Principles and Practice of Finite Volume Method

Principles and Practice of Finite Volume Method

Edited by **Haley Adison**

Clanrye International

New Jersey

Published by Clanrye International,
55 Van Reypen Street,
Jersey City, NJ 07306, USA
www.clanryeinternational.com

Principles and Practice of Finite Volume Method
Edited by Haley Adison

International Standard Book Number: 978-1-63240-417-6 (Hardback)

Printed in the United States of America.

Contents

Preface

This book aims to highlight the current researches and provides a platform to further the scope of innovations in this area. This book is a product of the combined efforts of many researchers and scientists, after going through thorough studies and analysis from different parts of the world. The objective of this book is to provide the readers with the latest information of the field.

The principles and practice of finite volume method are highlighted in this book. In this book, readers will find a subject which will increase their interest and involve them to further explore a challenge and work further on the presented solutions. This book can serve the purposes of both; a textbook and a practical guide. It presents a vast variety of ideas in FVM which is a result of the efforts of scientists from across the world. The major topics covered in this book are novel techniques and algorithms in FVM, solution of particular problems through FVM and application of FVM in medicine and engineering. This book is for anyone who wants to grow, improve and explore.

I would like to express my sincere thanks to the authors for their dedicated efforts in the completion of this book. I acknowledge the efforts of the publisher for providing constant support. Lastly, I would like to thank my family for their support in all academic endeavors.

Editor

Part 1

Different Aspects in FVM, New Techniques and Algorithms

1

Numerical Schemes for Hyperbolic Balance Laws – Applications to Fluid Flow Problems

Marek Brandner, Jiří Egermaier and Hana Kopincová
The University of West Bohemia
Czech Republic

1. Introduction

Balance laws arise from many areas of engineering practice specifically from the fluid mechanics. Many numerical methods for the solution of these balanced laws were developed in recent decades. The numerical methods are based on two views: solving hyperbolic PDE with a nonzero source term (the obvious description of the central and central-upwind schemes; (Kurganov & Levy, 2002; LeVeque, 2004)) or solving the augmented quasilinear nonconservative formulation (Gosse, 2001; Le Floch & Tzavaras, 1999; Parès, 2006). Furthermore, the methods can be interpreted using flux-difference splitting (or flux-vector splitting), or by selecting adaptive intervals and the transformation to the semidiscrete form (for example (Kurganov & Petrova, 2000)). We prefer the augmented quasilinear nonconservative formulation solved by the flux-difference splitting in our text. We try to formulate the methods in the most general form. The range of this text does not give the complete overview of currently used methods.

2. Mathematical models

In this section we describe the specific mathematical models based on hyperbolic balanced laws. There are many models that describe fluid flow phenomena but we are interested in the two type of them: models described open channel flow and urethra flow.

2.1 Shallow water equations

We are interested in solving the problem related to the fluid flow through the channel with the general cross-section area described by

$$
\begin{aligned}
a_t + q_x &= 0, \qquad (1)\\
q_t + \left(\frac{q^2}{a} + gI_1\right)_x &= -gab_x + gI_2,
\end{aligned}
$$

where $a = a(x,t)$ is the unknown cross-section area, $q = q(x,t)$ is the unknown discharge, $b = b(x)$ is given function of elevation of the bottom, g is the gravitational constant and

$$
I_1 = \int_0^{h(x,t)} [h(x,t) - \eta]\sigma(x,\eta)d\eta, \qquad (2)
$$

$$I_2 = \int_0^{h(x,t)} (h(x,t) - \eta) \left[\frac{\partial \sigma(x,\eta)}{\partial x} \right] d\eta, \tag{3}$$

here η is the depth integration variable, $h(x,t)$ is the water depth and $\sigma(x,\eta)$ is the width of the cross-section at the depth η. The derivation can be found in e.g. (Cunge at al., 1980).

The first special case are the equations reflecting the fluid flow through the spatially varying rectangular channel

$$\begin{aligned} a_t + q_x &= 0, \\ q_t + \left(\frac{q^2}{a} + \frac{ga^2}{2l} \right)_x &= \frac{ga^2}{2l^2} l_x - gab_x, \end{aligned} \tag{4}$$

with $l = l(x)$ being the function describing the width of the channel, and second one the system for the constant rectangular channel

$$\begin{aligned} h_t + (hu)_x &= 0, \\ (hu)_t + \left(hu^2 + \frac{1}{2} gh^2 \right)_x &= -ghb_x. \end{aligned} \tag{5}$$

In the above equation, $h(x,t)$ is the water depth and $u(x,t)$ is the horizontal velocity. It also possible to add some friction term to the system described above. For example, we can write

$$\begin{aligned} h_t + (hv)_x &= 0, \\ (hv)_t + \left(hv^2 + \frac{1}{2} gh^2 \right)_x &= -ghB_x - gM^2 \frac{hv|hv|}{h^{7/3}}, \end{aligned} \tag{6}$$

where M is Manning's coefficient.

All of the presented systems can be written in the compact matrix form

$$\mathbf{q}_t + [\mathbf{f}(\mathbf{q}, x)]_x = \psi(\mathbf{q}, x), \tag{7}$$

with $\mathbf{q}(x,t)$ being the vector of conserved quantities, $\mathbf{f}(\mathbf{q}, x)$ the flux function and $\psi(\mathbf{q}, x)$ the source term. This relation represents the balance laws.

It is possible to use any augmented formulation, which is suitable for rewriting the system to the quasilinear homogeneous one. For example, we can obtain

$$\begin{aligned} h_t + (hu)_x &= 0, \\ (hu)_t + \left(hu^2 + \frac{1}{2} gh^2 \right)_x &= -ghb_x, \\ b_t &= 0, \end{aligned} \tag{8}$$

i.e.

$$\begin{bmatrix} h \\ hv \\ b \end{bmatrix}_t + \begin{bmatrix} 0 & 1 & 0 \\ -v^2 + gh & 2v & -gh \\ 0 & 0 & 0 \end{bmatrix} \begin{bmatrix} h \\ hv \\ b \end{bmatrix}_x = \begin{bmatrix} 0 \\ 0 \\ 0 \end{bmatrix}. \tag{9}$$

2.2 Urethra flow

We now briefly introduce a problem describing fluid flow through the elastic tube represented by hyperbolic partial differential equations with the source term. In the case of the male urethra, the system based on model in (Stergiopulos at al., 1993) has the following form

$$\begin{aligned} a_t + q_x &= 0, \\ q_t + \left(\tfrac{q^2}{a} + \tfrac{a^2}{2\rho\beta}\right)_x &= \tfrac{a}{\rho}\left(\tfrac{a_0}{\beta}\right)_x + \tfrac{a^2}{2\rho\beta^2}\beta_x - \tfrac{q^2}{4a^2}\sqrt{\tfrac{\pi}{a}}\lambda(Re), \end{aligned} \tag{10}$$

where $a = a(x,t)$ is the unknown cross-section area, $q = q(x,t)$ is the unknown flow rate (we also denote $v = v(x,t)$ as the fluid velocity, $v = \frac{q}{a}$), ρ is the fluid density, $a_0 = a_0(x)$ is the cross-section of the tube under no pressure, $\beta = \beta(x,t)$ is the coefficient describing tube compliance and $\lambda(Re)$ is the Mooney-Darcy friction factor ($\lambda(Re) = 64/Re$ for laminar flow). Re is the Reynolds number defined by

$$Re = \frac{\rho q}{\mu a}\sqrt{\frac{4a}{\pi}}, \tag{11}$$

where μ is fluid viscosity. This model contains constitutive relation between the pressure and the cross section of the tube

$$p = \frac{a - a_0}{\beta} + p_e, \tag{12}$$

where p_e is surrounding pressure.

3. Conservative and nonconservative problems and numerical schemes

3.1 Conservative problems and numerical schemes

We consider the conservation law in the conservative form

$$\begin{aligned} \mathbf{q}_t + [\mathbf{f}(\mathbf{q})]_x &= \mathbf{0},\ x \in \mathbb{R},\ t \in (0,T), \\ \mathbf{q}(x,0) &= \mathbf{q}_0(x),\ x \in \mathbb{R}, \end{aligned} \tag{13}$$

The numerical scheme based on finite volume discretization in the conservation form can be written as follows,

$$\mathbf{Q}_j^{n+1} = \mathbf{Q}_j^n - \frac{\Delta t}{\Delta x}(\mathbf{F}_{j+1/2}^n - \mathbf{F}_{j-1/2}^n). \tag{14}$$

We use also the semidiscrete version of (14)

$$\frac{\partial \mathbf{Q}_j}{\partial t} = -\frac{1}{\Delta x}(\mathbf{F}_{j+1/2} - \mathbf{F}_{j-1/2}). \tag{15}$$

The relation (14) can be derived as the approximation of the integral conservation law at the interval $\left\langle x_{j-1/2}, x_{j+1/2}\right\rangle$ from time level t_n to t_{n+1}

$$\begin{aligned} \tfrac{1}{\Delta x}\int_{x_{j-1/2}}^{x_{j+1/2}} \mathbf{q}(x,t_{n+1})dx &= \tfrac{1}{\Delta x}\int_{x_{j-1/2}}^{x_{j+1/2}} \mathbf{q}(x,t_n)dx - \\ &- \tfrac{1}{\Delta t}\left[\int_{t_n}^{t_{n+1}} \mathbf{f}(\mathbf{q}(x_{j+1/2},t))dt - \int_{t_n}^{t_{n+1}} \mathbf{f}(\mathbf{q}(x_{j-1/2},t))dt\right]. \end{aligned} \tag{16}$$

The previous relations lead to the following approximations of integral averages

$$\mathbf{Q}_j^n \approx \frac{1}{\Delta x}\int_{x_{j-1/2}}^{x_{j+1/2}} \mathbf{q}(x,t_n)dx,$$
$$\mathbf{F}_{j+1/2}^n \approx \frac{1}{\Delta t}\int_{t_n}^{t_{n+1}} \mathbf{f}(\mathbf{q}(x_{j+1/2},t))dt. \tag{17}$$

Numerical fluxes $\mathbf{F}_{j+1/2}^n$ are usually defined by the approximate solution of the Riemann problem between states $\mathbf{Q}_{j+1}^n$ and $\mathbf{Q}_j^n$ (this technique is called the flux difference splitting; it will be described in the following parts) or by the Boltzmann approach (flux vector splitting; it will be described later). In what follows, we use the notation $\mathbf{Q}_{j+1/2}^+$ and $\mathbf{Q}_{j+1/2}^-$ for the reconstructed values of unknown function. Reconstructed values represent the approximations of limit values at the points $x_{j+1/2}$. The most common reconstructions are based on the minmod function (see for example (Kurganov & Tadmor, 2000)) or ENO and WENO techniques (Črnjarič-Zič at al., 2004).

If the exact solution of the problem has a compact support in the interval $\langle 0,T\rangle$ then it is possible to show, that the scheme (15) is conservative, i.e.

$$\sum_{j=-\infty}^{\infty} \mathbf{Q}_j^{n+1} = \sum_{j=-\infty}^{\infty} \mathbf{Q}_j^n. \tag{18}$$

The uniqueness of discontinuous solutions to the conservation laws is not guaranteed. Therefore the additional conditions, based on physical considerations, are required to isolate the physically relevant solution. The most common condition is called entropy condition. The unique entropy satisfying weak solution q holds

$$[\eta(q)]_t + [\varphi(q)]_x \leq 0, \tag{19}$$

for the convex entropy functions $\eta(q)$ and corresponding entropy fluxes $\varphi(q)$ (for example see (LeVeque, 2004)).

If the conservation law (13) is solved by the consistent method in conservation form (15) the Lax-Wendroff theorem is valid (LeVeque, 2004): if the approximate function $Q(x,t)$ converged to the function $q(x,t)$ for the $\Delta x, \Delta t \to 0$, the function $q(x,t)$ is the weak solution of the problem (13).

Many theoretical results in the field of hyperbolic PDEs can be found in the literature. For example, for scalar problems with the convex flux function, the convergence of the some method to the entropy-satisfying weak solutions, is proven. It means that if the solution q^ϵ of the problem

$$\begin{aligned} q_t + [f(q)]_x &= \epsilon q_{xx}, \quad x \in \mathbb{R}, 0 < t < T, \epsilon > 0, \\ q(x,0) &= q_0(x), \; x \in \mathbb{R}. \end{aligned} \tag{20}$$

exists and the limit $q^* = \lim_{\epsilon \to 0_+} q^\epsilon$ exists too than q^* is the entropy-satisfying weak solutions of the problem (20).

It is possible to define the appropriate properties of the methods. For example, in the case of scalar problem, TVD (Total Variation Diminishing) property, which ensures that the total variation of the solution is non-increasing, i.e.

$$\sum_{j=-\infty}^{\infty} |\mathbf{Q}_{j+1}^{n+1} - \mathbf{Q}_j^{n+1}| \leq \sum_{j=-\infty}^{\infty} |\mathbf{Q}_{j+1}^{n} - \mathbf{Q}_j^{n}| \tag{21}$$

for all time layers t_n. This property is important for limitation of the oscillations in the solution.

3.2 Nonconservative problems

We consider the nonlinear hyperbolic problem in nonconservative form

$$\begin{aligned} \mathbf{q}_t + \mathbf{A}(\mathbf{q})\mathbf{q}_x &= \mathbf{0},\ x \in \mathbf{R},\ t \in (0,T), \\ \mathbf{q}(x,0) &= \mathbf{q}_0(x),\ x \in \mathbf{R}. \end{aligned} \tag{22}$$

The numerical schemes for solving problems (22) can be written in fluctuation form

$$\frac{\partial \mathbf{Q_j}}{\partial t} = -\frac{1}{\Delta x}[\mathbf{A}^-(\mathbf{Q}_{j+1/2}^-, \mathbf{Q}_{j+1/2}^+) + \mathbf{A}(\mathbf{Q}_{j+1/2}^-, \mathbf{Q}_{j-1/2}^+) + \mathbf{A}^+(\mathbf{Q}_{j-1/2}^-, \mathbf{Q}_{j-1/2}^+)], \tag{23}$$

where $\mathbf{A}^{\pm}(\mathbf{Q}_{j+1/2}^-, \mathbf{Q}_{j+1/2}^+)$ are so called fluctuations. They can be defined by the sum of waves moving to the right or to the left. The directions are dependent on the signs of the speeds of these waves, which are related to the eigenvalues of matrix $\mathbf{A}(\mathbf{q})$.

When the problem (22) is derived from the conservation form (13), i.e. $\mathbf{f}'(\mathbf{q}) = \mathbf{A}(\mathbf{q})$ is the Jacobi matrix of the system, fluctuations can be defined as follows

$$\begin{aligned} \mathbf{A}(\mathbf{Q}_{j+1/2}^-, \mathbf{Q}_{j-1/2}^+) &= \mathbf{f}(\mathbf{Q}_{j+1/2}^-) - \mathbf{f}(\mathbf{Q}_{j-1/2}^+), \\ \mathbf{A}^-(\mathbf{Q}_{j+1/2}^-, \mathbf{Q}_{j+1/2}^+) &= \mathbf{F}_{j+1/2}^- - \mathbf{f}(\mathbf{Q}_{j+1/2}^-), \\ \mathbf{A}^+(\mathbf{Q}_{j-1/2}^-, \mathbf{Q}_{j-1/2}^+) &= \mathbf{f}(\mathbf{Q}_{j-1/2}^+) - \mathbf{F}_{j-1/2}^+. \end{aligned} \tag{24}$$

4. Riemann problem

4.1 Riemann problem for conservative systems

The Riemann problem is the special problem based on finite volume discretization with the discontinuous initial condition. In the nonlinear case it has the form

$$\begin{aligned} \mathbf{q}_t + [\mathbf{f}(\mathbf{q})]_x &= \mathbf{0},\ x \in \mathbb{R},\ t \in (0,T), \\ \mathbf{q}(x,0) &= \begin{cases} \mathbf{Q}_j^n, & x < x_{j+1/2}, \\ \mathbf{Q}_{j+1}^n, & x > x_{j+1/2}. \end{cases} \end{aligned} \tag{25}$$

We solve the transitions between two states, but this solution could not exists in general. These transitions can be rarefaction waves, shock waves or contact discontinuities. Rarefaction wave is case of a continuous solution when the following equality holds

$$\mathbf{q}(x,t) = \tilde{\mathbf{q}}(\xi(x,t)), \tag{26}$$

where

$$\tilde{\mathbf{q}}'(\xi) = \alpha(\xi)\mathbf{r}^p(\xi) \tag{27}$$

for any function $\xi(x,t)$, where $\alpha(\xi)$ is a coefficient dependent on the function ξ and $\mathbf{R}^p(\xi)$ is the corresponding p-th eigenvector of the Jacobi matrix $\mathbf{f}'(\mathbf{u})$.

Shock waves and contact discontinuities are special cases of discontinuous solutions. The requirement that this solution should be (see (LeVeque, 2004)) a weak solution of the problem (25) leads to the following relation

$$s(\mathbf{q}^+ - \mathbf{q}^-) = \mathbf{f}(\mathbf{q}^+) - \mathbf{f}(\mathbf{q}^-), \tag{28}$$

where s is the speed of the propagation of the discontinuities. The relation (28) is known as the Rankine-Hugoniot jump condition. In some cases it is possible to construct the solution of Riemann problem as the sequence of the transitions between the discontinuous states (in the case of strong nonlinearity the discontinuous solution consists of the shock waves, in the case of linear degeneration solution consists of the contact discontinuities).

In the case of a linear problem $\mathbf{f}(\mathbf{q}) = \mathbf{A}\mathbf{q}$ it is known that the initial state of each characteristic variable w^p is moving at a speed that corresponds to the eigenvalue λ^p of the matrix $\mathbf{A}$. The solution is a system of constant states separated by discontinuities that move at speeds correspondent to eigenvalues. Therefore, the jump $\mathbf{Q}_{j+1} - \mathbf{Q}_j$ over p-th discontinuity can be expressed as,

$$(w^p_{j+1} - w^p_j)\mathbf{r}^p = \alpha^p \mathbf{r}^p, \tag{29}$$

where $\mathbf{r}^p$ is the p-th eigenvector of matrix $\mathbf{A}$. The relation (29) represents the initial jump in the characteristic variable w^p and at the same time $\mathbf{q} = \mathbf{R}\mathbf{w}$. Therefore, the solution of linear Riemann problem can be defined by the decomposition of the initial jump of the unknown function to the eigenvectors $\mathbf{r}^p$ of the Jacobi matrix $\mathbf{A} = \mathbf{f}'(\mathbf{q})$

$$\mathbf{Q}_{j+1} - \mathbf{Q}_j = \sum_{p=1}^{m} \alpha^p \mathbf{r}^p. \tag{30}$$

The discontinuities $\mathbf{W}^p = \alpha^p \mathbf{r}^p$ are called waves and they are propagated by the speeds λ^p. For details see (LeVeque, 2004).

4.2 Riemann problem for nonconservative systems

In this section we are interested in nonlinear systems in nonconservative form

$$\begin{aligned} \mathbf{q}_t + \mathbf{A}(\mathbf{q})\mathbf{q}_x &= \mathbf{0}, \quad x \in \mathbb{R}, t > 0, \\ \mathbf{q}_0 &\in [\mathbb{BV}(\mathbb{R})]^m, \end{aligned} \tag{31}$$

where $\mathbf{q} \in \mathbb{R}^m$, $\mathbf{q} \to \mathbf{A}(\mathbf{q})$ is smooth locally bounded matrix-valued map, matrix $\mathbf{A}(\mathbf{q})$ is strictly hyperbolic (diagonalizable, with real and different eigenvalues). We suppose that $\mathbf{A}(\mathbf{q})$ is not Jacobi matrix so it is not possible to rewrite nonconservative system in conservative form (13). Here, $\mathbb{BV}[(\mathbb{R})]^m$ is function space contains functions with bounded total variation.

Above mentioned system makes sense only if $\mathbf{q}$ is differentiable. In the case when $\mathbf{q}$ admits discontinuities at a point, $\mathbf{A}(\mathbf{q})$ may admits discontinuities as well and $\mathbf{q}_x$ contains delta function with singularity at this point. Then $\mathbf{A}(\mathbf{q})\mathbf{q}_x$ is product of Heaviside function with delta function and in general is not unique. There is possibility to smooth out" of this function over width ϵ, for example by adding viscosity or diffusion. Then we get well defined product of continuous functions. The limiting behavior for $\epsilon \to 0$ is strongly depend on "smoothing out."

Under a special assumption, the nonconservative product $\mathbf{A}(\mathbf{q})\mathbf{q}_x$ can be understood as a Borel measure. In the following we introduce basic theorems and definitions, for details see (Gosse, 2001; Le Floch, 1989; Le Floch & Tzavaras, 1999; Parès, 2006).

Definition 4.1. *A path ϕ in $\Omega \in \mathbb{R}^m$ is a family of smooth maps $\langle 0,1\rangle \times \Omega \times \Omega \to \Omega$ satisfying:*

- $\phi(0;\mathbf{q}_l,\mathbf{q}_r) = \mathbf{q}_l$ *and* $\phi(1;\mathbf{q}_l,\mathbf{q}_r) = \mathbf{q}_r, \quad \forall \mathbf{q}_l,\mathbf{q}_r \in \Omega, \forall s \in \langle 0,1\rangle,$
- *for each bounded set $\mathcal{O} \in \Omega$, there exists a constant $k > 0$ such that*

$$\left|\frac{\partial \phi}{\partial s}(s;\mathbf{q}_l,\mathbf{q}_r)\right| \le k\,|\mathbf{q}_r - \mathbf{q}_l|$$

  for any $\mathbf{q}_l$, $\mathbf{q}_r \in \mathcal{O}$ and almost all $s \in \langle 0,1\rangle$,
- *for each bounded set $\mathcal{O} \in \Omega$, there exists a constant $K > 0$ such that*

$$\left|\frac{\partial \phi}{\partial s}(s;\mathbf{q}_l^1,\mathbf{q}_r^1) - \frac{\partial \phi}{\partial s}(s;\mathbf{q}_l^2,\mathbf{q}_r^2)\right| \le K\left(\left|\mathbf{q}_l^1 - \mathbf{q}_l^2\right| + \left|\mathbf{q}_r^1 - \mathbf{q}_r^2\right|\right)$$

  for any $\mathbf{q}_l^1$, $\mathbf{q}_r^1$, $\mathbf{q}_l^2$, $\mathbf{q}_r^2 \in \mathcal{O}$ and almost all $s \in \langle 0,1\rangle$.

Theorem 4.1 (Dal Maso, Le Floch, Murat). *Let $\mathbf{q} : (a,b) \to \mathbb{R}^m$ be a function with bounded variation and $\mathbf{A} : \mathbb{R}^m \to \mathbb{R}^{m\times m}$ a locally bounded function. There exists a unique signed Borel measure μ on (a,b) characterized by following properties:*

1. *if $x \to \mathbf{q}(x)$ is continuous on an open set $o \in (a,b)$ then*

$$\mu(o) = \int_o \mathbf{A}(\mathbf{q})\frac{\partial \mathbf{q}}{\partial x}\,dx,$$

2. *if $x_0 \in (a,b)$ is a discontinuity point of $x \to \mathbf{q}(x)$, then*

$$\mu(x_0) = \int_0^1 \mathbf{A}\left(\phi(s;\mathbf{q}(x_0^-),\mathbf{q}(x_0^+))\right)\frac{\partial \phi}{\partial s}(s;\mathbf{q}(x_0^-),\mathbf{q}(x_0^+))\,ds,$$

   where we denote $\mathbf{q}(x_0^-) = \lim_{x\to x_0^-}\mathbf{q}(x)$ and $\mathbf{q}(x_0^+) = \lim_{x\to x_0^+}\mathbf{q}(x)$.

Remark 4.1. *Borel measure μ is called nonconservative product and is usually written $[\mathbf{A}(\mathbf{q})\mathbf{q}_x]_\phi$.*

Remark 4.2. *In the case where $\mathbf{A}(\mathbf{q}) = \mathbf{f}'(\mathbf{q})$ then Borel measure, $[\mathbf{A}(\mathbf{q})\mathbf{q}_x]_\phi = \mathbf{f}(\mathbf{q})_x$ is independent of the path ϕ.*

Definition 4.2 (Weak solution). *Let ϕ be a family of paths in the sense of definition 4.1. A function $\mathbf{q} \in [L^\infty(\mathbb{R}\times\mathbb{R}^+) \cap \mathbb{BV}_{loc}(\mathbb{R}\times\mathbb{R}^+)]^m$ is a weak solution of system (31), if it satisfies*

$$\mathbf{q}_t + [\mathbf{A}(\mathbf{q})\mathbf{q}_x]_\phi = 0, \tag{32}$$

as a bounded Borel measure on $\mathbb{R}\times\mathbb{R}^+$.

Definition 4.3 (Entropy solution). *Given an entropy pair (η, φ)(entropy, entropy flux) for (31), i.e. a pair of regular functions $\Omega \to \mathbb{R}$, such that*

$$\nabla\varphi(\mathbf{q}) = \nabla\eta(\mathbf{q}) \cdot \mathbf{A}(\mathbf{q}), \quad \forall \mathbf{q} \in \Omega. \tag{33}$$

A weak solution is said to be entropic if it satisfies the inequality

$$\frac{\partial\eta(\mathbf{q})}{\partial t} + \frac{\partial\varphi(\mathbf{q})}{\partial x} \leq 0 \tag{34}$$

in the sense of distribution.

Function $\mathbf{q}(x,t)$ is weak solution if and only if, across a discontinuity with speed ζ, it satisfies generalized Rankine-Hugoniot condition,

$$-\zeta(\mathbf{q}_r - \mathbf{q}_l) + \int_0^1 \mathbf{A}\left(\phi(s; \mathbf{q}_l, \mathbf{q}_r)\right) \frac{\partial\phi}{\partial s}(s; \mathbf{q}_l, \mathbf{q}_r)\, ds = 0. \tag{35}$$

Now we define the Riemann problem for nonconservative strictly hyperbolic system:

$$\mathbf{q}_t + \mathbf{A}(\mathbf{q})\mathbf{q}_x = \mathbf{0}, \quad x \in \mathbb{R},\, t > 0 \tag{36}$$

with an initial condition

$$\mathbf{q}(x,0) = \mathbf{q}_0(x) = \begin{cases} \mathbf{q}_l \text{ pro } x < 0, \\ \mathbf{q}_r \text{ pro } x > 0, \end{cases} \tag{37}$$

where $\mathbf{q}_l, \mathbf{q}_r \in \mathbb{R}^m$ are vectors of constants.

Theorem 4.2. *Let ϕ be a family of path in the sense of definition 4.1. Assume that system (36) is strictly hyperbolic with genuinely nonlinear or linearly degenerate characteristic field and the family of path ϕ satisfies*

$$\frac{\partial\phi}{\partial\mathbf{q}_1}(1; \mathbf{q}_0, \mathbf{q}_0) - \frac{\partial\phi}{\partial\mathbf{q}_1}(0; \mathbf{q}_0, \mathbf{q}_0) = \mathbf{I}, \quad \forall \mathbf{q}_0 \in \mathbb{R}^m. \tag{38}$$

Then, for $|\mathbf{q}_r - \mathbf{q}_l|$ small enough, the Riemann problem (36) and (37) has a solution $\mathbf{q}(x,t)$ with bounded variation, which depends only on $\frac{x}{t}$ and has Lax's structure. That is $\mathbf{q}(x,t)$ consist of $m+1$ constant states separated by shock waves, rarefaction waves or contact discontinuities.

The solution of the Riemann problem for nonconservative system is related to the solution of the Riemann problem for conservative system. The only difference is in the case of shock wave, precisely in Rankine-Hugoniot condition formulation.

Before we define a class of useful numerical methods, we introduce a brief motivation, which can be in details found in (Gosse, 2001). Suppose nonconservative system

$$\mathbf{q}_t + \mathbf{A}(\mathbf{q})\mathbf{q}_x = \mathbf{0}, \quad x \in \mathbb{R},\, t > 0 \tag{39}$$

and suppose family of path in the sense of definition 4.1. If $\mathbf{q}(x,t)$ is piecewice regular weak solution then for given time t, the Borel measure can be written in following form

$$\mu^{\phi}(o) = \mu_a^{\phi}(o) + \mu_s^{\phi} = \int_o \mathbf{A}(\mathbf{q})\mathbf{q}_x\, dx + \sum_k \left[\int_0^1 \mathbf{A}(\phi(s;\mathbf{q}_k^-,\mathbf{q}_k^+)) \frac{\partial \phi}{\partial s}(s;\mathbf{q}_k^-,\mathbf{q}_k^+)\, ds \right] \delta_{x=x_k(t)}, \quad (40)$$

where index k represents number of discontinuities in solution, $x_k(t)$ are discontinuity points in the time $t > 0$ and $\mathbf{q}_k^- = \lim_{x \to x_k(t)^-} \mathbf{q}(x,t)$, $\mathbf{q}_k^+ = \lim_{x \to x_k(t)^+} \mathbf{q}(x,t)$, $\delta_{x=x_k(t)}$ is Dirac measure at the point $x = x_k(t)$. Then we can get

$$\bar{\mathbf{Q}}_j^{n+1} = \bar{\mathbf{Q}}_j^n - \frac{1}{\Delta x} \int_{t_n}^{t_{n+1}} \mu^{\phi}(I_j)\, dt. \quad (41)$$

The amount of the quantity on the cell boundaries $x_{j+1/2}$ can be splitted into contribution to cell I_{j+1} and the contribution to the cell I_j. In other words, we split it into two terms $\mathbf{A}_{j+1/2}^{\pm}$. So the sum of this terms can be understood as a discrete representation of the $\int_{t_n}^{t_{n+1}} \mu_s^{\phi} dt$ and $\mathbf{A}_j \approx \int_{t_n}^{t_{n+1}} \mu_a^{\phi}(I_j)dt$. By this way we get generalisation of classical conservative finite volume methods, see (Parès, 2006), i.e.

$$\bar{\mathbf{Q}}_j^{n+1} = \bar{\mathbf{Q}}_j^n - \frac{\Delta t}{\Delta x}[\mathbf{A}_{j-1/2}^{+} + \mathbf{A}_j + \mathbf{A}_{j+1/2}^{-}]. \quad (42)$$

Definition 4.4. *Given a family of path ϕ, a numerical scheme is said to be path-conservation if it can be written in form (42), where*

$$\mathbf{A}_{j+1/2}^{\pm} = \mathbf{A}^{\pm}(\bar{\mathbf{q}}_{j-p}, \ldots, \bar{\mathbf{q}}_{j+q}),$$

$$\mathbf{A}_j = \mathbf{A}(\bar{\mathbf{q}}_{j-p}, \ldots, \bar{\mathbf{q}}_{j+q}),$$

$\mathbf{A}^{\pm}, \mathbf{A} : \Omega^{p+q+1} \to \Omega$ *being continuous functions satisfying*

1.

$$\mathbf{A}^{\pm}(\mathbf{q}, \ldots, \mathbf{q}) = \mathbf{0}, \quad \forall \mathbf{q} \in \Omega, \quad (43)$$

2.

$$\mathbf{A}^{-}(\mathbf{q}_{-p}, \ldots, \mathbf{q}_q) + \mathbf{A}^{+}(\mathbf{q}_{-p}, \ldots, \mathbf{q}_q) = \int_0^1 \mathbf{A}(\phi(s;\mathbf{q}_l,\mathbf{q}_r)) \frac{\partial \phi}{\partial s}(s;\mathbf{q}_l,\mathbf{q}_r)\, ds \quad (44)$$
$$\forall \mathbf{q}_j \in \Omega, j = -p, \ldots, q.$$

3.

$$\mathbf{A}(\mathbf{q}, \ldots, \mathbf{q}) = \int_{x_{j-1/2}}^{x_{j+1/2}} \mathbf{A}(\mathbf{q})\mathbf{q}_x\, dx. \quad (45)$$

Remark 4.3. *If $\mathbf{A}(\mathbf{q}) = \mathbf{f}'(\mathbf{q})$ then path-conservation scheme is consistent and conservative (that is if the nonconservative system can be written as conservative one, path-conservative method become classical conservative and consistent). Notice here that $\mathbf{A}^{\pm}$ are fluctuations, see 3.2.*

4.3 The properties of exact solution

Consider a system in conservative form with special right hand side (this system agrees with shallow water system (5))

$$\mathbf{q}_t + \mathbf{f}(\mathbf{q})_x = \mathbf{S}(\mathbf{q})\sigma_x. \tag{46}$$

This system can be rewritten in a following homogeneous nonconservative form

$$\mathbf{u}_t + \mathbf{A}(\mathbf{u})\mathbf{u}_x = \mathbf{0}, \tag{47}$$

where

$$\mathbf{u} = \begin{bmatrix} \mathbf{q} \\ \sigma \end{bmatrix} \quad \text{and} \quad \mathbf{A}(\mathbf{u}) = \begin{bmatrix} \frac{\partial \mathbf{f}}{\partial \mathbf{u}}(\mathbf{u}) & -\mathbf{S}(\mathbf{u}) \\ 0 & 0 \end{bmatrix},$$

where $\frac{\partial \mathbf{f}}{\partial \mathbf{u}}(\mathbf{w})$ denotes Jacobian matrix of flux $\mathbf{f}$. In order to define weak solutions of the system (47) we have to choose family of paths (see definition 4.1 and theorem 4.2). And following requirement become natural: if $\mathbf{u} = [\mathbf{q}, \sigma]^T$ is a weak solution of the nonconservative system (47) and σ is constant, then $\mathbf{q}$ must be weak solution of the homogeneous system of conservation laws, i.e. (46) with zero right hand side. By this requirement we impose addition condition of family of paths:

$$\text{if } \mathbf{u}_l \text{ and } \mathbf{u}_r \text{ are such that } \sigma_l = \sigma_r = \bar{\sigma}, \text{ then}$$
$$\mathbf{\Phi}(s; \mathbf{u}_l, \mathbf{u}_r) = \bar{\sigma}, \quad s \in \langle 0, 1 \rangle\,. \tag{48}$$

For these systems, there are no difficulties with convergence. Moreover, the shock waves propagating in regions where σ is continuous are correctly captured independently of the choice of path. For details see (Le Floch at al., 2008).

5. Approximate Riemann solvers

There are many numerical schemes for solving (7) with different properties and possibilities of failing. The main types of the finite volume schemes are the central, upwind and central-upwind schemes. All these schemes relate together in variety of ways. We can interpret them as schemes with different deep knowledge about structure of the solution of Riemann problem. For example, the same scheme can be interpreted as the HLL solver (an approximate Riemann solver) or as the central–upwind scheme.

The main requirements on the numerical schemes are the consistency (in the finite volume sense, i.e. consistency with the flux function), the conservativity (if there is possibility to rewrite the problem to the conservative form it is required to have conservative numerical scheme), positive semidefiniteness (the scheme preserves nonnegativity of some quantities, which are essentially nonnegative from their physical fundamental) and the well-balancing (the schemes maintain some or all steady states which can occur). The next properties are the order of the schemes and stability. From the computational point of view there are other properties such as robustness, simplicity and computational efficiency. The second and third of these properties are typical for the central methods, but they are not common for scheme based on Roe's solver.

The steady states mean that the unknown quantities do not change in the time, i.e. $\mathbf{q_t} = \mathbf{0}$ in (7), and the flux function must balance the right hand side, i.e. $[\mathbf{f}(\mathbf{q})]_x = \psi(\mathbf{q}, x)$. Some schemes are constructed to preserve some special steady states for example called "lake at rest" in open channel flow problems, where there is no motion and the free surface is constant.

5.1 Central schemes, local Lax-Friedrichs method

These schemes use estimate of upper bound of maximal local speed of the propagating discontinuities (for example see (Kurganov & Petrova, 2000)). They are based on the following decomposition

$$\mathbf{f}(\mathbf{Q}^+_{j+1/2}) - \mathbf{f}(\mathbf{Q}^-_{j+1/2}) - \mathbf{\Psi}^{\pm}_{j+1/2} = s_{j+1/2}(\mathbf{Q}^+_{j+1/2} - \mathbf{Q}^*_{j+1/2}) - s_{j+1/2}(\mathbf{Q}^*_{j+1/2} - \mathbf{Q}^-_{j+1/2}), \quad (49)$$

where the approximation of local speed is defined by

$$s_{j+1/2} = \max_p \{\max\{|\lambda^p(\mathbf{Q}^-_{j+1/2})|, |\lambda^p(\mathbf{Q}^+_{j+1/2})|\}\}. \quad (50)$$

The local waves can be based on the decomposition (49). It means

$$\mathbf{Z}^1_{j+1/2} = -s_{j+1/2}(\mathbf{Q}^*_{j+1/2} - \mathbf{Q}^-_{j+1/2}), \qquad \mathbf{Z}^2_{j+1/2} = s_{j+1/2}(\mathbf{Q}^+_{j+1/2} - \mathbf{Q}^*_{j+1/2}). \quad (51)$$

We can use the scheme in conservation form based on numerical fluxes (15) or in nonconservative form based on the fluctuations (23). It can be used the scheme in the form

$$\frac{\partial \mathbf{Q}_j}{\partial t} = -\frac{1}{\Delta x}(\mathbf{F}_{j+1/2} - \mathbf{F}_{j-1/2}) + \mathbf{\Psi}_j \quad (52)$$

for the nonhomogeneous problems, where $\mathbf{\Psi}_j$ is a suitable approximation of the source term. The numerical fluxes can be defined by the following relation as

$$\mathbf{F}_{j+1/2} = \frac{1}{2}[\mathbf{f}(\mathbf{Q}^-_{j+1/2} + \mathbf{f}(\mathbf{Q}^+_{j+1/2})] - \frac{1}{2}s_{j+1/2}(\mathbf{Q}^+_{j+1/2} - \mathbf{Q}^-_{j+1/2}) \quad (53)$$

and the fluctuations as

$$\begin{aligned} \mathbf{A}^-(\mathbf{Q}^-_{j+1/2}, \mathbf{Q}^+_{j+1/2}) &= \mathbf{Z}^1_{j+1/2}, \\ \mathbf{A}(\mathbf{Q}^-_{j+1/2}, \mathbf{Q}^+_{j-1/2}) &= \mathbf{f}(\mathbf{Q}^-_{j+1/2}) - \mathbf{f}(\mathbf{Q}^+_{j-1/2}) - \mathbf{\Psi}^{\mp}_j, \\ \mathbf{A}^+(\mathbf{Q}^-_{j+1/2}, \mathbf{Q}^+_{j+1/2}) &= \mathbf{Z}^2_{j+1/2}. \end{aligned} \quad (54)$$

The advantage of central schemes is the fact, that it is not necessary to solve a full Riemann problem.

5.1.1 Steady states

These schemes do not preserve general steady states for the problems with source terms, but only special one for shallow water equations. The steady state called "lake at rest", where $v = 0, h + b = \text{const}$. If we use the scheme in fluctuation form, we need the equalities $\mathbf{A}^-_{j+1/2} = \mathbf{A}^+_{j+1/2} = \mathbf{A}_j = \mathbf{0}$. Here we present only the equality $\mathbf{A}^+_{j+1/2} = \mathbf{0}$, the $\mathbf{A}^-_{j+1/2} = \mathbf{0}$ can be derived by the similar way. The first component of decomposition (49) is

$$\underbrace{(HV)^+_{j+1/2} - (HV)_{j+1/2}}_{0} = \underbrace{s_{j+1/2}(H^+_{j+1/2} - H^*_{j+1/2})}_{A^{+,1}_{j+1/2}} \underbrace{- s_{j+1/2}(H^*_{j+1/2} - H^-_{j+1/2})}_{A^{-,1}_{j+1/2}}. \quad (55)$$

Because there is no motion of the fluid, we have

$$(HV)^+_{j+1/2} = (HV)^-_{j+1/2} = 0 \Rightarrow H^*_{j+1/2} = \frac{1}{2}(H^+_{j+1/2} + H^-_{j+1/2}). \quad (56)$$

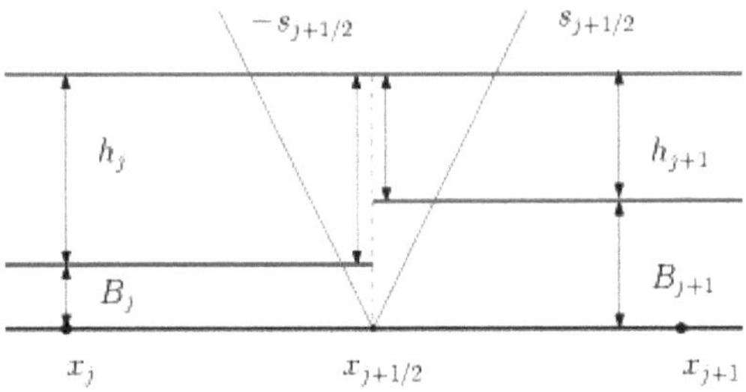

Fig. 1. The central methods construct only one middle state.

We consider nonhomogeneous problem, it means, that the $b_x \neq 0$. The constant water level then includes $h_x \neq 0$. Therefore the fluctuation are not equal to zero in general.

$$H^+_{j+1/2} \neq H^-_{j+1/2} \Rightarrow H^+_{j+1/2} - H^*_{j+1/2} \neq 0 \Rightarrow A^{+,1}_{j+1/2} \neq 0. \tag{57}$$

This situation is illustrated at the Fig.1. It can be seen that one middle state is in contradiction with physical model with source terms. This method could preserve only steady states that the reconstruction of unknown function is constant. This leads to the following modification of the system (5). We define new unknown function representing water level $y = h + b$. The shallow water equation can be modified to the following form

$$\begin{bmatrix} y \\ (y-B)v \end{bmatrix}_t + \begin{bmatrix} (y-B)v \\ (y-B)v^2 + \frac{1}{2}g(y-B)^2 \end{bmatrix}_x = \begin{bmatrix} 0 \\ -g(y-B)B_x \end{bmatrix}. \tag{58}$$

Then the first component of decomposition (49) is

$$\underbrace{((Y-B)V)^+_{j+1/2} - ((Y-B)V)^-_{j+1/2}}_{0} = \underbrace{s_{j+1/2}(Y^+_{j+1/2} - Y^*_{j+1/2})}_{A^{+,1}_{j+1/2}} \underbrace{-s_{j+1/2}(Y^*_{j+1/2} - Y^-_{j+1/2})}_{A^{-,1}_{j+1/2}}. \tag{59}$$

Again, no motion of the water ensures

$$((Y-B)V)^+_{j+1/2} = ((Y-B)V)^-_{j+1/2} = 0 \Rightarrow Y^*_{j+1/2} = \frac{1}{2}(Y^+_{j+1/2} + Y^-_{j+1/2}). \tag{60}$$

Constant water level means $y_x = 0$ and then

$$Y^+_{j+1/2} = Y^-_{j+1/2} \Rightarrow Y^+_{j+1/2} - Y^*_{j+1/2} = 0 \Rightarrow A^{+,1}_{j+1/2} = 0. \tag{61}$$

The second component of decomposition (49) is

$$\underbrace{f^2(\mathbf{Q}^+_{j+1/2}) - f^2(\mathbf{Q}^-_{j+1/2}) - \Psi^{2,\pm}_{j+1/2}}_{0\text{ (for suitable approximation)}} =$$
$$= \underbrace{s_{j+1/2}(Q^{+,2}_{j+1/2} - Q^{*,2}_{j+1/2})}_{A^{+,2}_{j+1/2}} \underbrace{-s_{j+1/2}(Q^{*,2}_{j+1/2} - Q^{-,2}_{j+1/2})}_{A^{-,2}_{j+1/2}}. \tag{62}$$

The second component of this middle state is defined by

$$Q^{*,2}_{j+1/2} = \frac{f^2(Q^-_{j+1/2}) - f^2(Q^+_{j+1/2}) - \Psi^{2,\pm}_{j+1/2}}{2s_{j+1/2}} + \frac{1}{2}(Q^{-,2}_{j+1/2} + Q^{+,2}_{j+1/2}), \tag{63}$$

therefore

$$Q^{*,2}_{j+1/2} = Q^{-,2}_{j+1/2} = Q^{+,2}_{j+1/2} \Rightarrow A^{+,2}_{j+1/2} = 0. \tag{64}$$

The term $\mathbf{A}_j = \mathbf{0}$ for suitable approximation $\mathbf{\Psi}^{\mp}_j$ so that

$$\mathbf{f}(\mathbf{Q}^-_{j+1/2}) - \mathbf{f}(\mathbf{Q}^+_{j-1/2}) = \mathbf{\Psi}^{\mp}_j. \tag{65}$$

We have seen that this method preserves steady state "lake at rest" for the modificate system (58) when the suitable approximation of the source terms is used. The common way how to construct this approximation is based on the equality with the flux difference.

Now we turn to general steady state. It means hv = const. We need again the equalities $\mathbf{A}^-_{j+1/2} = \mathbf{A}^+_{j+1/2} = \mathbf{A}_j = \mathbf{0}$. First component of the flux decomposition is

$$\underbrace{((Y-B)V)^+_{j+1/2} - ((Y-B)V)^-_{j+1/2}}_{0} = \underbrace{s_{j+1/2}(Y^+_{j+1/2} - Y^*_{j+1/2})}_{A^{+,1}_{j+1/2}} \underbrace{-s_{j+1/2}(Y^*_{j+1/2} - Y^-_{j+1/2})}_{A^{-,1}_{j+1/2}}. \tag{66}$$

The left and right values of the flux discharge are still equal

$$((Y-B)V)^+_{j+1/2} = ((Y-B)V)^-_{j+1/2} = 0 \Rightarrow Y^*_{j+1/2} = \frac{1}{2}(Y^+_{j+1/2} + Y^-_{j+1/2}), \tag{67}$$

but the reconstructed values of the water level are not equal in general

$$Y^+_{j+1/2} \neq Y^-_{j+1/2} \Rightarrow Y^+_{j+1/2} - Y^*_{j+1/2} \neq 0 \Rightarrow A^{+,1}_{j+1/2} \neq 0. \tag{68}$$

Therefore, the general steady state is not preserved.

5.1.2 Positive semidefiniteness

We recall that positive semidefiniteness is a property that consists in preserving nonnegativity of some unknown functions. In the cases of shallow water equations (water height h) and urethra flow modelling (cross-section of the tube a), the central scheme preserves this property. We illustrate this by the preserving the nonnegativity of the water level. The middle state is defined

$$H^*_{j+1/2} = \frac{s_{j+1/2}H^+_{j+1/2} + s_{j+1/2}H^-_{j+1/2} - H^+_{j+1/2}V^+_{j+1/2} + H^-_{j+1/2}V^-_{j+1/2}}{2s_{j+1/2}}. \tag{69}$$

Because the speed s is defined by (50), we can use the following inequality (for better display lower indexes $j+1/2$ are neglected),

$$H^* \geq \frac{(V^+ + \sqrt{gH^+})H^+ - (V^- - \sqrt{gH^-})H^- - H^+V^+ + H^-V^-}{2s} = \tag{70}$$

$$= \frac{\sqrt{gH^+}H^+ + \sqrt{gH^-_{j+1/2}}H^-}{2s} \geq 0. \tag{71}$$

5.2 Central-upwind schemes, HLL scheme

We will see that central-upwind scheme and HLL scheme can be understood as equivalent. The central-upwind scheme is based on the decomposition in the form

$$\mathbf{f}(\mathbf{Q}^+_{j+1/2}) - \mathbf{f}(\mathbf{Q}^-_{j+1/2}) - \mathbf{\Psi}^\pm_{j+1/2} = s^2_{j+1/2}(\mathbf{Q}^+_{j+1/2} - \mathbf{Q}^*_{j+1/2}) - s^1_{j+1/2}(\mathbf{Q}^*_{j+1/2} - \mathbf{Q}^-_{j+1/2}), \quad (72)$$

where $s^1_{j+1/2}$ and $s^1_{j+1/2}$ are the approximations of maximal and minimal speeds of the local waves. Furthemore

$$\mathbf{Z}^1_{j+1/2} = -s^1_{j+1/2}(\mathbf{Q}^*_{j+1/2} - \mathbf{Q}^-_{j+1/2}), \qquad \mathbf{Z}^2_{j+1/2} = s^2_{j+1/2}(\mathbf{Q}^+_{j+1/2} - \mathbf{Q}^*_{j+1/2}). \quad (73)$$

For the scheme in conservative form (13) (or in the form with approximation of the source term (52)), we can define the following numerical fluxes

$$\mathbf{F}_{j+1/2} = \frac{s^1_{j+1/2}\mathbf{f}(\mathbf{Q}^+_{j+1/2}) - s^2_{j+1/2}\mathbf{f}(\mathbf{Q}^-_{j+1/2})}{s^1_{j+1/2} - s^2_{j+1/2}} + \frac{s^1_{j+1/2}s^2_{j+1/2}}{s^1_{j+1/2} - s^2_{j+1/2}}(\mathbf{Q}^-_{j+1/2} - \mathbf{Q}^+_{j+1/2}). \quad (74)$$

If we use the scheme in fluctuation form (23), the fluctuations can be defined as follows

$$\begin{aligned}
\mathbf{A}^-(\mathbf{Q}^-_{j+1/2}, \mathbf{Q}^+_{j+1/2}) &= \sum_{p=1, s^p_{j+1/2}<0}^{2} \mathbf{Z}^p_{j+1/2}, \\
\mathbf{A}(\mathbf{Q}^-_{j+1/2}, \mathbf{Q}^+_{j-1/2}) &= \mathbf{f}(\mathbf{Q}^-_{j+1/2}) - \mathbf{f}(\mathbf{Q}^+_{j-1/2}) - \mathbf{\Psi}^\pm_{j+1/2}, \\
\mathbf{A}^+(\mathbf{Q}^-_{j+1/2}, \mathbf{Q}^+_{j+1/2}) &= \sum_{p=1, s^p_{j+1/2}>0}^{2} \mathbf{Z}^p_{j+1/2}.
\end{aligned} \quad (75)$$

As we mentioned before, the HLL solver can be identified with the central-upwind scheme. This solver depends on the choice of wave speeds. This solver does not use an explicit linearization of the Jacobi matrix, but the solution is constructed by consideration of two discontinuities (independent on the system dimension) propagating at speeds $s^1_{j+1/2}$ and $s^2_{j+1/2}$. These speeds approximate minimal and maximal local speeds of the system. The middle state $\mathbf{Q}^*_{j+1/2}$ between states $\mathbf{Q}_{j+1}$ and $\mathbf{Q}_j$ is determined by conservation law

$$s^1_{j+1/2}(\mathbf{Q}^*_{j+1/2} - \mathbf{Q}_j) + s^2_{j+1/2}(\mathbf{Q}_{j+1} - \mathbf{Q}^*_{j+1/2}) = \mathbf{f}(\mathbf{Q}_{j+1}) - \mathbf{f}(\mathbf{Q}_j). \quad (76)$$

Properties of this solver are strongly tied to the choice of the speeds $s^1_{j+1/2}$ and $s^2_{j+1/2}$. Speeds are determined by initial conditions and properties of the exact Riemann solver. Furthermore, it can be proved that in the case of the system of two equations when the wave speeds $s^1_{j+1/2}$ and $s^2_{j+1/2}$ are equal to Roe's speeds $\lambda^1_{j+1/2}$ and $\lambda^2_{j+1/2}$, the HLL solver is equal to Roe's solver (described in the following part).

5.2.1 Steady states

These methods preserve steady states by the similar way like the central schemes. It is important to use the suitable approximation of the source term function based on the flux diference. For example in (Kurganov & Levy, 2002) is presented approximation for preserving steady state "lake at rest" in the form

$$\Psi_j^{(2)}(t) \approx -g\frac{B(x_{j+1/2}) - B(x_{j-1/2})}{\Delta x} \cdot \frac{\left(Y_{j+1/2}^- - B(x_{j+1/2})\right) + \left(Y_{j-1/2}^+ - B(x_{j-1/2})\right)}{2}. \quad (77)$$

This approximation (77) supposes continuous approximation of function b describing bottom topography i.e. $B_{j+1/2}^+ = B_{j+1/2}^- = B_{j+1/2}$.

5.2.2 Positive semidefiniteness

The nonnegativity of the unknown function can be shown by a similar way as in the part 5.1.2. But if we use the scheme (Kurganov & Levy, 2002) and solve the system (58), the unknown function is the water level $y = h + B$ rather that the water height h. The nonnegativity of water height can be ensured for the system (5).

The scheme that ensures both properties at the same time is presented for example in (Audusse at al., 2004). It is based on special reconstruction of functions h and b

$$H_{j+1/2}^- = \max(0, H_j + B_j - B_{j+1/2}), \qquad H_{j+1/2}^+ = \max\{0, H_{j+1} + B_{j+1} - B_{j+1/2}\}, \quad (78)$$

where

$$B_{j+1/2} = \max\{B_j, B_{j+1}\}. \quad (79)$$

Again, as in previous cases, the source term approximation is equal to the flux difference.

5.3 Roe's solver

First, we assume the homogeneous system. Roe's solver is the approximate Riemann solver based on the local approximation of the nonlinear system $\mathbf{q}_t + [\mathbf{f}(\mathbf{q})]_x \equiv \mathbf{q}_t + \mathbf{A}(\mathbf{q})\mathbf{q}_x = \mathbf{0}$, where $\mathbf{A}(\mathbf{q})$ is the Jacobi matrix, by the linear system $\mathbf{q}_t + \mathbf{A}_{j+1/2}\mathbf{q}_x = \mathbf{0}$, where $\mathbf{A}_{j+1/2}$ is the Roe-averaged Jacobi matrix, which is defined by suitable combination of $\mathbf{A}(\mathbf{Q}_j)$ and $\mathbf{A}(\mathbf{Q}_{j+1})$. For details see (LeVeque, 2004).

For the conservation form (13) of the scheme we can define numerical fluxes

$$\mathbf{F}_{j+1/2} = \frac{1}{2}[\mathbf{f}(\mathbf{Q}_{j+1/2}^-) + \mathbf{f}(\mathbf{Q}_{j+1/2}^+)] - \frac{1}{2}|\mathbf{A}_{j+1/2}|(\mathbf{Q}_{j+1/2}^+ - \mathbf{Q}_{j+1/2}^-). \quad (80)$$

Fluctuation form is based on the following fluctuations

$$\begin{aligned} \mathbf{A}^-(\mathbf{Q}_{j+1/2}^-, \mathbf{Q}_{j+1/2}^+) &= \sum_{p=1}^{m} \lambda_{j+1/2}^{-,p} \mathbf{r}_{j+1/2}^p \Delta\gamma_{j+1/2}^p, \\ \mathbf{A}(\mathbf{Q}_{j+1/2}^-, \mathbf{Q}_{j+1/2}^+) &= \mathbf{f}(\mathbf{Q}_{j+1/2}^-) - \mathbf{f}(\mathbf{Q}_{j+1/2}^+), \\ \mathbf{A}^+(\mathbf{Q}_{j+1/2}^-, \mathbf{Q}_{j+1/2}^+) &= \sum_{p=1}^{m} \lambda_{j+1/2}^{+,p} \mathbf{r}_{j+1/2}^p \Delta\gamma_{j+1/2}^p, \end{aligned} \quad (81)$$

where $\mathbf{r}_{j+1/2}^p$ are eigenvectors of the Roe matrix $\mathbf{A}_{j+1/2}$, $\lambda_{j+1/2}^p$ are eigenvalues called Roe's speeds and $\Delta\gamma_{j+1/2} = \mathbf{R}_{j+1/2}^{-1}(\mathbf{Q}_{j+1/2}^+ - \mathbf{Q}_{j+1/2}^-)$.

5.3.1 Steady states

Roe's solver is based on upwind technique. For preserving steady states, it is possible to upwind the source terms too. See (Bermudez & Vasquez, 1994) for details. It is based on approximate Jacobi matrix constructed by the following relation (for simplicity the lower indexes are neglected)

$$\mathbf{f}(\mathbf{Q}^+) - \mathbf{f}(\mathbf{Q}^-) - \mathbf{\Psi}(\mathbf{Q}^+, \mathbf{Q}^-) = \sum_{p=1}^{m} \mathbf{Z}^p = \sum_{p=1}^{m} \lambda^p \alpha^p \mathbf{r}^p, \tag{82}$$

where λ^p and $\mathbf{r}^p$ are the eigenvalues and eigenvectors of matrix $\mathbf{A}$. Since the matrix $\mathbf{A}$ is diagonalizable we have

$$\mathbf{A} = \mathbf{R}\mathbf{\Lambda}\mathbf{R}^{-1}, \tag{83}$$

where $\mathbf{\Lambda}$ is the diagonal matrix of the eigenvalues of $\mathbf{A}$ and the columns of matrix $\mathbf{R}$ are the corresponding eigenvectors of $\mathbf{A}$. Then we can derive the decomposition of the source term

$$\mathbf{\Psi}^+ = \mathbf{R}\mathbf{\Lambda}^+\mathbf{\Lambda}^{-1}\sigma = \frac{1}{2}(\mathbf{I} + |\mathbf{A}|\mathbf{A}^{-1})\mathbf{\Psi}, \tag{84}$$

$$\mathbf{\Psi}^- = \mathbf{R}\mathbf{\Lambda}^-\mathbf{\Lambda}^{-1}\sigma = \frac{1}{2}(\mathbf{I} - |\mathbf{A}|\mathbf{A}^{-1})\mathbf{\Psi}, \tag{85}$$

where $\mathbf{\Lambda}^+$ and $\mathbf{\Lambda}^-$ are the positive and negative parts of $\mathbf{\Lambda}$ so that $\mathbf{\Lambda}^+ + \mathbf{\Lambda}^- = \mathbf{\Lambda}$. The jump of unknown function is decomposed to the eigenvectors of matrix $\mathbf{A}$

$$\mathbf{Q}^+ - \mathbf{Q}^- = \mathbf{R}\gamma. \tag{86}$$

Because

$$\mathbf{A} = \mathbf{R}(\mathbf{\Lambda}^+ + \mathbf{\Lambda}^-)\mathbf{R}^{-1} = \mathbf{A}^+ + \mathbf{A}^- \tag{87}$$

and

$$\gamma = \mathbf{R}^{-1}(\mathbf{Q}^+ - \mathbf{Q}^-) \tag{88}$$

we get

$$\begin{aligned} &\mathbf{A}^+(\mathbf{Q}^+ - \mathbf{Q}^-) + \mathbf{A}^-(\mathbf{Q}^+ - \mathbf{Q}^-) + \mathbf{R}\mathbf{\Lambda}^+\mathbf{\Lambda}^{-1}\sigma + \mathbf{R}\mathbf{\Lambda}^-\mathbf{\Lambda}^{-1}\sigma = \\ &= \mathbf{R}\mathbf{\Lambda}^+\gamma + \mathbf{R}\mathbf{\Lambda}^-\gamma + \mathbf{R}\mathbf{\Lambda}^+\mathbf{\Lambda}^{-1}\sigma + \mathbf{R}\mathbf{\Lambda}^-\mathbf{\Lambda}^{-1}\sigma = \\ &= \mathbf{R}\left[\mathbf{\Lambda}^-(\gamma + \mathbf{\Lambda}^{-1}\sigma) + \mathbf{\Lambda}^-(\gamma + \mathbf{\Lambda}^{-1}\sigma)\right] = \sum_{p=1}^{m} \lambda^{+,p}\alpha^p\mathbf{r}^p + \sum_{p=1}^{m} \lambda^{-,p}\alpha^p\mathbf{r}^p. \end{aligned} \tag{89}$$

The decomposition (89) is used for construction of fluctuation by the same way as in the following part 5.4 in the case of flux-difference splitting method. It can be shown (Bermudez & Vasquez, 1994), that such approximation preserves steady state "lake at rest" for the appropriate approximation of the source terms.

5.3.2 Positive semidefiniteness

Roe's solver is not positive semidefinite in general. The main reason is in using the linearization to construct the approximate Jacobi matrix.

5.4 Flux-difference splitting scheme

The main idea of this method is to decompose the flux difference into the combination of linearly independent vectors - in fact, this approach is a generalization of the decomposition used in Roe's solver. The suitable choice of these vectors are required because of consistency with the model. Such a suitable choice can be approximation of the eigenvectors of the Jacobi matrix of solved system. Sometimes it is preferable to add the approximation of the source term $\mathbf{\Psi}_{j+1/2}$ to this decomposition

$$\mathbf{f}(\mathbf{Q}^{+}_{j+1/2}) - \mathbf{f}(\mathbf{Q}^{-}_{j+1/2}) - \mathbf{\Psi}_{j+1/2} = \sum_{p=1}^{m} \gamma^{p}_{j+1/2}\mathbf{r}^{p}_{j+1/2}. \tag{90}$$

The fluctuation are defined based on this decomposition as

$$\begin{aligned} \mathcal{A}^{-}(\mathbf{Q}^{-}_{j+1/2}, \mathbf{Q}^{+}_{j+1/2}) &= \sum_{p=1, s^{p}_{j+1/2}<0}^{m} \gamma^{p}_{j+1/2}\mathbf{r}^{p}_{j+1/2}, \\ \mathcal{A}(\mathbf{Q}^{+}_{j-1/2}, \mathbf{Q}^{-}_{j+1/2}) &= \mathbf{f}(\mathbf{Q}^{-}_{j+1/2}) - \mathbf{f}(\mathbf{Q}^{+}_{j-1/2}) - \mathbf{\Psi}_{j}, \\ \mathcal{A}^{+}(\mathbf{Q}^{-}_{j+1/2}, \mathbf{Q}^{+}_{j+1/2}) &= \sum_{p=1, s^{p}_{j+1/2}>0}^{m} \gamma^{p}_{j+1/2}\mathbf{r}^{p}_{j+1/2}. \end{aligned} \tag{91}$$

This idea will be used in decompositions based on augmented system, which will be described later. In fact, the idea is very similar to the approach in (Bermudez & Vasquez, 1994)

5.4.1 Steady states

The preserving of steady states can be described very easily. If the left side of the relation (90) is equal to zero (it is condition for steady state) then all coefficients $\gamma^{p}_{j+1/2} = 0$ because of linearly independence of the vectors $\mathbf{r}^{p}_{j+1/2}$. Therefore the fluctuations are equal to zero too.

5.5 HLLE scheme

When the special choice of the characteristic speeds called the Einfeldt speeds is used, the HLL solver is called HLLE. The Einfeldt speeds are defined by

$$\begin{aligned} s^{1}_{j+1/2} &= \min_{p}\{\min\{\lambda^{p}_{jl}, \lambda^{p}_{j+1/2}, 0\}\}, \\ s^{2}_{j+1/2} &= \max_{p}\{\max\{\lambda^{p}_{jr}, \lambda^{p}_{j+1/2}, 0\}\}, \end{aligned} \tag{92}$$

where λ^{p}_{jl} are eigenvalues of the matrix $\mathbf{A}_{jl} = \mathbf{f}'(\mathbf{Q}^{-}_{j+1/2})$ and λ^{p}_{jr} are eigenvalues of the matrix $\mathbf{A}_{jr} = \mathbf{f}'(\mathbf{Q}^{+}_{j+1/2})$, $\lambda^{p}_{j+1/2}$ are the Roe's speeds. This choice of the speeds leads to smaller amount of numerical diffusion, for details see (Einfeldt, 1988).

5.6 Decompositions based on augmented system

This procedure is based on the extension of the system (10) by other equations (for simplicity we omit viscous term). This was derived in (George, 2008) for the shallow water flow. The advantage of this step is in the conversion of the nonhomogeneous system to the homogeneous one. In the case of urethra flow we obtain the system of four equations, where the augmented vector of unknown functions is $\mathbf{w} = [a, q, \frac{a_0}{\beta}, \beta]^T$. Furthermore we formally augment this system by adding components of the flux function $\mathbf{f}(\mathbf{u})$ to the vector of the unknown functions. We multiply balance law (7) by Jacobian matrix $\mathbf{f}'(\mathbf{u})$ and obtain following relation

$$\mathbf{f}'(\mathbf{q})\mathbf{q}_t + \mathbf{f}'(\mathbf{q})[\mathbf{f}(\mathbf{q})]_x = \mathbf{f}'(\mathbf{q})\psi(\mathbf{q}, x). \tag{93}$$

Because of $\mathbf{f}'(\mathbf{q})\mathbf{q}_t = [\mathbf{f}(\mathbf{q})]_t$ we obtain hyperbolic system for the flux function

$$[\mathbf{f}(\mathbf{q})]_t + \mathbf{f}'(\mathbf{q})[\mathbf{f}(\mathbf{q})]_x = \mathbf{f}'(\mathbf{q})\psi(\mathbf{q}, x). \tag{94}$$

In the case of the urethra fluid flow modelling we add only one equation for the second component of the flux function i.e. $\phi = av^2 + \frac{a^2}{2\rho\beta}$ (the first component q is unknown function of the original balance law), which has the form

$$\phi_t + (-v^2 + \frac{a}{2\rho\beta})(av)_x + 2v\phi_x - \frac{2av}{\rho}\left(\frac{a_0}{\beta}\right)_x - \frac{a^2 v}{\rho\beta^2}\beta_x = 0. \tag{95}$$

Finally augmented system can be written in the nonconservative form

$$\begin{bmatrix} a \\ q \\ \phi \\ \frac{a_0}{\beta} \\ \beta \end{bmatrix}_t + \begin{bmatrix} 0 & 1 & 0 & 0 & 0 \\ -\frac{q^2}{a^2} + \frac{a}{\rho\beta} & 2\frac{q}{a} & 0 & -\frac{a}{\rho} & -\frac{a^2}{\rho\beta^2} \\ 0 & -\frac{q^2}{a^2} + \frac{a}{\rho\beta} & 2\frac{q}{a} & 2\frac{q}{\rho} & -\frac{aq}{\rho\beta^2} \\ 0 & 0 & 0 & 0 & 0 \\ 0 & 0 & 0 & 0 & 0 \end{bmatrix} \begin{bmatrix} a \\ q \\ \phi \\ \frac{a_0}{\beta} \\ \beta \end{bmatrix}_x = \mathbf{0}, \tag{96}$$

briefly $\mathbf{w}_t + \mathbf{B}(\mathbf{w})\mathbf{w}_x = \mathbf{0}$, where matrix $\mathbf{B}(\mathbf{w})$ has following eigenvalues

$$\lambda^1 = v - \sqrt{\frac{a}{\rho\beta}}, \lambda^2 = v + \sqrt{\frac{a}{\rho\beta}}, \lambda^3 = 2v, \lambda^4 = \lambda^5 = 0 \tag{97}$$

and corresponding eigenvectors

$$\mathbf{r}^1 = \begin{bmatrix} 1 \\ \lambda^1 \\ (\lambda^1)^2 \\ 0 \\ 0 \end{bmatrix}, \mathbf{r}^2 = \begin{bmatrix} 1 \\ \lambda^2 \\ (\lambda^2)^2 \\ 0 \\ 0 \end{bmatrix}, \mathbf{r}^3 = \begin{bmatrix} 0 \\ 0 \\ 1 \\ 0 \\ 0 \end{bmatrix}, \mathbf{r}^4 = \begin{bmatrix} \frac{-a}{\rho\lambda^1\lambda^2} \\ 0 \\ \frac{a}{\rho} \\ 1 \\ 0 \end{bmatrix}, \mathbf{r}^5 = \begin{bmatrix} \frac{-a^2}{\rho\beta^2\lambda^1\lambda^2} \\ 0 \\ \frac{a^2}{2\rho\beta^2} \\ 0 \\ 1 \end{bmatrix}. \tag{98}$$

We have five linearly independent eigenvectors. The approximation is chosen to be able to prove the consistency and provide the stability of the algorithm. In some special cases this scheme is conservative and we can guarantee the positive semidefiniteness, but only under the additional assumptions (see (Brandner at al., 2009)).

The fluctuations are then defined by

$$\begin{aligned} \mathbf{A}^-(\mathbf{Q}^-_{j+1/2}, \mathbf{Q}^+_{j+1/2}) &= \begin{bmatrix} 0\,1\,0\,0\,1 \\ 0\,1\,0\,0\,1 \end{bmatrix} \cdot \sum_{p=1, s^{p,n}_{j+1/2}<0}^{m} \gamma^p_{j+1/2}\mathbf{r}^p_{j+1/2}, \\ \mathbf{A}^+(\mathbf{Q}^-_{j+1/2}, \mathbf{Q}^+_{j+1/2}) &= \begin{bmatrix} 0\,1\,0\,0\,1 \\ 0\,1\,0\,0\,1 \end{bmatrix} \cdot \sum_{p=1, s^{p,n}_{j+1/2}>0}^{m} \gamma^p_{j+1/2}\mathbf{r}^p_{j+1/2}, \\ \mathbf{A}(\mathbf{Q}^+_{j-1/2}, \mathbf{Q}^-_{j+1/2}) &= \mathbf{f}(\mathbf{Q}^-_{j+1/2}) - \mathbf{f}(\mathbf{Q}^+_{j-1/2}) - \mathbf{\Psi}(\mathbf{Q}^-_{j+1/2}, \mathbf{Q}^+_{j-1/2}), \end{aligned} \tag{99}$$

where $\mathbf{\Upsilon}(\mathbf{Q}_{j+1/2}^{-}, \mathbf{Q}_{j-1/2}^{+})$ is a suitable approximation of the source term and $\mathbf{r}_{j+1/2}^{p}$ are suitable approximations of the eigenvectors (98).

5.6.1 Steady states

The steady state for the augmented system means $\mathbf{B}(\mathbf{w})\mathbf{w}_x = \mathbf{0}$, therefore $\mathbf{w}_x$ is a linear combination of the eigenvectors corresponding to the zero eigenvalues. The discrete form of the vector $\Delta\mathbf{w}$ corresponds to the certain approximation of these eigenvectors. It can be shown (Brandner at al., 2009) that

$$\Delta \begin{bmatrix} A \\ Q \\ \Phi \\ \frac{a_0}{\beta} \\ \beta \end{bmatrix} = \begin{bmatrix} \frac{\bar{A}}{\rho} \frac{1}{\overline{\overline{\lambda^1 \lambda^2}}} \\ 0 \\ \frac{\bar{A}}{\rho} \frac{\widetilde{\lambda^1 \lambda^2}}{\overline{\overline{\lambda^1 \lambda^2}}} \\ 1 \\ 0 \end{bmatrix} \Delta \left(\frac{a_0}{\beta} \right) + \begin{bmatrix} \frac{\bar{A}^2}{\rho \beta_{j+1} \beta_j} \frac{1}{\overline{\overline{\lambda^1 \lambda^2}}} \\ 0 \\ \frac{\bar{A}^2}{\rho \beta_{j+1} \beta_j} \frac{\widetilde{\lambda^1 \lambda^2}}{\overline{\overline{\lambda^1 \lambda^2}}} - \frac{\tilde{A}^2}{2\rho \beta_{j+1} \beta_j} \\ 0 \\ 1 \end{bmatrix} \Delta \beta. \tag{100}$$

Therefore we use vectors on the RHS of (100) as approximations of the fourth and fifth eigenvectors of the matrix $\mathbf{B}(\mathbf{w})$ to preserves general steady state.

5.6.2 Positive semidefiniteness

Positive semidefiniteness of this scheme is shown in (George, 2008) for the case of shallow water equation. It is based on a special choice of approximations of the eigenvectors (98). This, in the case of urethra flow, is more complicated because of structure of the eigenvectors. Some necessary conditions for approximation of these eigenvectors are presented in (Brandner at al., 2009).

5.7 The other methods

Many other methods exist that are suitable for solving nonhomogeneous hyperbolic PDEs. These methods are often derived from the ideas described above. Some approaches are very different. However, we describe at least briefly the two of them.

5.7.1 ADER schemes

ADER scheme is an approach for constructing conservative nonlinear finite volume type methods of arbitrary accuracy in space and time. The first step in ADER algorithm is the reconstruction of point-wise values of solution from cell averages at time t_n via high-order polynomials. To design non-oscillatory schemes we can use for example WENO reconstruction. After this step we solve High-order Riemann problem (Derivative or Generalized Riemann problem). This problem is defined as a clasical Riemann problem with polynomial-wise initial condition (polynomials arise in reconstruction step). The solution of High-order Riemann problem is used for numerical fluxes evaluation. For details see for example (Toro & Titarev, 2006) and (Toro & Titarev, 2002).

5.7.2 Flux-vector splitting

This approach is based on solving Riemann problem for the augmented formulation of the system (for example (8)) and nonconservative reformulation of the zero-order terms of the right-hand-side of the equations in the form (46). This solution is decomposed to the stationary contact discontinuity described by the suitable Rankine-Hogoniot condition and the remaining part of the solution solved by the flux splitting for conservation systems. For details see for example (Gosse, 2001).

6. Numerical experiments

Now we present several numerical experiments of the urethra flow based on mathematical model (10) and shallow water flow based on model (5). The experiments are presented only for illustration of properties of some described methods. Detailed comparison of all presented methods can be found in using literature.

6.1 Steady urethra flow

This experiment simulates the urethra flow with no viscosity. The parameter $p_e = 1200$ and parameter β is illustrated at the figure 2. The initial condition is defined by the following

$$p(x,0) = \begin{cases} 3000 \text{ Pa}, & \text{for } x = 0 \\ 500 \text{ Pa}, & \text{else}, \end{cases}, v(x,0) = \begin{cases} 1 \text{ m/s}, & \text{for } x = 0 \\ 0 \text{ m/s}, & \text{else}. \end{cases} \tag{101}$$

It is used the constant input discharge at the point $x = 0$ during the whole simulation. The functions described discharge and cross section of the tube are extrapolated from the boundary st the outflow ($x = 0.19$). At the last figures 2 and 3 we can see the difference between the methods. While the method based on decomposition of augmented system (Sec.5.6) preserve steady state ($q \equiv const.$) the central-upwind method (Sec.5.2) do not preserve such general steady state.

6.2 Shallow water flow

These experiments compare solution computed by the method based on decomposition of augmented system (see section 5.6) using piecewise constant reconstruction (first order method) and the piecewise linear reconstruction (second order method). The preservation of general steady state and positive semidefiniteness of the scheme is shown.

In all experiments we used 200 grid points and the following bottom topography

$$b(x) = \begin{cases} \frac{1}{4}\left(1 + \cos\frac{\pi(x-0.5)}{0.1}\right) & \text{if } x \in \langle 0.4, 0.6 \rangle, \\ 0 & \text{otherwise.} \end{cases} \tag{102}$$

6.2.1 Small perturbations from the steady state

This experiment simulates propagation of the small perturbation from the steady state "lake at rest". The initial condition is given by

$$h + b = \begin{cases} 1 + 10^{-5}, & x \in \langle 0.1, 0.2 \rangle, \\ 1, & \text{otherwise.} \end{cases} \tag{103}$$

Water height and discharge are extrapolated at the boundaries. The propagation of the perturbation is illustrated at the Fig. 4

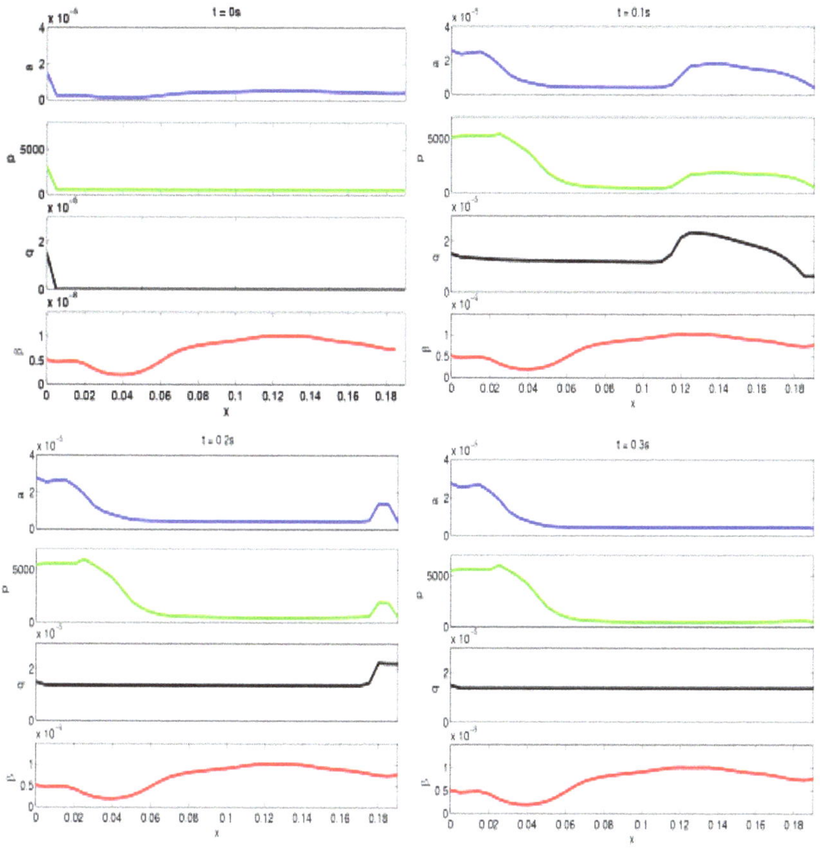

Fig. 2. Urethra flow simulation solved by the augmented system decomposition

6.2.2 Drainage of the reservoir

Here the initial condition is defined by

$$h + b = 0.8, \qquad q = 0. \tag{104}$$

This simulates reservoir initially at rest, draining onto a dry bed through its boundaries. So we use open boundary conditions at the right boundary, and reflecting boundary conditions at the left boundary. More specifically it is implemented by zero discharge $Q(0,t) = 0$ during the whole simulation, while the water height is extrapolated from the domain. At the outflow (right boundary), the boundary conditions are implemented as follows: if the flow is supercritical, $H + B$ and Q are extrapolated from the interior of the domain, while if the flow is subcritical, the water height H is prescribed $H(1,t) = 10^{-16}$ and Q is extrapolated. Solution is depicted in Fig. 5.

This flow demonstrates also the positive semidefinitness of the scheme which insures the water depth remains non-negative.

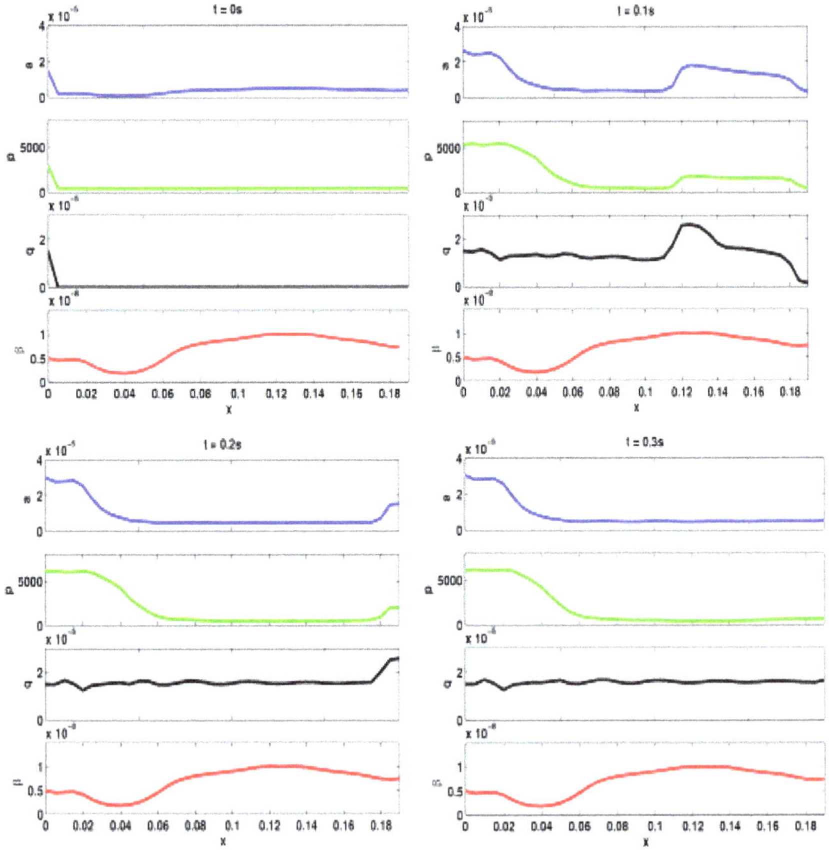

Fig. 3. Urethra flow simulation solved by the central-upwind scheme

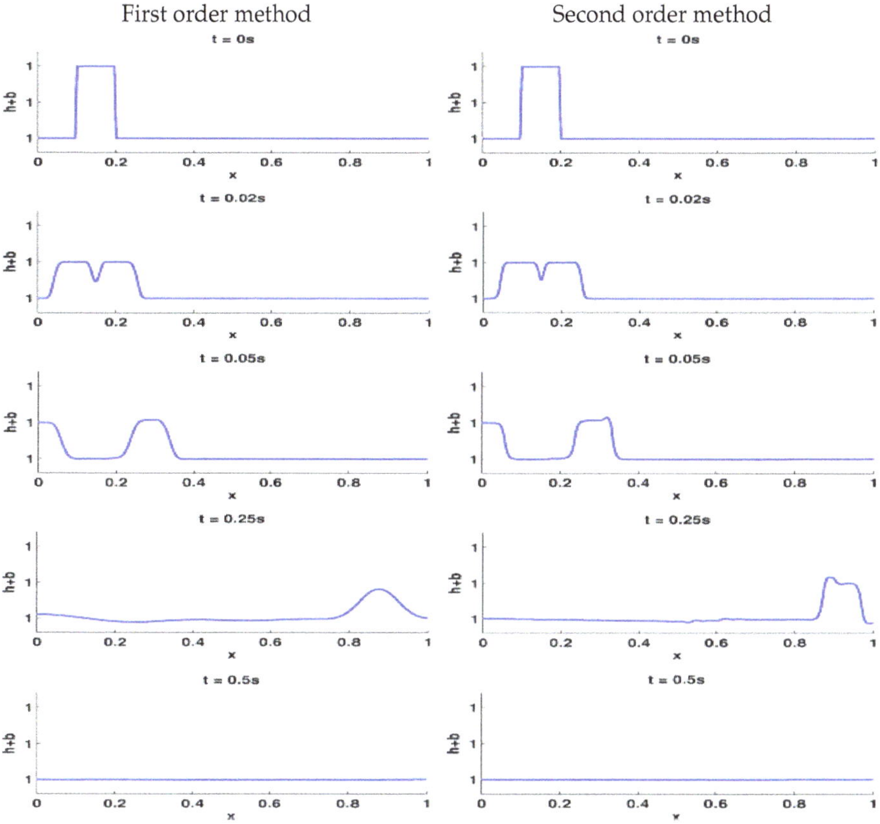

Fig. 4. Propagation of small perturbation of the water height from the steady state solution through the centered contracted channel.

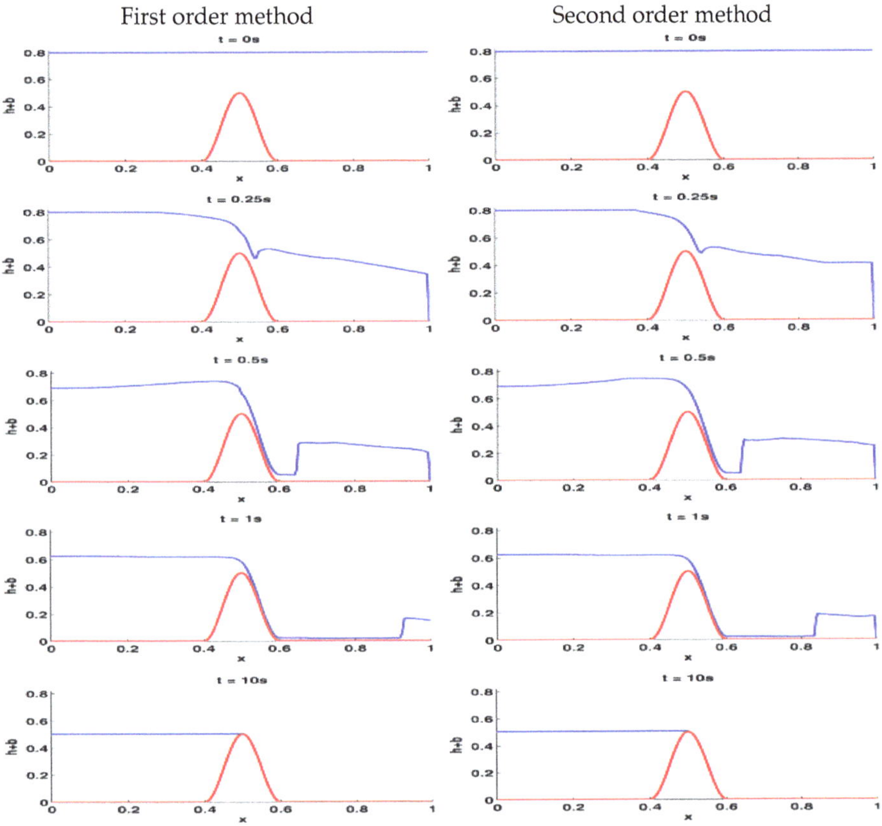

Fig. 5. Time evolution of the water height through the centered contracted channel from the initial condition to the general steady state solution.

7. Conclusion

In this chapter we have tried to summarize the basic approaches for solving hyperbolic equations with non-zero right hand side. We described all methods using the flux-difference or flux-vector splitting. We paid special attention to approximations that maintain steady states. Other desired property of the proposed schemes is positive semidefiniteness. It should be noted that it is very difficult to construct schemes that meet multiple properties simultaneously. For example, the central and central-upwind schemes are positive semidefinite, but maintain only the special steady state lake at rest. Similarly, we can explore the links between the just mentioned two properties and the order of schemes, the amount of numerical diffusion, etc. It can not be unambiguously decided which method is the best one. Before we choose the method that we have to decide which properties are essential for the concrete simulation. We hope that our overview provides to readers at least partial guidance on how to choose the most appropriate numerical approach to solve nonhomogeneous hyperbolic partial differential equations by finite volume methods.

8. References

Audusse, E.; Bouchut, F.; Bristeau, M.-O.; Klein, R. & Perthame, B. (2004). A fast and stable well-balanced scheme with hydrostatic reconstruction for shallow eater flows. *SIAM Journal on Scientific Computing*, Vol. 25, No. 6, pp. 2050-2065

Bermudez, A. & Vasquez, E. (1994). Upwind Methods for Conservation Laws with Source Terms. *Computers Fluids*, Vol. 23. No. 8, pp. 1049-1071

Brandner, M.; Egermaier, J. & Kopincová, H. (2009). Augmented Riemann solver for urethra flow modelling. *Mathematics and Computers in Simulations*, Vol. 80, No. 6, pp. 1222-1231

Črnjarič-Zič, N.; Vukovič, S. & Sopta, L. (2004). Balanced finite volume WENO and central WENO schemes for the shallow water and the open-channel flow equations. *Journal of Computational Physics*, Vol. 200, No. 2, pp. 512-548,

Cunge, J.; Holly, F. & Verwey, A. (1980). Practical Aspects of Computational River Hydraulics. Pitman Publishing Ltd,

Einfeldt, B. (1988). On Godunov-Type Methods for Gas Dynamics. *SIAM Journal on Numerical Analysis*, Vol. 25, pp. 294-318.

George, D., L. (2008). Augmented Riemann Solvers for the Shallow Water Equations over Variable Topography with Steady States and Inundation. *Journal of Computational Physics*, Vol. 227, pp. 3089-3113.

Gosse, L. (2001). A Well-Balanced Scheme Using Non-Conservative Products Designed for Hyperbolic Systems of Conservation Laws With Source Terms. *Mathematical Models and Methods in Applied Science*, Vol. 11, No. 2, pp. 339-365.

Kurganov, A. & Petrova, G. (2000). Central Schemes and Contact Discontinuities. *Mathematical Modelling and Numerical Analysis*, Vol. 34, pp. 1259-1275

Kurganov, A. & Tadmor, E. (2000). New High-Resolution Central Schemes for Nonlinear Conservation Laws and Convection-Diffusion Equations. *Journal of Computational Physics*, Vol. 160, No. 1, pp. 241-282

Kurganov, A. & Levy, D. (2002). Central-Upwind Schemes for the Saint-Venant System. *Mathematical Modelling and Numerical Analysis*, Vol. 36, No. 3, pp. 397-425

Le Floch, P., G. (1989). Shock Waves for Nonlinear Hyperbolic Systems in Nonconservative Form. IMA Preprint Series

Le Floch, P., G. & Tzavaras, A., E. (1999). Representation of Weak Limits and Definition of Nonconservative Products. *SIAM Journal on Mathematical Analysis*, Vol. 30, No. 6, pp. 1309-1342.

Le Floch, P., G.; Castro, M., J.; Muñoz-Ruiz & M., L., Parès, C. (2008). Why Many Theories of Shock Waves Are Necessary: Convergence Error in Formally Path-Consistent Schemes. *Journal of Computational Physics*, Vol. 227, No. 17, pp. 8107-8129.

LeVeque, R., J. (2004). *Finite Volume Methods for Hyperbolic Problems*, Cambridge Texts in Applied Mathematics (No. 31), Cambridge University Press

LeVeque, R., J. & George, D., L. (2004). High-Resolution Finite Volume Methods for Shallow Water Equations with Bathymetry and Dry States. *Proceedings of Third International Workshop on Long-Wave Runup Models*, Catalina

Parès, C. (2006). Numerical Methods for Nonconservative Hyperbolic Systems: a Theoretical Framework. *SIAM Journal on Numerical Analysis*, Vol. 44, No. 1, pp. 300-321.

Stergiopulos, N.; Tardy, Y. & Meister, J.-J. (1993). Nonlinear Separation of Forward and Backward Running Waves in Elastic Conduits. *Journal of Biomechanics*, Vol. 26, pp. 201-209

Toro, E., F. & Titarev V., A. (2006). Derivative Riemann Solvers for Systems of Conservation Laws and ADER Methods. *Journal of Computational Physics*, Vol. 212, No. 1, pp. 150-165.

Toro, E., F. & Titarev V., A. (2002). Solution of the Generalized Riemann Problem for Advection-Reaction Equations, *Proceedings of the Royal Society A*, Vol. 458, No. 2018, pp. 271-281.

2

The Complete Flux Scheme for Conservation Laws in Curvilinear Coordinates

J.H.M. ten Thije Boonkkamp and M.J.H. Anthonissen
Department of Mathematics and Computer Science,
Eindhoven University of Technology, Eindhoven
The Netherlands

1. Introduction

Conservation laws are ubiquitous in continuum physics, they occur in disciplines like fluid mechanics, combustion theory, plasma physics, etc. These conservation laws are often of advection-diffusion-reaction type, describing the interplay between different processes such as advection or drift, diffusion or conduction and chemical reactions or ionization. Examples are the conservation equations for reacting flow [8] or plasmas [9]. Sometimes, these conservation laws hold in spherical or cylindrical geometries, and in such cases it is convenient to reformulate the conservation laws in the corresponding coordinate system. In combustion theory, for example, the study of spherical and cylindrical flames is useful for finding parameters such as burning velocity or flame curvature [1].

For space discretization of these conservation laws we consider the finite volume method in combination with the complete flux scheme to approximate the fluxes at the cell interfaces. The complete flux scheme for Cartesian coordinates is introduced in [13]. The purpose of this contribution is to generalize the complete flux scheme to conservation laws in spherical or cylindrical coordinates.

The development of the complete flux scheme is inspired by papers by Thiart [10, 11]. The basic idea of the complete flux scheme is to compute the numerical flux at a cell interface from a local (one-dimensional) boundary value problem for the *entire* equation, including the source term. As such, the scheme is a generalization of the exponential scheme, where the flux is determined from a local, constant coefficient, homogeneous equation [4, 6]. Our approach is to first derive an integral representation for the flux, and subsequently apply suitable quadrature rules to obtain the numerical flux. As a consequence, the numerical flux is the superposition of a homogeneous and inhomogeneous flux, corresponding to the advection-diffusion operator and the source term, respectively. The resulting discretization has a three-point coupling in each spatial direction, shows uniform second order convergence and virtually never generates spurious oscillations [13]. The purpose of this chapter is to extend this approach to conservation laws, where the advection-diffusion operation is formulated in spherical or cylindrical coordinates. Another important issue is the extension to time-dependent problems. The key idea is then to consider the time derivative as a source term, and to include it in the inhomogeneous flux. The resulting implicit ODE system often has small dissipation and dispersion errors [15].

We have organised our paper as follows. The finite volume method for conservation laws in spherical and cylindrical coordinates is outlined in Section 2. In Section 3 we briefly repeat the complete flux scheme for stationary, one-dimensional conservation laws in Cartesian coordinates. The extension to spherical coordinates is presented in Section 4, and the next logical extension to cylindrical coordinates, is discussed in Section 5. How to deal with time dependent conservation laws is demonstrated in Section 6 for spherical coordinates. As an example, we present in Section 7 the numerical solution of a premixed, spherical flame, and finally in Section 8, we give a summary and formulate conclusions.

2. Finite volume discretization

In this section we outline the finite volume method (FVM) for a generic conservation law of advection-diffusion-reaction type, defined on a domain in $\mathbb{R}^d$ $(d = 1, 2, 3)$. Therefore, consider the following model equation

$$\frac{\partial \varphi}{\partial t} + \nabla\cdot(\boldsymbol{u}\varphi - \varepsilon\nabla\varphi) = s, \tag{2.1}$$

where $\boldsymbol{u}$ is a mass flux or (drift) velocity, $\varepsilon \geq \varepsilon_{\min} > 0$ a diffusion coefficient, and s a source term describing, e.g., chemical reactions or ionization. The unknown φ is then the mass fraction of one of the constituent species in a chemically reacting flow or a plasma. The parameters ε and s are usually (complicated) functions of φ whereas the vector field $\boldsymbol{u}$ has to be computed from (flow) equations corresponding to (2.1). However, for the sake of discretization, we will consider these parameters as given functions of the spatial coordinates $\boldsymbol{x}$ and the time t. Moreover, in the derivation of the numerical flux, we assume that the vector field $\boldsymbol{u}$ is incompressible, i.e.,

$$\nabla\cdot\boldsymbol{u} = 0. \tag{2.2}$$

Equation (2.1) is a prototype of a conservation law for a mixture, defining the mass balance for φ, and equation (2.2) is a simplified version of the corresponding continuity equation, describing conservation of mass or charge in the mixture.

Associated with equation (2.1) we introduce the flux vector $\boldsymbol{f}$, defined by

$$\boldsymbol{f} := \boldsymbol{u}\varphi - \varepsilon\nabla\varphi. \tag{2.3}$$

Consequently, equation (2.1) can be concisely written as $\partial\varphi/\partial t + \nabla\cdot\boldsymbol{f} = s$. Integrating this equation over a fixed domain $\Omega \subset \mathbb{R}^d$ and applying Gauss' theorem we obtain the integral form of the conservation law, i.e.,

$$\frac{\mathrm{d}}{\mathrm{d}t}\int_\Omega \varphi\,\mathrm{d}V + \oint_\Gamma \boldsymbol{f}\cdot\boldsymbol{n}\,\mathrm{d}S = \int_\Omega s\,\mathrm{d}V, \tag{2.4}$$

where $\boldsymbol{n}$ is the outward unit normal on the boundary $\Gamma = \partial\Omega$. This equation is the basic conservation law, which reduces to (2.1) provided φ is smooth enough.

In the FVM we cover the domain with a finite number of disjunct control volumes or cells Ω_{j} and impose the integral form (2.4) for each of these cells. The index j is an index vector for multi-dimensional problems. We restrict ourselves to uniform tensor product grids for an orthogonal, curvilinear coordinate system $\boldsymbol{\xi} = (\xi^1, \xi^2, \xi^3)$ and adopt the vertex-centred approach [16], i.e., we first choose the grid points $\boldsymbol{\xi}_{\mathrm{j}} = (\xi^1_j, \xi^2_k, \xi^3_l)$ with $\mathrm{j} = (j,k,l)$, where the unknown φ has to be approximated and subsequently choose the control volume $\Omega_{\mathrm{j}} = [\xi^1_{j-1/2}, \xi^1_{j+1/2}] \times [\xi^2_{k-1/2}, \xi^2_{k+1/2}] \times [\xi^3_{l-1/2}, \xi^3_{l+1/2}]$ with $\xi^1_{j\pm 1/2} := \frac{1}{2}(\xi^1_j + \xi^1_{j\pm 1})$ etc. The boundary $\Gamma_{\mathrm{j}} = \partial\Omega_{\mathrm{j}}$ is then the union of six interface surfaces $\Gamma_{\mathrm{j},\mathrm{j}\pm\mathrm{e}^i}$ $(i = 1,2,3)$ where, e.g., $\Gamma_{\mathrm{j},\mathrm{j}+\mathrm{e}^1} := \{\xi^1_{j+1/2}\} \times [\xi^2_{k-1/2}, \xi^2_{k+1/2}] \times [\xi^3_{l-1/2}, \xi^3_{l+1/2}]$ is the interface through $(\xi^1_{j+1/2}, \xi^2_k, \xi^3_l)$ and perpendicular to the line segment connecting ξ_{j} and $\xi_{\mathrm{j}+\mathrm{e}^1}$. The (integral) conservation law for such a control volume reads

$$\frac{\mathrm{d}}{\mathrm{d}t} \int_{\Omega_{\mathrm{j}}} \varphi \,\mathrm{d}V + \sum_{\mathrm{k}\in\mathcal{N}(\mathrm{j})} \int_{\Gamma_{\mathrm{j,k}}} \boldsymbol{f}\cdot\boldsymbol{n}\,\mathrm{d}S = \int_{\Omega_{\mathrm{j}}} s \,\mathrm{d}V, \tag{2.5}$$

where $\mathcal{N}(\mathrm{j}) = \{\mathrm{j} \pm \mathrm{e}^i \mid i = 1,2,3\}$ is the index set of neighbouring grid points of $\boldsymbol{\xi}_{\mathrm{j}}$ and where $\Gamma_{\mathrm{j,k}}$ is the face of the boundary Γ_{j} connecting the adjacent cells Ω_{j} and Ω_{k}. The unit normal $\boldsymbol{n}$ on $\Gamma_{\mathrm{j,k}}$ is directed from $\boldsymbol{\xi}_{\mathrm{j}}$ to $\boldsymbol{\xi}_{\mathrm{k}}$. Obviously, the volume element $\mathrm{d}V$ and the surface elements $\mathrm{d}S$ have to be expressed in terms of the curvilinear coordinates $\boldsymbol{\xi}$. Approximating the volume and surface integrals in (2.5) by the midpoint rule, we obtain the following semi-discrete conservation law for $\varphi_{\mathrm{j}}(t) \approx \varphi(\boldsymbol{\xi}_{\mathrm{j}}, t)$, i.e.,

$$\dot{\varphi}_{\mathrm{j}}(t) V_{\mathrm{j}} + \sum_{\mathrm{k}\in\mathcal{N}(\mathrm{j})} (\boldsymbol{F}\cdot\boldsymbol{n})_{\mathrm{j,k}} A_{\mathrm{j,k}} = s_{\mathrm{j}}(t) V_{\mathrm{j}}, \tag{2.6}$$

where V_{j} is the volume of Ω_{j}, $A_{\mathrm{j,k}}$ the area of $\Gamma_{\mathrm{j,k}}$, $\dot{\varphi}_{\mathrm{j}}(t) \approx \partial\varphi/\partial t(\boldsymbol{\xi}_{\mathrm{j}}, t)$ and $s_{\mathrm{j}}(t) = s(\boldsymbol{\xi}_{\mathrm{j}}, t)$. Furthermore, $(\boldsymbol{F}\cdot\boldsymbol{n})_{\mathrm{j,k}}$ is the normal component on $\Gamma_{\mathrm{j,k}}$, at the interface point $\boldsymbol{\xi}_{\mathrm{j,k}} := \frac{1}{2}(\boldsymbol{\xi}_{\mathrm{j}} + \boldsymbol{\xi}_{\mathrm{k}})$ of the numerical flux vector $\boldsymbol{F}$, approximating $(\boldsymbol{f}\cdot\boldsymbol{n})(\boldsymbol{\xi}_{\mathrm{j,k}}, t)$. Obviously, for stationary problems the time derivatives in (2.5) and (2.6) can be discarded.

In this paper we consider the formulation of the conservation law (2.1) in terms of the spherical coordinates (r, ϕ, θ) and the cylindrical coordinates (r, θ, z). In the first case, we assume spherical symmetry, i.e., $\varphi = \varphi(r,t)$ and $\boldsymbol{f} = f(r,t)\boldsymbol{e}_r$. As a typical example we mention a spherical flame; see Section 7. A control volume is then given by the spherical shell $\Omega_j = [r_{j-1/2}, r_{j+1/2}] \times [0, \pi] \times [0, 2\pi)$ and the surface integral over $\Gamma_j = \partial\Omega_j$ can be written as

$$\begin{aligned} \oint_{\Gamma_j} \boldsymbol{f}\cdot\boldsymbol{n}\,\mathrm{d}S &= \int_{r=r_{j+1/2}} \boldsymbol{f}\cdot\boldsymbol{e}_r\,\mathrm{d}S - \int_{r=r_{j-1/2}} \boldsymbol{f}\cdot\boldsymbol{e}_r\,\mathrm{d}S \\ &= 4\pi\big(r^2_{j+1/2} f(r_{j+1/2}, t) - r^2_{j-1/2} f(r_{j-1/2}, t)\big), \end{aligned} \tag{2.7}$$

where we used the shorthand notation $r = r_{j+1/2}$ to denote the sphere $\{r_{j+1/2}\} \times [0, \pi] \times [0, 2\pi)$. Note that this expression for the surface integral of the flux is exact and replaces the second term in (2.5). For the approximation of the volume integrals in (2.5) we apply the midpoint rule, so we find

$$\int_{\Omega_j} s\,\mathrm{d}V \doteq \tfrac{4}{3}\pi(r^3_{j+1/2} - r^3_{j-1/2})\, s_j(t). \tag{2.8}$$

Combining (2.5), (2.7) and (2.8) and using the relation $x^3 - y^3 = (x-y)(x^2+xy+y^2)$ we obtain the semidiscrete conservation law

$$\Delta r\big(r_j^2 + \tfrac{1}{12}\Delta r^2\big)\,\dot{\varphi}_j(t) + r_{j+1/2}^2 F_{j+1/2}(t) - r_{j-1/2}^2 F_{j-1/2}(t) = \Delta r\big(r_j^2 + \tfrac{1}{12}\Delta r^2\big)\,s_j(t), \qquad (2.9)$$

where $F_{j+1/2}(t)$ denotes the numerical flux approximating $f(r_{j+1/2},t)$ etc..

Next, for cylindrical coordinates, we assume cylindrical symmetry, i.e., $\varphi = \varphi(r,z,t)$ and $\boldsymbol{f} = \boldsymbol{f}_r(r,z,t)\boldsymbol{e}_r + \boldsymbol{f}_z(r,z,t)\boldsymbol{e}_z$. In this case a control volume is the cylindrical shell $\Omega_{j,l} = [r_{j-1/2}, r_{j+1/2}] \times [0,2\pi) \times [z_{l-1/2}, z_{l+1/2}]$. The surface integral of the flux over the boundary $\Gamma_{j,l} = \partial\Omega_{j,l}$ contains four terms and is given by

$$\begin{aligned}\oint_{\Gamma_{j,l}} \boldsymbol{f}\cdot\boldsymbol{n}\,\mathrm{d}S &= \int_{r=r_{j+1/2}} f_r\,\mathrm{d}S - \int_{r=r_{j-1/2}} f_r\,\mathrm{d}S + \int_{z=z_{l+1/2}} f_z\,\mathrm{d}S - \int_{z=z_{l-1/2}} f_z\,\mathrm{d}S \\ &\doteq 2\pi\Delta z\big(r_{j+1/2}f_{r,j+1/2,l}(t) - r_{j-1/2}f_{r,j-1/2,l}(t)\big) + \\ &\quad 2\pi\Delta r\, r_j\big(f_{z,j,l+1/2}(t) - f_{z,j,l-1/2}(t)\big),\end{aligned} \qquad (2.10)$$

where for example $r = r_{j+1/2}$ denotes the interface $\{r_{j+1/2}\} \times [0,2\pi) \times [z_{l-1/2}, z_{l+1/2}]$, and likewise for all other interfaces. For the approximation of the volume integrals in (2.5) we use once more the midpoint rule, giving the approximation

$$\int_{\Omega_{j,l}} s\,\mathrm{d}V \doteq 2\pi\Delta r\Delta z\, r_j s_{j,l}(t). \qquad (2.11)$$

Analogous to the previous case, combining (2.5), (2.10) and (2.11) we obtain the semidiscrete conservation law

$$\begin{aligned}&\Delta r\Delta z\, r_j\,\dot{\varphi}_{j,l}(t) + \Delta z\big(r_{j+1/2}F_{r,j+1/2,l}(t) - r_{j-1/2}F_{r,j-1/2,l}(t)\big) + \\ &\quad \Delta r\, r_j\big(F_{z,j,l+1/2}(t) - F_{z,j,l-1/2}(t)\big) = \Delta r\Delta z\, r_j\, s_{j,l}(t),\end{aligned} \qquad (2.12)$$

where $F_{r,j+1/2,l}(t)$ is the numerical flux approximating $f_r(r_{j+1/2}, z_l, t)$ and likewise for $F_{z,j,l+1/2}(t)$. In the following we suppress the explicit dependence on t.

The FVM has to be completed with expressions for the numerical flux. We require that $(\boldsymbol{F}\cdot\boldsymbol{n})_{\mathrm{j,k}}$ depends on φ and a modified source term $\tilde{s}$ in the neighbouring grid points $\boldsymbol{x}_{\mathrm{j}}$ and $\boldsymbol{x}_{\mathrm{k}}$, i.e., we are looking for an expression of the form

$$(\boldsymbol{F}\cdot\boldsymbol{n})_{\mathrm{j,k}} = \alpha_{\mathrm{j,k}}\varphi_{\mathrm{j}} - \beta_{\mathrm{j,k}}\varphi_{\mathrm{k}} + d_{\mathrm{j,k}}\big(\gamma_{\mathrm{j,k}}\tilde{s}_{\mathrm{j}} + \delta_{\mathrm{j,k}}\tilde{s}_{\mathrm{k}}\big), \qquad (2.13)$$

where $d_{\mathrm{j,k}} := |\boldsymbol{x}_{\mathrm{j}} - \boldsymbol{x}_{\mathrm{k}}|$. The variable $\tilde{s}$ includes the source term and an additional terms like the cross flux or time derivative, when appropriate. The derivation of expressions for the numerical flux is detailed in the next sections.

3. Numerical flux for Cartesian coordinates

In this section we outline the derivation of the complete flux scheme for the steady, one-dimensional conservation laws in Cartesian coordinates, which is based on the integral representation of the flux. The derivation is a summary of the theory in [3, 13].

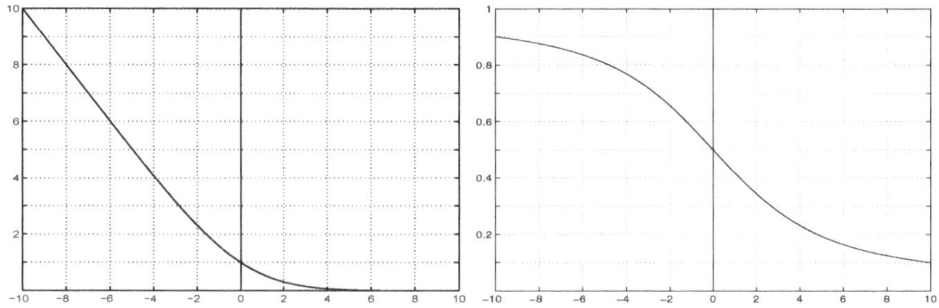

Fig. 1. The Bernoulli function B (left) and the function W (right).

The conservation law can be written as $\mathrm{d}f/\mathrm{d}x = s$ with $f = u\varphi - \varepsilon\,\mathrm{d}\varphi/\mathrm{d}x$. The integral representation of the flux $f_{j+1/2} := f(x_{j+1/2})$ at the cell edge $x_{j+1/2}$ located between the grid points x_j and x_{j+1} is based on the following model boundary value problem (BVP) for the variable φ:

$$\frac{\mathrm{d}}{\mathrm{d}x}\left(u\varphi - \varepsilon\frac{\mathrm{d}\varphi}{\mathrm{d}x}\right) = s, \quad x_j < x < x_{j+1}, \tag{3.1a}$$

$$\varphi(x_j) = \varphi_j, \quad \varphi(x_{j+1}) = \varphi_{j+1}. \tag{3.1b}$$

In accordance with (2.13), we derive an expression for the flux $f_{j+1/2}$ corresponding to the solution of the *inhomogeneous* BVP (3.1), implying that $f_{j+1/2}$ not only depends on u and ε, but also on the source term s. It is convenient to introduce the variables a, P, A and S for $x \in (x_j, x_{j+1})$ by

$$a := \frac{u}{\varepsilon}, \quad P := a\Delta x, \quad A(x) := \int_{x_{j+1/2}}^{x} a(\xi)\,\mathrm{d}\xi, \quad S(x) := \int_{x_{j+1/2}}^{x} s(\xi)\,\mathrm{d}\xi. \tag{3.2}$$

Here, P and A are the Peclet function and Peclet integral, respectively, generalizing the well-known (numerical) Peclet number. Integrating the differential equation $\mathrm{d}f/\mathrm{d}x = s$ from $x_{j+1/2}$ to $x \in (x_j, x_{j+1})$ we get the integral balance $f(x) - f_{j+1/2} = S(x)$. Using the definition of A in (3.2), it is clear that the flux can be rewritten as $f = -\varepsilon\,\mathrm{e}^{A}\,\mathrm{d}(\mathrm{e}^{-A}\,\varphi)/\mathrm{d}x$. Substituting this into the integral balance, isolating the derivative $\mathrm{d}(\mathrm{e}^{-A}\varphi)/\mathrm{d}x$, and integrating from x_j to x_{j+1} we obtain the following expressions for the flux:

$$f_{j+1/2} = f^{\mathrm{h}}_{j+1/2} + f^{\mathrm{i}}_{j+1/2}, \tag{3.3a}$$

$$f^{\mathrm{h}}_{j+1/2} = \left(\mathrm{e}^{-A_j}\varphi_j - \mathrm{e}^{-A_{j+1}}\varphi_{j+1}\right) \Big/ \int_{x_j}^{x_{j+1}} \varepsilon^{-1}\mathrm{e}^{-A}\,\mathrm{d}x, \tag{3.3b}$$

$$f^{\mathrm{i}}_{j+1/2} = -\int_{x_j}^{x_{j+1}} \varepsilon^{-1}\mathrm{e}^{-A}S\,\mathrm{d}x \Big/ \int_{x_j}^{x_{j+1}} \varepsilon^{-1}\mathrm{e}^{-A}\,\mathrm{d}x, \tag{3.3c}$$

where $f^{\mathrm{h}}_{j+1/2}$ and $f^{\mathrm{i}}_{j+1/2}$ are the homogeneous and inhomogeneous part, corresponding to the homogeneous and particular solution of (3.1), respectively.

In the following we assume that u and ε are constant; extension to variable coefficients is discussed in [3, 13]. In this case we can determine all integrals involved. Moreover, substituting the expression for $S(x)$ in (3.3c) and changing the order of integration, we can derive an alternative expression for the inhomogeneous flux. This way we obtain

$$f^{\mathrm{h}}_{j+1/2} = \frac{\varepsilon}{\Delta x}\big(B(-P)\varphi_j - B(P)\varphi_{j+1}\big), \tag{3.4a}$$

$$f^{\mathrm{i}}_{j+1/2} = \Delta x \int_0^1 G(\sigma; P, \sigma_{j+1/2}) s(x(\sigma))\,\mathrm{d}\sigma, \quad x(\sigma) = x_j + \sigma\Delta x. \tag{3.4b}$$

Here $B(z) := z/\left(\mathrm{e}^z - 1\right)$ is the generating function of the Bernoulli numbers, in short Bernoulli function, see Figure 1, $P := u\Delta x/\varepsilon$ is the Peclet number, and $G = G(\sigma; P, \sigma_{\mathrm{b}})$ is the Green's function for the flux, given by

$$G(\sigma; P, \sigma_{\mathrm{b}}) := \begin{cases} \dfrac{1-\mathrm{e}^{-P\sigma}}{1-\mathrm{e}^{-P}} & \text{for} \quad 0 \le \sigma \le \sigma_{\mathrm{b}}, \\[2ex] -\dfrac{1-\mathrm{e}^{P(1-\sigma)}}{1-\mathrm{e}^{P}} & \text{for} \quad \sigma_{\mathrm{b}} < \sigma \le 1, \end{cases} \tag{3.5}$$

see Figure 2. Note that G relates the *flux* to the source term and is different from the usual Green's function, which relates the *solution* to the source term. G is a function of the normalized coordinate $\sigma = (x - x_j)/\Delta x$ $(0 \le \sigma \le 1)$ between x_j and x_{j+1} and depends on the parameters P and σ_{b}, the σ-coordinate of the cell boundary. Obviously, $\sigma_{j+1/2} = \sigma(x_{j+1/2}) = \frac{1}{2}$. For the constant coefficient homogeneous flux we introduce the function

$$f^{\mathrm{h}}_{j+1/2} = \mathcal{F}^{\mathrm{h}}(\varepsilon/\Delta x, P; \varphi_j, \varphi_{j+1}) := \frac{\varepsilon}{\Delta x}\big(B(-P)\varphi_j - B(P)\varphi_{j+1}\big), \tag{3.6}$$

to denote the dependence of $f^{\mathrm{h}}_{j+1/2}$ on the parameter values $\varepsilon/\Delta x$ and P and on the function values φ_j and φ_{j+1}; cf. (2.13). The homogeneous flux (3.6) is the well-known exponential flux [7].

Next, we give the numerical flux $F_{j+1/2}$. For the homogeneous component $F^{\mathrm{h}}_{j+1/2}$ we obviously take (3.6), i.e., $F^{\mathrm{h}}_{j+1/2} = \mathcal{F}^{\mathrm{h}}(\varepsilon/\Delta x, P; \varphi_j, \varphi_{j+1})$. The approximation of the inhomogeneous component $f^{\mathrm{i}}_{j+1/2}$ depends on P. For dominant diffusion ($|P| \ll 1$) the average value of $G(\sigma; P)$ is small, which implies that the inhomogeneous flux is of little importance. On the contrary, for dominant advection ($|P| \gg 1$), the average value of $G(\sigma; P)$

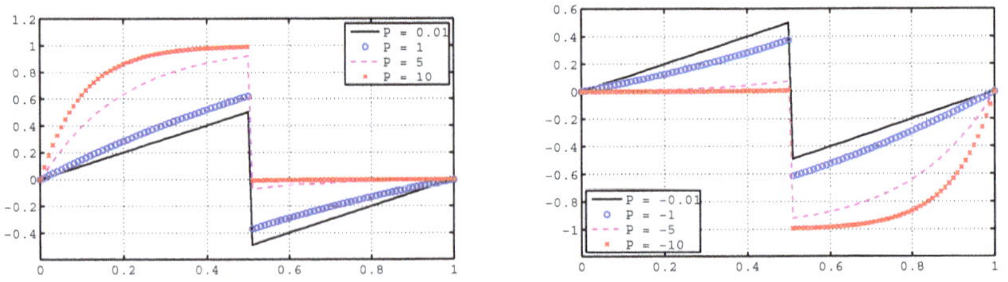

Fig. 2. Green's function for the flux for $P > 0$ (left) and $P < 0$ (right).

on the half interval upwind of $\sigma = \frac{1}{2}$, i.e., the interval $[0, \frac{1}{2}]$ for $u > 0$ and $[\frac{1}{2}, 1]$ for $u < 0$, is much larger than the average value on the downwind half. This means that for dominant advection the upwind value of s is the relevant one, and therefore we replace $s(x(\sigma))$ in (3.4b) by its upwind value $s_{u,j+1/2}$, i.e., $s_{u,j+1/2} = s_j$ if $u \geq 0$ and $s_{u,j+1/2} = s_{j+1}$ if $u < 0$, and evaluate the resulting integral exactly. This way we obtain

$$F_{j+1/2} = \mathcal{F}^{\mathrm{h}}(\varepsilon/\Delta x, P; \varphi_j, \varphi_{j+1}) + \Delta x\left(\tfrac{1}{2} - W(P)\right)s_{u,j+1/2}, \tag{3.7}$$

where $W(z) := (\mathrm{e}^z - 1 - z)/(z(\mathrm{e}^z - 1))$; see Figure 1. From this expression it is once more clear that the inhomogeneous component is only of importance for dominant advection. We refer to (3.7) as the complete flux (CF) scheme, as opposed to the homogeneous flux (HF) scheme for which we omit the inhomogeneous component.

4. Numerical flux for spherical coordinates

Our objective in this section is to extend the derivation in the previous section to spherical coordinates, assuming spherical symmetry.

The stationary conservation law can be written as $\mathrm{d}(r^2 f)/\mathrm{d}r = r^2 s$ with $f = u\varphi - \varepsilon \mathrm{d}\varphi/\mathrm{d}r$. The expression for the flux $f_{j+1/2} := f(r_{j+1/2})$ at the cell boundary $r_{j+1/2}$ is based on the following model BVP for the unknown φ:

$$\frac{1}{r^2}\frac{\mathrm{d}}{\mathrm{d}r}\left(r^2\left(u\varphi - \varepsilon\frac{\mathrm{d}\varphi}{\mathrm{d}r}\right)\right) = s, \quad r_j < r < r_{j+1}, \tag{4.1a}$$

$$\varphi(r_j) = \varphi_j, \quad \varphi(r_{j+1}) = \varphi_{j+1}, \tag{4.1b}$$

where ε and s are sufficiently smooth functions of r. Moreover, we assume that $u > 0$ and, in view of (2.2), u satisfies the relation

$$U := r^2 u = \mathrm{Const} \quad \text{for} \quad r \in (r_j, r_{j+1}). \tag{4.2}$$

Analogous to the flux in Cartesian coordinates, we derive an integral relation for the flux that is the superposition of the homogeneous flux, depending on the advection-diffusion operator, and the inhomogeneous flux, taking into account the effect of the source term s. Approximating all integrals involved gives us the expression for the numerical flux $F_{j+1/2}$.

Analogous to (3.2) we introduce the variables D, a, P, A and S, defined by

$$\begin{aligned} &D := r^2\varepsilon, \quad a := \frac{U}{D}, \quad P := a\Delta r, \\ &A(r) := \int_{r_{j+1/2}}^{r} a(\eta)\,\mathrm{d}\eta, \quad S(r) := \int_{r_{j+1/2}}^{r} \eta^2 s(\eta)\,\mathrm{d}\eta. \end{aligned} \tag{4.3}$$

We refer to P and A as the Peclet function and Peclet integral, respectively. Integrating the conservation law from $r_{j+1/2}$ to $r \in (r_j, r_{j+1})$, we obtain the relation

$$r^2 f(r) - (r^2 f)(r_{j+1/2}) = S(r). \tag{4.4}$$

Using the definitions of D and A in (4.3), it is clear that the expression for the flux can be rewritten as

$$r^2 f = U\varphi - D\frac{\mathrm{d}\varphi}{\mathrm{d}r} = -D\,\mathrm{e}^{A}\frac{\mathrm{d}}{\mathrm{d}r}\left(\varphi\,\mathrm{e}^{-A}\right). \tag{4.5}$$

Inserting this expression in (4.4), isolating the derivative $\mathrm{d}(\varphi\,\mathrm{e}^{-A})/\mathrm{d}r$, and integrating the resulting equation from r_j to r_{j+1} we obtain the following expressions for the flux:

$$\left(r^2 f\right)_{j+1/2} = \left(r^2 f^{\mathrm{h}}\right)_{j+1/2} + \left(r^2 f^{\mathrm{i}}\right)_{j+1/2}, \tag{4.6a}$$

$$\left(r^2 f^{\mathrm{h}}\right)_{j+1/2} = \left(\mathrm{e}^{-A_j}\varphi_j - \mathrm{e}^{-A_{j+1}}\varphi_{j+1}\right) \Big/ \int_{r_j}^{r_{j+1}} D^{-1}\mathrm{e}^{-A}\,\mathrm{d}r, \tag{4.6b}$$

$$\left(r^2 f^{\mathrm{i}}\right)_{j+1/2} = -\int_{r_j}^{r_{j+1}} D^{-1}\mathrm{e}^{-A}\,S\,\mathrm{d}r \Big/ \int_{r_j}^{r_{j+1}} D^{-1}\mathrm{e}^{-A}\,\mathrm{d}r, \tag{4.6c}$$

where $\left(r^2 f^{\mathrm{h}}\right)_{j+1/2}$ and $\left(r^2 f^{\mathrm{i}}\right)_{j+1/2}$ are the homogeneous and inhomogeneous part of $\left(r^2 f\right)_{j+1/2}$, corresponding to the homogeneous and particular solution of (4.1), respectively; cf. (3.3).

To elaborate the expressions in (4.6) we introduce some notation. $\langle a, b\rangle$ denotes the usual inner product of two functions a and b defined on (r_j, r_{j+1}), i.e.,

$$\langle a, b\rangle := \int_{r_j}^{r_{j+1}} a(r)b(r)\,\mathrm{d}r. \tag{4.7}$$

For a generic variable $v > 0$ defined on (r_j, r_{j+1}) we indicate the average, geometric average (of v_j and v_{j+1}) and the harmonic average by $\bar{v}_{j+1/2}$, $\tilde{v}_{j+1/2}$ and $\hat{v}_{j+1/2}$, respectively, i.e.,

$$\begin{aligned} \bar{v}_{j+1/2} &:= \tfrac{1}{2}\left(v_j + v_{j+1}\right), \\ \tilde{v}_{j+1/2} &:= \sqrt{v_j v_{j+1}}, \\ \frac{1}{\hat{v}_{j+1/2}} &:= \frac{\langle v^{-1}, 1\rangle}{\Delta r}. \end{aligned} \tag{4.8}$$

Consider the expression for the homogeneous flux. Assume first that $\varepsilon(r) = \mathrm{Const}$ on (r_j, r_{j+1}). In this case expression (4.6b) can be evaluated as

$$\left(r^2 f^{\mathrm{h}}\right)_{j+1/2} = \mathcal{F}^{\mathrm{h}}\left(\tilde{D}_{j+1/2}/\Delta r, \tilde{P}_{j+1/2}; \varphi_j, \varphi_{j+1}\right), \quad \tilde{P}_{j+1/2} := \frac{U\Delta r}{\tilde{D}_{j+1/2}}, \tag{4.9}$$

with $\mathcal{F}^{\mathrm{h}}$ defined in (3.6) and $\tilde{P}_{j+1/2}$ the geometric average of P, which we refer to as the constant coefficient homogeneous flux, i.e., $U = r^2 u = \mathrm{Const}$ and $\varepsilon = \mathrm{Const}$. In general, when ε is an arbitrary function of r, we can derive the following expression

$$(r^2 f^{\mathrm{h}})_{j+1/2} = \mathcal{F}^{\mathrm{h}}\big(\hat{D}_{j+1/2}/\Delta r, \langle a,1\rangle; \varphi_j, \varphi_{j+1}\big). \tag{4.10}$$

In the derivation we used that $D^{-1} = A'/U$ to evaluate the integral in (4.6b). Thus, the flux can be written as the constant coefficient flux (4.9) with $\tilde{D}_{j+1/2}$ and $\tilde{P}_{j+1/2}$ replaced by $\hat{D}_{j+1/2}$ and $\langle a,1\rangle$, respectively. Note that $\langle a,1\rangle$ can be interpreted as the average value of the Peclet function P over (r_j, r_{j+1}).

We consider next the expression for the inhomogeneous flux, and first take $\varepsilon(r) = \text{Const}$ on (r_j, r_{j+1}). Substituting the expression for $S(r)$ in (4.6c) and changing the order of integration, we can derive the representation

$$(r^2 f^{\mathrm{i}})_{j+1/2} = \Delta r \int_0^1 G(\sigma; \tilde{P}_{j+1/2}, \sigma_{j+1/2})\, r^2(\sigma) s(r(\sigma)) \left(\frac{r(\sigma)}{\tilde{r}_{j+1/2}}\right)^2 \mathrm{d}\sigma, \tag{4.11a}$$

with $\tilde{P}_{j+1/2}$ defined in (4.9) and with G the Green's function for the flux defined in (3.5), provided the normalized coordinate $\sigma(r)$ and the coordinate of the cell boundary $\sigma_{j+1/2}$ are chosen as

$$\sigma(r) = \frac{r - r_j}{\Delta r}\frac{r_{j+1}}{r}, \quad \sigma_{j+1/2} = \sigma(r_{j+1/2}). \tag{4.11b}$$

For arbitrary ε we can generalize (4.11) as follows

$$(r^2 f^{\mathrm{i}})_{j+1/2} = \Delta r \int_0^1 G(\sigma; \langle a,1\rangle, \sigma_{j+1/2})\, r^2(\sigma) s(r(\sigma)) \frac{D(r(\sigma))}{\hat{D}_{j+1/2}}\, \mathrm{d}\sigma, \tag{4.12a}$$

$$\sigma(r) = \int_{r_j}^{r} a(\eta)\,\mathrm{d}\eta / \langle a,1\rangle, \tag{4.12b}$$

where the correction factor $D(r(\sigma))/\hat{D}_{j+1/2}$ in (4.12a) is a consequence of the relation $\hat{D}_{j+1/2}\mathrm{d}r = \Delta r\, D(r(\sigma))\mathrm{d}\sigma$. Note that $a(r) > 0$ implies that $\sigma(r)$ defined in (4.12b) is monotonically increasing from 0 to 1 on the interval (r_j, r_{j+1}). Summarizing, the flux $f_{j+1/2}$ is the superposition of the homogeneous and inhomogeneous flux, defined in (4.10) and (4.12), respectively.

To derive expressions for the numerical flux, we need approximations for $\hat{D}_{j+1/2}$, $\langle a,1\rangle$, and $\sigma_{j+1/2} = \sigma(r_{j+1/2})$ with $\sigma(r)$ defined in (4.12b). Moreover, we need to evaluate the integral in (4.12a). A straightforward evaluation gives $\langle a,1\rangle = U\Delta r/\hat{D}_{j+1/2}$. To determine the harmonic average $\hat{D}_{j+1/2}$ from (4.8) we replace ε in the integrand by its average $\bar{\varepsilon}_{j+1/2}$ and evaluate the resulting integral exactly. This way we obtain the approximation $\hat{D}_{j+1/2} \approx \bar{\varepsilon}_{j+1/2} r^2_{j+1/2}$. Using the same approximation for ε in the evaluation of the integral in (4.12b) we obtain $\sigma_{j+1/2} = r_{j+1}/(2r_{j+1/2})$. Since $D(r(\sigma))/\hat{D}_{j+1/2} = 1 + \mathcal{O}(\Delta r)$, we omit the term altogether in (4.12a), resulting in an $\mathcal{O}(\Delta r^2)$ error for the inhomogeneous flux. Moreover, since for dominant advection G has a distinct bias toward the upwind end of (r_j, r_{j+1}), we replace

$r^2(\sigma)s(r(\sigma))$ by its upwind value $r_j^2 s_j$. The resulting integral can be evaluated as

$$\int_0^1 G(\sigma; \langle a,1\rangle, \sigma_{j+1/2})\,\mathrm{d}\sigma = \sigma_{j+1/2} - W(\langle a,1\rangle).$$

Then, applying all approximations mentioned above, we obtain the numerical flux

$$\left(r^2F\right)_{j+1/2} = \left(r^2F^{\mathrm{h}}\right)_{j+1/2} + \left(r^2F^{\mathrm{i}}\right)_{j+1/2}, \tag{4.13a}$$

$$\left(r^2F^{\mathrm{h}}\right)_{j+1/2} = \mathcal{F}^{\mathrm{h}}\left(D_{j+1/2}/\Delta r, P_{j+1/2}; \varphi_j, \varphi_{j+1}\right), \tag{4.13b}$$

$$\left(r^2F^{\mathrm{i}}\right)_{j+1/2} = \Delta r\left(\sigma_{j+1/2} - W(P_{j+1/2})\right) r_j^2 s_j, \tag{4.13c}$$

with coefficients $D_{j+1/2}$, $P_{j+1/2}$ and $\sigma_{j+1/2}$ given by

$$\begin{aligned} D_{j+1/2} &:= \bar{r}_{j+1/2}^2 \bar{\varepsilon}_{j+1/2}, \\ P_{j+1/2} &:= \frac{U\Delta r}{D_{j+1/2}}, \\ \sigma_{j+1/2} &= \frac{r_{j+1}}{2r_{j+1/2}}. \end{aligned} \tag{4.13d}$$

We refer to (4.13) as the complete flux (CF) scheme for spherical coordinates, with as special case the homogeneous flux (HF) scheme (4.13b).

5. Numerical flux for cylindrical coordinates

In this section we present the complete flux scheme for conservation laws in cylindrical coordinates, assuming rotational symmetry about the z-axis. Consequently, the problem does not depend on the azimuthal coordinate θ. We proceed in two steps. First, we derive the r-component of the flux in polar coordinates, so we solve an essentially one-dimensional problem, and second, we extend the scheme by including the z-component, to derive the full two-dimensional scheme.

The stationary, rotationally symmetric conservation law in polar coordinates reads $\mathrm{d}(rf)/\mathrm{d}r = rs$ with $f = u\varphi - \varepsilon\mathrm{d}\varphi/\mathrm{d}r$. We give a very concise derivation of the CF scheme, since it is quite similar to the CF scheme in spherical coordinates. To determine the integral relation for the flux $f_{j+1/2} := f(r_{j+1/2})$, we consider the one-dimensional model BVP:

$$\frac{1}{r}\frac{\mathrm{d}}{\mathrm{d}r}\left(r\left(u\varphi - \varepsilon\frac{\mathrm{d}\varphi}{\mathrm{d}r}\right)\right) = s, \quad r_j < r < r_{j+1}, \tag{5.1a}$$

$$\varphi(r_j) = \varphi_j, \quad \varphi(r_{j+1}) = \varphi_{j+1}, \tag{5.1b}$$

where ε and s are sufficiently smooth functions of r and where, because of (2.2), u satisfies the relation $U := ru = \text{Const}$. The definitions of the variables a, P and A in (4.3) still hold, whereas the definitions of D and S change to

$$D := \varepsilon r, \quad S(r) := \int_{r_{j+1/2}}^{r} \eta s(\eta)\,\mathrm{d}\eta. \tag{5.2}$$

We can essentially repeat the derivation in the previous section: integrate the conservation law from the cell boundary $r_{j+1/2}$ to $r \in (r_j, r_{j+1})$, rewrite the flux in terms of its integrating factor, substitute the flux in the integral relation and subsequently integrate over the interval (r_j, r_{j+1}), to arrive at the following expressions:

$$(rf)_{j+1/2} = (rf^{\mathrm{h}})_{j+1/2} + (rf^{\mathrm{i}})_{j+1/2}, \tag{5.3a}$$

$$(rf^{\mathrm{h}})_{j+1/2} = \left(\mathrm{e}^{-A_j}\varphi_j - \mathrm{e}^{-A_{j+1}}\varphi_{j+1}\right) \Big/ \int_{r_j}^{r_{j+1}} D^{-1}\mathrm{e}^{-A}\,\mathrm{d}r, \tag{5.3b}$$

$$(rf^{\mathrm{i}})_{j+1/2} = -\int_{r_j}^{r_{j+1}} D^{-1}\mathrm{e}^{-A}\,S\,\mathrm{d}r \Big/ \int_{r_j}^{r_{j+1}} D^{-1}\mathrm{e}^{-A}\,\mathrm{d}r, \tag{5.3c}$$

thus, as anticipated, the flux $f_{j+1/2}$ is the superposition of the homogeneous flux $f^{\mathrm{h}}_{j+1/2}$ and the inhomogeneous flux $f^{\mathrm{i}}_{j+1/2}$; cf. (4.6).

Next, we have to elaborate (5.3b) and (5.3c). Evaluating all integrals involved, we recover relation (4.10) for the homogeneous flux. Substituting the expression for S in (5.3c) and changing the order of integration, we obtain the expression

$$(rf^{\mathrm{i}})_{j+1/2} = \Delta r \int_0^1 G(\sigma; \langle a,1\rangle, \sigma_{j+1/2})\, r(\sigma)s(r(\sigma))\, \frac{D(r(\sigma))}{\hat{D}_{j+1/2}}\,\mathrm{d}\sigma, \tag{5.4}$$

where the normalized coordinate σ is defined in (4.12b). Finally, to derive expressions for the numerical flux, we need approximations for $\hat{D}_{j+1/2}$, $\langle a,1\rangle$, $\sigma_{j+1/2}$ and for the integral in the right hand side of (5.4). For the latter, we replace the term $r\,s(r)$ in the integrand by its upwind value $(r\,s(r))_{\mathrm{u},j+1/2}$, i.e., $(r\,s(r))_{\mathrm{u},j+1/2} = r_j\,s_j$ if $\bar{u}_{j+1/2} \geq 0$ and $(r\,s(r))_{\mathrm{u},j+1/2} = r_{j+1}\,s_{j+1}$ if $\bar{u}_{j+1/2} < 0$. Approximating ε by its average $\bar{\varepsilon}_{j+1/2}$, we obtain similar results as in Section 4, except that the harmonic average $\hat{D}_{j+1/2}$ is now approximated as

$$\hat{D}_{j+1/2} \approx \bar{\varepsilon}_{j+1/2}\hat{r}_{j+1/2}, \quad \hat{r}_{j+1/2} = \frac{r_{j+1} - r_j}{\ln(r_{j+1}/r_j)}.$$

From straightforward Taylor expansions we conclude that $\hat{r}_{j+1/2} = r_{j+1/2} + \mathcal{O}(\Delta r^2)$. Putting everything together, we obtain the following version of the complete flux scheme:

$$(rF)_{j+1/2} = (rF^{\mathrm{h}})_{j+1/2} + (rF^{\mathrm{i}})_{j+1/2}, \tag{5.5a}$$

$$(rF^{\mathrm{h}})_{j+1/2} = \mathcal{F}^{\mathrm{h}}(D_{j+1/2}/\Delta r, P_{j+1/2}; \varphi_j, \varphi_{j+1}), \tag{5.5b}$$

$$(rF^{\mathrm{i}})_{j+1/2} = \Delta r\,\big(\sigma_{j+1/2} - W(P_{j+1/2})\big)(r\,s)_{\mathrm{u},j+1/2}, \tag{5.5c}$$

where the coefficients $D_{j+1/2}$, $P_{j+1/2}$ and $\sigma_{j+1/2}$ are given by

$$D_{j+1/2} := r_{j+1/2}\,\bar{\varepsilon}_{j+1/2}, \quad P_{j+1/2} := \frac{U\Delta r}{D_{j+1/2}}, \quad \sigma_{j+1/2} = \frac{\ln(r_{j+1/2}/r_j)}{\ln(r_{j+1}/r_j)}; \tag{5.5d}$$

cf. (4.13). Note that $P_{j+1/2}$ is the average of the Peclet function P over the interval (r_j, r_{j+1}).

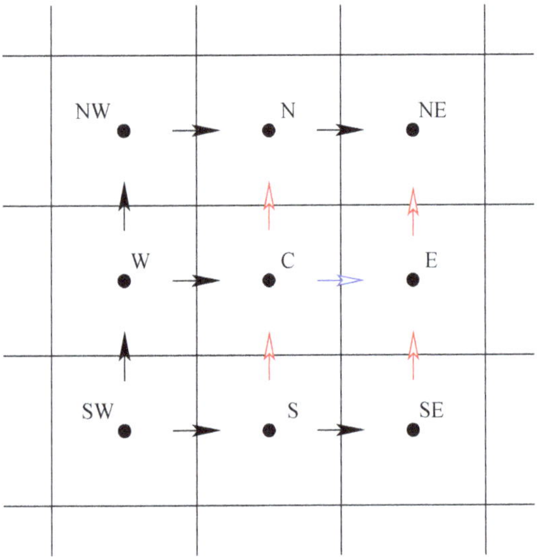

Fig. 3. Control volume Ω_C and corresponding stencil.

Next, we extend the derivation to two-dimensional conservation laws, including the z-component of the flux. In particular, we derive the expression for the r-component of the numerical flux. For ease of notation, we use both index notation and the compass notation; see Figure 3. Thus, φ_C should be undersood as $\varphi_{j,l}$ and $f_{r,e}$ as $f_{r,j+1/2,l}$ etc.

The flux corresponding to equation (2.1) is given by

$$\boldsymbol{f} = f_r \boldsymbol{e}_r + f_z \boldsymbol{e}_z = \left(u_r\varphi - \varepsilon\frac{\partial\varphi}{\partial r}\right)\boldsymbol{e}_r + \left(u_z\varphi - \varepsilon\frac{\partial\varphi}{\partial z}\right)\boldsymbol{e}_z. \tag{5.6}$$

We outline the derivation of the r-component of the numerical flux $F_{r,j+1/2,l}$ at the eastern edge of the control volume $\Omega_{j,l}$; see Figure 3. The derivation of the z-component $F_{z,j,l+1/2}$ of the numerical flux at the northern edge is completely analogous and is therefore omitted. The key idea is to include the cross flux term $\partial f_z/\partial z$ in the evaluation of the flux. Therefore we determine the numerical flux $F_{r,j+1/2,l}$ from the quasi-one-dimensional BVP:

$$\frac{1}{r}\frac{\partial}{\partial r}\left(r\left(u_r\varphi - \varepsilon\frac{\partial\varphi}{\partial r}\right)\right) = s_r, \quad r_j < r < r_{j+1},\ z = z_l, \tag{5.7a}$$

$$\varphi(\boldsymbol{x}_{j,l}) = \varphi_{j,l}, \quad \varphi(\boldsymbol{x}_{j+1,l}) = \varphi_{j+1,l}, \tag{5.7b}$$

where the modified source term s_r is defined by

$$s_r := s - \frac{\partial f_z}{\partial z}. \tag{5.7c}$$

The derivation of the expression for the numerical flux is essentially the same as for (5.5), the main difference being the inclusion of the cross flux term $\partial f_z/\partial z$ in the source term. In the computation of s_r we replace $\partial f_z/\partial z$ by its central difference approximation and for f_z we take the homogeneous numerical flux. A similar procedure applies to the z-component of the flux, which is actually the Cartesian flux from Section 3, albeit with nonconstant coefficients, [13]. Putting everything together, we obtain the following two-dimensional complete flux scheme.

---

two-dimensional CF scheme

Peclet function

$$P_r := U_r\Delta r/D \qquad\qquad P_z := u_z\Delta z/\varepsilon$$

$$U_r = ru_r, \quad D := r\varepsilon$$

(weighted) average

$$\bar{v}_{\mathrm{e}} := \tfrac{1}{2}(v_{\mathrm{C}} + v_{\mathrm{E}}) \qquad\qquad \bar{v}_{\mathrm{n}} := \tfrac{1}{2}(v_{\mathrm{C}} + v_{\mathrm{N}})$$

$$v_{\mathrm{n}}^* := W(-\bar{P}_{z,\mathrm{n}})v_{\mathrm{C}} + W(\bar{P}_{z,\mathrm{n}})v_{\mathrm{N}}$$

homogeneous flux

$$\left(rF_r^{\mathrm{h}}\right)_{\mathrm{e}} = \mathcal{F}^{\mathrm{h}}\left(D_{\mathrm{e}}/\Delta r, P_{r,\mathrm{e}}; \varphi_{\mathrm{C}}, \varphi_{\mathrm{E}}\right) \qquad\qquad F_{z,\mathrm{n}}^{\mathrm{h}} = \mathcal{F}^{\mathrm{h}}\left(\mathcal{E}_{\mathrm{n}}/\Delta z, \bar{P}_{z,\mathrm{n}}; \varphi_{\mathrm{C}}, \varphi_{\mathrm{N}}\right)$$

$$D_{\mathrm{e}} = r_{\mathrm{e}}\bar{\varepsilon}_{\mathrm{e}}, \quad P_{r,\mathrm{e}} = \bar{U}_{r,\mathrm{e}}\Delta r/D_{\mathrm{e}} \qquad\qquad \mathcal{E}_n = P_{z,\mathrm{n}}^*\varepsilon_{\mathrm{n}}^*/\bar{P}_{z,\mathrm{n}}$$

source term with cross wind diffusion

$$s_{r,\mathrm{C}} = s_{\mathrm{C}} - \frac{1}{\Delta z}\left(F_{z,\mathrm{n}}^{\mathrm{h}} - F_{z,\mathrm{s}}^{\mathrm{h}}\right) \qquad\qquad s_{z,\mathrm{C}} = s_{\mathrm{C}} - \frac{1}{r_{\mathrm{C}}}\frac{1}{\Delta r}\left(\left(rF_r^{\mathrm{h}}\right)_{\mathrm{e}} - \left(rF_r^{\mathrm{h}}\right)_{\mathrm{w}}\right)$$

upwinded source term

$$s_{r,\mathrm{u,e}} = \begin{cases} s_{r,\mathrm{C}} & \text{if} \quad \bar{u}_{r,\mathrm{e}} \geq 0 \\ s_{r,\mathrm{E}} & \text{if} \quad \bar{u}_{r,\mathrm{e}} < 0 \end{cases} \qquad\qquad s_{z,\mathrm{u,n}} = \begin{cases} s_{z,\mathrm{C}} & \text{if} \quad \bar{u}_{z,\mathrm{n}} \geq 0 \\ s_{z,\mathrm{N}} & \text{if} \quad \bar{u}_{z,\mathrm{n}} < 0 \end{cases}$$

inhomogeneous flux

$$\left(rF_r^{\mathrm{i}}\right)_{\mathrm{e}} = \Delta r\left(\sigma_{\mathrm{e}} - W(P_{r,\mathrm{e}})\right)\left(rs_r\right)_{\mathrm{u,e}} \qquad\qquad F_{z,\mathrm{n}}^{\mathrm{i}} = \Delta z\left(\tfrac{1}{2} - W(\bar{P}_{z,\mathrm{n}})\right)s_{z,\mathrm{u,n}}$$

$$\sigma_{\mathrm{e}} = \frac{\ln\left(r_{\mathrm{e}}/r_{\mathrm{C}}\right)}{\ln\left(r_{\mathrm{E}}/r_{\mathrm{C}}\right)}$$

complete flux

$$\left(rF_r\right)_{\mathrm{e}} = \left(rF_r^{\mathrm{h}}\right)_{\mathrm{e}} + \left(rF_r^{\mathrm{i}}\right)_{\mathrm{e}} \qquad\qquad F_{z,\mathrm{n}} = F_{z,\mathrm{n}}^{\mathrm{h}} + F_{z,\mathrm{n}}^{\mathrm{i}}.$$

The stencil of the flux approximation for $F_{r,\mathrm{e}}$ is depicted in Figure 3. Assume first that $\bar{u}_{r,\mathrm{e}} > 0$. Then $F_{r,\mathrm{e}}$ depends on φ in the grid points $\boldsymbol{x}_{\mathrm{C}}$ and $\boldsymbol{x}_{\mathrm{E}}$, on s in the central point $\boldsymbol{x}_{\mathrm{C}}$ and on the homogeneous fluxes $F_{z,\mathrm{n}}^{\mathrm{h}}$ and $F_{z,\mathrm{s}}^{\mathrm{h}}$ and through these fluxes again on φ in $\boldsymbol{x}_{\mathrm{N}}$ and $\boldsymbol{x}_{\mathrm{S}}$. For $\bar{u}_{r,\mathrm{e}} < 0$, $F_{r,\mathrm{e}}$ again depends on φ_{C} and φ_{E}, but this time on the source term s_{E} and the homogeneous fluxes $F_{z,\mathrm{En}}^{\mathrm{h}}$ and $F_{z,\mathrm{Es}}^{\mathrm{h}}$, inducing a further dependency on φ_{NS} and φ_{SE}. Thus, in addition to the direct neighbours, $F_{r,e}$ depends on a few other values of φ, determined by the local upwind direction.

6. Aspects of time integration

Next, we extend the derivation to time-dependent conservation laws, restricting ourselves to spherically symmetric conservation laws; for Cartesian coordinates see [13, 15].

The semidiscrete conservation law for $\varphi_j(t) \approx \varphi(r_j, t)$ can be written as

$$\left(r^2F\right)_{j+1/2}(t) - \left(r^2F\right)_{j-1/2}(t) = \Delta r\left(r_j^2 + \tfrac{1}{12}\Delta r^2\right)\left(s_j(t) - \dot{\varphi}_j(t)\right), \tag{6.1}$$

where $\dot{\varphi}_j(t) \approx \partial\varphi/\partial t(r_j, t)$ and $s_j(t) = s(r_j, t)$. In the following we shall omit the explicit dependence on the variable t.

For the numerical flux $F_{j+1/2}$ in (6.1) we have two options. We can simply take the flux (4.13) derived from the corresponding BVP (4.1), and henceforth referred to as the stationary complete flux (SCF) scheme. Alternatively, we can include $\partial\varphi/\partial t$ in the numerical flux, if we determine $\left(r^2F\right)_{j+1/2}$ from the quasi-stationary BVP:

$$\frac{1}{r^2}\frac{\partial}{\partial r}\left(r^2\left(u\varphi - \varepsilon\frac{\partial\varphi}{\partial r}\right)\right) = s - \frac{\partial\varphi}{\partial t}, \quad r_j < r < r_{j+1}, \tag{6.2a}$$

$$\varphi(r_j) = \varphi_j, \quad \varphi(r_{j+1}) = \varphi_{j+1}, \tag{6.2b}$$

thus subtracting the time derivative from the source term. We can repeat the derivation in Section 4, to arrive at the following expression for the numerical flux

$$\left(r^2F\right)_{j+1/2} = \frac{D_{j+1/2}}{\Delta r}\left(B(-P_{j+1/2})\varphi_j - B(P_{j+1/2})\varphi_{j+1}\right) + \Delta r\left(\sigma_{j+1/2} - W(P_{j+1/2})\right) r_j^2\left(s_j - \dot{\varphi}_j\right), \tag{6.3}$$

referred to as the transient complete flux (TCF) scheme; cf. (4.13). This numerical flux can be written in the desired form (2.13) as

$$\left(r^2F\right)_{j+1/2} = \alpha_{j+1/2}\,\varphi_j - \beta_{j+1/2}\,\varphi_{j+1} + \Delta r\left(\gamma_{j+1/2}\,\tilde{s}_j + \delta_{j+1/2}\,\tilde{s}_{j+1}\right), \tag{6.4a}$$

with $\tilde{s} := s - \partial\varphi/\partial t$ and where the coefficient $\alpha_{j+1/2}$ etc. are defined by

$$\begin{aligned}
\alpha_{j+1/2} &:= \frac{D_{j+1/2}}{\Delta r}B^-_{j+1/2},\\
\beta_{j+1/2} &:= \frac{D_{j+1/2}}{\Delta r}B^+_{j+1/2},\\
B^\pm_{j+1/2} &:= B(\pm P_{j+1/2}),\\
\gamma_{j+1/2} &:= \sigma_{j+1/2} - W(P_{j+1/2}),\\
\delta_{j+1/2} &:= 0.
\end{aligned} \tag{6.4b}$$

A similar expression holds for the numerical flux $F_{j-1/2}$. Substituting the TCF approximations in (6.1) we obtain the finite volume TCF semidiscretisation, given by

$$b_{W,j}\dot{\varphi}_{j-1} + b_{C,j}\dot{\varphi}_j - a_{W,j}\varphi_{j-1} + a_{C,j}\varphi_j - a_{E,j}\varphi_{j+1} = b_{W,j}s_{j-1} + b_{C,j}s_j, \tag{6.5a}$$

where the coefficients $a_{W,j}$ etc. are defined by

$$\begin{aligned} a_{W,j} &:= \alpha_{j-1/2}, \quad a_{E,j} := \beta_{j+1/2}, \quad a_{C,j} := \alpha_{j+1/2} + \beta_{j-1/2}, \\ b_{W,j} &:= \Delta r\, \gamma_{j-1/2}, \quad b_{C,j} := \Delta r\big(r_j^2 + \tfrac{1}{12}\Delta r^2 - \gamma_{j+1/2}\big). \end{aligned} \tag{6.5b}$$

The semidiscretization in (6.5) defines an implicit ODE system, for which we require an A-stable, one-step time integrator. Our choise is the trapezoidal rule. In [15] we have shown that the Cartesian TCF scheme has usually much smaller dissipation and dispersion errors than the corresponding SCF scheme, provided the solution is smooth.

7. Numerical example

In this section we apply the complete flux scheme to a model problem, describing a premixed spherical flame stabilized by a point source of combustible mixture.

A point source at the origin issues a mass flux of $4\pi U$ of combustible mixture. After ignition, a stable spherical flame is formed, provided the value of U is in the proper range. The governing equations for this system are given by [2, 12]:

$$\frac{\partial C}{\partial t} + \frac{1}{r^2}\frac{\partial}{\partial r}\Big(UC - r^2\frac{1}{\text{Le}}\frac{\partial C}{\partial r}\Big) = \omega, \quad r > 0, t > 0, \tag{7.1a}$$

$$\frac{\partial T}{\partial t} + \frac{1}{r^2}\frac{\partial}{\partial r}\Big(UT - r^2\frac{\partial T}{\partial r}\Big) = \omega, \tag{7.1b}$$

where C and T are the dimensionless concentration of combustion product and temperature, respectively, and where ω is the (dimensionless) reaction rate. The radial coordinate r and the time t are dimensionless as well. Parameters in (7.1) are the mass flux (per solid angle) U and the Lewis number Le. The reaction rate ω depends on C and T as follows

$$\omega = \frac{1}{2\text{Le}}\beta^2(1-Y)\mathrm{e}^{\beta(T-1)}, \tag{7.2}$$

with β the dimensionless activation energy. In the unburnt gas mixture, far ahead of the flame front, there is no combustion product and the temperature equals the temperature of the unburnt gas. In the burnt gas, beyond the flame, we assume that the reaction is completed, and consequently the combustion product is the only species and the temperature is equal to the adiabatic temperature of the burnt gas mixture. These conditions lead to the following boundary conditions

$$C(0,t) = T(0,t) = 0, \quad C(\infty,t) = T(\infty,t) = 1, \quad t > 0. \tag{7.3}$$

As initial conditions, we take the linear profiles $C(r,0) = r/r_{\max}$ and $T(r,0) = r/r_{\max}$ on the truncated domain $(0, r_{\max})$ and let the solution evolve to its steady state. We take $r_{\max} = 120$.

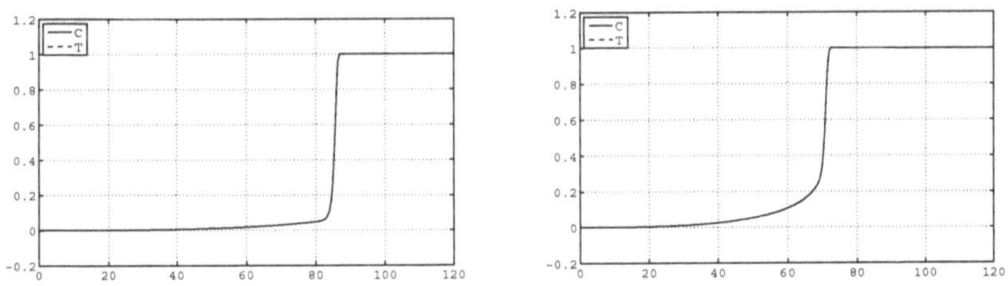

Fig. 4. Numerical solutions of the thermo-diffusive model (7.1) for $\beta = 10$ (left) and $\beta = 8$ (right). Other parameters are: $U = 1.0475 \times 10^4$ and $\mathrm{Le} = 1$.

For space discretisation of (7.1) we employ the TCF scheme (6.4a) in combination with the θ-method for time integration [5]. The resulting nonlinear system at each time step is solved applying one Newton iteration step. Moreover, to enhance the robustness of the method, we bound the numerical solutions according to $0 \leq C_j, T_j \leq 1$, followed by a smoothing step as follows: $C_j := \frac{1}{4}(C_{j-1} + 2C_j + C_{j+1})$, and likewise for T_j.

As an example, the numerical solutions at $t = 100$ for $U = 1.0475 \times 10^4$, $\mathrm{Le} = 1$ and $\beta = 10, 8$ are shown in Figure 4, computed with grid size $\Delta r = 0.4$ and time step $\Delta t = 0.25$. The solutions exhibit a steep interior layer, the so-called flame front, connecting the (virtually) constant unburnt and burnt states. Since $\mathrm{Le} = 1$, the numerical solutions for C and T are identical. The solution for $\beta = 10$ is very close to the asymptotic solution [12]

$$C(r,0) = \begin{cases} \exp\left(\mathrm{Le}U\left(\frac{1}{r_\mathrm{f}} - \frac{1}{r}\right)\right) & \text{if} \quad r \leq r_\mathrm{f}, \\ 1 & \text{if} \quad r \geq r_\mathrm{f} \end{cases},$$

$$T(r,0) = \begin{cases} \exp\left(U\left(\frac{1}{r_\mathrm{f}} - \frac{1}{r}\right)\right) & \text{if} \quad r \leq r_\mathrm{f}, \\ 1 & \text{if} \quad r \geq r_\mathrm{f}, \end{cases},$$

with $r_\mathrm{f} = 93.4$ the radius of the flame. For decreasing β the reaction slows down, resulting in a slightly wider flame front and a location of the flame front closer to the source. We define $e_C := \left|\left|(\boldsymbol{C}^{n+1} - \boldsymbol{C}^n)/\Delta t\right|\right|_1 / N$ with $\boldsymbol{C}^n = \left(C_j^n\right)^\mathrm{T}$ and N the number of grid points, and likewise e_T. The time histories of e_C and e_T corresponding to the numerical solutions in Figure 4 are shown in Figure 5. We observe a regular convergence to the steady state. Finally, in order to study the effect of preferential diffusion, the numerical simulations are repeated for $\mathrm{Le} = 0.3$, and the results are shown in Figure 6. As expected, the interior layer for C is slightly wider than for T.

8. Conclusions and future research

In this paper we have derived complete flux schemes for spherically or cylindrically symmetric conservation laws of advection-diffusion-reaction type. An integral relation for the flux is derived from a local one-dimensional BVP for the *entire* equation, including

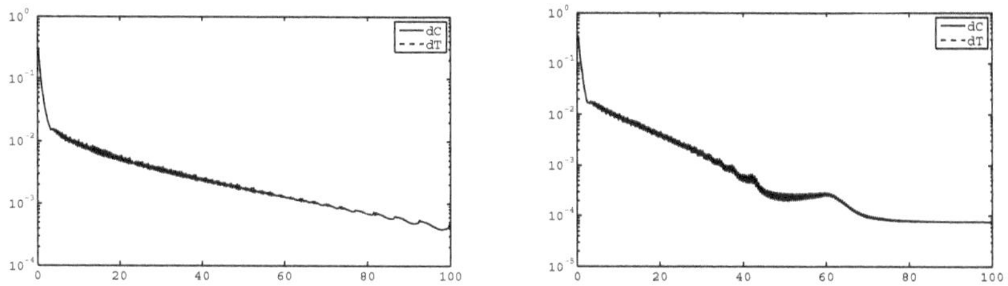

Fig. 5. Time history of the (discrete) time derivatives. Parameter values are: $\beta = 10$ (left), $\beta = 8$ (right), $U = 1.0475^4$ and $\mathrm{Le} = 1$.

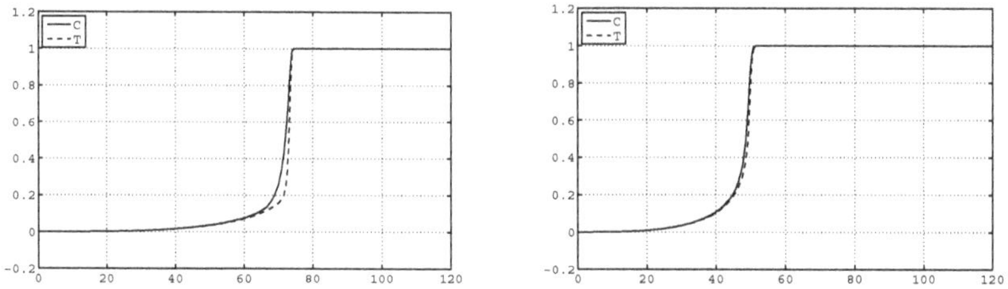

Fig. 6. Numerical solutions of the thermo-diffusive model (7.1) for $\beta = 10$ (left) and $\beta = 8$ (right). Other parameters are: $U = 1.0475 \times 10^4$ and $\mathrm{Le} = 0.3$.

the source term. Applying suitable quadrature rules, we derived expressions for the numerical flux. As a result of this procedure, we obtained a numerical flux that is the superposition of a homogeneous flux, corresponding to the advection-diffusion operator, and an inhomogeneous flux, corresponding to the reaction term. For time-dependent conservation laws, we included the time derivative in the inhomogeneous flux, resulting in an implicit ODE system. The CF-scheme has been applied to a thermo-diffusive model for a spherical flame.

Possible directions of future research are the following: first, a rigourous convergence analysis of the (stationary) CF-schemes for spherical and cylindrical coordinates, and second, a dispersion analysis of time-dependent CF scheme; cf. [15] where such analysis is presented for Cartesian coordinates. Finally, from a more fundamental point of view, it would be very interesting to base the derivation of the time-dependent CF scheme on a local initial boundary-value problem for a truly time-dependent equation, rather than computing the flux from a quasi-stationary BVP.

9. References

[1] Groot, G.R.A.: *Modelling of Propagating Spherical and Cylindrical Premixed Flames*. PhD Thesis, Eindhoven University of Technology (2003).

[2] Guillard, H. and Peyret, R.: On the use of spectral methods for the numerical solution of stiff problems, *Comp. Meth. in Appl. Mech. and Eng.* 66, 17-43, 1988.

[3] Van 't Hof, B., Ten Thije Boonkkamp, J.H.M., and Mattheij, R.M.M.: Discretisation of the Stationary Convection-Diffusion-Reaction Equation. *Numer. Meth. for Part. Diff. Eq.* 14, 607-625 (1998).

[4] Il'in, A.M.: Differencing scheme for a differential equation with a small parameter affecting the highest derivative. *Mat. Zametki* 6, 237-248. (in Russian).

[5] Mattheij, R.M.M., Rienstra, S.W. and Ten Thije Boonkkamp, J.H.M.: *Partial Differential Equations, Modeling, Analysis, Computation*, SIAM, Philadelphia, 2005.

[6] Morton, K.W.: *Numerical Solution of Convection-Diffusion Problems, Applied Mathematics and Mathematical Computation 12*, Chapman & Hall, London, 1991.

[7] Patankar, S.V.: *Numerical Heat Transfer and Fluid Flow; Series in Computational Methods in Mechanics and Thermal Sciences*, Hemisphere Publishing Corporation, New York, 1980.

[8] Poinsot, T. and Veynante, D.: *Theoretical and Numerical Combustion, Second Edition*, Edwards, Philadelphia, 2005.

[9] Raizer, Yu.P.: *gas Discharge Physics*, Springer, Berlin, 1991.

[10] Thiart, G.D.: Finite difference scheme for the numerical solution of fluid flow and heat transfer problems on nonstaggered grids. *Numerical Heat Transfer, Part B*, 17, 43–62 (1990).

[11] Thiart, G.D.: Improved finite difference scheme for the solution of convection-diffusion problems with the SIMPLEN algorithm. *Numerical Heat Transfer, Part B*, 18, 81–95 (1990).

[12] Sivashinsky, G.I.: On self-turbulization of a laminar flame, *Acta Astronautica* 6, 569-591, 1979.

[13] Ten Thije Boonkkamp, J.H.M. and Anthonissen, M.J.H.: The Finite Volume-Complete Flux Scheme for Advection-Diffusion-Reaction Equations, *J. Sci. Comput.* 46: 47-70 (2011).

[14] Ten Thije Boonkkamp, J.H.M. and Anthonissen, M.J.H.: The Complete Flux Scheme for Spherically Symmetric Conservation Laws, in: M. Bubak et al. (Eds.) ICCS 2008, Part I, LNCS 5101, pp. 651-660, 2008. Springer-Verlag Berlin Heidelberg 2008.

[15] Ten Thije Boonkkamp, J.H.M. and Anthonissen, M.J.H.: Extension of the Complete Flux Scheme to Time-Dependent Conservation Laws, in: G. Kreiss et al. (eds.) Numerical Mathematics and Advanced Applications 2009, Proceedings ENUMATH 2009, Springer-Verlag, Berlin, 2010, p. 865-873.

[16] Wesseling, P., *Principles of Computational Fluid Dynamics, Springer Series in Computational Mathematics 29*, Springer, Berlin, 2001.

3

Use of Proper Closure Equations in Finite Volume Discretization Schemes

A. Ashrafizadeh, M. Rezvani and B. Alinia
K. N. Toosi University of Technology
Iran

1. Introduction

Finite Volume Method (FVM) is a popular field discretization approach for the numerical simulation of physical processes described by conservation laws. This article presents some advances in finite volume discretization of heat and fluid flow problems.

In the case of thermo-fluid problems, decision regarding the choice of unknown variables needs to be taken at the continuous level of formulation. A popular choice, among many available options, is the pressure-based primitive variable formulation (Acharya, 2007), in which the flow continuity equation is considered as a constraint on fluid pressure. Pressure-based finite volume methods, in turn, can be categorized based on the computational grid arrangement used for the discretization of solution domain. In a staggered grid arrangement, velocity components and other unknown scalars are defined at different nodal points (Patankar, 1980). On the contrary, all unknown variables are defined at the same nodal locations in a co-located grid arrangement. The discussion of finite volume discretization schemes in this article is limited to pressure-based formulations on co-located grids.

At the continuous level of formulation in a finite volume method, an intensive variable φ is constrained by an integral balance equation for a control volume CV bounded by the control surface CS as follows:

$$\frac{\partial}{\partial t}\iiint_{CV} \rho\varphi dV + \oint\!\!\oint_{CS} \vec{J}_{\varphi} \bullet \vec{n} dA = \iiint_{CV} \dot{S}_{\varphi} dV \tag{1}$$

In Eq. (1) $\vec{J}_{\varphi}$ represents the flux of φ across the control surface, $\dot{S}_{\varphi}$ stands for the rate of generation of φ within the control volume and $\vec{n}$ is the outward unit normal vector to the control surface. In most of the engineering problems of interest, there are two mechanisms for the transport of φ, i.e. advection and diffusion, and the flux function is mathematically described as follows:

$$\vec{J}_{\varphi} \equiv \left(\rho\vec{V}\right)\varphi - \left(\Gamma_{\varphi}\right)\vec{\nabla}\varphi \tag{2}$$

In Eq. (2) $\rho\vec{V}$ is the advective mass flux and Γ_{φ} is the diffusion coefficient associated with variable φ. By properly defining φ, $\vec{J}_{\varphi}$ and $\dot{S}_{\varphi}$, Eq. (1) can be used as a generic equation to mathematically describe the transport of mass, momentum and energy in fluid flow problems (Patankar, 1980).

A typical two dimensional discrete domain covered by contiguous control volumes (cells) is shown in Fig. 1. Each cell represents a nodal point and is bounded by a number of panels which comprise the control surface. Typically, each panel includes one integration point. There isn't any limitation on the shape and the number of panels of a cell. Here we use a background finite element mesh to define the control volumes. Figure 1 shows structured and unstructured two-dimensional grids in an element-based finite volume method (EBFVM) context (Ashrafizadeh et al., 2011).

The objective of any finite volume discretization scheme is to use the CV balance equations, similar to Eq. (1), to obtain a set of algebraic equations which constrains the unknown nodal values throughout the domain. This objective is achieved by carrying out the discretization in two stages.

At the first stage of discretization, here called level-1 approximation, Eq. (1) is converted to a semi-discrete form comprising of both nodal and integration point variables. Level-1 approximate form of Eq. (1) for an internal finite volume in Fig. 1 can be written as follows:

$$\frac{\Delta(\rho\varphi)_{np}}{\Delta t}\delta V+\sum_{ip=1}^{N_{ip}}(\rho\varphi\vec{V}\bullet\vec{n}dA)_{ip}=\sum_{ip=1}^{N_{ip}}(\Gamma_{\varphi}\frac{\partial\varphi}{\partial n}A)_{ip}+\dot{S}\varphi\delta V \tag{3}$$

In Eq. (3), subscripts np and ip stand for nodal and integration point variables respectively. Closure equations are now needed to relate ip variables to np variables.

At the second stage of discretization, here called level-2 approximation, Eq. (3) is converted to a fully-discrete form, i.e. an algebraic equation which constrains the variable φ at nodal point P. The computational molecule obtained at this stage of discretization is written in the following convenient form:

$$a_P\varphi_P=\sum_{nb=1}^{N_{nb}}a_{nb}\varphi_{nb}+S_P^{\varphi} \tag{4}$$

Here, subscript nb stands for the neighbor nodes and superscript N_{nb} refers to the number of influential neighbor nodes. Influence coefficients a_P and a_{nb} and the source term S contain terms and parameters which model all physical effects relevant to the value of φ at the nodal point P.

If Eq. (1) is a nonlinear equation, linearization also becomes necessary and should be carried out during the discretization. In principle, the equation can be linearized before or after each of the approximation levels and this might affect the properties of the final discrete model.

In most engineering problems, including heat and fluid flow ones, a coupled set of nonlinear partial differential equations needs to be discretized and solved. In such cases, the inter-equation couplings should also be numerically modeled. Two-dimensional, incompressible Navier-Stokes equations, for example, can be written in a form similar to Eq. (1) in which φ is equal to 1 in the mass balance equation, u in the x-momentum balance equation, v in the y-momentum balance equation and e (specific energy) in the energy

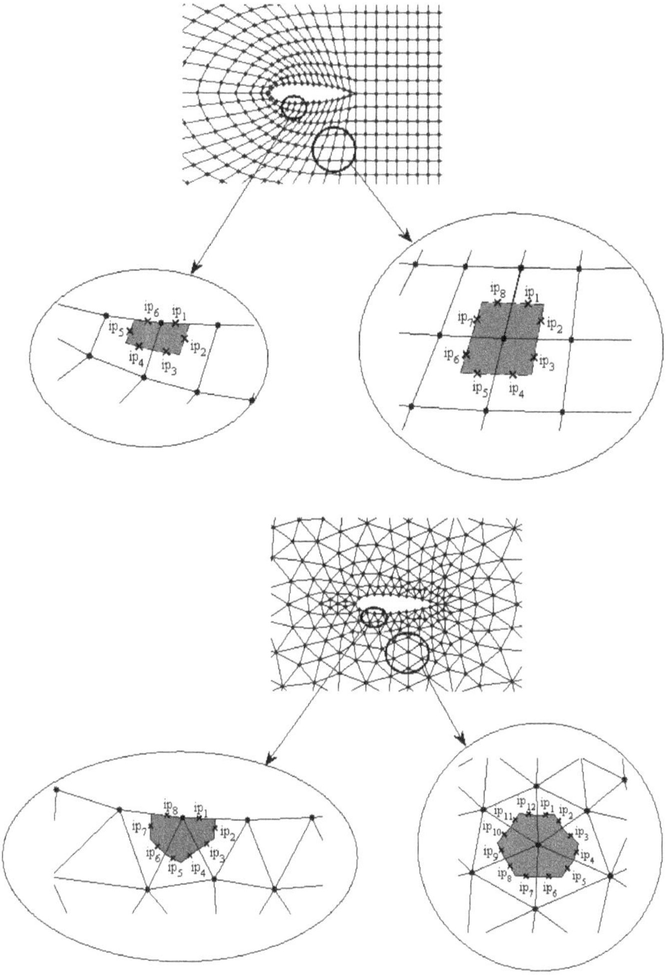

Fig. 1. Structured and unstructured element-based finite volume grids.

balance equation. In general, the computational molecule associated with each equation of a coupled set includes more than one variable to reflect the fact that the equations and the variables are inter-related. Therefore, in the case of a set of equations with two coupled variables φ and ψ, the following modified form of Eq. (4) is used to constrain φ at the nodal point P:

$$a_P^{\varphi}\varphi_P = \sum_{nb=1}^{N_{nb}} a_{nb}^{\varphi}\varphi_{nb} + a_P^{\psi}\psi_P + \sum_{nb=1}^{N_{nb}} a_{nb}^{\psi}\psi_{nb} + S_P^{\varphi} \tag{5}$$

In this article, we focus on level-2 approximation of Navier-Stokes equations in the context of co-located, pressure-based finite volume method. This level of approximation is particularly important because it deals with the modeling of physical influences such as upwind and downwind effects and the couplings between variables. A number of famous level-2 approximations are reviewed and the method of proper closure equations (MPCE), proposed by the first author, is introduced. To avoid the complexities associated with multi-dimensional problems and to focus on the nature of the approximations in a simple setting, one dimensional equations and test cases are discussed in details. Two-dimensional results are only briefly presented to show the applicability of the MPCE in multi-dimensional problems.

2. The semi-discrete form of equations in a 1D context

A one-dimensional variable-area duct is shown in Fig. 2. The complete set of governing equations for unsteady 1D-inviscid compressible flow in this duct, using the momentum variable formulation, is as follows (Chterental, 1999):

$$\frac{\partial}{\partial \tau}\begin{bmatrix} \rho \\ f \\ \rho e \end{bmatrix} + \frac{\partial}{\partial x}\begin{bmatrix} f \\ fu + p \\ fe + pu \end{bmatrix} = \begin{bmatrix} -f\dfrac{1}{a}\dfrac{da}{dx} + S^m \\ -fu\dfrac{1}{a}\dfrac{da}{dx} + S^u \\ -pu\dfrac{1}{a}\dfrac{da}{dx} \end{bmatrix} \tag{6}$$

Where τ is time, $f = \rho u$ is the momentum variable, u is velocity, p is pressure and a is the cross sectional area. Note that no artificial energy source term is used in the energy equation.

In order to close the above set of equations, an auxiliary equation is needed. Here, the ideal gas equation of state is used:

$$p = \rho R t \tag{7}$$

Where t is temperature and R is the gas constant. Neglecting the potential energy term, the total energy per unit mass (e) of an ideal gas can be written as follows:

$$e = c_v t + \frac{1}{2}u^2 \tag{8}$$

Where c_v is the specific heat at constant volume.

By integrating Eq. (6) over the P control volume in Fig. 2 and after using divergence theorem, the semi-discrete forms of the governing equations are obtained as follows:

$$\begin{bmatrix} \dfrac{V_P}{\Delta\tau}\left(\rho - \rho^{\circ}\right)_P + f_e a_e - f_w a_w \\ \dfrac{V_P}{\Delta\tau}\left(F_P - F_P^{\circ}\right) + [fu + p]_e a_e - [fu + p]_w a_w + P_P(a_w - a_a) \\ \dfrac{V_P}{\Delta\tau}\left((\rho E) - (\rho E)^{\circ}\right)_P + [fe + pu]_e a_e - [fe + Pu]_w a_w \end{bmatrix} = \begin{bmatrix} S_P^m V_P \\ S_P^p V_P \\ 0 \end{bmatrix} \tag{9}$$

Note that capital letters (like F and P) are used for the nodal values and small letters (like f and p) are used for integration point variables. Here, V_P stands for the volume of cell P. Closure equations for the integration point variables are discussed next.

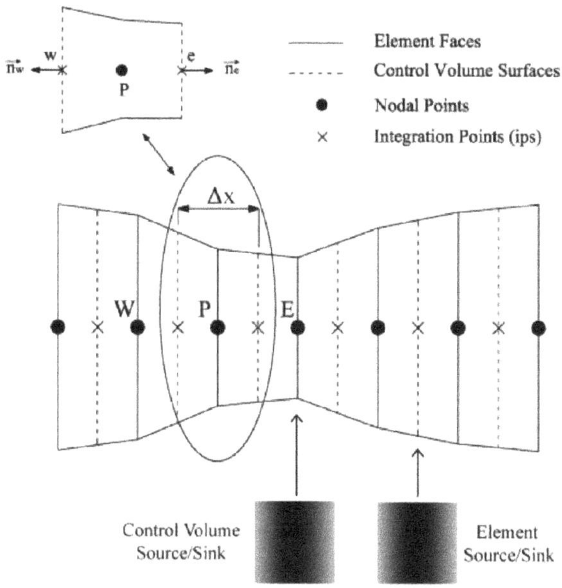

Fig. 2. Grid arrangement in a one-dimensional variable area duct.

3. Closure equations for integration point variables

After obtaining the semi-discrete form of governing equations and linearizing them, closure equations are needed to approximate integration point (ip) quantities in terms of the nodal values. By employing these closure equations, the semi-discrete equations are converted to fully-discrete equations. These ip-equations play a critical role in the robustness and accuracy of a collocated scheme. Therefore, different solution strategies for defining ip-equation are briefly described in this section. Here, we only discuss the closure equations for "east" integration point of a typical cell around node P. The closure equations at the "west" ip are obtained similarly.

3.1 Methods based on geometrical interpolation

Simple symmetric interpolation is the most trivial choice for an ip-value in a uniform grid:

$$\varphi_e = \frac{\phi_P + \phi_E}{2} \tag{10}$$

On non-uniform grids, weighted interpolation is used as follows:

$$\varphi_e = \lambda_e \phi_E + (1 - \lambda_e)\phi_P \tag{11}$$

Where λ_e is the weight factor for "east" ip, calculated by the following formula:

$$\lambda_e = \frac{x_e - X_P}{X_E - X_P} \tag{12}$$

This scheme is 2nd-order accurate, but is unbounded so that non-physical oscillations may appear in regions with high gradients. To show the problem associated with this scheme, consider the 1-D inviscid incompressible flow in a constant area duct shown in Fig. 3. The semi-discrete form of continuity and momentum equations in this case are as follows:

$$\rho A\left(u_e - u_w\right) = 0 \tag{13}$$

$$\bar{\dot{m}}\left(u_e - u_w\right) - A\left(p_e - p_w\right) = 0 \tag{14}$$

Where A is the cross sectional area of the duct and $\bar{\dot{m}}$ is the mass flow rate based on the most recent available value of velocity.

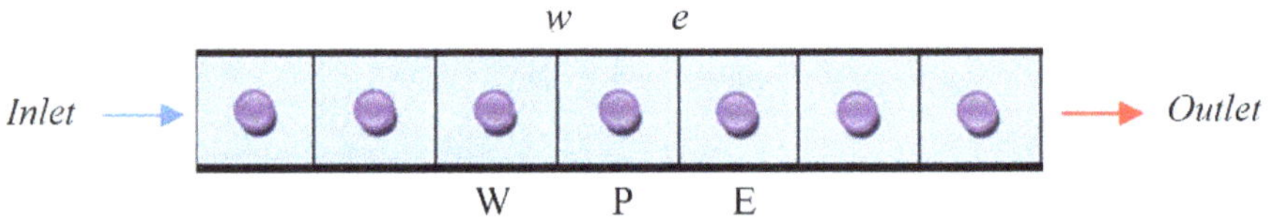

Fig. 3. A 1-D Co-located grid in a constant cross sectional area duct.

Using simple symmetric interpolation, Eqs. (13) and (14) are converted to the following fully-discrete forms:

$$\frac{\rho A}{2}\left(U_E - U_W\right) = 0 \tag{15}$$

$$\frac{\bar{\dot{m}}}{2}\left(U_E - U_W\right) - \frac{A}{2}\left(P_E - P_W\right) = 0 \tag{16}$$

Combining the above equations, one obtains the following constraint on the pressure at nodal point P:

$$P_E = P_W \tag{17}$$

Note that pressure at node P is absent in this equation and this constraint cannot distinguish between a uniform pressure field and the wiggly pressure field shown in Fig. 4.

3.2 Methods based on taylor-series expansion

It is logical to assume that convected quantities in a thermo-fluid problem are influenced by the upstream condition. Therefore, assuming the flow direction from P to E, the following truncated Taylor series expansions are both valid approximations for φ_e :

$$\varphi_e \approx \phi_P \tag{18}$$

$$\varphi_e \approx \phi_P + \left.\frac{\partial \varphi}{\partial x}\right|_P \left(x_e - X_P\right) \tag{19}$$

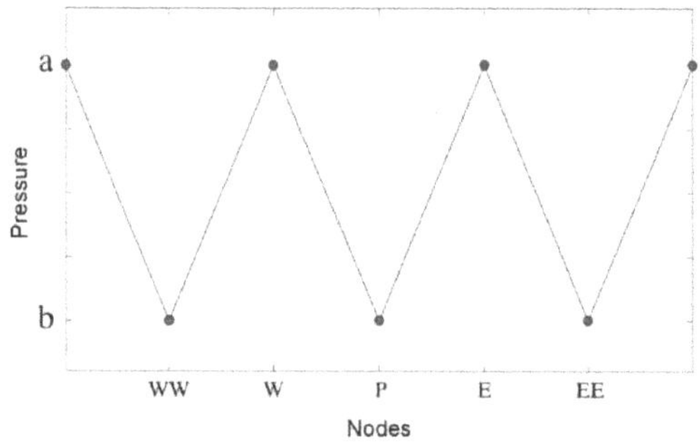

Fig. 4. A typical wiggly pressure filed.

Equation (18) is called the simple or Fist Order Upwinding (FOU) and provides highly stable and unconditionally bounded numerical schemes. However, it is only first order accurate. Equation (19) provides a Second Order Upwinding (SOU) strategy. One can show that a numerical scheme which employs Eq. (10) for integration point pressures and Eqs. (18) or (19) for integration point velocities in Eqs. (13) and (14), would not be able to detect the checkerboard pressure fields. Such numerical schemes are prone to unphysical pressure-velocity decoupling symptom.

3.3 Methods based on momentum interpolation

Rhie and Chow (Rhie, 1983) employed a co-located grid to solve the flow field around an airfoil. To avoid nonphysical, wiggly numerical solutions, they came to the conclusion that different closure equations had to be used for the convecting or mass-carrying ($\hat{u}$) and convected or transported (u) velocity components at the integration points. The convecting velocity, used in the semi-discrete continuity equation, is linked to the pressure through a momentum interpolation procedure. Hence, this discretization scheme is called the Momentum Interpolation Method (MIM). As compared to the classical schemes on staggered grids, one may assume that nodal momentum balances are conceptually staggered to obtain closure equations for convecting velocity at integration points. The discrete momentum balance for node P can be written as follows:

$$a_P U_P - \sum a_{nb} U_{nb} + V_P \left.\frac{\partial P}{\partial x}\right|_P = 0 \quad \Rightarrow \quad U_P = \frac{\sum b_{nb} U_{nb}}{a_P} - \frac{V_P}{a_P}\left.\frac{\partial P}{\partial x}\right|_P \tag{20}$$

The convecting velocity at "e" is obtained by staggering the stencil of the "P" computational molecule to the "e" integration point:

$$\hat{u}_e = \left.\frac{\sum a_{nb} U_{nb}}{a}\right|_e \quad \left.\frac{V_e}{a_e}\frac{\overline{\partial P}}{\partial x}\right|_e \tag{21}$$

Which results in:

$$\hat{u}_e = \frac{U_P + U_E}{2} + \frac{\bar{d}_e}{2}\left(\left.\frac{\overline{\partial P}}{\partial x}\right|_P + \left.\frac{\overline{\partial P}}{\partial x}\right|_E - 2\left.\frac{\partial P}{\partial x}\right|_e \right) \tag{22}$$

Where

$$\bar{d}_e \equiv \frac{1}{2}\left(\frac{V_P}{a_P}+\frac{V_E}{a_E}\right) \tag{23}$$

The closure for the convected velocity, and any other transported variable, comes from an upwind scheme. The simplest option at the integration point e would be the FOU:

$$u_e = U_P \quad \text{if} \quad \vec{m}_e > 0 \tag{24}$$

The interpolation formula, Eq. (22), is not unique as explained in (Acharya, 2007) and one might successfully implement the idea using modified forms of this approach. The key point, however, is to use different closure equations for the convected and convecting velocity components. The convecting velocity should be properly related to the pressure field.

The closure equation for p_e has not been a matter of concern, at least for incompressible flows. Taking into account the elliptic behavior of the pressure, the following closure equation is used in the MIM:

$$P_e = \frac{P_E + P_P}{2} \tag{25}$$

This technique guarantees the required physical pressure-velocity coupling in the numerical model and prevents any wiggly solutions (Ashrafizadeh et al., 2009).

3.4 Methods based on physical influences

The pleasure of working with all variables on a single nodal position, as opposed to the pain of working with staggered grids limited to simple geometries, has made the MIM very popular. Therefore, attempts have been made to employ and improve the MIM for the solution of flows at all speeds. Schneider and Raw (Schneider & Raw, 1987a, 1987b) proposed the physical influence scheme to unify the integration point velocity interpolation formulas in co-located grids. They retained a unified definition for the integration-point velocity components, but argued that the closure for u_e should not be obtained solely by purely mathematical upwind schemes. They suggested the non-conservative form of the momentum equation to provide a closure for u_e :

$$\left[u\frac{\partial(\rho u)}{\partial x}+\rho u\frac{\partial u}{\partial x}+\frac{\partial P}{\partial x}\right]=0 \tag{26}$$

The method, called the physical influence scheme, provides the following closure equation in the context of the 1-D test case:

$$u_e = \hat{u}_e = U_p + \frac{\Delta x}{2\rho\bar{u}_e}\frac{P_P - P_E}{\Delta x} \tag{27}$$

Where the over bar refers to the most recent available value in the nonlinear iterations. Equation (25) is used as the closure for p_e in this scheme.

While successfully implemented in a number of multidimensional viscous problems, the physical influence scheme fails to prevent pressure checkerboard problem in inviscid incompressible flows (Bakhtiari, 2008). This failure reinforces the belief that two definitions for convected and convecting velocities at integration points are necessary when the calculations are carried out on a co-located grid.

3.5 Methods based on modified physical influences

To fix the problem associated with the physical influence scheme, Karimian and Schneider (Karimian, 1994a) accept the notion of dual velocities at the integration points and employ a combination of momentum and continuity equations to provide a closure for the convecting velocity ($\hat{u}$):

$$\hat{u}_e = \frac{U_P + U_E}{2} - \frac{\Delta x}{2\rho\bar{u}_e}\left[(Momentum\ Error) - \bar{u}_e(Continuity\ Error)\right]_e \tag{28}$$

Which results in:

$$\hat{u}_e = \frac{U_P + U_E}{2} - \frac{P_P - P_E}{2\rho\bar{u}_e} \tag{29}$$

They also use Eq. (27) for convected velocity (u_e) and Eq. (25) as the closure for p_e. Numerical experiments with the 1-D test case show that this method works well and successfully suppresses any oscillatory numerical solution even in time-depended all speed flows (Karimian, 1994b). This method has also been extended to 2D flows on structured (Alisadeghi et al., 2011a) and unstructured (Alisadeghi et al., 2011b) grids.

Darbandi (Darbandi et al., 1997) proposed a pressure-based all speed method similar to the Karimian's method, in which the momentum components ($f = \rho u$) are used as the flow dependent variables instead of the velocity components (u). They used the following form of the momentum equation as the physical constraint on the integration point convected velocity:

$$\left[\frac{\partial f}{\partial \tau} + \bar{u}\frac{\partial f}{\partial x} + \bar{f}\frac{\partial u}{\partial x} + \frac{\partial P}{\partial x}\right]_e = 0 \tag{30}$$

Where $f = \rho u$ is the momentum variable. Appropriate discretization of Eq.(30) leads to:

$$f_e = F_P + \frac{1}{4\bar{u}_e}(P_P - P_E) \tag{31}$$

In this method, a combination of momentum and continuity equations is used to constrain the integration point convecting velocity:

$$\left[\frac{\partial f}{\partial \tau} + \bar{u}\frac{\partial f}{\partial x} + \bar{f}\frac{\partial u}{\partial x} + \frac{\partial P}{\partial x}\right]_e - \alpha\bar{u}_e\left[\frac{\partial \bar{\rho}}{\partial \tau} + \frac{\partial f}{\partial x}\right]_e = 0 \tag{32}$$

Where α is a user-defined weight factor. This equation can be discretized to obtain the following formula:

$$\hat{f}_e = F_P + \frac{1}{2\bar{u}_e}(P_P - P_E) \tag{33}$$

In this method, like all other methods discussed so far, Eq. (25) is used as the closure for p_e. Numerical results have shown that this set of closure equations removes the possibility of a checkerboard solution and provides strong coupling between pressure and velocity fields. Another analysis has shown that the use of momentum-variable formulation, instead of primitive variable formulation, improves the stability and accuracy of the solution especially in the solution of compressible flows with shock waves (Darbandi, 2004).

3.6 Method of proper closure equations (MPCE)

3.6.1 One-dimensional incompressible flow

Ashrafizadeh et al. (Ashrafizadeh et al., 2009) have shown that it is possible to develop a co-located numerical scheme without resorting to the convecting and convected velocity concepts, originally proposed by Rhie and Chow. The proposed method, called the method of proper closure equations (MPCE), employs a proper set of physically relevant equations to constrain the velocity and pressure at integration points. It has already been successfully implemented and used in the solution of steady 1-D inviscid incompressible and compressible flow problems (Ashrafizadeh et al., 2008, 2009) and steady 2-D viscous incompressible flow problems on both structured and unstructured grids (Alinia, 2011; Bakhtiari, 2008).

In this method, following the physical-based approach proposed by Schneider and Raw (Schneider et al., 1987) and modifications proposed by Karimian and Darbandi (Darbandi, 1997; Karimian, 1994a), a combination of non-conservative form of momentum and continuity equations is used to obtain an interpolation formula for the ip-velocity:

$$[momentum\ Eq.]_e - \alpha_u u_e [continuity\ Eq.]_e = 0 \tag{34}$$

which results in:

$$\left[\underbrace{\rho u \frac{\partial u}{\partial x}}_{UDS} + \underbrace{\rho u \frac{\partial u}{\partial x}}_{1} + \rho u^2 \frac{1}{A} \underbrace{\frac{da}{dx}}_{CDS} + \underbrace{\frac{\partial P}{\partial x}}_{CDS} - S^u \right]_e - \alpha_u u_e \left[\rho \underbrace{\frac{\partial u}{\partial x}}_{2} + \rho u \frac{1}{A} \underbrace{\frac{da}{dx}}_{CDS} - S^m \right]_e = 0 \tag{35}$$

Here S^u and S^m are artificial momentum and mass source terms, and a and α_u are the ip-cross sectional area and a scheme control parameter respectively. There are options available to the user at this stage. Using UDS or CDS Schemes for terms 1 and 2 and $\alpha_u = 0, 1, 2$, as discussed in (Rezvani, 2008), different formulas for u_e can be obtained. The most common formula is as follows:

$$u_e = U_P + \frac{P_P - P_E}{2\rho\bar{u}_e} + \frac{\Delta x}{2\rho\bar{u}_e}(S_e^u - \bar{u}_e S_e^m) \tag{36}$$

As mentioned earlier, the classical closure choice for the integration point pressure has been the linear approximation, i.e. Eq.(25). However, the MPCE employs relevant governing equations to obtain the required closures for all integration point quantities, including the pressure. The momentum equation is a natural choice for obtaining a proper closure equation for p_e. However, only the pressure gradient appears in the momentum equation and CDS-based discretization, physically proper for the pressure, results in wiggly solutions. As discussed in (Rezvani, 2010), by taking the divergence of the momentum equation, a pressure poisson equation for the pressure is obtained. The pressure Poisson equation is an appropriate constraint equation for the pressure discretization. Therefore, as compared to Eq. (34), the following closure equation is proposed for the integration point pressure:

$$\begin{bmatrix} Divergence\ of \\ Momentum\ Eq. \end{bmatrix}_e - \alpha_P \frac{4\bar{u}_e}{\Delta x}\left[Continuity\ Eq.\right]_e = 0 \tag{37}$$

Which results in:

$$P_e = \frac{P_P + P_E}{2} + \alpha_P \frac{\rho \bar{u}_e}{2}\left(U_P - U_E\right) + \alpha_P \frac{\rho \bar{u}_e^2}{2a_e}\left(A_P - A_E\right) + \frac{\Delta x^2}{4}\left[\left.\frac{\partial u}{\partial x}\right|_e + 2\alpha_P \frac{\bar{u}_e}{\Delta x} S_e^m\right] \tag{38}$$

Where $\alpha_P > 0$ is a scheme control parameter. Numerical test results have shown that the method works well with any nonzero positive value for α_P (Rezvani-2010).

3.6.2 One-dimensional compressible flow

Extension of the MPCE to compressible flow problems requires two additional tasks. First, the energy equation should also be solved in order to find the temperature filed. Second, depending on the chosen unknown variables, proper linearization is crucially important. Following approaches presented in (Darbandi et al., 2007; Karimian, 1994a), the Newton-Raphson linearization strategy is employed to linearize the nonlinear terms in the balance equations. For example, term pu in the energy equation is linearized as follows:

$$pu = (\rho Rt)u = (\rho u)Rt = R(ft) = R\left(\bar{f}t + \bar{t}f - \bar{f}\bar{t}\right) \tag{39}$$

Closure equation for f_e can be easily obtained using the unsteady-compressible form of Eq. (34). First order upwind difference scheme is used for the discretization of momentum variable gradient terms. Use of $\alpha_u = 1$ in the general form of the closure for the momentum variable at the east face of the cell, results in the following formula for f_e:

$$f_e = \frac{2C\dfrac{\bar{\rho}_e}{\bar{\rho}_P}}{1+2C}F_P + \frac{C\left(P_P - P_E\right)}{\bar{u}_e(1+2C)} + \frac{f_e^\circ}{1+2C} + \frac{C\Delta x}{1+2C}\left.\frac{\partial \bar{\rho}}{\partial \tau}\right|_e + \frac{C\Delta x}{\bar{u}_e(1+2C)}\left(S_e^u - \bar{u}_e S_e^m\right) \tag{40}$$

Where, C is the Courant number defined as $C = \bar{u}\Delta\tau/\Delta x$ and superscript $^\circ$ stands for the values from previous time step.

Similarly, Closure equation for p_e can be obtained using the unsteady-compressible form of Eq. (37) as follow:

$$p_e = \frac{P_P + P_E}{2} + \frac{\alpha_p}{2}\bar{u}_e\left(F_P - F_E\right) + \frac{\alpha_p}{2}\frac{\bar{f}_e\bar{u}_e}{a_e}\left(A_P - A_E\right) + \frac{\Delta x^2}{4}\left[\left.\frac{\partial u}{\partial x}\right|_e + 2\alpha_p\frac{\bar{u}_e}{\Delta x}\right]S_e^m - \frac{\alpha_p}{2}\bar{u}_e\Delta x\left.\frac{\partial\bar{\rho}}{\partial\tau}\right|_e \tag{41}$$

The non-conservative form of the continuity equation in Eq. (6), in the absence of its source term, is proposed as a proper closure equation for the density (Darbandi et al., 1997; Karimian, 1994a). Here, in order to control the stability and accuracy of the code in both subsonic and supersonic regimes, a smart blending factor (β) is used which makes the equation hyperbolic in supersonic flow regions [Karimian, 1994a]:

$$\rho_e = \rho_P - \beta_e\frac{\Delta x}{2\bar{u}_e}\left[\frac{\partial\bar{F}}{\partial x} + \frac{\bar{f}}{a}\frac{\partial A}{\partial x} + \frac{\partial\bar{\rho}}{\partial\tau}\right]_e + \left(1 - \beta_e\right)\left.\frac{\partial\bar{\rho}}{\partial x}\right|_e\frac{\Delta x}{2} \tag{42}$$

$$\beta_e = \frac{M_e^m}{M_e^m + 1} \tag{43}$$

where M_e is local Mach number at the "east" integration point and m=2 is suggested for the superscript m.

The final required closure equation is an interpolation formula for the integration point temperature. Here, the non-conservative form of the energy equation in the absence of the source term is a natural candidate (Darbandi et al., 1997) and one obtains the following expression:

$$t_e = \frac{2C}{1+2C}T_P + \frac{C\bar{p}_e}{\bar{f}_e c_v\left(1+2C\right)}\left(\frac{F_P}{\bar{\rho}_P} - \frac{F_E}{\bar{\rho}_E}\right) + \frac{C}{1+2C}\frac{\bar{p}_e\left(A_P - A_E\right)}{\bar{\rho}_e c_v a_e} + \frac{t_e^\circ}{1+2C} \tag{44}$$

The final discrete computational molecules for solving velocity, pressure and temperature fields are obtained after closure equations are substituted in Eq. (9):

$$a_P^{pp}P_P + \sum a_{nb}^{pp}P_{nb} + a_P^{pf}F_P + \sum a_{nb}^{pf}F_{nb} + a_P^{pt}T_P + \sum a_{nb}^{pt}T_{nb} = b_P^p \tag{45}$$

$$a_P^{ff}F_P + \sum a_{nb}^{ff}F_{nb} + a_P^{fp}P_P + \sum a_{nb}^{fp}P_{nb} + a_P^{ft}T_P + \sum a_{nb}^{ft}T_{nb} = b_P^f \tag{46}$$

$$a_P^{tt}T_P + \sum a_{nb}^{tt}T_{nb} + a_P^{tf}F_P + \sum a_{nb}^{tf}F_{nb} + a_P^{tp}P_P + \sum a_{nb}^{tp}P_{nb} = b_P^t \tag{47}$$

In this linear algebraic set, $a_P^{ff}, a_P^{fp}, \ldots$ are the influence coefficients and $b_P^p, b_P^f, \ldots$ are right hand side constant terms. Subscript nb stands for the immediate neighbors of node P (i.e. W and E). The first superscript refers to the relevant equation and the second superscript points to the relevant physical variable. The assembled linear equations can be solved using any direct or iterative solver. Here, the Gauss elimination method is used to solve the set of simultaneous equations. The coupled set of the algebraic equations might also be solved using a semi-implicit (segregated) solution strategy.

3.6.3 Two-dimensional incompressible flow

The MPCE has also been used to solve 2D, incompressible laminar flows on both structured and unstructured grids (Bakhtiari, 2008; Alinia, 2011). The implementation follows the basic philosophy already described in a 1D context. However, more complex interpolation formulas are required to take into account the effect of cell shape and orientation on balance equations. Implementation details are not discussed here for the sake of brevity.

4. One-dimensional test cases

4.1 Incompressible flow

Performance of the MPCE for solving incompressible and compressible flows is examined here through two different test cases. Steady incompressible flow in a constant area duct is the first test case which is used to examine the pressure-velocity coupling in MPCE. Quasi 1D flow in a convergent-divergent nozzle is another test case used to check the accuracy of the proposed method in inviscid incompressible flow problems.

4.1.1 Steady incompressible flow in a constant area duct

The 1D incompressible flow in a constant area duct is used to see if oscillatory solutions appear when the MPCE equations govern the discrete flow field. For this purpose, artificial element and volume sources/sinks are used in continuity and momentum equations. In a domain with 30 nodes, the artificial element source terms are located at the 10th and 20th cell-faces and control volume source terms are activated at the 10th and 20th nodal positions.

Numerical results show that the effects of sharp variations in velocity and pressure due to the artificial source terms are quickly damped out and MPCE works well even with $\alpha_p = 0$. Figures 5 and 6 show the effects of control-volume artificial sources on the numerical solution. These results show that numerical oscillations are developed in both velocity and pressure fields for $\alpha_p = 0$. However, oscillations are strongly damped for any non-zero value of α_p. Note that the notation "N-D" in these figures stands for "Non-Dimensional" and velocity is normalized with the inlet velocity and pressure is normalized using the exit pressure.

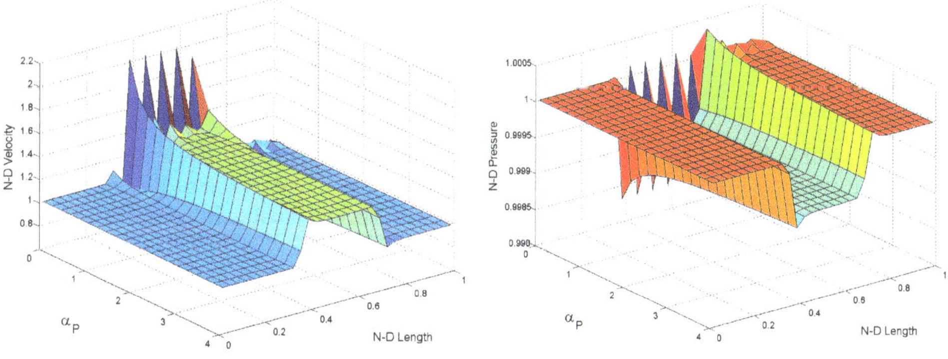

Fig. 5. Velocity and pressure fields in a constant area duct with control-volume mass source/sink.

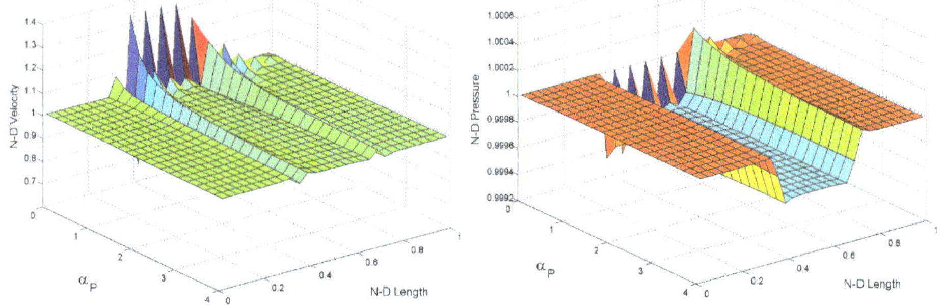

Fig. 6. Velocity and pressure fields in a constant area duct with control-volume momentum source/sink.

4.1.2 The quasi-1-D test case

Ideal flow in a converging-diverging nozzle is another test case to examine the performance of MPCE. Figure 7 compares the non-dimensional velocity and pressure distributions along the nozzle with the analytical solutions. Excellent agreement is observed.

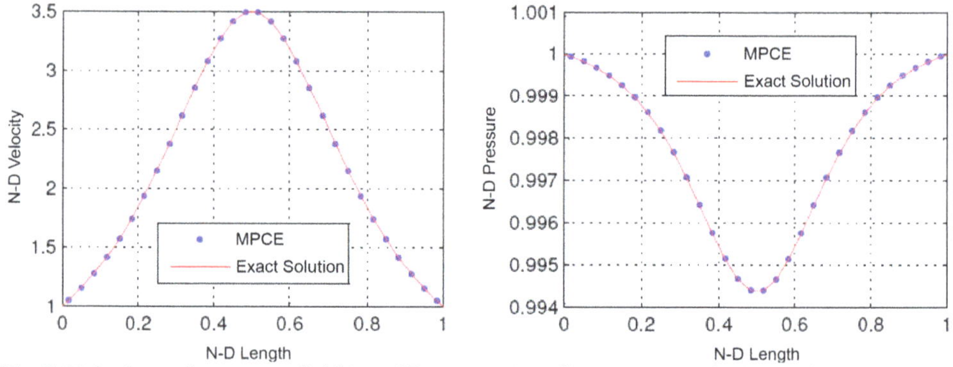

Fig. 7. Velocity and pressure field in a 1D converging-diverging nozzle ($\alpha_p = 1$).

4.2 Compressible flow

For compressible flow, performance of MPCE is examined in three different test problems. Steady subsonic compressible flow in convergent-divergent nozzle, steady compressible flow with normal shock in convergent-divergent nozzle and unsteady compressible flow with normal shock and expansion waves in a shock tube.

4.2.1 Steady subsonic compressible flow in a convergent-divergent nozzle

Results of solving subsonic compressible flow in a symmetric convergent-divergent nozzle with aspect ratio 2 using 51 nodes is shown in this section. For the boundary condition implementation, mass flow rate and the static temperature are specified at the inlet and the static pressure is provided at the outlet. In order to assess the accuracy of the proposed method, numerical solutions corresponding to inlet Mach numbers of 0.05, 0.1, 0.15, 0.2, and 0.25 are presented in Fig. 8. Flow with the inlet Mach number of 0.3059, for which the nozzle is choked, is also considered. In this test case, $\alpha_p = 1$ is used and pressure, temperature, and density are nondimensionalized with the outlet pressure, inlet temperature, and inlet density, respectively.

4.2.2 Steady flow with shock waves

In the divergent section of the previous test case, normal shock waves appear for certain inlet boundary conditions while the position and the strength of the shock can be controlled by regulating the back pressure. Figure 9 presents the Mach number, non-dimensional pressure, temperature and density distributions along the nozzle for the pressure ratios (P_{outlet}/P_{inlet}) of 0.65, 0.75 and 0.85 using 81 nodes. It is seen that the normal shock is well captured using a few nodes in all cases and no numerical oscillation is developed along the nozzle. Note that this results are computed using $m = 2$ in Eq.(43) and α_p takes values lower than 1 in the supersonic parts of the solution domain and values higher than 1 in the subsonic regions.

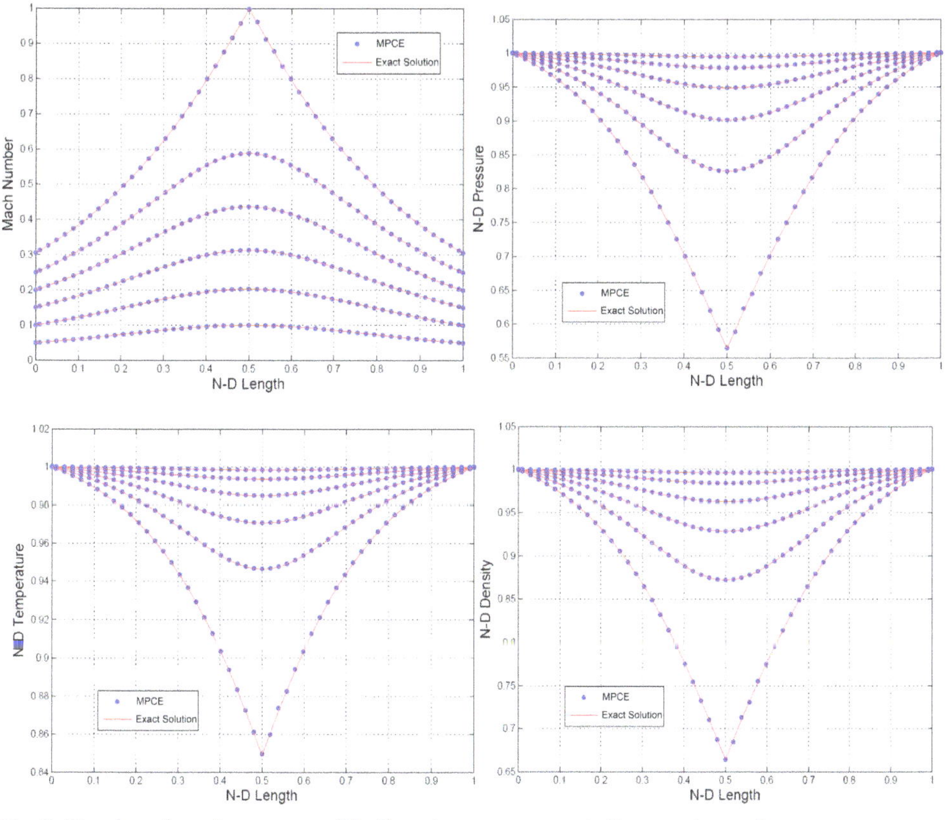

Fig. 8. Steady subsonic compressible flow in a convergent-divergent nozzle.

4.2.3 Unsteady compressible flow in a shock tube

As the final unsteady 1D test case with both compression and expansion waves, shock tube problem is discussed. The geometry shown in Fig.10 is a constant area duct divided into two regions separated by a diaphragm at the middle of the duct. At $\tau = 0$ the gas temperature is 298 K, the pressure at the right side of the membrane is 100 KPa and the pressure at the left side of the membrane is 1000 KPa. The time step for the marching numerical solution is equal to $\Delta\tau = 4\times10^{-6}$ seconds. Figure 11 shows the numerically calculated distributions of Mach number, non-dimensional pressure, temperature, and density after 0.42×10^{-3} seconds. Here, the Courant number is 0.229 which is computed using the maximum magnitude of velocity in the domain. The pressure, temperature and density are non-dimensionalized using the low pressure side values. Good agreement with the exact solution is observed.

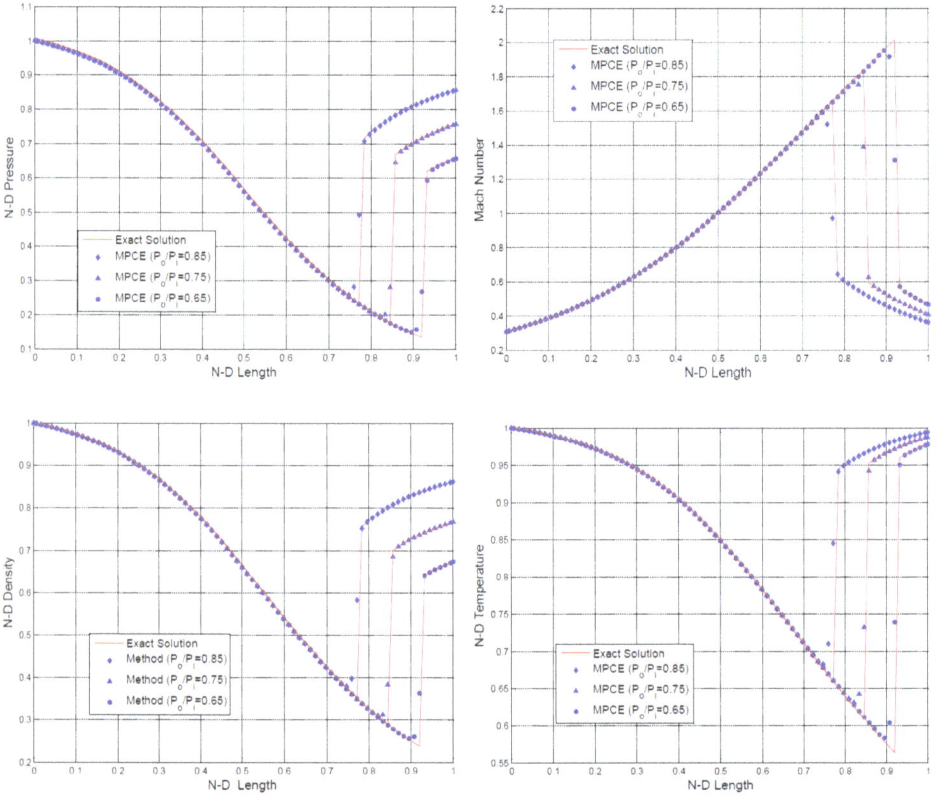

Fig. 9. Steady flow with shock waves in a convergent-divergent nozzle.

5. Two-dimensional incompressible flow test cases

The MPCE has also been used to solve many 2D benchmark test problems. Here, only the results for two standard test cases are presented.

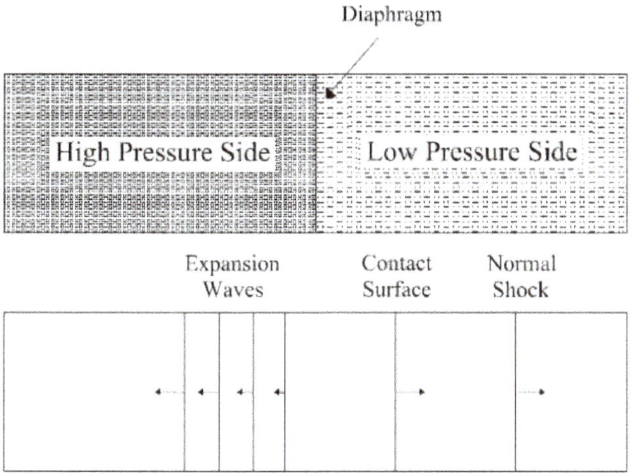

Fig. 10. Geometry and the flow pattern for the shock tube problem.

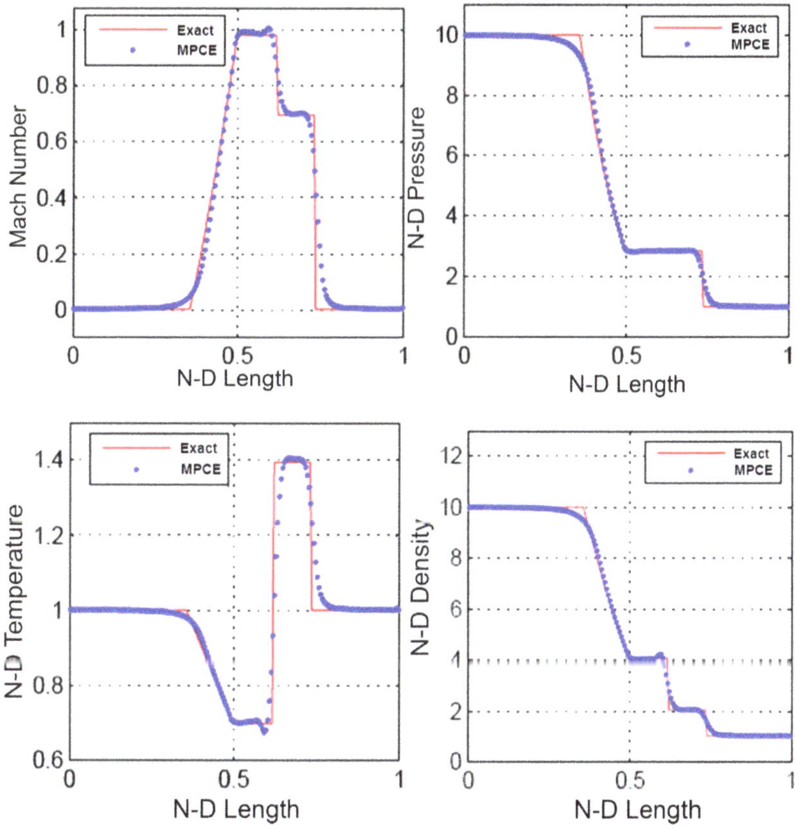

Fig. 11. Results of unsteady compressible flow in shock tube

5.1 Flow in a 2D lid-driven cavity

Flow in a lid-driven cavity, shown in Fig. 12, is a widely used test case for 2D steady incompressible flows governed by Navier-Stokes equations. Numerical solution via MPCE on an unstructured grid with 12656 triangular elements, shown in Fig. 13, is compared to the numerical results of Erturk (Erturk, 2005). Flow patterns at different Reynolds numbers are shown in Fig. 14. Excellent agreement between the calculated velocities at cavity centerlines and the benchmark results in Fig. 14 is also observed.

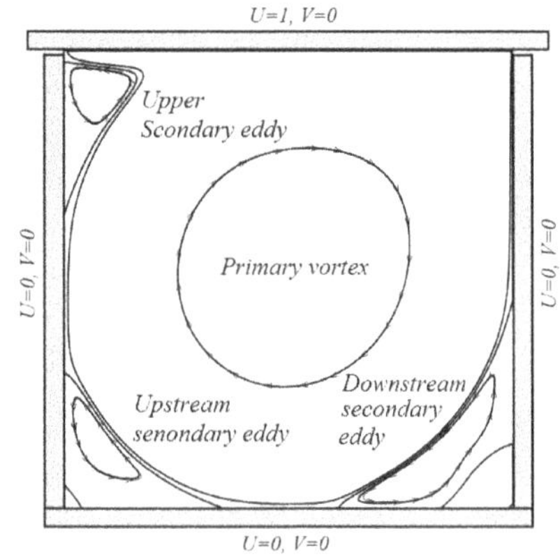

Fig. 12. Lid-driven cavity boundary conditions with an schematic of flow features.

Fig. 13. Generated unstructured mesh for lid-driven cavity test case.

5.2 Flow over a backward facing-step

The backward step flow is another commonly used test case for numerical solution algorithms. The geometry of this test case is shown in Fig. 15. Forty nodes are used across the channel, 300 nodes are used from the step to 30h and 50 nodes are used from 30h to 50h along the channel. Nodes are distributed uniformly. Figure 16 shows the calculated streamlines at Re=800, in which a recirculation zone appears at the upper wall. Normalized locations of the re-attachment point at the lower wall (X1), the separation point at the upper wall (X2), and the re-attachment point at the upper wall (X3) are compared with the results obtained by Gartling (Gartling, 1990) for flow at Re = 800. Comparison results are shown in Table 1 and excellent agreement is obvious (Erturk, 2008). A comparison has also been made between numerical results via MPCE and the benchmark results in (Gartling, 1990) and shown in Fig. 18. Again, excellent agreement is observed. Similar calculations and comparisons have also been made for a backward-step flow model with an inlet channel. Excellent results, not shown here, are also obtained in this test case.

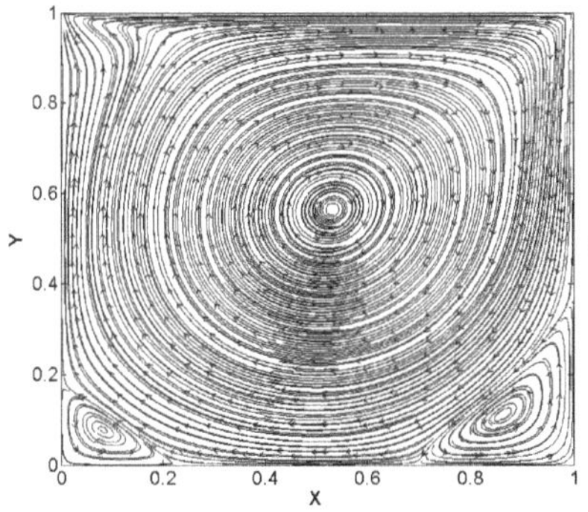

(a) Streamlines at Re = 1000.

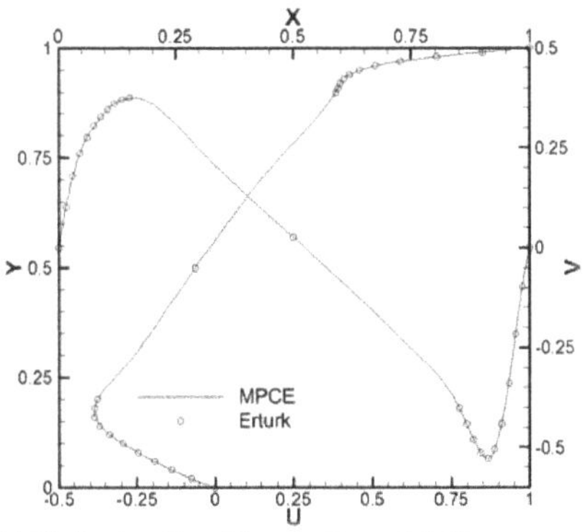

(b) Velocity Profiles at Re = 1000.

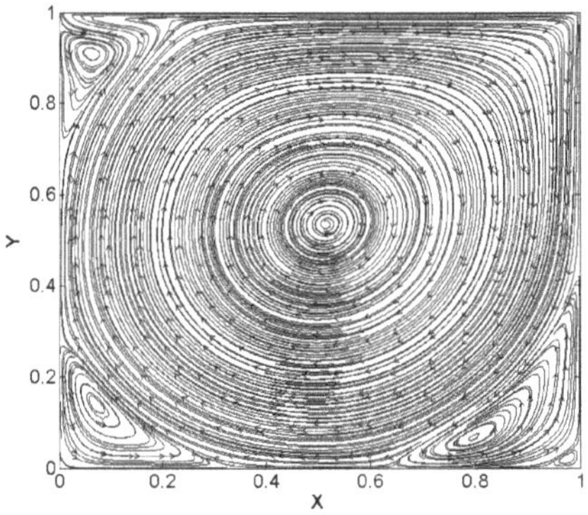

(c) Streamlines at Re = 5000.

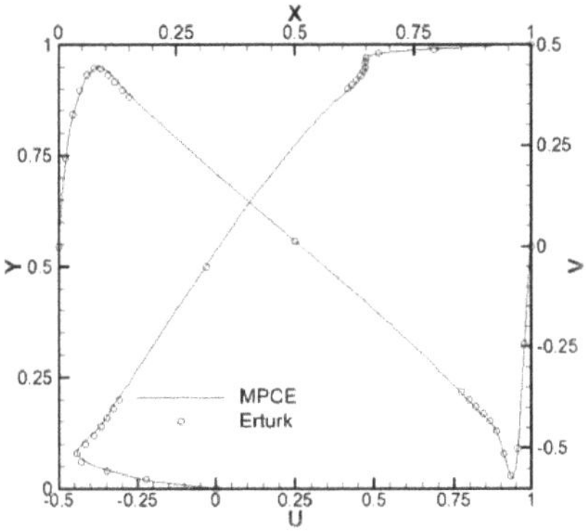

(d) Velocity Profiles at Re = 5000.

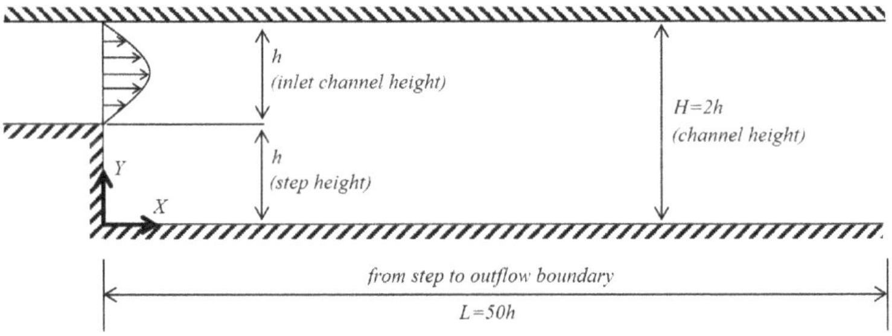

Fig. 14. Schematic view of the backward facing-step flow problem.

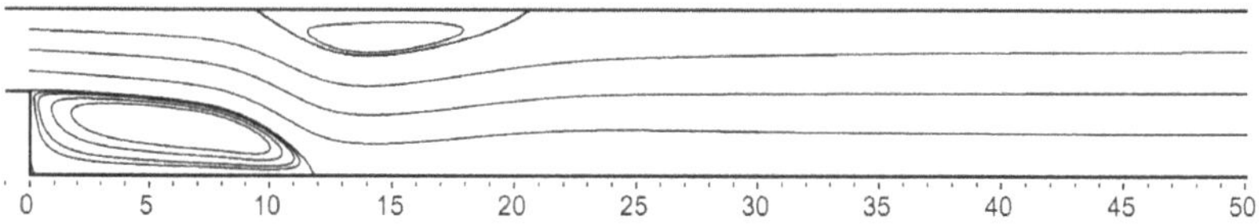

Fig. 15. Calculated streamlines in the backward-step flow problem at Re = 800.

	X1	X2	X3	X3 - X2
MPCE	12.0419	9.5528	20.8093	11.2565
Erturk	12.20	9.70	20.96	11.26

Table 1. Comparison of normalized X1, X2 and X3 locations and length of the first recirculating region on the upper wall (X3 - X2) for Re = 800

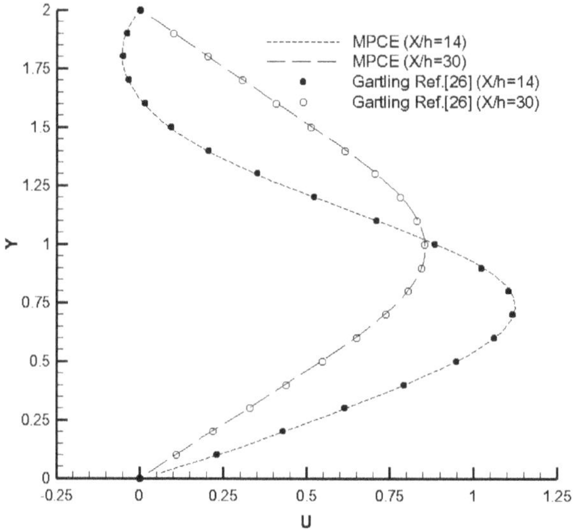

Fig. 16. Horizontal velocity profiles across the channel at X/h = 14 and 30 for Re = 800.

6. Conclusion

Different pressure-based finite volume discretization schemes for the solution of fluid flow problems are explained and compared in the context of co-located grid arrangements and a new scheme, called the Method of Proper Closure Equations (MPCE), is proposed. The proposed method is the only discretization scheme on co-located grids which employs only one definition for the integration point velocity and successfully solves incompressible as well as compressible flow problems. Simple one-dimensional test cases are presented to discuss different available schemes and the building blocks of MPCE. Implementation of the MPCE on 2D structured and unstructured grids has also been carried out but the details are not reported here for the sake of brevity. Numerical solutions via MPCE, however, are presented for both 1D and 2D problems which clearly show the satisfactory performance of the MPCE in the numerical solution of fluid flow problems.

7. References

Acharya, S; Baliga, B. R; Karki, K.; Murthy, J. Y.; Prakash, C. & Vanka, S. P. (2007). Pressure-Based Finite-Volume Methods in Computational Fluid Dynamics. *Journal of Heat Transfer*, Vol. 129, pp. 407-424

Alinia, B. (2011). Implementation of the Method of Proper Closure Equations on Two-Dimensional Unstructured Grids for Steady, Incompressible Viscous Problems, *M.Sc. Thesis*, Department of Mechanical Engineering, K.N.Toosi University of Technology, Tehran, Iran, (In Persian)

Alisadeghi, H. & Karimian, S. M. H. (2011a). Different Modelings of Cell-Face Velocities and Their Effects on the Pressure-Velocity Coupling, Accuracy and Convergence of Solution, *International Journal for Numerical Methods in Fluids*, Vol. 65, pp.969-988

Alisadeghi, H. & Karimian, S. M. H. (2011b). Comparison of Different Solution Algorithms for Collocated Method of MCIM to Calculate Steady and Unsteady Incompressible Flows on Unstructured Grids, *Computers and Fluids*, Vol. 46, 94-100

Ashrafizadeh, A.; Okhovat, S.; Pourbagian, M. & Raithby, G. D. (2011). A Semi-Coupled Solution Algorithm in Aerodynamic Inverse Shape Design, *Inverse Problems in Science and Engineering*, Vol. 19, No. 4, pp. 509-528, ISSN 1741-5985

Ashrafizadeh, A.; Rezvani M.; & Bakhtiari, B. (2008). A New Co-located Pressure-Based Finite Volume Method, *16th Annual Conference of the CFD Society of Canada*, Saskatoon, Saskatchewan, Canada

Ashrafizadeh, A.; Rezvani M.; & Bakhtiari, B. (2009). Pressure-Velocity Coupling On Co-Located Grids Using the Method of Proper Closure Equations, *Numerical Heat Transfer; Part B*, Vol. 56, pp.259–273

Bakhtiari, B. (2008). Decoupling of Pressure and Velocity in the Numerical Simulation of Incompressible Flows: A review. *M.Sc. Thesis*, K.N.Toosi University of Technology, Tehran, Iran, (In Persian)

Chterental, I. (1999). Two-Dimensional Euler Equations Solver, *M.Sc. Thesis*, Graduate Department of Aerospace Science and Engineering, University of Toronto, Toronto, Canada

Darbandi, M. & Schneider, G. E. (1997). Momentum Variable Procedure for Solving Compressible and Incompressible Flows. *AIAA Journal*, Vol. 35, pp. 1801-1805

Darbandi, M. & Mokarizadeh, V. (2004). A Modified Pressure-Based Algorithm to Solve Flow Fields With Shock and Expansion Waves, *Numerical Heat Transfer, Part B*, Vol. 46, pp.497–504

Darbandi, M.; Roohi, E. & Mokarizadeh, V. (2007). Conceptual Linearization of Euler Governing Equations to Solve High Speed Compressible Flow Using a Pressure-Based Method, *Numerical Methods for Partial Differential Equations*, DOI 10.1002, pp 583-604

Erturk, E.; Corke, T. C. & Gökçöl, C. (2005). Numerical Solutions of 2-D Steady Incompressible Driven Cavity Flow at High Reynolds Numbers, *Int. J. Numer. Meth.* Fluids, Vol. 48, pp. 747-774

Erturk, E. (2008). Numerical Solutions of 2-D Steady Incompressible Flow Over a Backward Facing Step, Part I: High Reynolds Number Solutions, *Int. J. Computers & Fluids*, Vol. 37, pp. 633-655

Gartling, D. K. (1990). A Test Problem for Outflow Boundary Conditions – Flow Over a Backward Facing Step, *Int. J. Numer. Meth. Fluids*, Vol. 11, pp. 953-969

Karimian, S. M. H. & Schneider, G. E. (1994a). Application of a Control-Volume-Based Finite-Element Formulation to the Shock Tube Problem, *AIAA Journal*, Vol.33 No.1, pp. 165-167

Karimian, S. M. H. & Schneider, G. E. (1994b). Pressure-based Computational Method for Compressible and Incompressible Flows, *Journal of Thermophysics and Heat Transfer*, Vol.8, No.2 , pp. 267-274

Patankar, S. V. (1980). *Numerical Heat Transfer and Fluid Flow*, Hemisphere Publishing Corporation, Washington, D. C.

Rezvani, M. (2008). A New Pressure-Based Computational Method for Solving Fluid flow at all speeds, M.Sc. Thesis, Department of Mechanical Engineering, K.N.Toosi University of Technology, Tehran, Iran, (In Persian)

Rezvani, M. & Ashrafizadeh, A. (2010). Numerical Simulation of the Inter-Equation Couplings in All-Speed Flows via the Method of Proper Closure Equations, *Numerical Heat Transfer; Part A*, Vol. 58, pp. 1–20

Rhie, C.M. & Chow, W. L. (1983). A Numerical Study of the Turbulent Flow Past an Isolated airfoil With Trailing Edge Separation. *AIAA Journal*, Vol. 21, pp. 1525-1532

Schneider, G. E. & Raw, M. J. (1987a). Control Volume Finite-Element Method For heat Transfer and Fluid Flow Using Colocated Variables-1. Computational Procedure. *Numerical Heat Transfer, part B*, Vol. 11, pp. 363-390

Schneider, G. E. & Raw, M. J. (1987b). Control Volume Finite-Element Method For heat Transfer and Fluid Flow Using Colocated Variables-2. Application and Validation. *Numerical Heat Transfer, part B*, Vol. 11, pp. 391-400

4

Application of Finite Volume Method in Fluid Dynamics and Inverse Design Based Optimization

Árpád Veress and József Rohács
Budapest University of Technology and Economics
Hungary

1. Introduction

The Euler and Navier-Stokes (NS) equations are derived by applying Newton's second law to an infinitely small inviscid and viscous control volume respectively, and with the extension of the mass and energy conservation laws, they provide the highest level of approximations for the flow physics within approaching of continuum-mechanics based flow regime. The mentioned governing equations[1] are the system of the nonlinear partial differential equations and they do not have general closed form solution as yet. However, in consideration of increasing expectations arisen from the industry and the high level evolution of the computer technology, the different numerical and optimization methods are developed and implemented in the complex framework of CFD (Computational Fluid Dynamics) programs. These software are widely spread in the engineering practice and they provide efficient solutions for different industrial problems. Beside the advantages of the virtual prototyping, the wide range of visualization tools and optimization, significant number of measurements can be replaced by the validated numerical analyses, which reduce time, capacity, cost and strongly contribute to the benefit of the company. Hence, followed by the short overview of different numerical methods generally used in CFD, a complete physical and mathematical interpretation is presented for a compressible NS and Euler solvers with validation and extension for coupling of inviscid flow solver with inverse design based optimization algorithm in a framework of the finite volume method.

1.1 Numerical methods for fluid flow

The continuum mechanics based NS equations describe flow physics in continuum and slip flow regime defined by Knudsen number as $Kn<0.01$ and $0.01<Kn<0.1$ respectively (Zucrow & Hoffman, 1976). In case of Euler equations, in which the viscous (diffusion) effects are not considered, $Re\rightarrow\infty$ and $Kn\rightarrow 0$. The continuum approach does not count the individual molecules and instead, considers the substance of interest as being continuously distributed in space (Wassgren, 2010). The continuum based method requires that the smallest length

---

[1] In modern fluid dynamics literature, the NS or Euler equations refer to the complete system of equations beside momentum, including both mass and energy conservation laws.

scale to be much larger than the microscopic length scale, typically the mean free path of a molecule (for gases). The Knudsen number for a generic engineering flow, in which the mean free path (λ) is around 1*10E-5 cm and the macroscopic length of interest (L) is 1 mm, is $Kn=\lambda/L=0.0001$ (Wassgren, 2010). The other expectation is that the highest length scale of the spatial resolution must be small enough to accurately capture the parameters (e.g. density) to be approximately constant in the control volume and do not be affected by the physics or geometry. The continuum assumption is valid in the vast majority of the industrial applications and so the NS (and Euler) equations can be used in the most of the industrial applications. However, in general, they do not have closed form solution till now. Hence, different discretization methods were developed from the second part of the last century till now to have approximate results for the nonlinear partial differential system of the equations in each point of the temporal (in case of transient problem) and spatial discretization. Beside fulfilling consistency, stability and convergence characteristics as a measure of the mathematical correctness of the discretization methods, the final results of the numerical analyses are evoluted as a function of the applied governing equations, boundary conditions and the geometry.

The three most frequent discretization methods (in the percent of the available commercial CFD codes) are the finite difference (**FDM**) (~ 2 %), finite element (**FEM**) (~ 15 %) and finite volume (**FVM**) (~ 80 %) methods. The rest 3 % are consist mostly of Spectral, Boundary element, Vorticity type and Lattice gas or Lattice Boltzmann methods.

The most traditional and oldest methods applied for numerical solution of partial differential equations are the **FDM**. They are essentially based upon the properties of the Taylor series expansion and the method is only applicable to structured grids in practice (Manna, 1992). The accuracy of FDM method strongly depends upon the mesh size and its properties as stretch ratio, aspect ratio and skewness for example. Although the increasing number of mesh point improves the accuracy, it can lead to difficulties in solution procedures due to the matrix inversion at the algebraic system of the equations (Manna, 1992).

The **FEM** historically originated from structural mechanics. The physical domain is divided by cells or elements, they form the numerical mesh, which can be structured or unstructured providing higher flexibility for handling complex geometry than FDM (Manna, 1992). The field variables are approximated by linear combinations of known basis functions, which can be quite general with varying degrees of continuity at the inter-element boundaries (Hirsch, 2007). Its mathematical rigor and elegance makes the FEM algorithms very attractive and widely researched area in the CFD community (Manna, 1992).

The **FVM** is originally introduced by McDonald 1971 and they are based on the observation that the conservation laws have to be interpreted in integral form to preserve discontinuous solution as vortex sheets, contact discontinuities or shock waves. Similarly to the FEM, the physical domain is subdivided by cells. The flow field variables are evaluated in some discrete points on each cell and they are interpreted as average value over the finite volumes. The finite volumes are constructed as parts of one or more cells, which can, moreover, be either overlapped or non-overlapped. The conservation laws are then applied to the finite volumes to obtain the discrete equations. The possibility of modifying the shape and location of the control volumes associated to a given mesh allows large freedom in the choice of the function representation of the flow field. This property is not shared by either the FDM or FEM and it is mostly the reason of higher popularity of the FVM in the engineering applications (all paragraph from Manna, 1992).

Turn back to the physical interpretations of modelling, there are two main approaches can be distinguished. The Pressure- and the Density-based classes of methods have evolved distinct strategies for the discretization, non-linear relaxation and linear solution aspects underlying the computational schemes. Historically, the pressure-based approach was developed for low-speed incompressible flows, while the density-based approach was mainly used for high-speed compressible flows. However, recently both methods have been extended and reformulated to solve and operate for a wide range of flow conditions beyond their traditional or original intent. In the **Pressure-based methods** the velocity is obtained from the momentum equations. The pressure field is extracted by solving a pressure or pressure correction equation, which is derived by manipulating continuity and momentum equations. The one of the most widespread methods is the SIMPLE (Semi-Implicit Pressure Linked Equations) family of schemes. Karki and Patankar developed the SIMPLER method for compressible flows, applicable for a wide range of speeds (Karki, Patankar 1989). Munz et al. extended the SIMPLE scheme for low Mach number flow employing multiple pressure variables, each being associated with different physical response (Munz et al. 2003). The time marching **Density-based methods** represent a large class of schemes adopted for compressible flows and applied widely in computational fluid dynamics for modelling steady and transient, transonic, supersonic and hypersonic flows. The continuity equation is used to determine the density field while the pressure distribution is obtained from the equation of state. The velocity field is computed also from the momentum equations. Both approaches are now applicable to a broad range of flows (from incompressible to highly compressible), but the origins of the density-based formulation may give it an accuracy (i.e. shock resolution) advantage over the pressure-based solver for high-speed compressible flows (ANSYS, Inc. 2010). Hence, this method is used in followings due to the high speed aeronautical applications.

1.2 Turbulence modelling

Concerning the physical level of modelling turbulent flows in CFD, the three most frequent approaches are the DNS (Direct Numerical Simulation), LES (Large Eddy Simulation) and RANS (Reynolds-averaged Navier-Stokes equations) based simulations, meanwhile for modelling laminar flows, there are no special treatment of the original NS equations are required. The closest model to the real phenomenon is the DNS, in which the Navier-Stokes equations are numerically solved without any turbulence model. It means that the whole range of spatial and temporal scales of the turbulence is resolved. The computational cost is extremely high, it is proportional to Re^3 (Pope, 2000). This method can not be used in the most part of the engineering practice due to the economical aspects. In case of LES, the large scales of turbulence are resolved directly providing more accurate results compared with RANS, meanwhile the geometry-independent, small scales and expensive structures are modelled. Although this approach is less computationally expensive compared with DNS, the industrial use is confined to low Reynolds number from purely computational consideration. The RANS based methods are the highest feasible level of approximation for the turbulent flows in general engineering applications. In this case, the parameters in the governing equations are averaged over a characteristic time interval in order to eliminate the influence of the turbulent fluctuations, meanwhile the unsteadiness of the other physical phenomenon are preserved (Manna, 1992). The resulting equations are formally identical to the laminar Navier-Stokes equation, except for some extra terms, so called Reynolds stresses, which results from the non linear terms of the governing equations (Manna, 1992). The main goal of the different turbulence modelling is to provide functional relations, expressions and/or closure equations related with these new terms in order to couple and so make definite the governing equations (in discretized form). In the last forty years several turbulence models have been developed. Following the Boussinesq approximation (which expects the similarity of the mechanism of laminar and turbulent stresses) the Reynolds stresses can be expressed in terms of eddy viscosity, which is considered as the turbulent counterpart of the laminar molecular viscosity. Opposite to the molecular viscosity, which is a fluid property, the eddy viscosity is a function of the flow properties (Manna, 1992). The direct aim of the turbulence models is to identify the functional relations between the flow properties and the turbulent eddy viscosity. At the lowest level, they are based upon the mixing length concept, introduced by Prandtl in 1925, which effectively relates the turbulent shear stresses to the mean velocity gradients. These algebraic turbulence models are often called zero equation models. A higher degree of approximation is reached by solving additional equations written for the turbulence variables, as the transport equations for the turbulent kinetic energy and its rate of dissipation. This class of turbulence models is usually classified according to the number of additional equations applied for the turbulence variables, i.e. one equations models, two equation models or higher order closure like the Reynolds stress models (Manna, 1992). The two-equation models, especially the k-ε and k-ω based models, are the widest spread applications in the industry due to the best compromise of the physical accuracy and computational cost. However, beside the advantages, they have disadvantages also.

The first low Reynolds number k-ε model has been developed by Jones and Launder (1973) and suppose that the flow is fully turbulent. It is a computationally cheap and provides reasonable accuracy for a wide range of flows. However, the k-ε model performs poorly for complex flows involving severe pressure gradient, separation and strong streamline curvature. From the standpoint of aerodynamics, the most disturbing problem is the lack of sensitivity to adverse pressure-gradients. Under those conditions, the model predicts significantly too high shear-stress levels and thereby delays (or completely prevents) separation. Furthermore, it requires the application of damping mechanism for stabilization when the equations are integrated through the viscous sublayer (Menter, 1994). The standard k-ε model has been modified by many authors.

There are a significant number of alternative models that have been developed to overcome the shortcomings of the k-ε model. One of the most successful, with respect to the accuracy and the robustness, is the k-ω model of Wilcox (1988). It solves one equation for the turbulent kinetic energy k and a second equation for the specific turbulent dissipation rate ω. The most prominent advantages is that the equations can be integrated without additional terms through the viscous sublayer, which makes them y+ insensitive and provides straightforward application of the boundary conditions. This leads to significant advantages in numerical stability. The model performs significantly better under adverse pressure-gradient conditions and separation than the k-ε model and suitable for the complex boundary layer flows (e.g. external aerodynamics and turbomachinery). However, the k-ω model also has some shortcomings. The model depends strongly on the free stream values of ω that are specified outside the shear-layer. Another point of concern is that the model predicts spreading rates that are too low for free shear-layers, if the correct values are specified for ω (Menter, 1994).

In order to improve both the k-ε and the k-ω models Menter (1994) suggested to combine the two models called the SST (Shear Stress Transport) k-ω turbulence model. The use of a k-ω formulation in the inner parts of the boundary layer makes the model directly usable all the way down to the wall through the viscous sub-layer, hence the SST k-ω model can be used as a Low-Re turbulence model without any extra damping functions. The SST formulation also switches to the k-ε behaviour in the free-stream and thereby avoids the common k-ω problem that the model is too sensitive to the free-stream value of the turbulence variables (in particular ω). The further distinct of the SST turbulence model is the modified turbulence eddy-viscosity function. The purpose is to improve the accuracy of prediction of flows with strong adverse pressure gradients and pressure-induced boundary layer separation. The modification accounts for the transport of the turbulent shear stress, which is based on Bradshaw's assumption that the principal shear stress is proportional to the turbulent kinetic energy (Blazek, 2005). Due to the above mentioned characteristics, the Menter model has gained significant popularity in the aeronautical community and can be regarded as one of the standard approaches today. Despite of the improvements of the original SST model as SST-2003, SST-sust, SST-Vsust variants for example, there are also some complaints. The one of them is that the distance to the nearest wall has to be known explicitly. This requires special provisions on multiblock structured or on unstructured grids (Blazek, 2005). Also, the k-ε to k-ω switch can produce some unrealistic effective viscosity, which may not affect the results. Meanwhile the SST turbulence model provides similar benefits as standard k-ω (except for the free stream sensitivity), the dependency on wall distance can make it less suitable for the free shear flows compared to standard k-ω and it requires mesh resolution near to the wall (CFD Online Discussion Forum, 2010).

The Wilcox's improved k-ω model (Wilcox, 1998) predicts free shear flow spreading rates more accurately than version 1988. The results of the benchmark simulations are in close agreement with flow measurements on far wakes, mixing layers, and plane, round, and radial jets, and it thus applicable to wall-bounded flows and free shear flows (ANSYS, 2010). Hence, this turbulence model has been implemented.

2. Finite volume method based compressible flow solver

Nowadays, in spite of disadvantages of turbulence closure models for RANS (Reynolds Averaged Navier-Stokes equations), they are at present the only tools available for the computation of complex turbulent flows of practical relevance. Their popularity comes from high efficiency in terms of accuracy and computational cost, which makes them widely used in commercial codes and related multidisciplinary applications. Hence, a detailed description of the physical and mathematical aspects of a RANS based compressible flow solver is presented in followings.

The governing equations in conservative form are derived by using density weighted averaging coupled with the time averaging of RANS. The code is based on structured, density-based cell centered finite volume method, in which the convective terms are discretized by Roe approximated Riemann method. The method of Roe is highly non-dissipative and closely linked to the concept of characteristic transport. It is one of the most powerful linear Riemann solvers due to the excellent discontinuity-capturing property including shear waves. However, it is well-known that flux function mentioned above can produce non-physical expansion shocks that violate the entropy condition. This can be avoided by modifying the modulus of the eigenvalues for the non-linear fields. The method of Yee is used and discussed at the present case to cure the problem. Central discretization is applied for diffusive terms on a shifted mesh. MUSCL (Monotone Upstream Schemes for Conservation Laws) approach is implemented for higher order spatial reconstruction with Mulder limiter for monotonicity preserving. Wilcox k-ω two equations turbulence model is adopted and used (Wilcox, 1998). The explicit system of the equations is solved by the 4th order Runge-Kutta method. The numerical boundary conditions are determined by the extrapolation technique for the NS solver and by the method of characteristics at the Euler solver. The interest is mostly in high speed industrial and aeronautical applications hence, the validation is completed for test cases are in the transonic, supersonic and subsonic flow regime as circular bump in the transonic channel and compression corner for the NS solver and flow over a wing profile and cascade for the Euler solver. The description of the benchmarks and the results are presented in Chapter 3.

2.1 Governing equations

In absence of external forces, heat addition, mass diffusion and finite rate chemical reaction, the unsteady two dimensional Navier-Stokes equations coupled with k-ω turbulence model-equations (Wilcox, 1998) in conservative, divergence and dimensional form are next:

$$\frac{\partial U}{\partial t} + \frac{\partial (F(U) - F_v(U))}{\partial x} + \frac{\partial (G(U) - G_v(U))}{\partial y} = S(U) \tag{1}$$

in $\Lambda(x,y)$ of the Cartesian coordinate system, where $x, y \in \mathbf{R}$ and $t \in \mathbf{R}^+$. The conservative variables and convective fluxes are given by

$$U = \begin{pmatrix} \bar{\rho} \\ \bar{\rho}\tilde{u} \\ \bar{\rho}\tilde{v} \\ \bar{\rho}\tilde{E} \\ \bar{\rho}k \\ \bar{\rho}\omega \end{pmatrix}, \quad F(U) = \begin{pmatrix} \bar{\rho}\tilde{u} \\ \bar{\rho}\tilde{u}^2 + p^* \\ \bar{\rho}\tilde{u}\tilde{v} \\ \tilde{u}(\bar{\rho}\tilde{E} + p^*) \\ \bar{\rho}\tilde{u}k \\ \bar{\rho}\tilde{u}\omega \end{pmatrix}, \quad G(U) = \begin{pmatrix} \bar{\rho}\tilde{v} \\ \bar{\rho}\tilde{v}\tilde{u} \\ \bar{\rho}\tilde{v}^2 + p^* \\ \bar{\rho}\tilde{v}\tilde{w} \\ \tilde{v}(\bar{\rho}\tilde{E} + p^*) \\ \bar{\rho}\tilde{v}k \\ \bar{\rho}\tilde{v}\omega \end{pmatrix} \tag{2}$$

while the viscous fluxes and source terms are following:

$$F_v(U) = \begin{pmatrix} 0 \\ \tau_{xx}^F \\ \tau_{xy}^F \\ \tilde{u}\tau_{xx}^F + \tilde{v}\tau_{xy}^F - q_x + \sigma_{kx} \\ \sigma_{kx} \\ \sigma_{\omega x} \end{pmatrix}, \quad G_v(U) = \begin{pmatrix} 0 \\ \tau_{yx}^F \\ \tau_{yy}^F \\ \tilde{u}\tau_{yx}^F + \tilde{v}\tau_{yy}^F - q_y + \sigma_{ky} \\ \sigma_{ky} \\ \sigma_{\omega y} \end{pmatrix}, \quad S(U) = \begin{pmatrix} 0 \\ 0 \\ 0 \\ 0 \\ S_k \\ S_\omega \end{pmatrix} \tag{3}$$

The bar over variables represents the Reynolds averaging over the characteristic time scale in order to separate and filter the small sized phenomena as turbulence fluctuation:

$$\bar{\phi} = \frac{1}{\Delta t}\int_{t_0}^{t_0+\Delta t} \phi \; dt \tag{4}$$

For a supersonic or hypersonic compressible flow the local density is not constant and in case of turbulent flow it fluctuates also due to the pressure diffusion, dilatation, work and turbulent transport/molecular diffusion of turbulent energy. Hence, the instantaneous density can also be separated by averaging and fluctuating part, which requires the introduction of Favre averaging given by (5).

$$\tilde{\phi} = \frac{1}{\bar{\rho}}\frac{1}{\Delta t}\int_{t_0}^{t_0+\Delta t} (\rho\phi) dt \tag{5}$$

The F in superscripts and tides (~) over the parameters in (2) and (3) and in followings mean the Favre averaged parameters. Other relationships in (2) and (3) are given by (6)-(9),

$$p^* = \bar{p} + \frac{2}{3}\bar{\rho}k \, , \; \tilde{E} = \tilde{e} + \frac{1}{2}\left(\tilde{u}^2 + \tilde{v}^2\right) + k \, , \; \tilde{e} = c_v \tilde{T} \, , \tag{6}$$

$$\tau_{ij}^F = (\mu + \mu_t)\left[\frac{\partial \tilde{u}_i}{\partial x_j} + \frac{\partial \tilde{u}_j}{\partial x_i} - \frac{2}{3}\delta_{ij}\sum_{k=1}^{2}\frac{\partial \tilde{u}_k}{\partial x_k}\right], \; q_j = -c_p\left(\frac{\mu}{Pr} + \frac{\mu_t}{Pr_t}\right)\frac{\partial \tilde{T}}{\partial x_j}, \; \sigma_{kj} = (\mu + \sigma^* \mu_t)\frac{\partial k}{\partial x_j} \tag{7}$$

$$\sigma_{\omega j} = (\mu + \sigma\mu_t)\frac{\partial \omega}{\partial x_j}, \; \mu = C_1 \frac{\tilde{T}^{\frac{3}{2}}}{\tilde{T} + C_2}, \; \mu_t = \bar{\rho}\frac{k}{\omega}, \; S_k = \sum_{i=1}^{2}\sum_{j=1}^{2}\left(\tau_{ijT}^F \frac{\partial \tilde{u}_i}{\partial x_j}\right) - \beta^* \bar{\rho} k \omega \tag{8}$$

$$S_\omega = \alpha\frac{\omega}{k}\sum_{i=1}^{2}\sum_{j=1}^{2}\left(\tau_{ijT}^F \frac{\partial \tilde{u}_i}{\partial x_j}\right) - \beta\bar{\rho}\omega^2, \; \tau_{ijT}^F = \mu_t\left[\frac{\partial \tilde{u}_i}{\partial x_j} + \frac{\partial \tilde{u}_j}{\partial x_i} - \frac{2}{3}\delta_{ij}\sum_{k=1}^{2}\frac{\partial \tilde{u}_k}{\partial x_k}\right] - \frac{2}{3}\delta_{ij}\bar{\rho}k \, , \tag{9}$$

in which p is the static pressure, ρ is the density, k is the turbulent kinetic energy, e is the internal energy u and v are the Cartesian components of velocity vector, c_v is the specific heat at constant volume, T is the static temperature, μ is the dynamic molecular viscosity, μ_t is the dynamic turbulent or eddy viscosity, δ_{ij} is the Kronecker's delta, c_p is the specific heat at constant pressure, $Pr(=0.72)$ and $Pr_t(=0.9)$ (for air) are the Prandtl number and turbulent Prandtl number respectively, $C_1 = 1.458 * 10E-6 \; kg/m/s/\sqrt{K}$ and $C_2 = 110.4 \, K$ are the constants in the Sutherland's formula to count the effect of temperature on dynamic viscosity and ω is the specific turbulent dissipation rate. The terms in the expressions, which are related to values of i, j and k in indexes, range from 1 to 2. The closure expression of the NS equations is the ideal gas law (10).

$$\bar{p} = \bar{\rho}R\tilde{T} \tag{10}$$

The not mentioned parameters and expressions in the turbulence model equations are given by (11)-(18) (Wilcox, 1998),

$$\alpha = \frac{13}{25},\ \sigma^* = \frac{1}{2},\ \sigma = \frac{1}{2} \tag{11}$$

$$\beta^* = \beta_0^* f_{\beta^*}\left[1 + \xi^* F(M_t)\right],\ \beta = \beta_0 f_\beta - \beta_0^* f_{\beta^*} \xi^* F(M_t) \tag{12}$$

$$\beta_0^* = \frac{9}{100},\ \beta_0 = \frac{9}{125} \tag{13}$$

$$f_{\beta^*} = \begin{cases} 1 & if \quad \chi_k \le 0 \\ \dfrac{1 + 680\chi_k^2}{1 + 400\chi_k^2} & if \quad \chi_k > 0 \end{cases},\ \chi_k = \frac{1}{\omega^3}\frac{\partial k}{\partial x_j}\frac{\partial \omega}{\partial x_j} \tag{14}$$

$$f_\beta = \frac{1 + 70\chi_\omega}{1 + 80\chi_\omega},\ \chi_\omega = \left|\frac{\Omega_{ij}\Omega_{jk}S_{ki}}{(\beta_0^*\omega)^3}\right|,\ (\chi_\omega = 0 \text{ for 2D flows}) \tag{15}$$

$$\xi^* = \frac{3}{2},\ M_{t0} = \frac{1}{4},\ M_t^2 = \frac{2k}{c^2} \tag{16}$$

$$F(M_t) = \left[M_t^2 - M_{t0}^2\right]H(M_t - M_{t0}),\ H(x) = \begin{cases} 0 & if \quad x \le 0 \\ 1 & if \quad x > 0 \end{cases} \tag{17}$$

$$S_{ij} = \frac{1}{2}\left(\frac{\partial \tilde{u}_i}{\partial x_j} + \frac{\partial \tilde{u}_j}{\partial x_i}\right),\ \Omega_{ij} = \frac{1}{2}\left(\frac{\partial \tilde{u}_i}{\partial x_j} - \frac{\partial \tilde{u}_j}{\partial x_i}\right), \tag{18}$$

where H is the Heaviside step function and c is the sound speed. The system of the nonlinear partial differential equations is already coupled. After discretization the system of algebraic equations can easily be solved.

Assuming a frictionless and isentropic flow, the NS equations - neglecting viscous and heat conducting terms - can be reduced to the Euler equations, which are the highest level approximation of the inviscid flow.

2.2 Boundary conditions

The numerical treatment of the boundary conditions strongly influences not only the convergence properties but the accuracy of the results in solving partial differential system of the equations. The physical boundary conditions secure the existence and uniqueness of the exact solution and numerical boundary conditions are supposed to ensure that various perturbations generated in the interior of the computational domain leave it without being reflected at the boundaries. Due to the convection dominated problem, the method of characteristic is used to determine the number and the exact values of the numerical boundary conditions in case of Euler equations, meanwhile extrapolation technique is applied for the NS equations. The direction of wave propagation (V_n, V_n, V_n+c and V_n-c) depends not only on the sign of the cell face normal velocity V_n but also on the local speed of sound c. At the boundary, the number of physical boundary condition to be imposed equals the number of negative eigenvalues, which correspond to the incoming characteristics from the outside (boundary) to the computational domain. The need for numerical boundary conditions comes from the fact that the actual problem to be solved is formulated in terms of the conservative variables rather than Riemann invariants. Therefore, it is hard to impose the Dirichlet boundary conditions in the usual way. It is common practice to recover the boundary values by switching to the characteristic variables, evaluating the incoming Riemann invariants from the physical boundary conditions and extrapolating the outgoing ones from the interior of the computational domain (Kuzmin & Möller, 2004).

Concerning the inlet, it is examined whether the flow is supersonic or subsonic. First, consider the supersonic case, at which only incoming characteristics are available. Hence, total pressure, static pressure, total temperature and flow angle are imposed as physical boundary conditions and no numerical boundary conditions are required for the Euler equations. If the flow is subsonic, one outgoing characteristic is appeared (V_n-c), so the two dimensional local Riemann problem belongs to that characteristic curve is solved by using physical and computed (existing) parameters and the total pressure, total temperature and flow angle are imposed as physical boundary conditions. The temperature and the components of the velocity vector are recovered by using ideal gas law and inlet flow angle, while the tangential velocity component is kept to be constant. Concerning the NS equations, additionally to the above mentioned specifications, the turbulent kinetic energy (k) and specific dissipation rate (ω) are imposed as physical boundary conditions and the static pressure is extrapolated from the computational domain in case of subsonic inlet.

If the outcoming flow is supersonic, there is no incoming characteristic hence, no physical boundary conditions are specified. If the flow is subsonic, there is one incoming characteristic hence, one parameter (static pressure) is imposed as physical boundary condition. The numerical boundary conditions are calculated by using characteristic variables (compatibility equations) in case of Euler equations, or they are extrapolated from the interior as the NS equations are implemented for viscous flow modelling. The static temperature is calculated by ideal gas law.

Concerning the Euler equations, the solid wall boundary conditions are considered as an outlet with the restriction of normal velocity is set to be zero across the wall. Hence, the numerical boundary conditions are calculated by using characteristic variables (compatibility equations) belongs to characteristic curves V_n, V_n and V_n+c. The static temperature is calculated by ideal gas law also. In case of NS equations, the no-slip boundary condition is implemented at the solid walls, the velocity vectors are set to be zero. Assuming zero pressure gradients, the pressure is set equal to the one at the cell centre nearest to the wall. Adiabatic wall condition is used to determine temperature. The turbulent kinetic energy is zero at the wall and the specific dissipation rate is computed by suggestion of (Wilcox, 1998) assuming a rough wall.

If the flow field has any kind of periodicity, the calculation time can be reduced significantly by using periodic boundary condition, by which the rotationally or translationally shifted parameters are used in the cells at the boundaries. Periodic boundary condition is used before and after the profile to recover infinite blade number in cascades.

Meanwhile the expected pressure distribution is imposed at the solid wall boundary in the inverse mode of the inviscid solver, the opening boundary is used instead of solid wall to allocate the local flow direction - determined by the pressure difference between the boundary and computational domain - and its velocity V_n (see Fig. 11.). The main outcome of the present mode is to have velocity profile over the geometry, which will be used for modifying mesh points in the wall modification module of the inverse design method (see Chapter 4.).

The detailed description of the presented boundary conditions for the Euler equations is found in (Veress et al., 2011).

2.3 Finite volume discretization

The finite volume method is a technique to handle the spatial derivatives that are appeared in the governing equations. The method is based on the integration of the equations over a finite volume. Then the integrals are transformed using the Gauss' divergence theorem where applicable. The physical meaning of the method is that fluxes flow through the faces of the finite volume while flux balance over the volume is satisfied. In the finite volume approach the first issue consists in evaluating the contour integral of the inviscid and viscous flux vectors in (1), hence the numerical flux functions are written in vector form given by (19). The $\vec{e}_x$ and $\vec{e}_y$ are the unit vectors in x and y directions.

$$\vec{H}(U) = F(U)\vec{e}_x + G(U)\vec{e}_y \text{ and } \vec{H}_v(U) = F_v(U)\vec{e}_x + G_v(U)\vec{e}_y \tag{19}$$

It is convenient to define total fluxes normal to the boundary of elementary control volumes rather than making use of the individual Cartesians components, using the rotational invariance of the governing equations. Integrating system eq. (1) over a control volume Ω, which is bounded by interface Γ and applying the Gauss' divergence theorem gives (20) and (21) (Manna, 1992),

$$\frac{\partial}{\partial t}\iint_{\Omega} U d\Omega + \int_{\Gamma}\left[\vec{H}(U)\cdot\vec{n}\right]d\Gamma = \int_{\Gamma}\left[\vec{H}_v(U)\cdot\vec{n}\right]d\Gamma + \iint_{\Omega}[S(U)]d\Omega \tag{20}$$

$$\frac{\partial}{\partial t}\iint_{\Omega} U d\Omega = \int_{\Gamma}[H_{vn}(U) - H_n(U)]d\Gamma + \iint_{\Omega}[S(U)]d\Omega \tag{21}$$

where $\vec{n} = (n_x, n_y)$ is the local outward pointing unit normal vector of the cell interface. The variables in conservative form, inviscid fluxes, source terms and viscous fluxes are the followings:

$$U = \begin{pmatrix} \bar{\rho} \\ \bar{\rho}\tilde{u} \\ \bar{\rho}\tilde{v} \\ \bar{\rho}\tilde{E} \\ \bar{\rho}k \\ \bar{\rho}\omega \end{pmatrix}, \quad H_n(U) = \begin{pmatrix} \bar{\rho}V_n \\ \bar{\rho}\tilde{u}V_n + p^* n_x \\ \bar{\rho}\tilde{v}V_n + p^* n_y \\ (\bar{\rho}\tilde{E} + p^*)V_n \\ \bar{\rho}V_n k \\ \bar{\rho}V_n\omega \end{pmatrix}, \quad S(U) = \begin{pmatrix} 0 \\ 0 \\ 0 \\ 0 \\ S_k \\ S_\omega \end{pmatrix}, \tag{22}$$

$$H_{vn}(U)=\begin{pmatrix} 0 \\ \tau_{xx}^{Fk}n_x+\tau_{yx}^{Fk}n_y \\ \tau_{xy}^{Fk}n_x+\tau_{yy}^{Fk}n_y \\ \left(\bar{u}\,\tau_{xx}^{Fk}+\bar{v}\,\tau_{xy}^{Fk}-q_x\right)n_x+\left(\bar{u}\,\tau_{yx}^{Fk}+\bar{v}\,\tau_{yy}^{Fk}-q_y\right)n_y+\left(\mu+\sigma^{*}\mu_t\right)\left[\frac{\partial k}{\partial x}n_x+\frac{\partial k}{\partial y}n_y\right] \\ \left(\mu+\sigma^{*}\mu_t\right)\left[\frac{\partial k}{\partial x}n_x+\frac{\partial k}{\partial y}n_y\right] \\ \left(\mu+\sigma\mu_t\right)\left[\frac{\partial \omega}{\partial x}n_x+\frac{\partial \omega}{\partial y}n_y\right] \end{pmatrix}, \tag{23}$$

where

$$V_n=V\vec{n}=\left(u\vec{e}_x+v\vec{e}_y\right)\cdot\left(n_x\vec{e}_x+n_y\vec{e}_y\right)=un_x+vn_y\,. \tag{24}$$

By means of finite volume discretization, in order to pass from a continuous to a discrete form, the unknown in a general finite volume of the partitioned computational domain is defined by (25),

$$U_{i,j}=\frac{1}{\Omega_{i,j}}\iint_{\Omega}U d\Omega \tag{25}$$

which corresponds to cell centre discretization. The vector $U_{i,j}$ has been interpreted as a mean value over the control volume. The fluxes are computed across quadrilateral cells in a structured grid, which can be seen in Fig. 1. The computational domain is divided by finite number of non overlapping finite surfaces or cells and the (21) is applied for each cell separately. It means that the second integral in (21) is replaced by summation over the all boundaries N_b of the cell i,j and so eq. (21) can be written in the general form of the semi-discrete expression over cell i,j as it is shown in (26).

$$\frac{d}{dt}U_{i,j}=\frac{1}{\Omega_{i,j}}\left(\sum_{k=1}^{N_b}\left[H_{vn}-H_n\right]_{ij,k}\Gamma_{ij,k}\right)+\left[S\right]_{i,j}=\Re_{i,j} \tag{26}$$

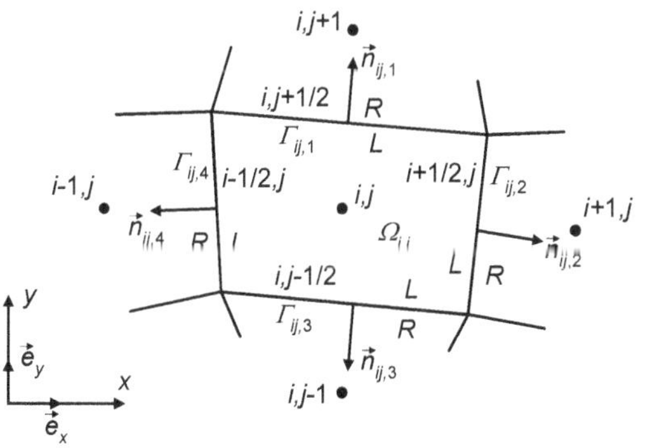

Fig. 1. Cell centred quadrilateral finite volume

$\left[H_n\right]_{ij,k}$ is the inviscid and $\left[H_{vn}\right]_{ij,k}$ is the viscous flux function normal to the cell boundary of cell interface k and $U_{i,j}$ is the vector of conservative variables (2). In present case, in 2D, $\Omega_{i,j}$ is the area of the cell surface and $\Gamma_{ij,k}$ is the length of a cell boundary k of $\Omega_{i,j}$.

The one of the key point in the convergence and the accuracy point of view is the correct determination of numerical flux function. It is especially true for the convective flux function, as it is expressed in the function of the left (L) and the right (R) side parameters of the cell interface (see Fig. 1.). In case of upstream differencing (or upwind) schemes, the quantity $\left[H_n\right]_{ij,k}$ are characterized by a flux function $H_n\left(U^L,U^R\right)$, which takes into account the sign of the Jacobian matrices, or in other words the relevant propagation directions between the L and R states (Manna, 1992). The $\hat{H}_n\left(U^L,U^R\right)$ can be evaluated by linear wave decomposition where an unique average state (which is denoted by a hat) of the left and right states exist (Roe, 1981):

$$\hat{H}_n\left(U^L,U^R\right)=\frac{1}{2}\left\{H_n\left(U^L\right)+H_n\left(U^R\right)-\left|\hat{D}_n\left(U^L,U^R\right)\right|\left(U^R-U^L\right)\right\}. \tag{27}$$

For the ideal gas, Roe has shown that the matrix $\hat{D}_n$ is equal to the Jacobian D_n when it is expressed as a function of the variables $\hat{\rho}$, $\hat{u}$, $\hat{v}$, and $\hat{h}_0$, which are weighted variables of the square root of density. h_0 is the total enthalpy. Detailed information about the Roe's method of the approximate Riemann solver is found in (Roe, 1981). The method of Roe is highly non-dissipative and closely linked to the concept of characteristic transport. It is one of the most powerful linear Riemann solvers due to the excellent discontinuity-capturing property including shear waves. However, it is well-known that flux function mentioned above can produce non-physical expansion shocks that violate the entropy condition. This can be avoided, by modifying the modulus of the eigenvalues for the non-linear fields. The method of Yee (1989) is used at the present case.

MUSCL (Monotone Upstream Schemes for Conservation Laws) approach is implemented for higher order spatial extension, by which the piece-wise constant distribution of the initial variables over the cell can be replaced by a piecewise linear or quadratic one. The mathematical deduction starts with the introduction of Taylor series expansion around point i. The results are found at (28) after discretization and integration.

$$U^R_{i+\frac{1}{2}}=U_{i+1}-\frac{1}{4}\left[(1-\kappa)\Delta_{i+\frac{3}{2}}+(1+\kappa)\Delta_{i+\frac{1}{2}}\right], \quad U^R_{i+\frac{1}{2}}=U_{i+1}-\frac{1}{4}\left[(1-\kappa)\Delta_{i+\frac{3}{2}}+(1+\kappa)\Delta_{i+\frac{1}{2}}\right] \tag{28}$$

$\Delta_{i-\frac{1}{2}}=U_i-U_{i-1}$, $\Delta_{i+\frac{1}{2}}=U_{i+1}-U_i$, $\Delta_{i+\frac{3}{2}}=U_{i+2}-U_{i+1}$ and the new left and right states next to the cell boundary $i+1/2$ (between points i and $i+1$) are denoted by U^L and U^R. The $\kappa=1/3$ in equation (28) corresponds to a third order accurate space discretization in one dimensional problem (Manna, 1992). The spurious oscillations (wiggles) can occur with high order spatial discretization schemes due to shocks, discontinuities or sharp changes in the solution domain. Hence, in this case, Mulder limiter is implemented in the high resolution schemes for monotonicity preserving (Manna, 1992):

$$U^R_{i+\frac{1}{2}} = U_{i+1} - \frac{1}{4}\psi^R\left[\left(1-\kappa\psi^R\right)\Delta_{i+\frac{3}{2}} + \left(1+\kappa\psi^R\right)\Delta_{i+\frac{1}{2}}\right],\ U^L_{i+\frac{1}{2}} = U_i + \frac{1}{4}\psi^L\left[\left(1-\kappa\psi^L\right)\Delta_{i-\frac{1}{2}} + \left(1+\kappa\psi^L\right)\Delta_{i+\frac{1}{2}}\right] \quad (29)$$

where

$$\psi^R = \frac{2\Delta_{j+\frac{1}{2}}\Delta_{j+\frac{3}{2}} + \varepsilon}{\Delta^2_{j+\frac{1}{2}} + \Delta^2_{j+\frac{3}{2}} + \varepsilon},\quad \psi^L = \frac{2\Delta_{j-\frac{1}{2}}\Delta_{j+\frac{1}{2}} + \varepsilon}{\Delta^2_{j-\frac{1}{2}} + \Delta^2_{j+\frac{1}{2}} + \varepsilon} \quad \text{and} \quad 10^{-7} \le \varepsilon \le 10^{-5}. \quad (30)$$

The same method was used for NS and for the turbulence model equations, however they were handled separately.

A simple central scheme is applied for the space discretization of the diffusive terms in (26) as follows:

$$H_{vn} = \frac{1}{2}\left[H_{vn}\left(U^L\right) + H_{vn}\left(U^R\right)\right] \quad (31)$$

U^L and U^R are the conservative variables at the cell centres. The derivatives in the diffusive terms are determined at the centre of the cell interface. Hence, two new types of cells are formed, which are shifted by $i+1/2$ and $j+1/2$ directions respectively. The centres of the boundary of such cells are coincident with the centres and vertices of the original cells. In case of former situation, the parameters are known, because they are stored at the cell centre of the original cells. If the centres of the boundaries of the new cells are coincident with cell vertex of the original cells, the flow variables at the new cell boundary centre are the simple averages of the four neighbouring cell centre values of the original cells in case of quadrilateral mesh. Then, the derivatives into the x and y directions of the viscous flux function (23) can be obtained by using Green-Gauss theorem:

$$\iint_{\Omega'} \nabla\phi d\Omega = \int_{\Gamma'} \phi\vec{n} d\Gamma \quad \rightarrow \quad \nabla\phi = \frac{1}{\Omega_{i,j}'}\int_{\Gamma'} \phi\vec{n} d\Gamma \quad \rightarrow \quad (32)$$

$$\rightarrow \quad \frac{\partial\phi}{\partial x} = \frac{1}{\Omega_{i,j}'}\sum_{k=1}^{N_b}\left(n'_{kx}\,\phi_{ij,k}\Gamma'_{ij,k}\right) \quad \text{and} \quad \frac{\partial\phi}{\partial y} = \frac{1}{\Omega_{i,j}'}\sum_{k=1}^{N_b}\left(n'_{ky}\,\phi_{ij,k}\Gamma'_{ij,k}\right), \quad (33)$$

where ϕ is an arbitrary flow variable, $\Omega_{i,j}'$ is the area of the shifted cell i,j, k runs through the number of the boundaries of the shifted cell till the N_b, which is the number of maximal boundary, n'_{kx} and n'_{ky} is the x and y components of the cell boundary normal unit vector of the interface k, $\phi_{ij,k}$ is the value of the flow variable at the given interface centre of the shifted cell i,j and $\Gamma'_{ij,k}$ is the length of the face k. At the boundaries of the computational domain, a series of the ghost cells are applied and filled with values, so no special treatment is necessary to determine the derivatives. The geometry of the ghost cells are extrapolated from the last two cells.

The integral of the source term in (21) are approximated as follows:

$$\iint_{\Omega} [S(U)] d\Omega = [S(U)]_{i,j}\,\Omega''_{i,j} \quad (34)$$

where $[S(U)]_{i,j}$ is an average value over the cell $\Omega''_{i,j}$. Derivatives in the expressions are determined at the cell centres by using similar treatment as in case of diffusive term discretization.

A widely used class of non linear multi-stage time integration techniques is given by the Runge-Kutta (RK) schemes. They are usually designed to obtain higher order temporal accuracy with minimum computational storage and the large stability range with the specific coefficients, even though it has been often used for steady state calculations as herein. The 4 stages RK method (RK4) is used to solve the time derivatives of the conservative variables in (26) with $U = U^{\cdot}_{i,j}$ and $\Re = \Re_{i,j}$ for simplicity in each cell given by:

$$\begin{aligned} &step\ 1: && U_0 = U^n \\ &step\ 2: && U_k = U_0 + \alpha_k \Delta t \Re(U_{k-1}), \quad k = 1,\ldots,4 \\ &step\ 3: && U^{n+1} = U_4 \end{aligned} \tag{35}$$

where $\alpha_1 = 1/4$, $\alpha_2 = 1/3$, $\alpha_3 = 1/2$, $\alpha_4 = 1$ for the NS equations and $\alpha_1 = 1/8$, $\alpha_2 = 0.306$, $\alpha_3 = 0.587$ and $\alpha_4 = 1$ for the Euler equations are the coefficients of RK4, n represents the parameters at previous time step and *n+1* at the next time step over a cell. The RK4 index is denoted by k and it runs from 1 to m with its maximum value of 4 in *step 2* (35). Due to the steady state assumption, the time accuracy is not required hence, the RK4 coefficients are applied to have high stability and smoothing properties of the upwind scheme with MUSCL reconstruction. In order to optimize the time step behind the stability criterion, the local time stepping has been used for every cells i,j as follows (Lefebvre, Arts, 1997):

$$\Delta t_{i,j} = \frac{\Omega_{i,j} \nu}{\sum_{k=1}^{N_b} \left(|V_n| + c\right)_{ij,k} \Gamma_{ij,k}} \tag{36}$$

where $\Omega_{i,j}$ is the area of the cell *i,j*, ν is the Courant number, $\Gamma_{ij,k}$ is the length of the cell boundary k of $\Omega_{i,j}$, V_n is the cell face normal velocity, N_b is the number of cell boundaries and c is the sound speed.

3. Validation of the flow solver

The goal of the validation – in case of any calculation methods – is to provide information about the correct mathematical and physical operation of the simulation by means of comparing the results with real tests or other benchmarks especially referring to the application of flow physics under investigation.

3.1 Validation of the viscous flow solver

In the following sections, the numerical results are presented for transonic channel over circular bump and compression corner for validating frictional and heat conducted flow simulations.

The first test case is transonic channel over circular bump, in which the flow enters into the channel with Mach number 0.85 and a shockwave develops over the circular bump. The bump has 4.2 % maximum thickness. At the inlet, the total pressure, total temperature, and flow angles are specified as physical boundary conditions. The static pressure corresponds to the isentropic flow at Mach=0.85 is imposed at the outlet. Under these conditions the flow expands in the rear part of the bump up to a Mach number 1.2 and ends up into a week shock wave to allow the recovery of the free stream conditions. The results of FLUENT and own code are compared to each other at the same solver settings however, the k-ε turbulence model was used in the commercial program. The Mach number iso-lines show reasonable deflections (see Fig. 2.), the present method predicts the shockwave earlier.

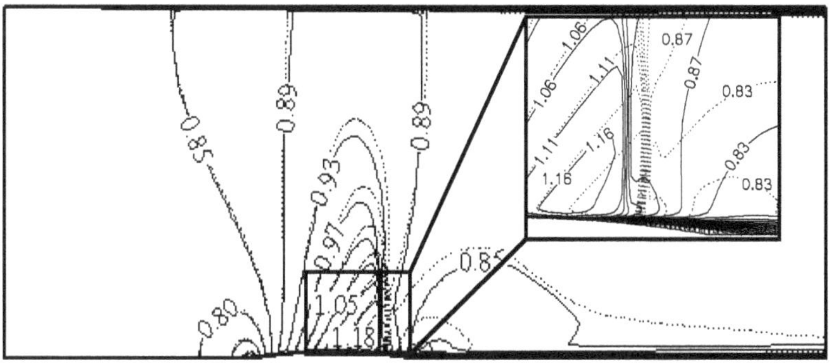

Fig. 2. Mach number distribution in transonic channel over circular bump test case (dotted line: FLUENT, continuous line: recent solver)

The shape of the geometry and the thickness of the boundary layer have a dominant effect on the shock wave evolution. Different numerical methods and turbulence models have different inherent mechanism to model boundary layer and shock wave–boundary layer interaction. The one of the criticisms against the k-ε turbulence is the lack of sensitivity to adverse pressure-gradients (see Subchapter 1.2.). The boundary layer seems to be thinner at the downstream of the circular bump in case of the commercial code compared to the own one. Hence, the shockwave triggered earlier in case of the present model. However, the differences between the two approaches in the entire computational domain are less than

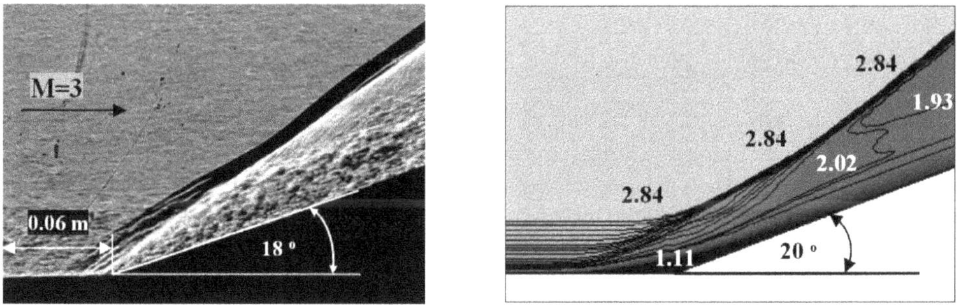

Fig. 3. Configuration and Schlieren photograph about compression corner at inlet Mach number 3 and with slope 18 ° (left side) (Settles, 1975) and Mach number distribution by the recent solver at inlet Mach number 2.85 and with slope 20 ° (right side)

5 % hence, based on the strongly validated commercial code, the accuracy of the in-house software can be accepted for this benchmark.

In the second test case a ramp with 20 degrees slope angle is located in a flow channel. The air enters into a channel with Mach number 2.85. Before the ramp an oblique shockwave develops. The geometry and Mach number distribution can be found in Fig. 3. The comparison of the measured and simulated results shows similar shock wave pattern, however they can not be compared with each other directly due to slight difference between the inlet conditions and slope values. The reason why the presented condition is used in the simulation is the available measured quantitative parameters found in the Gerolymos' publication (Gerolymos et al., 2003). The locations and directions of the coordinate systems, along which velocity distributions are measured is shown in Fig. 4.

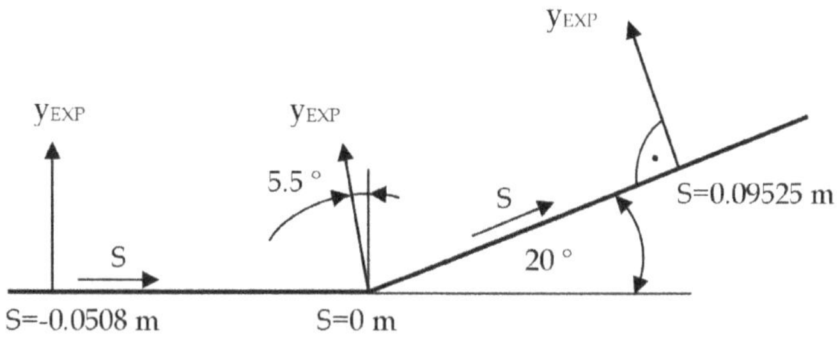

Fig. 4. Locations and directions of the coordinate systems, along which the velocity distributions of the measurements are compared with the results of simulation

y_{EXP} is the distance from the wall. The velocity profiles of the recent viscous flow solver with k-ω model and the experiments are found in Figs. 5-7. for comparison. The velocity profiles in Fig. 5 and 6. are even quantitatively agreed with each other, but the results shown in Fig. 7. are slightly far from the experiments. The reason of that can be caused by the fact, that the error, which is generated by the velocity profile at the beginning of the computational domain is growing along with the flow and so the small difference becomes larger at downstream. The other problem can be the free-stream sensitivity of the k-ω model described in Subchapter 1.2. The solution could be improved by further adjusting boundary layer at upstream and the ω.

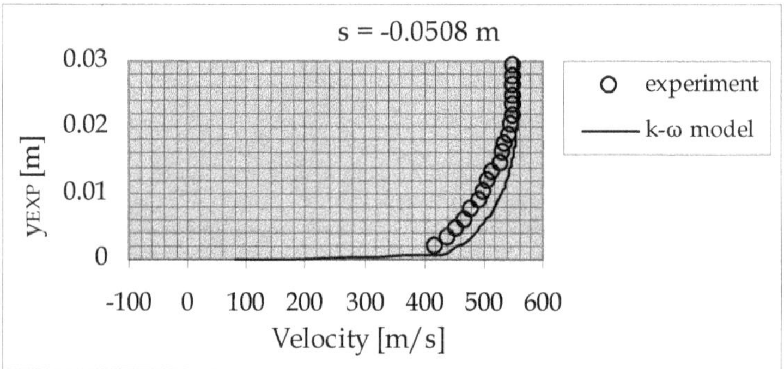

Fig. 5. Validation of the viscous flow solver. Velocity profiles at s = -0.0508 m (see Fig. 4.)

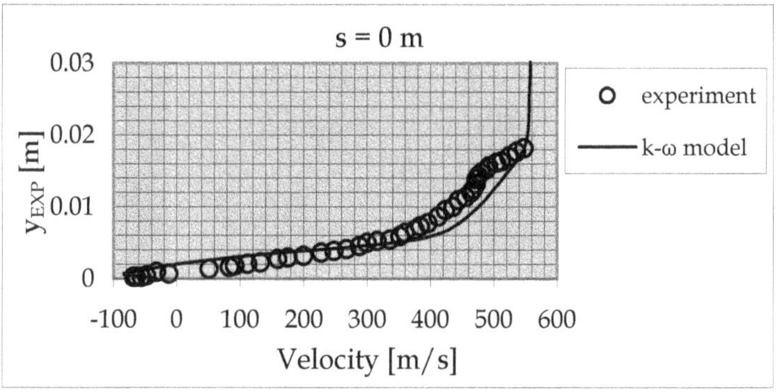

Fig. 6. Validation of the viscous flow solver. Velocity profiles at s = 0 m (see Fig. 4.)

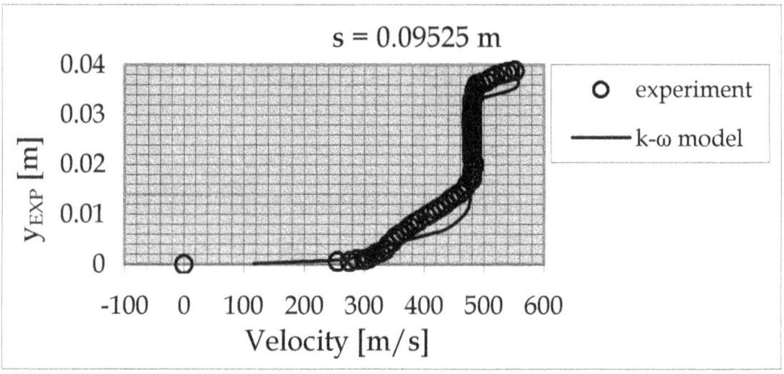

Fig. 7. Validation of the viscous flow solver. Velocity profiles at s = 0.09525 m (see Fig. 4.)

3.2 Validation of the inviscid flow solver

The compressible viscous flow solver presented in Chapter 2. requires relative high computational time due to the significant number of equations and fine mesh especially in the boundary layer. Hence, this approach can not be used economically for coupling with optimization methods in the explicit time marching manner. However, assuming a frictionless and convection dominated problems, the NS equations can be reduced to the Euler equations, which are the highest level approximation of inviscid flows. The Euler equations are valid for modelling compressible high speed flows outside of the boundary layer without separation. As most of the industrial process under the interest, as well as significant number of flow situations encountered in nature, are dominated by convective effects, and therefore, they are well approximated by the Euler equations as it will be seen in the validation also (Manna, 1992). Furthermore, the number of equations and the desired cell number are significantly reduced compared with NS based solver, hence it is more suitable for applying them in the optimization methods.

Although the mathematical characteristics of the presented numerical method for the Euler equations are investigated in many articles (e.g. Barth et al., 2004), a measurement of the wing profile NACA 65-410 has been used in the first case to check the accuracy of the calculation. The measurements are performed by Abbott et al. in 1945 and they include the experimental analysis of the lift, drag, pitching moment, etc. of the NACA 6 series airfoils.

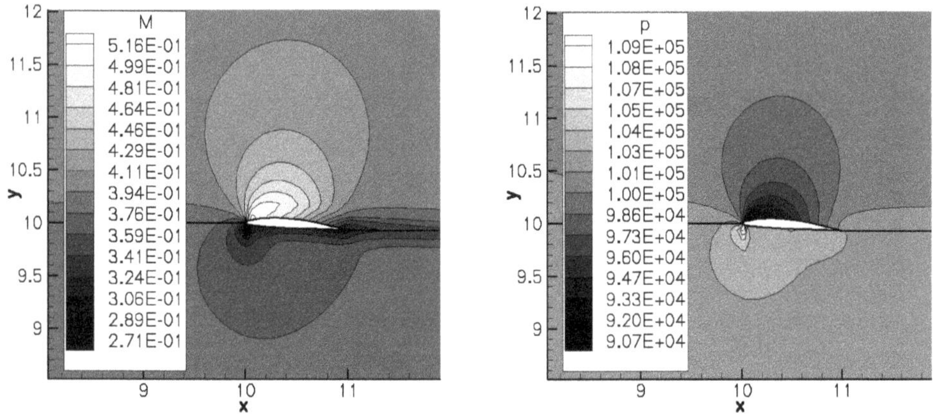

Fig. 8. Mach number and pressure distribution over the profile NACA 65-410 at α=4 degrees angle of attack

Most of the data on airfoil section characteristics were obtained in the Langley two-dimensional low-turbulence pressure tunnel with a rectangular test section (0.9144 meters wide and 2.286 meters high), in which usually 0.6 meters chord models were tested. The test models completely span the width of the tunnel has a maximum speed of about 70 m/s. More information about the experiments is found in (Abbott, 1945). The Mach number and pressure distribution around the profile is found in Fig. 8. as a result of the computation at α=4 degrees angle of attack (angle between the up stream flow and chord). The boundary conditions are the following: inlet total pressure: $p_{tot,in}$=112800 [Pa]; inlet total temperature: $T_{tot,in}$=293.15 [K]; outlet static pressure: $p_{stat,out}$=101325 [Pa]. The mesh size is 87×120. The result of the analysis and measurements are compared with each other and they are shown in Fig. 9. in the plot of the lift coefficient in the function of angle of attack. The results of the analysis are accepted in engineering point of view, the overall deviation is less then 5 % at the investigated range. However, it must be considered that the results depend on the mesh resolution and the mesh sensitivity analyses are indispensable to have.

The second test case for the validation is a compressor cascade analysis[2] based on the technical report by (Emery et al., 1958). The cascade has been constructed by using NACA 65-410 profile also with:

$$\sigma = \frac{c}{g} \tag{37}$$

[2] The original and full version of the present investigation is found in (Veress et al., 2010)

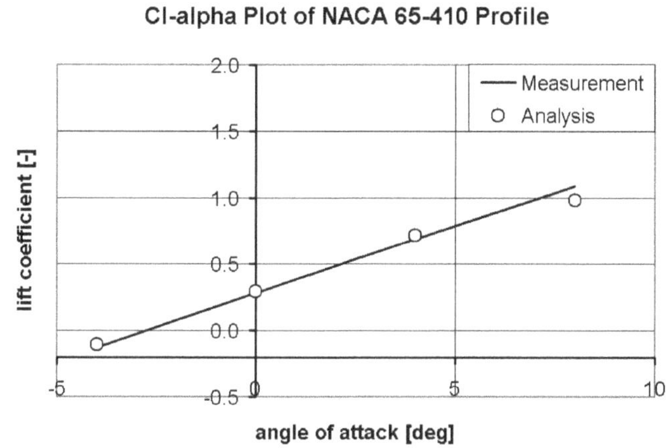

Fig. 9. Lift coefficient in the function of angle of attack at the profile NACA 65-410

where $\sigma = 1.5$ is the solidity, $c = 12.7cm$ is the chord and the g is the tangential spacing. The β_1 and α represents the angle between flow direction and the rotation of axis and the flow direction and chord (angle of attack) respectively in following. The boundary conditions are set to provide the same Reynolds number as in the experiment. The total inlet pressure is $p_{tot,in}$=101750 [Pa], the total inlet temperature is $T_{tot,in}$=293.15 [K] and the static outlet pressure is $p_{stat,out}$=101325 [Pa] over the H-type mesh (110×60). The pressure coefficients (38) along the profile are considered in the validation, where p_{tot} is the upstream total pressure, q_1 and $p_{stat,1}$ are the dynamic and static pressure respectively at wall surface position 1.

$$C_p = \frac{p_{tot} - p_{stat,1}}{q_1} \tag{38}$$

The measured parameters and the results of the simulation are compared with each other and the quantitative parameters of the pressure coefficients are shown in Fig. 10. The investigated variables of the calculation, at different angle of attack, are in a good correlation with the measurements (Emery et al., 1958). The difference between them is under the limit of the acceptance, the overall deviation is less then 8 %. The C_p values show higher dispersions near to the leading edge due to the geometrical inaccuracy (sharp edge).

Although the mathematical aspects of the applied methods as consistency, stability and convergence characteristics are strongly investigated and published in many articles (e.g. Barth et al., 2004), the validation of the Euler based CFD solver was completed in the present subchapter. The results of the analysis and measurements are compared with each other, for a 2D NACA 65-410 wing profile and its low speed cascade. The resembling shows acceptable agreement in engineering point of view. The average deviation between the real tests and the analyses is less then 8 % in both investigated cases, the accuracy of the numerical tool is reasonable, it can be used for further applications.

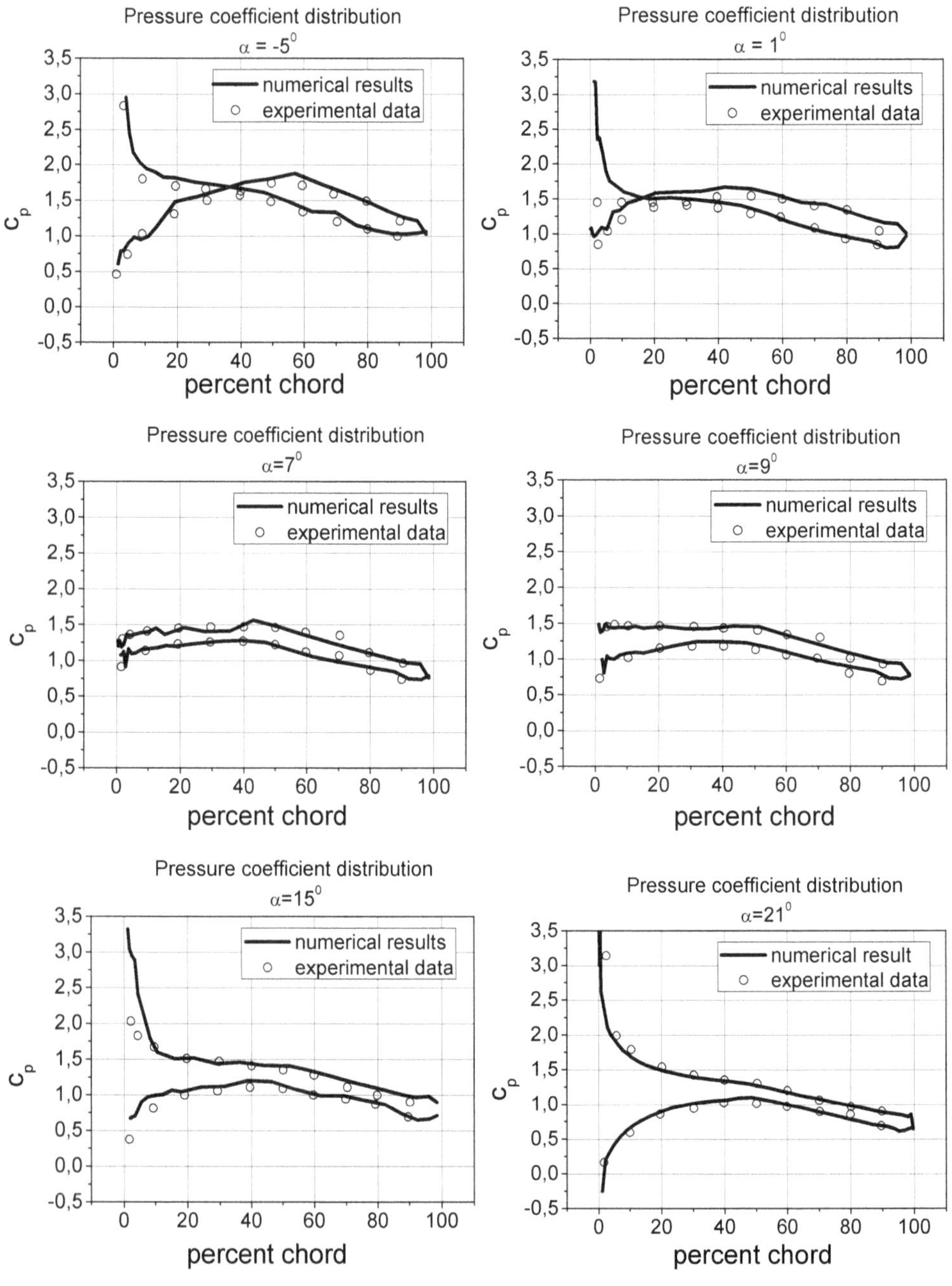

Fig. 10. Pressure coefficient distribution over the NACA 65-410 compressor cascade profile at different α (angle of attack) (β_1=30°, Re=2.45E5, and $v_{inf,\ inlet}$=29 m/s). The experimental data are from (Emery et al., 1958)

4. Finite volume method based optimization and inverse design for aeronautical applications

Today, beside the developments of the central core of the fluid dynamics solvers, the different optimization techniques, related with CFD, are also under intensive research. In case of direct optimization techniques, an attempt has been made to find the optimal solution. They typically utilize some sort of search technique (e.g. gradient-based optimizer), stochastic based algorithms (e.g. evolutionary strategies, genetic algorithms), artificial neural networks or some other optimization methods. These procedures can be computationally expensive because several flow solutions must be calculated to specify the direction of deepest descent, fitness of individuals in the population, etc. in order to determine the shape changes (Lane, 2010). Furthermore, the required number of flow solutions increases dramatically with the number of design variables.

In case of a specific set of the inverse design-type methods, the geometry modification is based on the prescribed set of the pre-defined variables at the wall by simple, fast and robust algorithms, which makes them especially attractive amongst other optimization techniques (Lane, 2010). The wall modification can be completed within much less flow solutions for inverse design techniques than for direct optimization methods. Hence, the inverse design methods typically being much more computationally efficient and they are very innovative to be used in practice. The main drawback of inverse design methods is that the designer should create target (optimum in a specific sense) pressure or velocity distributions that should correspond to the design goals and meet the required aerodynamic characteristics. However, it can be difficult to specify the expected pressure or velocity distribution that satisfies all design goals. Also, one cannot guarantee that an arbitrarily prescribed pressure/velocity distribution will provide mechanically correct surfaces or bodies (airfoils without trailing edge open or cross over for example). Hence, the one of the main goals of the following subchapters is to provide solutions for the above mention complaints.

The calculation process of the developed iterative type inverse design method is shown in Fig. 11. The procedure, first of all, requires an initial geometry and a required pressure distribution (p^{req}) along the wall to be modified. The prescribed distribution can be the goal function of an optimization method or it can come from the industrial experiences and/or theory. The iterative cycle starts with the direct solution of the inviscid Euler solver. Completing the convergence criteria, if the target conditions are not reached, a new (opening) boundary condition is applied at the solid boundary to be redesigned or optimized. The required pressure distribution (p^{req}) is imposed at the solid wall boundary, which is become locally opening as inlet or outlet, depends upon the evolved pressure differences between the boundary and computational domain. The outcome of this analysis is a velocity distribution along the wall, which is not necessarily parallel with it. The final step of the cycle is the wall modification. The wall becomes parallel with the local velocity vector corresponds to a new streamline of the flow field. The mentioned steps are repeated until the target distribution is reached by the direct analysis and so the new geometry is available (Leonard & Van den Braembussche, 1990).

All the contributions of the above presented procedure has been described in Chapter 2. except for the wall modification algorithm and the determination of the required pressure distribution (p^{req} in Fig. 11.), which are the topic of the following paragraph and subchapter respectively.

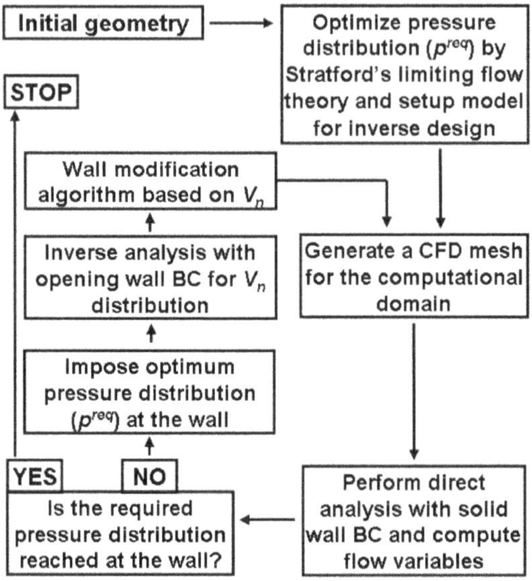

Fig. 11. Flowchart of the computational procedure of the iterative inverse design calculation

While the incoming and out coming velocity distribution (see V_n in Fig. 11.) is given at the solid wall, based on the inverse mode of the inviscid solver (see Subchapter 2.2 at opening wall boundary), the last step of the iterative design cycle is the modification of the geometry. The new position of the solid boundary coordinates is calculated by setting the wall to be parallel with the local velocity vector of the cell centre:

$$\Delta y_i(x_i) = \sum_{k=Le}^{i} \left(\frac{v_k}{u_k} \Delta x_k \right) \tag{39}$$

where u and v are the Cartesian component of the velocity vector at the wall. The geometry modification starts from the leading edge or inlet stagnation point till the trailing edge or the outlet stagnation point and completed in vertical directions (see Fig. 12.).

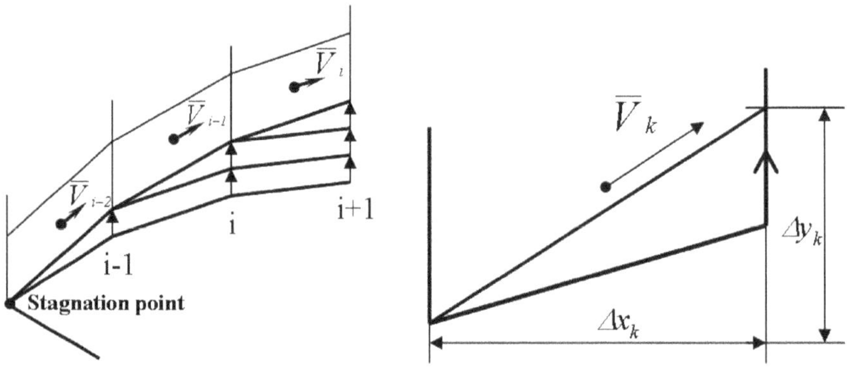

Fig. 12. Schematic view of the wall modification process based on the local velocity vector

In the following two subchapters, two case studies of the application of the inverse design method have been presented. In the first one, the lift force is a goal function of an optimization procedure over the NACA 65-410 profile in external flow, meanwhile the increased static pressure ratio is the target condition of redesigning NACA 65-410 cascade geometry in the second case. The pressure distribution should be as low as possible over the solid surface of the suction side at given operational conditions for maximal profile loading. However, in order to reach the downstream conditions, the pressure must increase after the location of the maximum velocity. Stratford's limiting flow theory is used and coupled with the SQP (Sequential Quadratic Programming) nonlinear constraint optimization to provide the target pressure distribution represents the maximum lift force close but certain distance far from the separation in case of external flow. Stratford's limiting flow theory is implemented also in case of redesigning cascade geometry. The presented inverse design method is used to complete wall surface modification till the previously defined target pressure distributions are reached by means of the corresponding sequence of the inverse, wall modification and direct algorithms. The Euler equations are used for modelling basic physics for both cases. The standard cell centred finite volume method has been applied with Roe's approximated Riemann solver, MUSCL approach and Mulder limiter, which are described in Chapter 2. The validation of the Euler solver is found in Subchapter 3.2.

4.1 Airfoil optimization for maximal lift force by means of inviscid inverse design method[3]

It has been pointed out in the introduction of the Chapter 4. that the inverse design methods require optimal pressure or velocity distributions to determine the adherent geometry. In order to maximise the lift force of the suction side of a profile at given and constant operational (boundary) conditions, the pressure distribution should be minimized. However, the adverse pressure gradient is appeared after the location of the maximum velocity (and minimum pressure) in order to recover downstream conditions. The adverse pressure gradient till the trailing edge should have limited in each discretized points to be just below the condition of causing separation. The maximum area bounded by the suction and the pressure side distributions in conjunction with the mentioned limited values of pressure gradients will provide the optimum solution as a target distribution to be specified for the inverse design method.

There are several existing methods for predicting separation as Goldschmied, Stratford, Head, and Cebeci-Smith for example (Smith, 1975). The accuracy these methods were examined several times. One of the output of these investigation shows that the operation of Goldschmied's method is unreliable. The other three are in reasonable agreement and Stratford's method tended to predict separation slightly early. The Cebeci-Smith method is appeared to be best and the Head method is a strong second one (Smith, 1975). Due to the good accuracy, simple expressions and conservative characteristics for predicting separation, Stratford's method has been used in followings (Veress et al., 2011).

Stratford has derived an empirical formula for predicting the point of separation in an arbitrary decelerating flow at the order of Re=10E6 (Stratford, 1959),

[3] The original and full version of the subchapter is found in (Veress et al., 2011).

$$\frac{\overline{C}_p\left[x\left(d\overline{C}_p/dx\right)\right]^{1/2}}{\left(10^{-6}\,\mathrm{Re}\right)^{1/10}}=S \tag{40}$$

where the canonical pressure distribution is

$$\overline{C}_p=\frac{p-p_0}{\frac{1}{2}\rho_0 u_0^2} \tag{41}$$

and if $d^2p/dx^2 \geq 0$ then S = 0.39 or if $d^2p/dx^2 < 0$ then S =0.35. Additionally, $\overline{C}_p \leq 4/7$. The flows under investigations consist first of a flat-plate flow. Hence, x is distance measured from the leading edge of the plate and $\mathrm{Re}=u_0 x/\nu$. If the flows begin the pressure rise at a point x_0 (it is the position of minimum pressure, p_0 and maximum velocity, u_0), left-hand side of eq. (40) starts from a zero value. The left-hand side then grows. When it reaches the limiting value of S, separation is said to occur. If S is held at its limiting value of 0.39 for $d^2p/dx^2 > 0$ eq. (40) amounts to an ordinary differential equation for $\overline{C}_p(x)$. It is evident from eq. (40) that the equation describes a flow that is ready everywhere to separate. Stratford presents the following solutions (Stratford, 1959),

$$\overline{C}_p=0.645\left\{0.435\,\mathrm{Re}_0^{\;1/5}\left[\left(x/x_0\right)^{1/5}-1\right]\right\}^{2/n} \quad for \quad \overline{C}_p \leq (n-2)/(n+1) \tag{42}$$

and

$$\overline{C}_p=1-\frac{a}{\left[\left(x/x_0\right)+b\right]^{1/2}} \quad for \quad \overline{C}_p \geq (n-2)/(n+1) \tag{43}$$

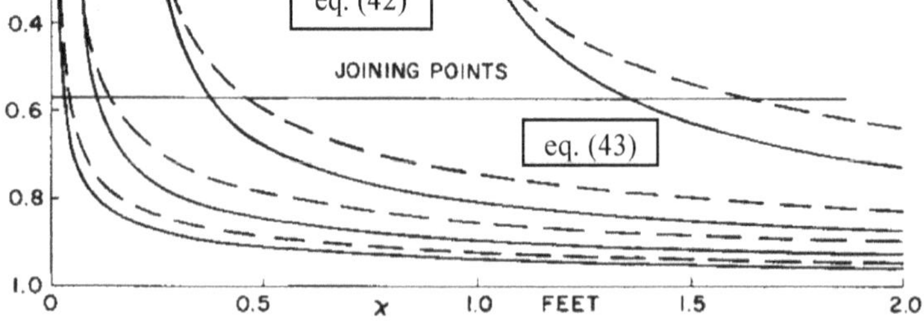

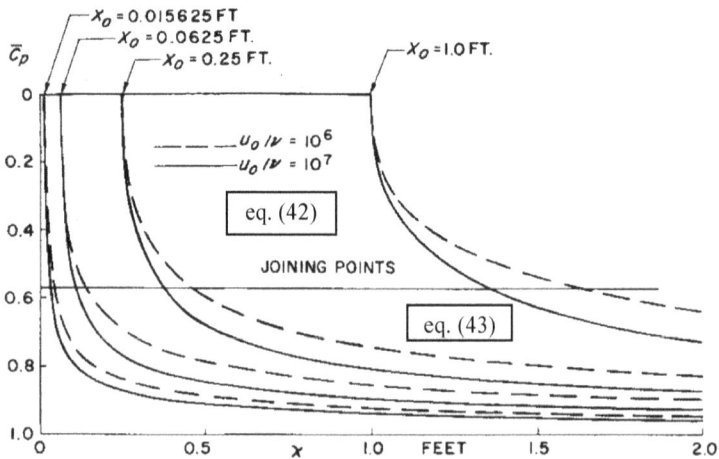

Fig. 13. Stratford limiting flows at two values of unit Reynolds number (Smith, 1975)

In that two-part solution, x_0 is the start of pressure rise, $\mathrm{Re}_0 = u_0 x_0/\nu$, x is the distance measured from the very start of the flow, which begins as flat-plate, turbulent flow. The number n is a constant that Stratford finds to be about 6. The quantities a and b are arbitrary constants used in matching values and slopes in the two equations at the joining point, $\overline{C}_p \geq (n-2)/(n+1)$. Of course, eq. (42) describes the beginning of the flow and eq. (43) the final part. The flow is an equilibrium flow that always has the same margin, if any, against separation. Two families of such flows have been computed; they are shown in Fig. 13. The features of the presented diagram, together with eq. (42) and (43) are found in (Smith, 1975).

The method presented above is included in determining the pressure distribution at maximum lift force and at the limit of separation on the suction side for given far field conditions:

$$\aleph\left(\overline{C}_p(x)\right) = \Im\left(C_p(x)\right) = \oint C_p(x)dx = \oint \frac{p - p_\infty}{\frac{1}{2}\rho_\infty u_\infty^2} dx = \oint \frac{p - p_\infty}{0.5\gamma p_\infty M_\infty^2} dx \tag{44}$$

where p is the static pressure at the given wall location and the other primitive variables correspond to free stream condition denoted by ∞ (downstream conditions are used in case of cascade design). The connection between $\overline{C}_p(x)$ and $C_p(x)$ (pressure coefficient) is given by:

$$\overline{C}_p = \frac{p - p_0}{\frac{1}{2}\rho_0 u_0^2} = \frac{C_p - C_{p,0}}{1 - C_{p,0}} = \frac{\frac{p - p_\infty}{\frac{1}{2}\rho_\infty u_\infty^2} - \frac{p_0 - p_\infty}{\frac{1}{2}\rho_\infty u_\infty^2}}{1 - \frac{p_0 - p_\infty}{\frac{1}{2}\rho_\infty u_\infty^2}} = \frac{p - p_0}{p^{total} - p_0} = \frac{p - p_0}{\frac{1}{2}\rho_0 u_0^2} \tag{45}$$

The objective function is to

$$\text{minimize } \frac{1}{\aleph\left(\overline{C}_p(x)\right)} \tag{46}$$

$$\text{subject to } p_{opt}^{TE} - p_{init}^{TE} = 0 \tag{47}$$

The reason of the constraint to be specified at the presented way is to fix trailing edge (*TE*) condition of Stratford's method and to minimize the disturbances of the optimal pressure distribution of the computational domain respect to the initial flow field. The posterior numerical test shows that the latter condition is not required, it can differ from zero.

The optimization procedure is divided by two sub steps. In the first sub step the physical connections between different parameters are described by Stratford's criteria to evaluate limiting pressure distribution. The pressure coefficient at the minimum pressure (p_0) is given by:

$$C_p = C_{p,0} = \frac{p_0 - p_\infty}{\frac{1}{2}\rho_\infty u_\infty^2} = \frac{p_0 - p_\infty}{0.5\gamma p_\infty M_\infty^2} = \frac{p_0 - p_\infty}{0.7 p_\infty M_\infty^2} \tag{48}$$

where p_0 and maximum velocity u_0 is supposed to be constant starting from the leading edge of the suction side till the starting of the positive pressure gradient (x_0). The Mach numbers M_0 at these points are calculated by:

$$C_p = C_{p,0} = \frac{p_0 - p_\infty}{0.7 p_\infty M_\infty^2} = \frac{1}{0.7 M_\infty^2}\left[\left(\frac{1+0.2M_\infty^2}{1+0.2M_0^2}\right)^{\frac{\gamma}{\gamma-1}} - 1\right] \tag{49}$$

The T_0, u_0 and ρ_0 are obtained by the energy equation of the isentropic flow and ideal gas law:

$$T_0 = T^{total}\left(1+\frac{\gamma-1}{2}M_0^2\right)^{-1} \tag{50}$$

$$u_0 = \sqrt{\frac{\gamma}{\gamma-1}R\left(T^{total} - T_0\right)2} \tag{51}$$

$$\rho_0 = \frac{p_0}{RT_0} \tag{52}$$

The *total* quantities correspond to the given operational (flight) or inlet boundary conditions.

A general way of determining pressure distribution starts with specifying a possible p_0. All parameter belongs to p_0 can be calculated by eqs. (48)-(52). The next step is to find location x_0, which gives back the required trailing edge static pressure by using Stratford's equations (42) and (43) over x. Hence, the location of starting flow deceleration (x_0) and the Stratford's limiting pressure distribution till the required trailing edge pressure is the output of the first sub step of the optimization procedure. There are infinite possible pressure distribution existing of the presented method hence, the second sub step of the optimization procedure is the constraint optimization in order to determine the corresponding flow parameters and location belongs to the minimum pressure and maximum velocity point on the suction surface, which provides the maximum area bounded by the pressure distribution of the suction and pressure side of the profile. p_0, T_0, u_0, ρ_0, x_0 and $p(x)$ (by Stratford's criteria) parameters will be modified in the second sub step to satisfy (46) and (47).

The pressure side distribution is also modified by means of constraint optimization to maximize the area under the function restricted to less than or equal to the maximum pressure gradient or higher than or equal to the minimum pressure gradient respect to the original distribution.

NACA 65-410 profile has been used to provide initial geometry and flow field for the optimization. The boundary conditions are the followings: inlet total pressure: $p_{tot,in}$=112800 [Pa]; inlet total temperature: $T_{tot,in}$=293.15 [K]; outlet static pressure: $p_{stat,out}$=101325 [Pa] over the mesh size of 87×120. The pressure distributions of the given geometry are shown in Fig. 14. and they are noted as init (initial). The inverse design program modifies initial geometry till the result pressure distribution over the geometry gives back exactly the target (optimum) one. The optimum pressure distribution belongs to the maximum area of the closed distribution of the pressure and suction side at the limit of separation in case of adverse pressure gradient flow conditions on the suction side. However, several points near to the leading edge of the suction side are modified to make the extremely high pressure gradient smoother. Moreover, an arbitrary (optimal) target pressure distribution often causes non-realistic geometry as negative thickness, trailing edge opening or cross over. Based on several theoretical investigation and computational tests, it can be noticed, that the expected pressure distribution can not be arbitrary in case of subsonic flow due to the information propagation into the upstream (leading edge) direction along the streamline bounded by the wall. If the required pressure is differ from the initial one at the certain representative part of the near wall region, the flow can be retarded or sucked depends on the local conditions. This effect has an influence on the flow evolution starting from the leading edge and the pressure should be redistributed by considering higher or lower local kinetic energy along the stream line especially at the first couple mesh points of the leading edge.

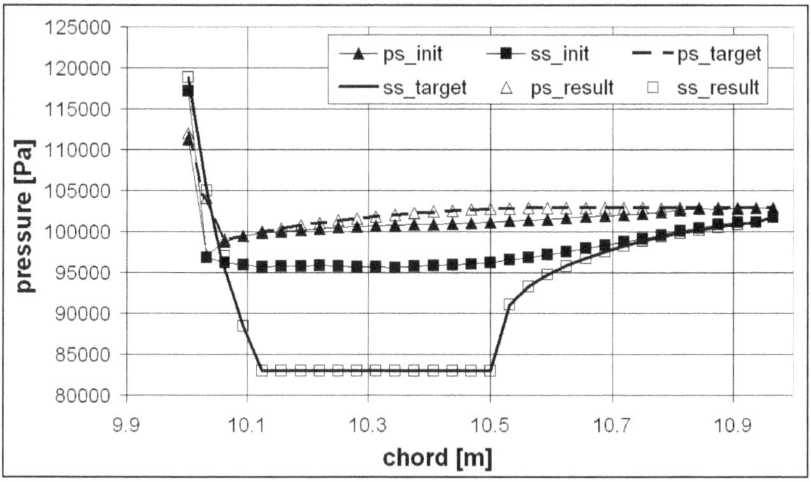

Fig. 14. Pressure distribution of the initial (init), optimum (target) and result (of the inverse design based optimization procedure) cases (ps: pressure side and ss: suction side)

The modified distributions have been imposed in the inverse design procedure to determine the geometry, which provides the optimal conditions. The inverse design method was converged after 10 iteration cycles of the inverse, wall modification and direct modes. The normal velocity distribution across the solid wall becomes near to zero at the last inverse subroutine, which represents that there is no need for any further steps, the pressure gradient is infinitesimally small (no flow) across the solid boundary. The corresponding results of the optimization procedure are found in Fig. 14. The target and optimized (result) pressure distribution are compared with each other and the deviation between them is negligible. The optimized geometry with Mach number and pressure distribution is shown in Fig. 15. The improvements are straightforward; the lower pressures at the suction side provide higher lift force in case of the optimum geometry. Further quantitative results are found in Fig. 16. The optimization was completed at zero angle of attack. However, the off-design conditions show also the same order of improvements as the optimization at design point. The lift force coefficient is increased significantly around by 100 % in the investigated range of the angle of attack. The effect of drag force should be analyzed and considered with the aim of including it in the optimization process.

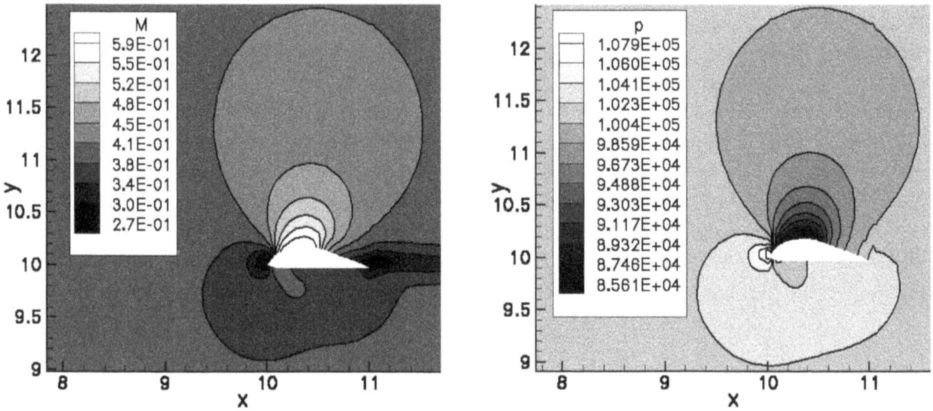

Fig. 15. Mach number and pressure distribution [Pa] of the result case (result of the inverse design based optimization procedure)

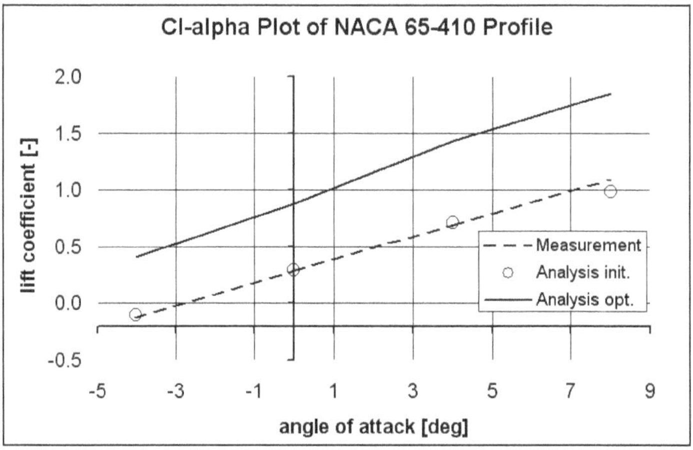

Fig. 16. Lift coefficient distribution in the function of angle of attack (the optimization has been completed at angle of attack 0)

4.2 Application of the inverse design based semi-optimization in cascade flows

Compressors and turbines are widely used in the vehicle engines as gas turbines for shaft power and jet engines. The axial compressors and turbines can be simplified to cascade geometry by extracting a cylindrical cut surface from them and laying out in 2D. The basic characteristics of the elementary flow field can be analysed and studied by this way.

NACA 65-410 profile has been used also in followings. The validation of the Euler solver for the cascade geometry is found in Subchapter 3.2.

The main goal of the first part of the subchapter is to show the outcome of the profile with maximum blade loading developed by the inverse design method coupling with SQP and Stratford's limiting flow theory (see Subchapter 4.1.). The result of the optimization was analyzed by the present Euler solver and ANSYS CFX. The same computational procedure has been used in both software except for the mesh at viscous mode of the CFX. y+ determines the first cell size next to the wall and has an influence on the cell number, which results 71X23 2D volumes. Concerning the boundary conditions, the following physical parameters were used: the inlet total pressure is $p_{tot,in}$= 107853 [Pa], the inlet total temperature is $T_{tot,in}$= 298.42 [K], the inlet flow angle is 30° and the outlet static pressure is $p_{stat,out}$=101325 [Pa].

The CFX was convergent after 110 iterations. The quantitative results of the analyses are shown in Fig. 17. The target and the result pressure distributions of the inviscid inverse design based optimization are presented over the final geometry beside the inviscid (Ansys-ss and Ansys-ps) and viscous (Ansys-ss-viscous and Ansys-ps-viscous) results of the CFX (ss: suction side and ps: pressure side). The average difference between the three approaches is less than 8 %, which is acceptable in engineering point of view. The main deviation is at the leading edge stagnation point; the static pressure in the Euler solver is higher compared with CFX. This unphysical feature is caused by the linear extrapolation of determining static pressure at the ghost cell of the solid wall boundaries without averaging procedure.

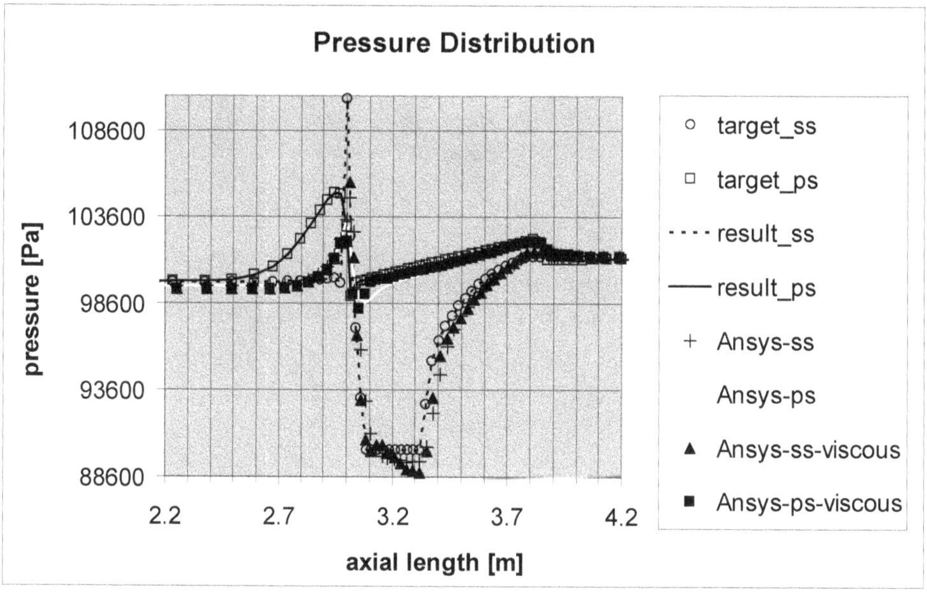

Fig. 17. Pressure distribution of the redesigned blade in case of recent Euler solver and the inviscid and viscous analysis of ANSYS CFX (ss: suction side and ps: pressure side)

It can be observed in Fig. 17 that the outlet static pressure is higher than the inlet one, so the cascade is working in a compressor mode, however, the static pressure ratio is negligible due to the unexpected thick profile, which causes chocking.

Of course, the lower outlet static pressure - or higher mass flow - can improve design specification by means of increasing static pressure rise over the cascade due to the energy conversion. Moreover, the higher blade loading can also increase the static pressure ratio, which is shown in the second part of the present subchapter. The blade geometry variants in the function of the blade loading at the same mesh, solver settings, initial and boundary conditions are found in Fig. 18. with the corresponding pressure coefficients and distributions, which are based on the Stratford's limiting flow theory. The pressure side pressure distribution was the same at the three investigated cases. The physical boundary conditions are the followings: the inlet total pressure is $p_{tot,in}$= 107853.4 [Pa], the inlet total temperature is $T_{tot,in}$= 298.4267 [K], the inlet flow angle is 30° and the outlet static pressure is $p_{stat,out}$= 83325 [Pa]. Although the static pressure ratio is increased from 1.034 to 1.14 proportionally with blade loading and the absolute value of the pressure coefficient, further numerical and experimental investigations are indispensable to have for well established conclusions in the field of cascade optimization.

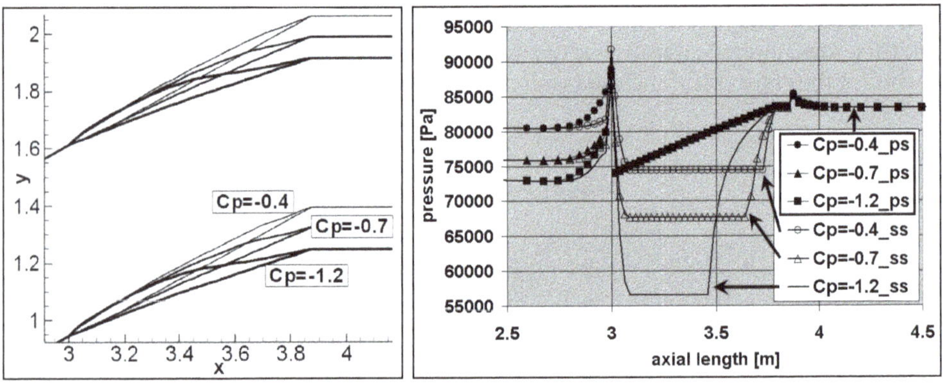

Fig. 18. Blade geometries (left side) at different Stratford's based pressure distributions (blade loadings) (right side) with the corresponding minimum pressure coefficients (ss: suction side and ps: pressure side)

5. Acknowledgment

This work has been supported by the Hungarian National Fund for Science and Research (OTKA) under the fund No. F 67555. The results reported in the chapter has been developed in the framework of the project "Talent care and cultivation in the scientific workshops of BME" project. This project is supported by the grant TÁMOP-4.2.2.B-10/1-2010-0009.

HUNGARY'S RENEWAL

The project is supported by the European Union.

NEW SZÉCHENYI PLAN

6. References

Abbott H.; Doenhoff A. E. & Stivers, L. S. (1945). *NACA REPORT No. 824., Summary of Airfoil Data,* USA

ANSYS, Inc. (2010). ANSYS Fluent User's guide, *ANSYS, Inc. Southpointe, 275 Technology Derive Canonsburg, PA 15317, ansysinfo@ansys.com, http://www.ansys.com, USA*

Barth, T. & Ohlberger, M. (2004). *Finite volume methods: foundation and analysis,* In: Stein, E.; Borst, R. & Hughes, T. (eds.) Encyclopedia of Computational Mechanics, John Wiley & Sons, New York, USA

Blazek, J. (2005). *Computational Fluid Dynamics: Principle and Applications,* Elsevier, ISBN-13: 978-0-08-044506-9, ISBN-10: 0-08-044506-3, United Kingdom

CFD Online Discussion Forum (2010). *http://www.cfd-online.com/Forums/main/75554-use-k-epsilon-k-omega-models.html*

Emery, J. C.; Herrig, L. J; Erwin, J. R. & Felix, A. R. (1958). *Systematic two-dimensional cascade tests of NACA 65-series compressor blades at low speeds,* NACA Technical Report 1368, USA

Gerolymos, G. A.; Sauret, E. & Vallet, I. (2003). *Oblique-Shock-Wave/Boundary-Layer Interaction using Near-Wall Reynolds-Stress Models,* Université Pierre-et-Marie-Curie, AIAA 2003-3466, 33rd Fluid Dynamics Conference, 23-26 June 2003 Orlando, Florida, USA

Hirsch, C. (1990, 2007). *Numerical Computation of Internal and External Flows,* Wiley, ISBN 978-0-7506-6594-0, United Kingdom

Jones, W. P. & Launder B. E. (1973).*The Calculation of Low-Reynolds-Number-Phenomena with a Two-Equation Model of Turbulence,* Int. J. Heat Mass Transl., Vol. 16, 1973, pp. 1119-1130

Karki, K. & Patankar, S. V. (1989). *Pressure based calculation procedure for viscous flows at all speeds in arbitrary configurations,* AIAA Journal, Vol. 27, No 9, pp. 1167-1174, 1989, USA

Kuzmin, D. & Möller, M. (2004). *Algebraic Flux Correction II. Compressible Euler Equations,* http://www.mathematik.tu-dortmund.de/papers/KuzminMoeller2004a.pdf, Germany

Lane, K. A. (2010). *Novel Inverse Airfoil Design Utilizing Parametric Equations,* MSC thesis, Faculty of California Polytechnic State University, San Luis Obispo.

Lefebvre, M. & Arts, T. (1997). *Numerical aero-thermal prediction of laminar/turbulent flows in a two-dimensional high pressure turbine linear cascade,* Second European Conference on Turbomachinery - Fluid Dynamics and Thermodynamics, Antwerp, Belgium, pp. 401-409

Leonard, O. & Van den Braembussche, R. (1990). *Subsonic and Transonic Cascade Design,* AGARD-VKI Special Course on Inverse Methods in Airfoil Design for Aeronautical and Turbomachinery Applications, May, 14-18, 1990, Belgium

Manna, M. (1992). *A Three Dimensional High Resolution Compressible Flow Solver,* PhD thesis, Catholic University of Leuven, Belgium

Menter, F. R. (1994). *Two-Equation Eddy-Viscosity Turbulence Models for Engineering Applications,* AIAA Journal, vol. 32, no. 8, pp. 1598-1605, USA

McDonald, P. M. (1971). *The Computation of Transonic Flow through two Dimensional Gas Turbine Cascades,* ASME paper 71-GT-89, USA

Munz, C.-D.; Roller, S.; Klein, R. & Geratz, K. J. (2003). *The extension of incompressible flow solvers to the weakly compressible regime*, Computers & Fluids, Vol. 32, No 2, pp. 173-196

Pope, S. B. (2000). *Turbulent Flows*, Cambridge University Press, ISBN 978-0521598866, United Kingdom

Roe, P. L. (1981). *Approximate Riemann Solvers, Parameter Vectors, and Difference Schemes*, Journal of Computational Physics, Vol. 43 pp. 357-372

Settles, G. (1975). *The shadowgram of Mach 3 airflow over an 18 degree compression corner*, http://www.efluids.com/efluids/gallery/gallery_pages/18degramp.htm, Gas Dynamics Lab, Penn State University (PHD Thesis, Princeton, 1975), USA

Smith, A. M. O. (1975), *High-Lift Aerodynamics*, Journal of Aircraft, Vol. 12 No. 6, pp. 501-530, USA

Stein, E.; Borst, R. & Hughes, T. (2004). *Finite volume methods: foundation and analysis*, http://weberknecht.uni-muenster.de/num/publications/2004/BO04a/finvol_script.pdf, Edited by John Wiley & Sons, Ltd., USA

Stratford, B. S. (1959), *The Prediction of Separation of the Turbulent Boundary Layer*, Journal of Fluid Mechanics, Vol. 5. pp 1-16, USA

Veress, Á.; Gallina, T. & Rohács, J. (2010). *Fast and Robust Inverse Design Method for Internal and Cascade Flows*, International Review of Aerospace Engineering (IREASE), ISSN 1973-7459 Vol. 3 N. 1. pp. 41-50.

Veress, Á.; Felföldi, A.; Gausz, T. & Palkovics, L. (2011). *Coupled Problem of the Inverse Design and Constraint Optimization*, Applied Mathematics and Computation, DOI: 10.1016/j.amc.2011.08.110, Paper In Press, Corrected Proof.

Wassgren, C. (2010). *Notes on Fluid Mechanics and Gas Dynamics*, lecture note, School of Mechanical Engineering, Purdue University, USA

Wilcox, D. C. (1998, (1988)). *Turbulence Modelling for CFD*, DCW Industries Inc. Second edition, ISBN-10: 0963605151, ISBN-13: 978-0963605153, USA

Yee, H. C. (1989). *A class of high-resolution explicit and implicit shock-capturing methods*, VKI lecture series 1989-04, March 6-10, 1989; NASA TM-101088, Feb. 1989, Belgium

Zucrow, M. J. & Hoffman, J. D. (1976). *Gas Dynamics Vol. 1*, Wiley, ISBN 0-471-98440-X, USA

5

Alternative Methods for Generating Elliptic Grids in Finite Volume Applications

A. Ashrafizadeh, M. Ebrahim and R. Jalalabadi
K. N. Toosi University of Technology
Iran

1. Introduction

Numerical solution of an engineering problem via finite volume method (FVM) requires the discretization of the solution domain and computational grid generation. While both structured and unstructured grids can be used, elliptic structured grid generation methods, when applicable, have favorable features in terms of both accuracy and computational cost.

Among the elliptic grid generation (EGG) methods, the most well known and widely used are the algebraic transfinite interpolation and differential methods which employ Poisson equations. In this chapter classical EGG methods are reviewed. It is then proposed that these methods can be classified based on the parameters being interpolated (i.e. interpolants), the interpolation method used and the grid generation equations being employed. The proposed unified view provides a framework for the development of new grid generation methods; some of which are introduced here for the first time.

Another major task in this chapter is to show that finite volume method, which employs the computational grid, can itself be used in the numerical grid generation process. In other words, FVM can be used for two different tasks; discretization of the differential equations which govern the coordinates of the computational grid points and discretization of the differential equations which govern the physical process of interest.

A typical 2D structured grid in the physical domain is shown in Fig. 1a and the corresponding logical or computational grid is shown in Fig. 1b. The classical structured grid generation methods provide equations which define, directly or indirectly, the mapping functions which describe the curvilinear coordinate lines in the physical domain, i.e. $\xi(x,y)$ and $\eta(x,y)$ curves. The grid point (i,j) in the physical domain is defined at the intersection of the curvilinear coordinate lines ξ_i and η_j as shown in Fig. 1a.

The so called algebraic grid generation methods directly specify the formulas used to calculate the physical coordinates (x,y) in terms of the logical coordinates (ξ,η) (Eiseman, 1979; Eiseman et al., 1992; Lehtimaki, 2000; Zhou, 1998). For example, in the Trans-Finite Interpolation (TFI) method (Eiseman et al., 1992), the generating equations employ the boundary nodal coordinates and some derivative terms to calculate the nodal coordinates throughout the solution domain. This method is often described as a Boolean sum of one dimensional interpolation functions U and V as follows:

$$\vec{R}(\xi,\eta) = (x,y) = U \oplus V = U + V - UV \tag{1}$$

$$U(\xi,\eta) = \sum_{i=1}^{L}\sum_{n=0}^{P} \alpha_i^n(\xi) \frac{\partial^n \vec{R}(\xi_i,\eta)}{\partial \xi^n} \tag{2}$$

$$V(\xi,\eta) = \sum_{j=1}^{M}\sum_{m=0}^{Q} \beta_j^m(\eta) \frac{\partial^m \vec{R}(\xi,\eta_j)}{\partial \eta^m} \tag{3}$$

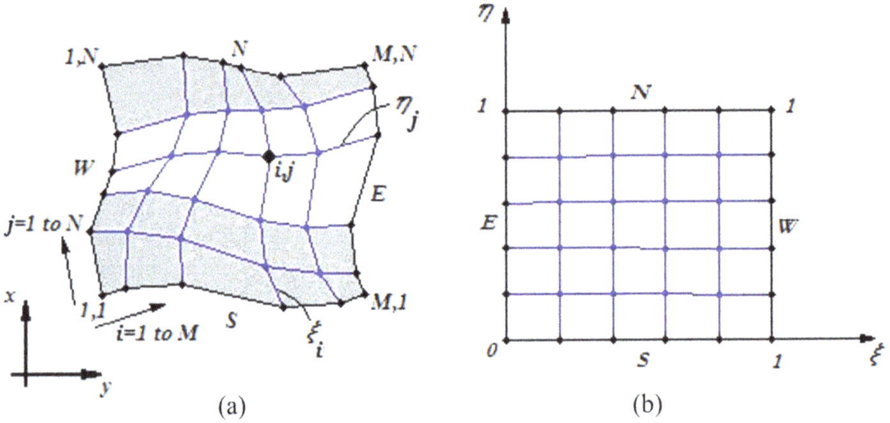

Fig. 1. (a) A physical grid, (b) the corresponding logical grid.

The highest orders of one-dimensional interpolation formulas in Eqs. (2) and (3) are specified by P and Q respectively, and L and M specify the number of auxiliary nodes used in these interpolations. For example, a zero order TFI computational molecule for generating a (M×N) grid is as follows:

$$\begin{aligned}\vec{R}_{i,j} = {} & C_{i,N}\vec{R}_{i,N} + C_{i,1}\vec{R}_{i,1} + C_{1,j}\vec{R}_{1,j} + C_{M,j}\vec{R}_{M,j} + \\ & C_{1,N}\vec{R}_{1,N} + C_{M,N}\vec{R}_{M,N} + C_{1,1}\vec{R}_{1,1} + C_{M,1}\vec{R}_{M,1}\end{aligned} \tag{4}$$

Coefficients $C_{i,N}$, $C_{i,1}$, ... in this nine-point computational molecule for the calculation of the coordinates of the nodal point (i,j) can be linear or nonlinear functions of the logical coordinates of this point, i.e. ξ_i, η_j.

In contrast to the TFI, the mapping functions are not explicitly provided in the so called differential grid generators. For example, the following differential constraints on the unknown mapping functions $\xi(x,y)$ and $\eta(x,y)$ are proposed by Thompson, Thames and Mastin (TTM) (Thompson, et al., 1974):

$$\xi_{xx} + \xi_{yy} = P(\xi,\eta) \tag{5}$$

$$\eta_{xx} + \eta_{yy} = Q(\xi,\eta) \tag{6}$$

Equations (5) and (6) are often analytically inverted and the calculations are carried out in the logical domain. The non-linear inverted equations are as follows:

$$g_{11}\, x_{\xi\xi} - 2 g_{12}\, x_{\xi\eta} + g_{22}\, x_{\eta\eta} = - J^2 (P x_\xi + Q x_\eta) \tag{7}$$

$$g_{11}\, y_{\xi\xi} - 2 g_{12}\, y_{\xi\eta} + g_{22}\, y_{\eta\eta} = - J^2 (P y_\xi + Q y_\eta) \tag{8}$$

In Eqs. (7) and (8), $g_{11} = x_\eta^2 + y_\eta^2$, $g_{22} = x_\xi^2 + y_\xi^2$, $g_{12} = x_\xi\, x_\eta + y_\xi\, y_\eta$ and J is the Jacobian of the transformation ($J = x_\xi\, y_\eta - y_\xi\, x_\eta$).

Ashrafizadeh and Raithby (Ashrafizadeh & Raithby, 2006) have shown that the TTM grid generation equations, i.e. Eqs. (5) and (6), can be discretized and solved in the physical domain. To apply the finite volume method to the solution of Eqs. (5) and (6) in the physical domain, an initial algebraic grid is generated first. Then, a control volume is associated with each node of the initial grid. Defining

$$\vec{q}^{\,\xi} = \vec{\nabla} \xi \; ; \qquad \vec{q}^{\,\eta} = \vec{\nabla} \eta \tag{9}$$

the integral of Eq. (5) over a control volume associated with node i, with volume V_i and surface S_i , is

$$\int_{V_i} \vec{\nabla} \bullet \vec{q}^{\,\xi}\, dV = \int_{S_i} \bullet\, d\vec{S} = \int_{V_i} P\, dV \tag{10}$$

The surface S_i consists of a number of panels, with an integration point ip located at the centre of each panel. The panel containing ip has area $\vec{S}_{ip}$. The integrals in Eq. (10) are approximated as follows

$$\sum_{ip} \vec{q}^{\,\xi}_{ip} \bullet \vec{S}_{ip} \equiv \sum_{ip} F^{\xi}_{ip} = (PV)_i \tag{11}$$

where F^{ξ}_{ip} can be thought of as a generalized "flow" across the panel ip driven by $\vec{\nabla}\xi$.

The final algebraic equation is obtained by approximating each term in Eq. (11) by an equation that involves nodal values of ξ, x, and y. This provides one constraint for ξ_i , x_i , and y_i . Applying a similar procedure, Eq. (6) leads to another algebraic equation relating η_i , x_i , and y_i for each interior node. But the values of ξ_i and η_i are known for all interior nodes, so that these two algebraic equations provide the necessary constraints for computing (x_i , y_i). The nodal values of $\xi_i , \eta_i\; x_i$, and y_i are all prescribed for boundary nodes, so the set of equations is closed throughout the solution domain and its boundary. This is called the Direct Design Method for solving the elliptic grid generation problem because no inversion of equations is required and the unknown nodal values of (x_i , y_i) appear explicitly (i.e. "directly") as dependent variables in the finite volume equations in the physical domain.

Calculation of the source or control functions at the right hand sides of Eqs. (5) and (6) is an important part of any method which uses this set of equations to generate the grid. In addition to the elementary method, proposed in (Thompson, et al., 1974), many researchers have proposed methods for the automatic calculation of the boundary values of control functions (Thomas & Middlecoff, 1980; Spekreijse, 1995; Steger & Sorenson, 1997; Kaul, 2003; Lee & Soni, 2004; Ashrafizadeh & Raithby, 2006; Kaul, 2010). Assuming that the P values are known at $(\xi_i, \eta = 1)$ and $(\xi_i, \eta = 0)$ boundaries in Fig. 1a, the P values at internal nodes can be obtained through the following one dimensional interpolation formula (Thomas & Middlecoff, 1980):

$$P(\xi,\eta) = C(\eta)\ P(\xi,0) + (1 - C(\eta))\ P(\xi,1) \tag{12}$$

The Q values at internal nodes can also be calculated similarly:

$$Q(\xi,\eta) = C(\xi)\ Q(0,\eta) + (1 - C(\xi))\ Q(1,\eta) \tag{13}$$

Coefficients $C(\xi)$ and $C(\eta)$ in Eqs. (12) and (13) can be linear or non-linear functions of the corresponding logical coordinates.

Another noticeable classical Grid Generation method, which is known as the Orthogonal Grid Generation (OGG) method and is elliptic in certain situations, is based on the assumptions of continuity and orthogonality of the coordinate lines. The final forms of the grid generation equations in the OGG method are as follows (Ryskin & Leal, 1983):

$$\frac{\partial}{\partial \xi}\left(f \frac{\partial x}{\partial \xi}\right) + \frac{\partial}{\partial \eta}\left(\frac{1}{f} \frac{\partial x}{\partial \eta}\right) = 0 \tag{14}$$

$$\frac{\partial}{\partial \xi}\left(f \frac{\partial y}{\partial \xi}\right) + \frac{\partial}{\partial \eta}\left(\frac{1}{f} \frac{\partial y}{\partial \eta}\right) = 0 \tag{15}$$

The orthogonality condition, $g_{12} = x_\xi x_\eta + y_\xi y_\eta = 0$, is implied in Eqs. (14) and (15). The scale factor, f, is defined based on the transformation metrics relevant to the magnification effects of the mapping in different logical directions as follows:

$$f \equiv \frac{\sqrt{g_{22}}}{\sqrt{g_{11}}} = \frac{\sqrt{\left(x_\eta^2 + y_\eta^2\right)}}{\sqrt{\left(x_\xi^2 + y_\xi^2\right)}} \tag{16}$$

Calculation of the scale factor near the boundaries and throughout the solution domain is a major step in the orthogonal grid generation and is discussed in a number of publications (Ryskin & Leal, 1983; Kang & Leal, 1992; Eca, 1996; A. Bourchetin & L. Bourchetin, 2006). Most commonly, boundary values of f are calculated first and then linear or non-linear interpolation techniques are used to obtain the internal values.

Imposition of the orthogonality constraint in some problems may be difficult or even impossible. Therefore, modifications on the OGG have also been proposed to generate nearly orthogonal grids (Akcelik et al., 2001; Zhang et al., 2004, Zhang et. Al, 2006a, Zhang et. Al, 2006b).

Based on the above brief review of some of the classical EGG methods, it can be argued that in each one of these methods a set of grid generation equations is developed to calculate the physical coordinates of the nodal points. The set of equations may directly introduce the mapping functions which transform the logical grid to the physical grid, e. g. the TFI, or they may provide differential constraints on the mapping functions and indirectly describe them, e. g. the TTM and the OGG methods. However, it is important to note that the governing equations in grid generation are radically different from equations which govern physical processes. While experimental observations provide a basis for the development of the physical governing equations, the grid generation equations are not expressions of natural phenomena and are developed based on analogy or mathematical considerations. An example of the use of physical analogy in determining the control functions in Eq. (5) and (6) is provided by Kaul (Kaul, 2003). Considering the arbitrariness in the development of elliptic grid generation equations, it is very desirable to have a clear, simple and systematic approach for proposing the governing equations in the context of structured grid generation.

In this paper, we propose a unifying rationale for the development of elliptic grid generation methods. Based on the proposed unifying view point, all existing EGG methods can be viewed as multi-dimensional geometrical interpolation techniques which employ different interpolants, interpolation methods and grid generation equations. Once the grid generation equations, the interpolants and interpolation techniques are chosen, there are various numerical solution methods to solve the algebraic equations and to calculate the nodal coordinates.

To explain the proposed framework, a number of applicable interpolants in structured grid generation are first introduced in the next section. Then, different applicable interpolation techniques are presented. Afterwards, the rationale behind the development of grid generation equations is discussed. Finally, a number of alternative EGG methods are introduced and examples of elliptic grid generation via the classical and proposed alternative methods are presented.

For the sake of simplicity and brevity, the grid generation examples in this chapter are limited to two dimensional solution domains, but the underlying ideas are clearly applicable in three-dimensional problems as well.

2. Interpolants

The logical grid, shown in Fig. 1b, is the simplest possible two-dimensional grid. The boundaries are straight lines and the nodes are distributed uniformly. This simplicity makes the logical grid generation trivial. In contrast, boundaries of the physical grid may be complex curves and the boundary nodes in this case can be distributed non-uniformly. For example, the N, S, W and E boundaries shown in Fig. 1a are different curves with different non-uniform distributions of nodes. It is exactly this complexity that makes the grid generation in the physical domain a rather difficult task as compared to the logical domain. The shape of the boundary of the domain and the distribution of the nodes along the boundary are the most important information which needs to be taken into consideration in the elliptic grid generation process. Quantities which provide information regarding the shape of the boundary coordinate lines, expected shape of the crossing coordinate lines near the boundaries and the distribution of nodes along the boundaries are here called the boundary data. Some or all of these data are the inputs required in an elliptic grid generation method.

The interpolants in an EGG problem are quantities related to the boundary data. The simplest, and most obvious, quantities which describe the boundary coordinate lines are the x and y coordinates of boundary nodal points. These are the interpolants used in many simple algebraic grid generators and we call them here the zero order boundary data.

It is possible to use higher order boundary data as interpolants as well. Quantities such as nodal values of x_ξ and y_ξ along the north boundary in Fig. 1a are tangential slope-related first order data. Similarly, $x_{\xi\xi}$ and $y_{\xi\xi}$ along the south boundary are tangential curvature-related second order data. These data provide information regarding the stretching of nodes along a boundary coordinate line.

By generating paving layers near boundaries or by making assumptions regarding the coordinate lines which cross the boundaries, it is also possible to generate normal first, second,......, and nth order boundary data. For example, using the shaded paving layer near the north boundary in Fig. 1a, it is possible to generate x_η and y_η data at the north boundary. Similarly, $x_{\eta\eta}$ and $y_{\eta\eta}$ can be generated near the south boundary using the data obtained from the two shaded paving layers near the south boundary in Fig. 1a. Therefore, using the paving layers, it is possible to generate boundary data which actually describe the boundary cell geometries or the shape of coordinate lines crossing the boundary. A simple algebraic method for generating high quality paving layers is introduced in (Ashrafizadeh & Raithby, 2006).

The first and second order boundary data can also be defined with the physical coordinates as the independent variables. For example quantities such as ξ_x, ξ_{xx}, ξ_y and ξ_{yy} fall in this category. However, in contrast to the logical coordinates, the physical coordinates of nodes are not univariate variables and the denominator in the discrete form of a quantity such as ξ_x can be zero, resulting in computational difficulties.

Boundary data of different orders, just described, can also be combined to provide more information regarding the boundary nodes and cells. Such combinations of the boundary data can be employed as the interpolants in the formulation of EGG problems. For example, in the TTM method, boundary values of $(\xi_{xx}+\xi_{yy})$, called P, and $(\eta_{xx}+\eta_{yy})$, called Q, are used as the interpolants. Figure 2 shows how the source term P provides information regarding the shape of the cells and the distribution of the nodes along the boundary coordinate line (ξ_i,η_N). In Fig. 2a, the boundary is a straight line and the nodes are distributed uniformly. The source term P is identically zero everywhere along the boundary in this case. Figure 2b shows a straight boundary with non-uniform distribution of nodes. It is seen that the source term P is not zero anymore at locations with contraction or expansion of the grid. Figures 2c and 2d show curved boundaries with uniform and non-uniform distribution of nodes respectively. It is clear that the source term P varies along the η_N coordinate line and carries some information regarding the boundary geometry in the latter two cases as well. Therefore, the source terms P and Q can be used as interpolants in an elliptic grid generation problem. The source values in these examples, which are Laplacians of the logical coordinates, have been calculated using the finite volume method as described in (Ashrafizadeh et al., 2002; Ashrafizadeh et al., 2003).

The scale factor f is used as the interpolant in the classical OGG method. The shape of the boundary and the distribution of nodes in Figs. 3a to 3d have been chosen similar to Figs. 2a to 2d respectively to study the effect of the boundary geometry on the boundary values of

f .For the straight uniform paving layer shown in Fig. 3a, $f = 1$ everywhere along the boundary. In all other cases in Fig. 3, the curved boundary and/or non-uniform distribution of nodes result in a corresponding change in the nodal f values. Therefore, the scale factor f can also be used as the interpolant in an elliptic grid generation problem.

Other combinations of the boundary data can also be used as the interpolants in an EGG problem and one can check the sensitivity of a chosen interpolant to the boundary specifications before actually using them in an elliptic grid generation algorithm as just explained. Since these computations are done on distorted and/or non-uniform grids in the physical domain, finite volume method is a suitable numerical solution choice as explained before.

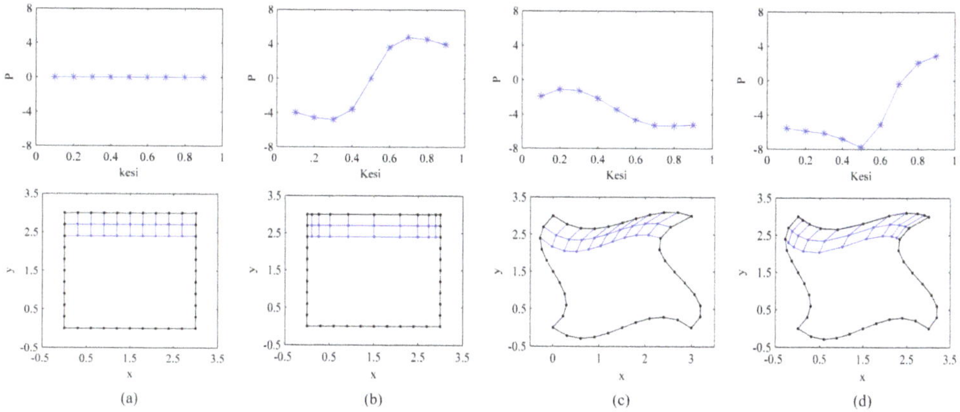

Fig. 2. Sensitivity of the source term, P, with respect to the boundary geometry and nodal distribution.

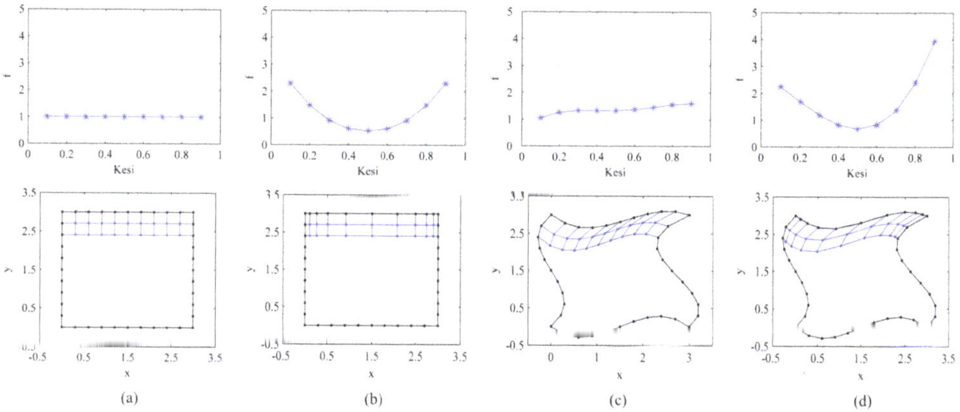

Fig. 3. Sensitivity of the scale factor, f, with respect to the boundary geometry and nodal distribution.

3. Interpolation techniques

Having chosen the interpolants, an interpolation technique is required to find the corresponding values at internal nodes. The idea is that the boundary data should be used to determine the geometrical properties of the internal cells and coordinate lines via interpolation techniques. There are three geometrical interpolation techniques that can be used in a multi-dimensional problem as follows.

3.1 One-dimensional interpolation

Univariate stretching functions provide the relations for the one dimensional interpolation. Equations (12) and (13) are examples of one dimensional interpolation formulas used in a 2D problem.

3.2 Quasi multi-dimensional interpolation

Quasi multi-dimensional interpolation techniques such as TFI, which employs the Boolean sum of 1D interpolations, can also be used to interpolate the chosen interpolants in an elliptic grid generation process. The interpolation coefficients in these algebraic methods can be constant, linear or non-linear functions of the logical or physical coordinates. Use of the physical coordinates as the independent variables in the interpolation coefficients worsens the nonlinearity of the interpolation and, therefore, this option has not gained any popularity. As an example, it will be shown later that the boundary values of the control functions in the TTM method can also be interpolated via the TFI.

3.3 Multi-dimensional interpolation

The interpolants in the EGG methods can also be interpolated by truly multi-dimensional methods, i.e. by the solution of boundary value problems. For example, it will be shown that the boundary values of the control functions in the TTM method can also be interpolated through the solution of Dirichlet boundary value problems to obtain the corresponding values for the internal nodes.

4. Grid generation equations

By viewing the elliptic grid generation problem as a multi-dimensional interpolation problem, the focus is obviously on the selection of interpolants and the interpolation techniques. However, depending on the chosen interpolants, it may also be necessary to develop grid generation equations, i.e. equations which ultimately provide the nodal coordinates throughout the domain.

The process of interpolation may actually play the role of the grid generation equations. In an algebraic grid generation method such as the zero order TFI, the interpolants are the nodal coordinates. By carrying out the interpolation, nodal coordinates are obtained throughout the domain and no additional grid generation equation is required. In other words, the interpolation equations in this case are themselves the grid generation equations.

The expressions used to define the interpolants may also be used to develop the grid generation equations. For example, in the classical TTM method, in which Eqs. (5) and (6) are used to define the interpolants, the same equations are also inverted to obtain the grid generation equations in the logical domain. In the method proposed by Ashrafizadeh and Raithby (Ashrafizdeh & Raithby, 2006), Eqs. (5) and (6) are used to obtain the grid generation equations in the physical domain.

The classical orthogonal grid generation method is a good representative example of the cases in which neither the interpolation formulas nor the definition of the interpolants can be used for the grid generation. In contrast to the TTM, the interpolated values of the scale factor are not directly used to calculate the coordinates of internal nodes in the OGG. First order differential interpolants, such as the scale factor f, are not appropriate for the calculation of the nodal coordinates. Therefore, second order differential equations are developed by imposing the continuity constraints on the mapping functions. Consequently, Eqs. (14) and (15) are obtained and used as the grid generation equations.

Now that the commonly used grid generation methods are explained in the framework of the proposed unifying view, a number of alternative elliptic grid generation methods are introduced in the next section. The objective is to show that how new elliptic grid generation methods can be developed in the context of the suggested vantage point.

5. Alternative grid generation methods

Based on the proposed view point, there are many possible alternatives for the development of elliptic grid generation methods. By focusing on the interpolants and the interpolation techniques, EGG methods can be divided into the following four categories:

- Algebraic interpolation of Algebraic interpolants (AA methods).
- Algebraic interpolation of Differential interpolants (AD methods).
- Differential interpolation of Algebraic interpolants (DA methods).
- Differential interpolation of Differential interpolants (DD methods).

As mentioned before, the selection of appropriate grid generation equations provides another degree of freedom in the development of elliptic grid generation methods. Here we present two alternative grid generation methods in each category. It is clear that there are other possibilities and one may develop new grid generation methods in the proposed context.

5.1 AA methods

5.1.1 The AA1 method

This method works with algebraic boundary data and employs an algebraic interpolation formula with constant coefficients. Grid generation equations here are algebraic interpolation formulas similar to Eq. (4) applied to a sub-domain close to the nodal point, shown in Fig. 4, as follows:

$$x_P = 0.5(x_W + x_E + x_S + x_N) - 0.25(x_{NW} + x_{NE} + x_{SW} + x_{SE}) \tag{17}$$

$$y_P = 0.5(y_W + y_E + y_S + y_N) - 0.25(y_{NW} + y_{NE} + y_{SW} + y_{SE}) \tag{18}$$

Note that here algebraic formulas are used to ultimately interpolate the coordinates of boundary nodes (the interpolants). The nine-point computational molecules provide two sets of simultaneous equations which have to be solved to obtain the coordinates of grid points. In the context of the classical EGG methods it is hard to call this method an algebraic grid generator. We prefer to avoid the confusion by simply associate the method to the selected interpolants and the mathematical nature of the interpolation technique.

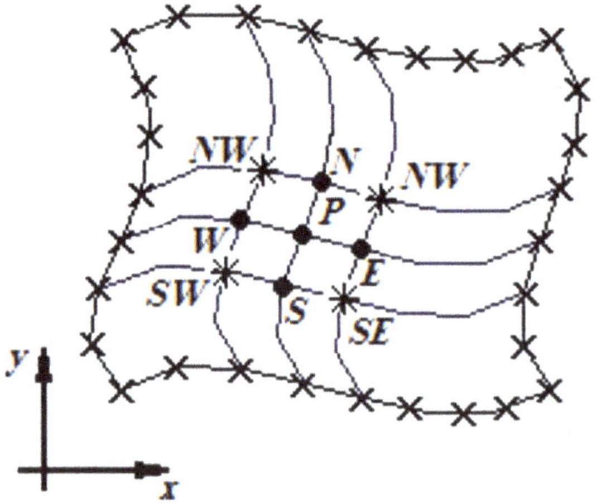

Fig. 4. Contributing nodes in the AA1, DA1 and DD1 methods.

5.1.2 The AA2 method

This method provides a combined local/global interpolation formula. Coordinates of some adjacent and neighbor boundary nodes are interpolated in a TFI-like interpolation procedure. To obtain the nodal coordinates, the following procedure is carried out:

- Coordinates of node P are calculated using zero order TFI in the shaded area in Fig. 5a. The contributing nodes are shown by solid dots in Fig. 5a. The calculated coordinates at this stage are called $(x_1, y_1)_P$.
- A similar procedure is carried out using the TFI in shaded areas shown in Figs 5b, 5c and 5d to obtain new coordinates $(x_2, y_2)_P$, $(x_3, y_3)_P$ and $(x_4, y_4)_P$.
- The final coordinates of node P , i. e. $(x, y)_P$, are obtained as follows:

$$(x,y)_P = C_1(x_1,y_1)_P + C_2(x_2,y_2)_P + C_3(x_3,y_3)_P + C_4(x_4,y_4)_P \tag{19}$$

The simplest choice for the weight factors would be $C_1 = C_2 = C_3 = C_4 = 0.25$. The weight factors may also be chosen taking into consideration the logical coordinates of point P.

Note that formula for $(x_1, y_1)_P$ includes coordinates of some boundary points as well as points P and P_1. Similarly, the $(x_2, y_2)_P$, $(x_3, y_3)_P$ and $(x_4, y_4)_P$ terms bring the coordinates of points P_2, P_3 and P_4 in the mix. Therefore, as compared to the traditional zero order TFI method, which takes the coordinates of 8 boundary nodes to calculate the coordinates of an internal node P, this method employs the information at 16 boundary nodes as well as 4 neighbor nodes to construct a computational molecule for nodal point P. Considering the fact that the coordinates of boundary nodes are known, Eq. (14) can be re-written as 5-point computational molecule for the coordinates of node P as follows:

$$(x,y)_P = C_{P_1}(x,y)_{P_1} + C_{P_2}(x,y)_{P_2} + C_{P_3}(x,y)_{P_3} + C_{P_4}(x,y)_{P_4} + C_P \quad (20)$$

Coefficient C_P in Eq. (20) includes the effects of the above mentioned 16 boundary nodes. In contrast to the classical TFI, a simultaneous set of equations needs to be solved to obtain the coordinates of internal nodes. If only two shaded areas shown in Figs. 5a and 5b are used to develop the interpolation formulas, the coordinates of internal nodes can be obtained in a marching calculation process starting from the south east corner of the domain.

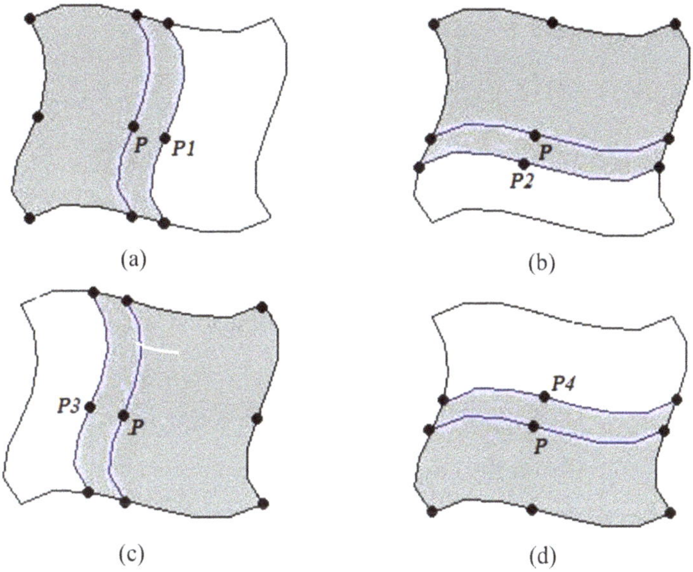

Fig. 5. Four sub-domains used to develop the computational molecule in the AA2 method.

5.2 AD methods

5.2.1 The AD1 method

In this method the interpolants are some differential boundary data interpolated by algebraic interpolation formulas. The contributing boundary nodes in each computational molecule are shown in Fig. 6 by × signs. Second order derivatives of boundary coordinates are interpolated by the algebraic TFI method:

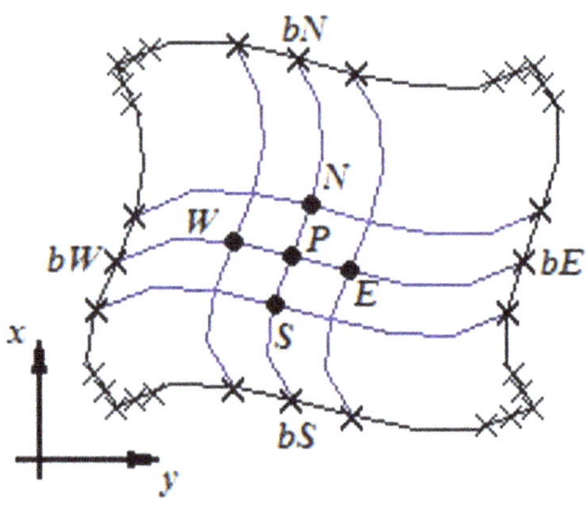

Fig. 6. Contributing nodes in the AD1 method.

$$
\begin{aligned}
(x_{\eta\eta}, y_{\eta\eta})_P &= (1-\xi_P)(x_{\eta\eta}, y_{\eta\eta})_{bW} + (\xi_P)(x_{\eta\eta}, y_{\eta\eta})_{bE} + (1-\eta_p)(x_{\eta\eta}, y_{\eta\eta})_{bS} \\
&+ (\eta_P)(x_{\eta\eta}, y_{\eta\eta})_{bN} - (1-\xi_P)(1-\eta_p)(x_{\eta\eta}, y_{\eta\eta})_{P\text{-}SW} \\
&- (1-\xi_P)(\eta_P)(x_{\eta\eta}, y_{\eta\eta})_{P\text{-}NW} - (\xi_P)(1-\eta_p)(x_{\eta\eta}, y_{\eta\eta})_{P\text{-}SE} \\
&- (\xi_P)(\eta_P)(x_{\eta\eta}, y_{\eta\eta})_{P\text{-}NE}
\end{aligned} \tag{21}
$$

$$
\begin{aligned}
(x_{\xi\xi}, y_{\xi\xi})_P &= (1-\xi_P)(x_{\xi\xi}, y_{\xi\xi})_{bW} + (\xi_P)(x_{\xi\xi}, y_{\xi\xi})_{bE} + (1-\eta_p)(x_{\xi\xi}, y_{\xi\xi})_{bS} \\
&+ (\eta_P)(x_{\xi\xi}, y_{\xi\xi})_{bN} - (1-\xi_P)(1-\eta_p)(x_{\xi\xi}, y_{\xi\xi})_{P\text{-}SW} \\
&- (1-\xi_P)(\eta_P)(x_{\xi\xi}, y_{\xi\xi})_{P\text{-}NW} - (\xi_P)(1-\eta_p)(x_{\xi\xi}, y_{\xi\xi})_{P\text{-}SE} \\
&- (\xi_P)(\eta_P)(x_{\xi\xi}, y_{\xi\xi})_{P\text{-}NE}
\end{aligned} \tag{22}
$$

Discrete forms of Eqs (21) and (22) result in the following formulas for the coordinates of node P:

$$(x_1, y_1)_P = C_{1S}(x, y)_{1S} + C_{1N}(x, y)_{1N} + C_{1P} \tag{23}$$

$$(x_2, y_2)_P = C_{2E}(x, y)_{2E} + C_{2W}(x, y)_{2W} + C_{2P} \tag{24}$$

Coefficients C_{1P} and C_{2P} include the contribution of boundary nodes corresponding to the nodal point P as shown in Fig. 6. The final computational molecule is obtained as follows:

$$(x, y)_P = \beta_{1P}(x_1, y_1)_P + \beta_{2P}(x_2, y_2)_P \tag{25}$$

Again the simplest choice for the weight factors would be $\beta_{1P} = \beta_{2P} = 0.5$, however, these factors may also be chosen taking into consideration the logical coordinates of point P.

Therefore, in this method the coordinates of each node are constrained directly by the coordinates of 4 neighbor nodes and indirectly by 32 boundary nodes. Once again all boundary nodes contribute to the calculation of the coordinates of node P through the solution of a set of algebraic equations similar to Eq. (25).

5.2.2 The AD2 method

In this method the interpolants are the source functions P and Q, defined in Eqs. (5) and (6). The zero-order TFI is used for the interpolation of the interpolants as follows:

$$\begin{aligned}(P)_P =& (1-\xi_P)(P)_{bW} + (\xi_P)(P)_{bE} + (1-\eta_p)(P)_{bS} + (\eta_P)(P)_{bN}\\ &-(1-\xi_P)(1-\eta_p)(P)_{P\text{-}SW} - (1-\xi_P)(\eta_P)(P)_{P\text{-}NW}\\ &-(\xi_P)(1-\eta_p)(P)_{P\text{-}SE} - (\xi_P)(\eta_P)(P)_{P\text{-}NE}\end{aligned} \tag{26}$$

$$\begin{aligned}(Q)_P =& (1-\xi_P)(Q)_{bW} + (\xi_P)(Q)_{bE} + (1-\eta_p)(Q)_{bS} + (\eta_P)(Q)_{bN}\\ &-(1-\xi_P)(1-\eta_p)(Q)_{P\text{-}SW} - (1-\xi_P)(\eta_P)(Q)_{P\text{-}NW}\\ &-(\xi_P)(1-\eta_p)(Q)_{P\text{-}SE} - (\xi_P)(\eta_P)(Q)_{P\text{-}NE}\end{aligned} \tag{27}$$

Equations (5) and (6) are used as the grid generation equations.

5.3 DA methods

5.3.1 The DA1 method

In this case Dirichlet boundary value problems are solved to interpolate algebraic boundary data, i.e. coordinates of boundary nodes. All boundary nodes, shown by × signs in Fig. 4, indirectly contribute to the calculation of coordinates of each internal node. The interpolation formulas, which are actually the grid generation equations, are mathematical expressions which imply the smoothness of functions $x(\xi,\eta)$ and $y(\xi,\eta)$ as follows:

$$x_{\xi\xi} + x_{\eta\eta} = 0 \tag{28}$$

$$y_{\xi\xi} + y_{\eta\eta} = 0 \tag{29}$$

Nodal points which contribute to the computational molecule for the calculation of the coordinates of node P are shown by • signs in Fig. 4. It is interesting to note that Eqs. (28) and (29) correspond also to the conformal mapping. These equations are also obtained by setting $f=1$ in Eqs. (14) and (15).

5.3.2 The DA2 method

In this case an initial grid in a simple domain, (x_0,y_0), is used to generate the grid in the physical domain. The interpolants are algebraic quantities $\delta x = x - x_0$ and $\delta y = y - y_0$, in which (x,y) are the corresponding boundary nodes of the target geometry. The target domain and the initial grid are both shown in Fig. 7. The interpolation formulas, which are actually the grid generation equations, are mathematical expressions for a two-dimensional interpolation of the boundary values of δx and δy as follows:

$$\nabla_0^2(\delta x) = \frac{\partial^2(\delta x)}{(\partial x_0)^2} + \frac{\partial^2(\delta x)}{(\partial y_0)^2} = 0 \tag{30}$$

$$\nabla_0^2(\delta y) = \frac{\partial^2(\delta y)}{(\partial x_0)^2} + \frac{\partial^2(\delta y)}{(\partial y_0)^2} = 0 \tag{31}$$

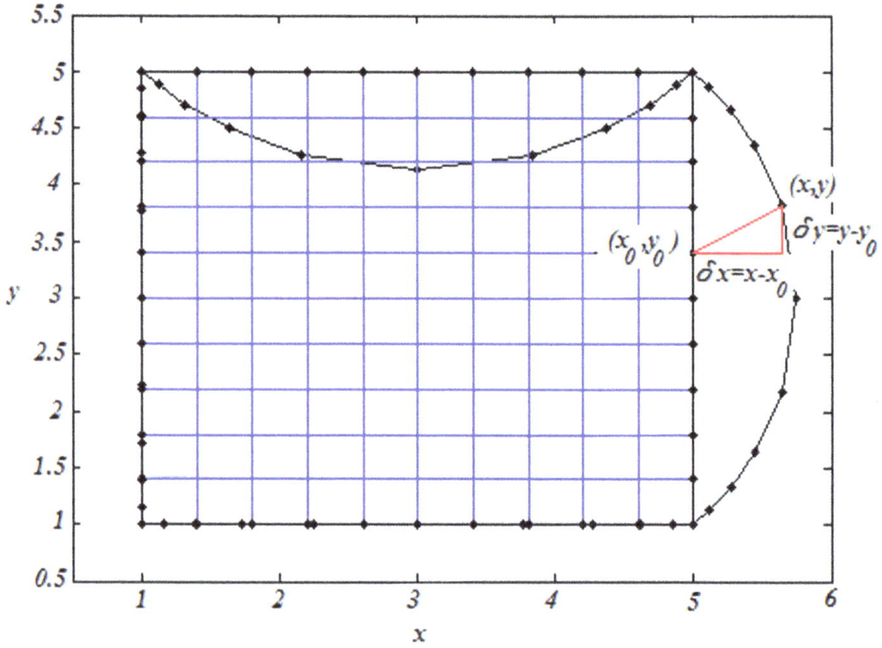

Fig. 7. Grid generation by interpolating the nodal boundary displacements (the DA2 method).

Both finite difference and finite volume methods can be used to numerically solve these equations. This method can also be used to re-mesh the computational domain in a moving boundary problem. More discussion on this method can be found in (Ashrafizadeh et al., 2009).

5.4 The DD methods

5.4.1 The DD1 method

In this example of a DD method, two differential quantities, $P = x_{\xi\xi} + x_{\eta\eta}$ and $Q = y_{\xi\xi} + y_{\eta\eta}$, are calculated at all nodes adjacent to the boundary using the paving layers as explained in (Ashrafizadeh & Raithby, 2006). These boundary data are then interpolated differentially as follows:

$$P_{\xi\xi} + P_{\eta\eta} = 0 \tag{32}$$

$$Q_{\xi\xi} + Q_{\eta\eta} = 0 \tag{33}$$

With the P and Q terms available at all internal nodes, the following grid generation equations are solved to obtain the internal nodal coordinates:

$$x_{\xi\xi} + x_{\eta\eta} = P \tag{34}$$

$$y_{\xi\xi} + y_{\eta\eta} = Q \tag{35}$$

Here again all boundary nodes, depicted by × signs in Fig. 4, contribute through the implementation of boundary conditions in the interpolation procedure for each internal node.

5.4.2 The DD2 method

Another alternative for the description of the boundary information is to use $P = \xi_{xx} + \xi_{yy}$ and $Q = \eta_{xx} + \eta_{yy}$ interpolants. Boundary values of these source functions are interpolated by a multi-dimensional interpolation technique, i.e. Eqs. (32) and (33), and the coordinates are generated by solving Eqs. (5) and (6). This method is similar to the TTM as employed in (Thomas & Middlecoff, 1980) except that a multi-dimensional interpolation method is used to calculate internal values of the source functions.

6. A brief discussion

It is worthwhile to mention few points here for further clarification:

1. The grid generation methods, just introduced, are few examples of many methods that can be developed based on the three main choices in the proposed unifying view, i. e. the choice of the interpolants, the choice of the interpolation technique and the choice of the grid generation equations. For example, a family of new methods, not discussed here, have also been developed by the authors which employ the transformation metrics at or near the boundary as interpolants. Such methods may be viewed as a continuation, and generalization, of the orthogonal grid generation method.
2. The smooth distribution of the source terms in the TTM method is a sign of grid smoothness. By properly choosing the interpolants, the interpolation technique and the grid generation equations, boundary coordinate lines are smoothly interpolated into the domain and a smooth distribution of source terms is obtained.
3. The possibility of folding exists in nearly all of the commonly used elliptic grid generation methods except for the TTM with P = Q = 0 on sufficiently fine grids. Therefore, the above new grid generation methods may result in folded grids for some geometrically complex domains. However, the objective in the development of new elliptic grid generation techniques is to obtain methods which generate high quality grids and are more resilient to folding as compared to the classical methods.
4. Many of the alternative grid generation methods presented here are executed much faster than the classical methods. They may also be used to generate the background or initial grids for other more expensive grid generators.

7. Grid generation examples

The performances of the proposed grid generation methods are now studied by solving various grid generation problems. In this section two geometries, often used to test the elliptic grid generators and here called the test cases, are chosen to examine some of the proposed methods and to also compare them with the classical elliptic and algebraic grid generation methods. The first test case is a quadrilateral domain, for which all four boundaries are distorted. The second test case is also a quadrilateral domain for which only two of the neighboring boundaries are distorted. A (11×11) grid is generated in all test cases. Finer, and nicer, grids can obviously be generated but we have chosen a rather coarse grid to be able to visualize the details of the performance of the methods.

The EGG methods can be compared in terms of the computational cost and the grid quality measures. Considering the fact that the grid generation cost depends mostly on the cost of the solution of the grid generation equations, the comparison in terms of the computational cost seems a trivial task in nearly all cases. As a general guideline, the solution of a nonlinear set of equations is computationally more expensive than the solution of a linear set. Regarding the grid quality, two parameters, i.e. the skewness and the aspect ratio, are chosen as the quality measures in this study. Skewness of a cell varies between 0 and 1 and measures the deviation from the orthogonality of the coordinate lines. Aspect ratio of a cell is defined as the ratio of the longest edge length to the shortest one and measures the deviation from a square cell. Cells in the logical space have zero skewness and aspect ratio equal to one.

Figure 8 shows the grids generated by the classical methods in the test domains. The generated grids by the zero-order TFI are shown in Figs. 8a and 8b. Figures 8c and 8d show the grids obtained from the TTM and Figs. 8e and 8f are orthogonal grids generated by the OGG. It is seen that the OGG method results in folded grids in both test cases. The corresponding grid quality measures are shown in Figs. 9a and 9b, 9c and 9d, and 9e and 9f respectively. Note that there are 10 cells along each coordinate line in the test grids. Each quality measure diagram shows the relevant quality measures for all 100 cells on a three-dimensional plot containing 10×10 data points.

Figures 10, 11, 12 and 13 show the grids in the test geometries obtained via the proposed new methods. It can be seen that all of the methods provide unfolded grids comparable to the grids obtained by the classical methods. Furthermore, and as expected, it is obvious that methods which employ differential interpolation techniques result in smoother grids.

The spatial distribution of the source functions $P=\xi_{xx}+\xi_{yy}$ and $Q=\eta_{xx}+\eta_{yy}$ using three different interpolation techniques are shown in Fig. 14. One dimensional interpolation is used to interpolate the control functions shown in Figs. 14a and 14b. The grids corresponding to these control functions are shown in Figs. 8c and 8d. Figures 14c and 14d show the distributions of control functions, which are obtained through a quasi-two dimensional method, i.e. the TFI. The grids corresponding to these control functions are shown in Figs. 11c and 11d. Finally, Figs. 14e and 14f show the interpolated control functions via a truly two-dimensional interpolation method, i.e. the solution of Dirichlet boundary value problems. The grids corresponding to these control functions are shown in

Figs. 13c and 13d. As expected, it is seen that the grids corresponding to the control functions shown in Figs. 14e and 14f are smoother than the other grids. A stretched grid with higher number of nodes is shown in Fig. 15 to show the applicability of AD1 in more complex domains. Similar results can be obtained via other proposed methods.

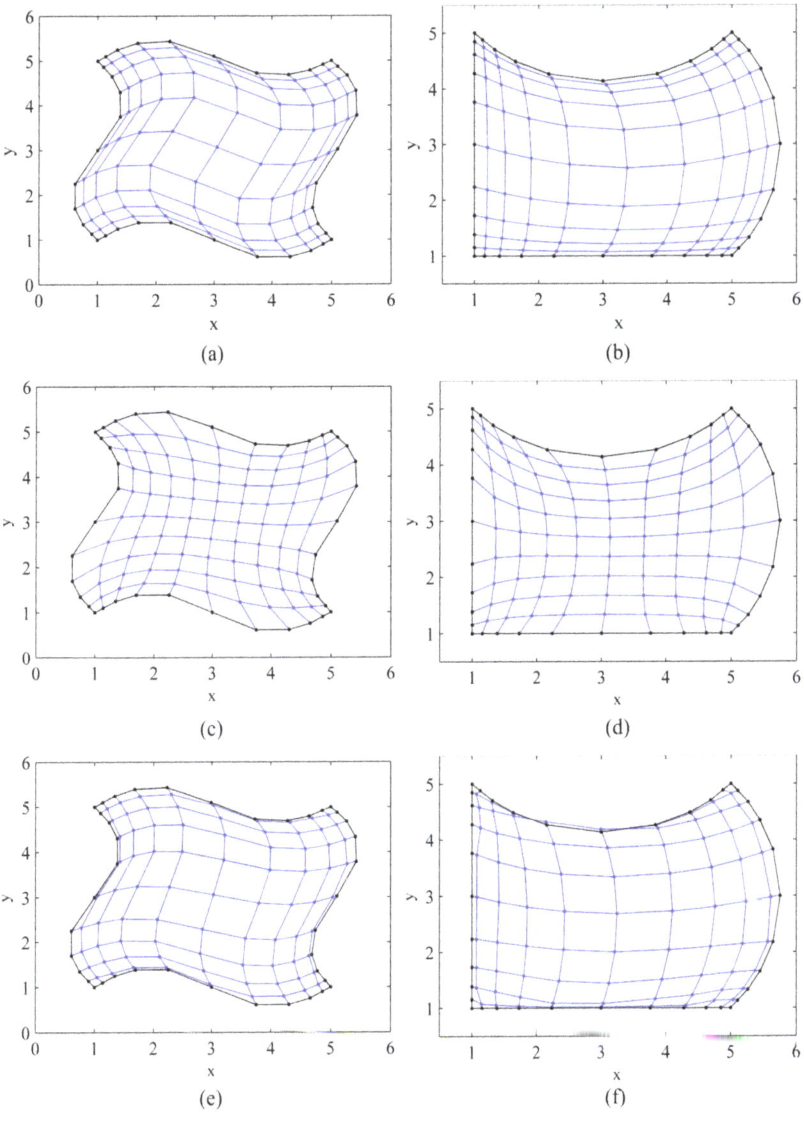

Fig. 8. Generated grids by TFI (a,b), TTM (c,d) and OGG (e,f).

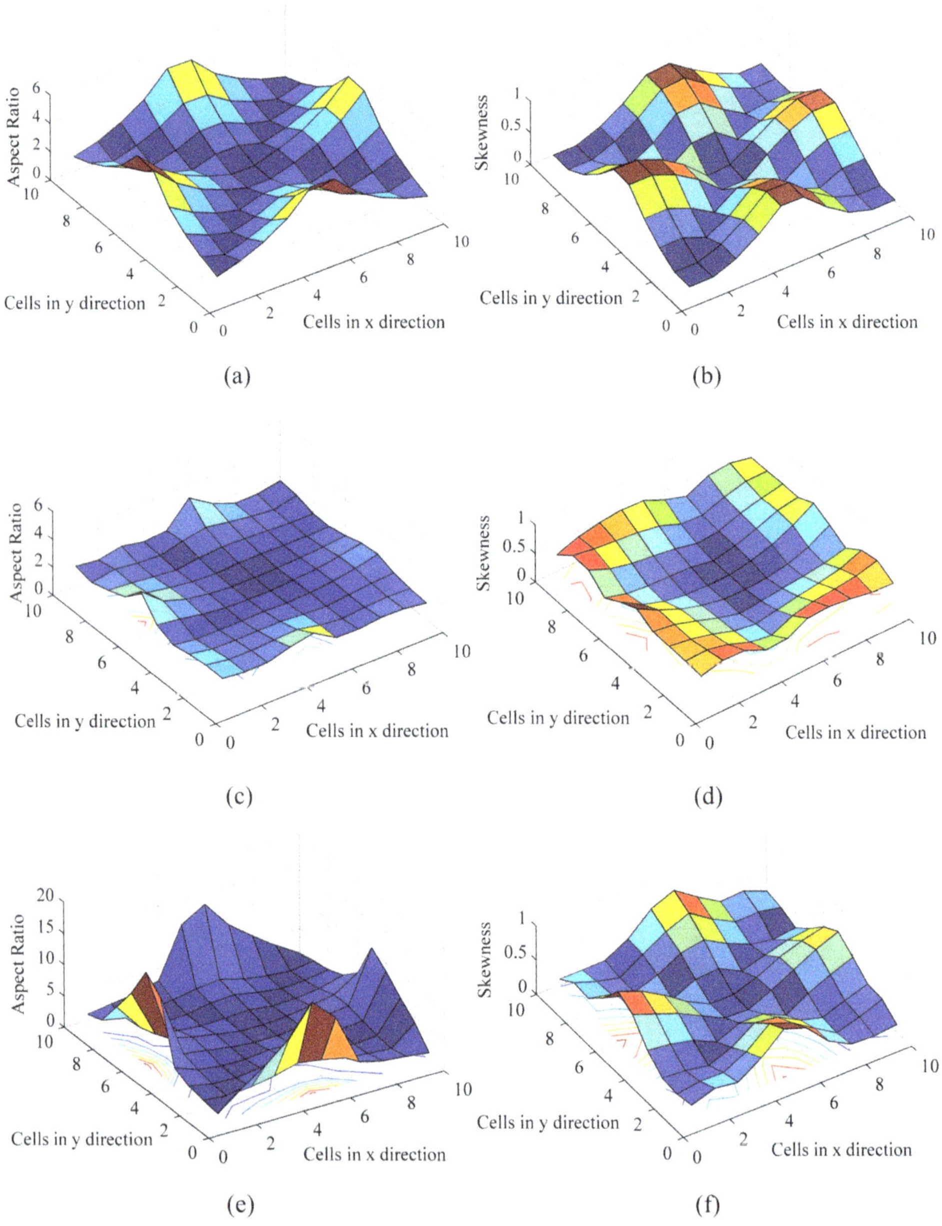

Fig. 9. Quality measures for the grids generated by TFI (a,b), TTM (c,d) and OGG (e,f).

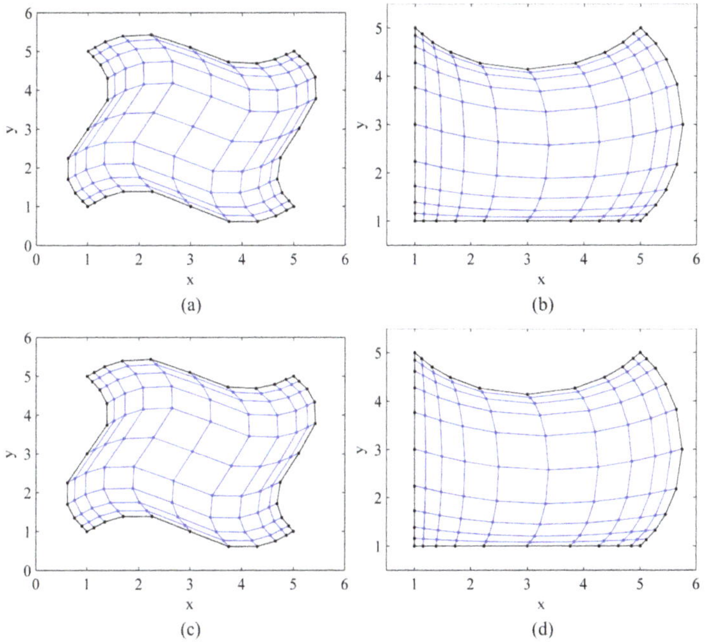

Fig. 10. Generated grids by AA1 (a, b), AA2 (c, d).

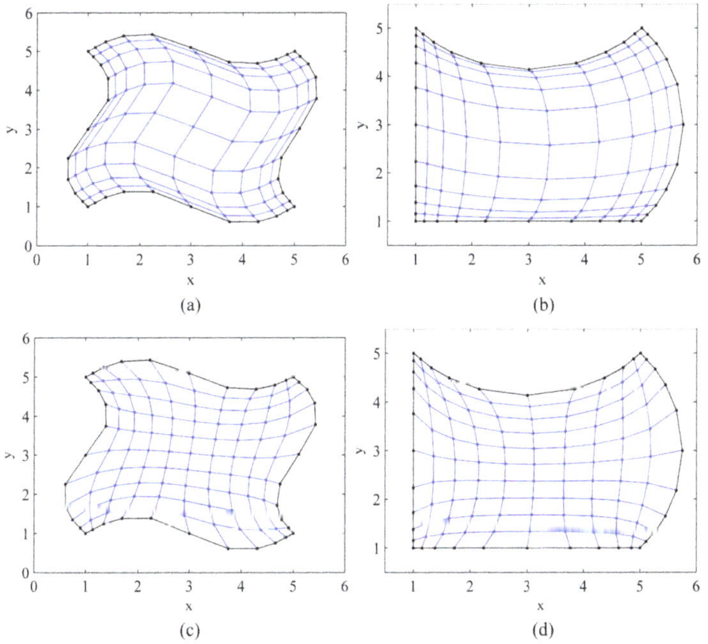

Fig. 11. Generated grids by AD1 (a, b) and AD2 (c, d).

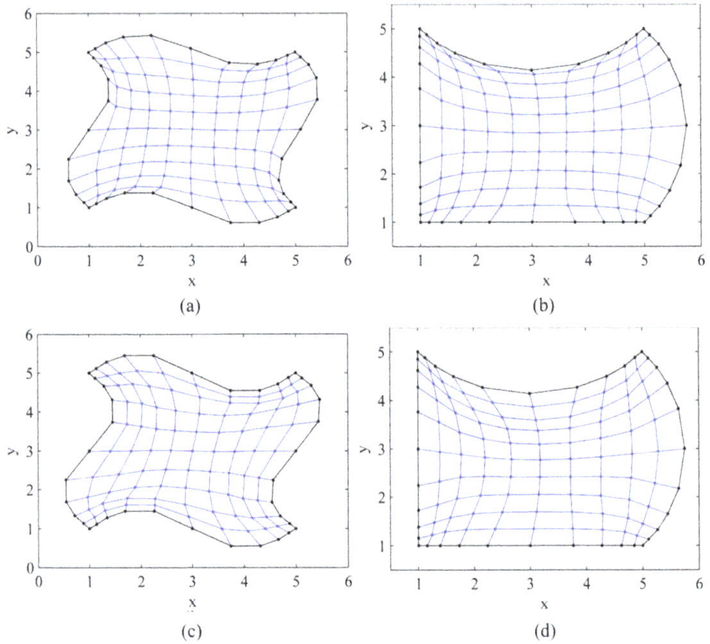

Fig. 12. Generated grids by DA1 (a, b) and DA2 (c, d).

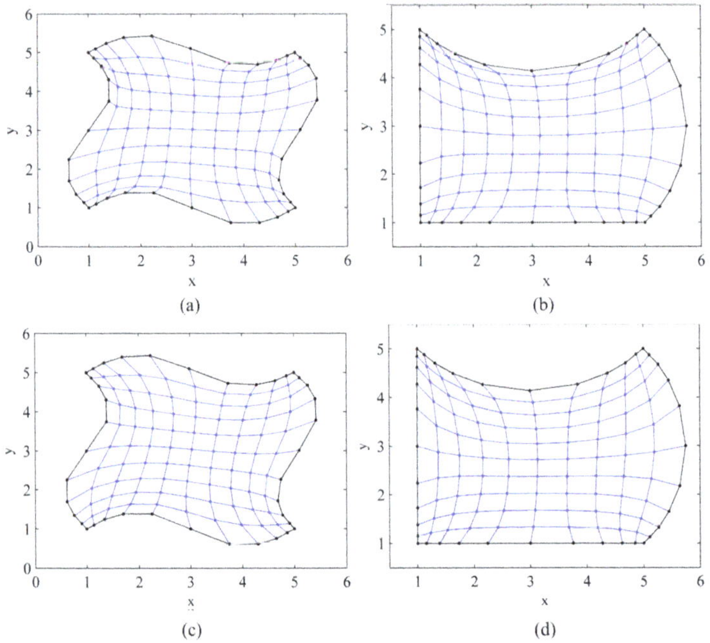

Fig. 13. Generated grids by DD1 (a, b) and DD2 (c, d).

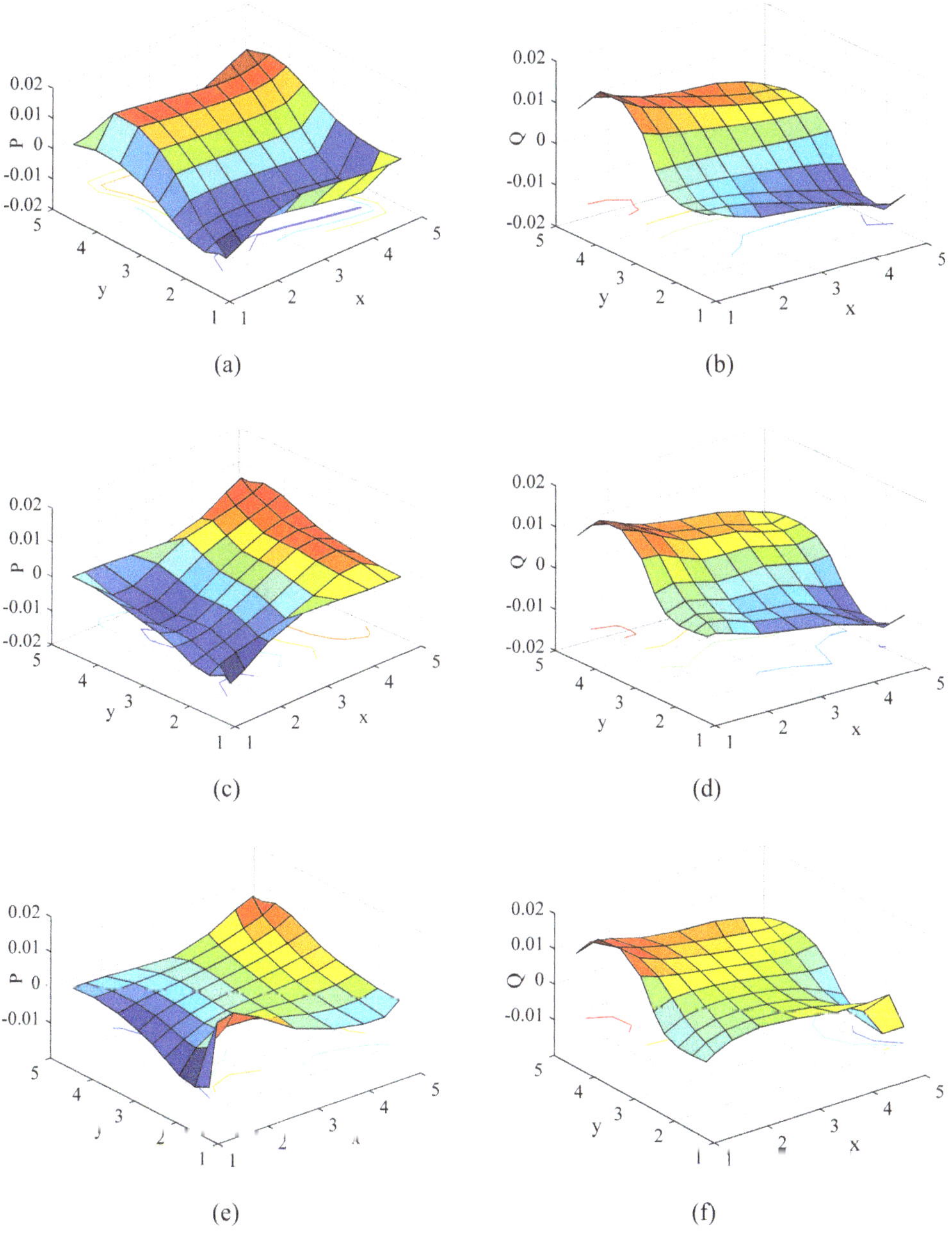

Fig. 14. Calculated P and Q by 1D interpolation (a, b), AD2 (c, d) and DD2 (e, f).

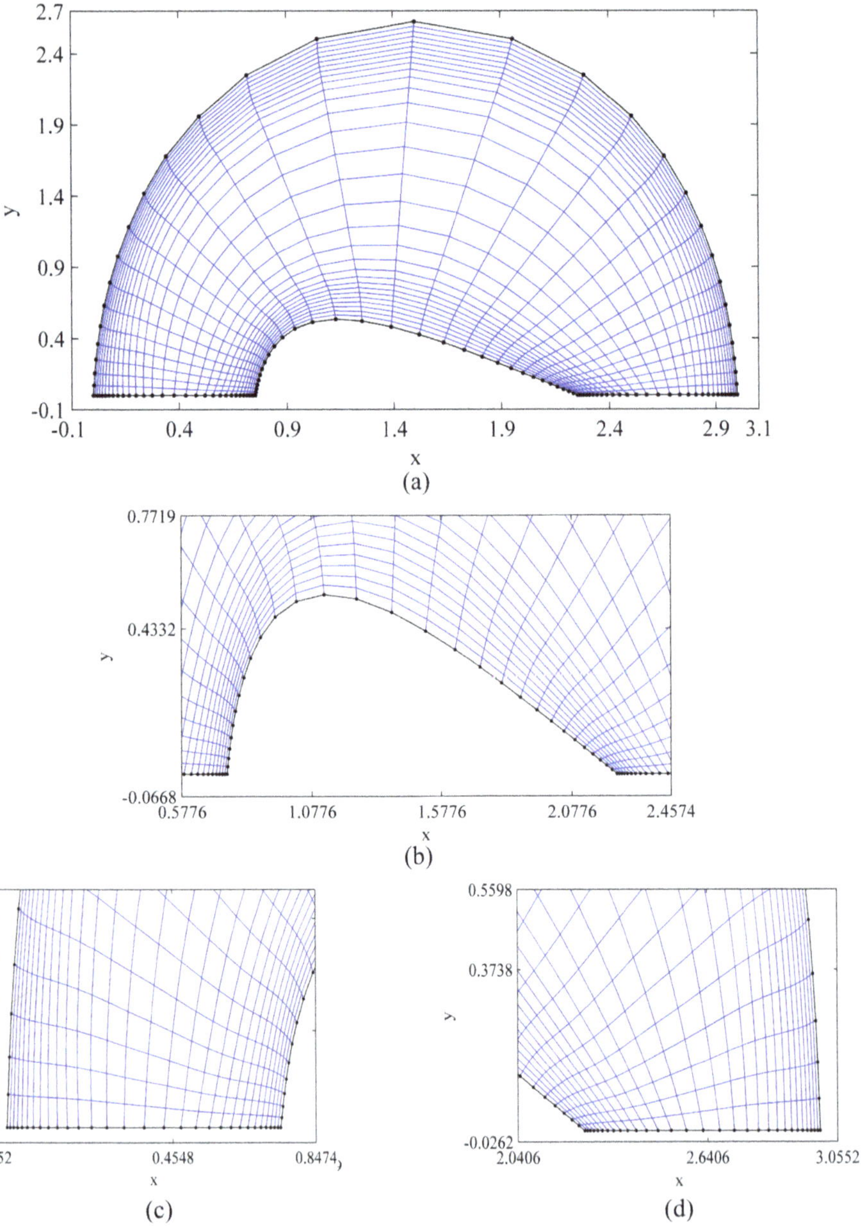

Fig. 15. A sample grid generated by AD1 (a), and a larger view of sections of the grid (b, c and d).

8. Conclusion

A unified view of the elliptic grid generation is proposed in this Chapter. It is argued that elliptic grid generation techniques are actually methods for multi-dimensional geometrical interpolation and can be described in terms of the interpolants, interpolation technique, and grid generation equations. Interpolants are used to describe the boundary shape and nodal distribution, interpolation technique is used to bring the boundary data into the domain, and grid generation equations are used to calculate the internal nodal coordinates. The most commonly used classical elliptic grid generation methods are explained in the context of the proposed unified view and new grid generation methods are also presented in the same context. A number of grid generation examples are chosen to show the applicability of the proposed methods. Authors believe that the proposed unified view provides a systematic and comprehensible approach to explain and develop a large class of elliptic grid generation methods. Some of these methods are computationally cheaper than the existing methods, yet provide grids with comparable qualities.

9. References

Akcelik V., Jaramaz B.and Ghattas O. (2001). Nearly Orthogonal Two Dimensional Grid Generation with Aspect Ratio Control. *Journal of Computational Physics,* Vol. 171, Issue 2, (August 2001), pp. 805-821, ISBN No. 0021-9991.

Ashrafizadeh, A.; Raithby, G. D. & Stubley, G. D. (2002). Direct Design of Shape. *Numerical Heat Transfer,* Part B, Vol. 41, (June 2002), pp. 501-520, ISSN: 1040-7790 print/ 1521-0626 online.

Ashrafizadeh, A.; Raithby, G. D. & Stubley, G. D. (2003). Direct Design of Ducts. *Journal of Fluids Engineering,* Transactions of ASME, Vol. 125, (January 2003), pp. 158-165, ISSN: 0098-2202 print/ 1528-901X online.

Ashrafizadeh, A.; & Raithby, G. D. (2006). Direct Design Solution of the Elliptic Grid Generation Equations. *Numerical Heat Transfer,* Part B, Vol. 50, (September 2006), pp. 217-230, ISSN: 1040-7790 print/ 1521-0626 online.

Ashrafizadeh, A.; Jalalabadi, R. & Bazargan, M. (2009). A New Structured Grid Generation Method. *Proceedings of 11th ISGG Conference,* Ecole Polytechnique: Montreal, Canada, May, 2009.

Bourchtein A. & Bourchtein. L. (2006). On Generation of Orthogonal Grids. *Journal of Applied Mathematics and Computation,* Volume 173, Issue 2, (February 2006), pp. 767-781, ISSN: 0096-3003.

Duraiswami R. & Prosperetti. A. (1992). Orthogonal Mapping in Two Dimensions. *Journal of Computational Physics,* Vol. 98, Issue 2, (February 1992), pp. 254-268, ISBN No. 0021-9991.

Eca. L. (1996). Two Dimensional Orthogonal Grid Generation with Boundary Point Distribution Control. *Journal of Computational Physics,* Vol. 125, Issue 2, (May 1996), pp. 440-453, ISBN No. 0021-9991.

Eiseman, P. R.; Choo, Y. K. & Smith, R. E. (1992). Algebraic. Grid Generation with Control Points. *Finite Elements in Fluids,* Vol. 8, (1992), pp. 97-116, ISBN No. 0-89116-850-8.

Eiseman, P. R. (1979). A Multi-Surface Method of Coordinate Generation. *Journal of Computational Physics,* Vol. 33, Issue 1, (October 1979), pp. 118-150, ISBN No. 0021-9991.

Kang. I.S. & Leal. L.G. (1992). Orthogonal Grid Generation in a 2D Domain via the Boundary Integral Technique. *Journal of Computational Physics,*Vol. 102, Issue 1, (September 1992), pp. 78-87, ISBN No. 0021-9991.

Kaul. U. K. (2003). New Boundary Constraints for Elliptic systems used in Grid Generation Problems. *Journal of Computational Physics,* Vol. 189, Issue 2, (August 2003), pp. 476-492, ISBN No. 0021-9991.

Kaul. U. K. (2010). Three-Dimensional Elliptic Grid Generation with Fully Automatic Boundary Constraints. *Journal of Computational Physics,* Vol. 229, Issue 17, (August 2010), pp. 5966-5979, ISBN No. 0021-9991.

Lee, S.H. & Soni, B.K. (2004). The Enhancement of an Elliptic Grid Using Appropriate Control Functions, *Journal of Applied Mathematics and Computation,* Vol. 159, Issue 3, (December 2004), pp. 809-821, ISSN: 0096-3003.

Lehtimaki. R. (2000). An Algebraic Boundary Orthogonalization Procedure for Structured Grids. *International Journal for Numerical Methods in Fluids,* Vol. 32, Issue 5, (March 2000), pp. 605–618, Online ISSN: 1097-0363.

Ryskin G. & Leal. L.G. (1983). Orthogonal Mapping. *Journal of Computational Physics,* Vol. 50, Issue 1, (April 1983), pp. 71-100, ISBN No. 0021-9991.

Spekreijse, S. P. (1995). Elliptic Grid Generation Based on Laplace Equations and Algebraic Transformations. *Journal of Computational Physics*, Vol. 118, Issue 1, (April 1995), pp. 38-61, ISBN No. 0021-9991.

Steger. J. L. & Sorensos. R. L. (1997). Automatic Mesh-Point Clustering Near Boundary in Grid Generation with Elliptic Partial Differential Equations. *Journal of Computational Physics,* Vol. 33, Issue 3, (December 1979), pp. 405-410, ISBN No. 0021-9991.

Thomas, P.D. & Middlecoff, J. F. (1980). Direct Control of the Grid Points Distribution in Meshes Generated by Elliptic Equations. *AIAA Journal,* (1980), Vol. 18, no.6, pp. 652-656, ISSN: 0001-1452 print/ 1533-385X.

Thompson, J. F.; Thames, F. C. & Mastin, C. W. (1974). Automatic Numerical Generation of Body-Fitted Curvilinear Coordinate System for Fields Containing Any Number of Arbitrary Two-Dimentional Bodies. *Journal of Computational Physics,* Vol. 15, Issue 3, (July 1974), pp. 299-319, ISBN No. 0021-9991.

Zhang Y.; Jia Y. & Wang S. S. Y. (2004). 2D Nearly Orthogonal Mesh Generation. *International Journal for Numerical Methods in Fluids,* Vol. 46, Issue 7, (November 2004), pp. 685–707, Online ISSN: 1097-0363.

Zhang Y.; Jia Y. & Wang S. S. Y. (2006). 2D Nearly Orthogonal Mesh Generation with Controls of Distortion Function. *Journal of Computational Physics*, Vol. 218, Issue 2, (November 2006), pp. 549-571, ISBN No. 0021-9991.

Zhang Y.; Jia Y. & Wang S. S. Y. (2006). Structured Mesh Generation with Smoothness Controls. *International Journal for Numerical Methods in Fluids,* Vol. 51, Issue 11, (August 20060, pp. 1255–1276, Online ISSN: 1097-0363.

Zhou. Q. (1998) A Simple Grid Generation Method. *International Journal for Numerical Methods in Fluids,* Volume 26, Issue 6, (March 1998), pp. 713–724, Online ISSN: 1097-0363.

6

An Alternative Finite Volume Discretization of Body Force Field on Collocated Grid

Jure Mencinger
LFDT, Faculty of Mechanical Engineering, University of Ljubljana
Slovenia

1. Introduction

Collocated grids are more suitable for the implementation on general geometries than the staggered counterparts, but their use requires the enhancement of the pressure-velocity ($p - \vec{v}$) field coupling. This is achieved by thoughtful interpolation of the velocities on finite volume faces. However, employing standard interpolation schemes, such as the well known Rhie-Chow scheme, can cause unphysical spikes in the velocity field when an abruptly changing body force field is present; an example is shown in Fig. 1. To understand the problem and find the remedy, proposed originally in (Mencinger & Žun, 2007), we should analyze the connection between p and $\vec{v}$ fields. This connection is highlighted in the following subsections.

1.1 The Navier-Stokes equation

The conservation of momentum, expressed with the Newton's second law, is in fluid dynamics represented through equation

$$\rho\left(\frac{\partial \vec{v}}{\partial t} + \vec{v}\cdot\nabla\vec{v}\right) \equiv \rho\frac{D\vec{v}}{Dt} = \vec{f} + \nabla\cdot\underline{\sigma} \tag{1}$$

where ρ is the density of an infinitesimal fluid particle and $D\vec{v}/Dt$ is its acceleration due to the presence of body force field $\vec{f}$ and stress $\underline{\sigma}$ in the considered fluid. The stress tensor $\underline{\sigma}$ contains both the pressure p and the viscous stress which is (for Newtonian fluid) proportional to the rate of strain. For incompressible fluids ($\nabla\cdot\vec{v} = 0$), considered in this text, it is written as

$$\underline{\sigma} = -p\underline{I} + \eta\left(\nabla\vec{v} + (\nabla\vec{v})^t\right) \tag{2}$$

where η denotes the viscosity and $\underline{I}$ the identity tensor. If η does not change substantially in the fluid, then (1) becomes the Navier-Stokes equation (Landau & Lifshitz, 1987)

$$\rho\frac{D\vec{v}}{Dt} = -\nabla p + \nabla\cdot(\eta\nabla\vec{v}) + \vec{f}. \tag{3}$$

The components of (3), i.e. the x-component in the Cartesian system (where $\vec{v} = (u, v, w)^t$)

$$\rho\frac{Du}{Dt} = -\frac{\partial p}{\partial x} + \nabla\cdot(\eta\nabla u) + f^x \tag{4}$$

can be written in the form of the general transport equation for a specific scalar quantity ϕ

$$\rho \frac{D\phi}{Dt} = \nabla \cdot (\Gamma \nabla \phi) + S^{\phi} \tag{5}$$

by setting $\phi = u$, $\Gamma = \eta$, $S^{\phi} = -\partial p / \partial x + f^x$ and can thus be straightforwardly discretized with the finite volume (FV) method. However, finding the solution of the obtained discretized momentum equations is not straightforward. Firstly, the unknown pressure field needs to be found. Secondly, the uninformed discretization of the terms in (4) leads to obtaining nonphysical numerical artifacts which can overwhelm the solution. One such typical artifact is the appearance of checkerboarding pressure field which originates from weak $p - \vec{v}$ coupling, explained in the next subsection.

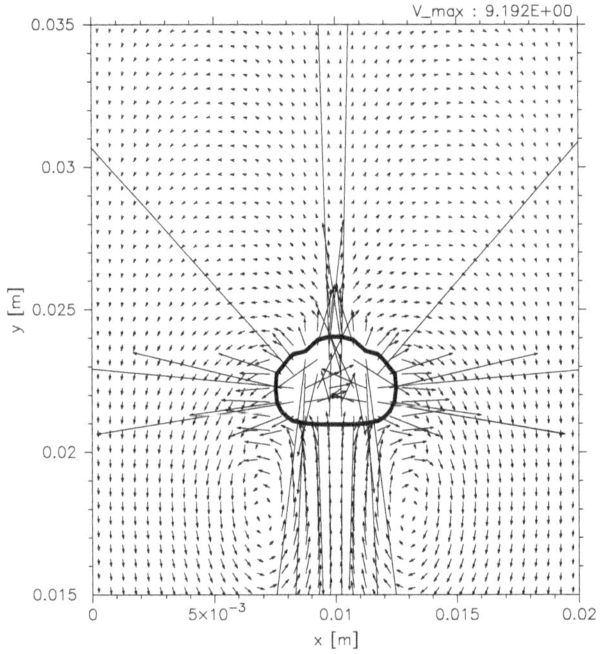

Fig. 1. Example of calculation of a rising bubble with VOF model; the unphysical velocity spikes appear in the vicinity of the interface indicated by the contour.

1.2 Origin of weak $p - \vec{v}$ coupling

The finite volume (FV) discretization of (4), obtained by its integration[1] over control volume P sized V_P (e.g., see Ferziger & Perić (2002)) and time interval Δt results in

$$\frac{\rho_P u_P - \rho_P^0 u_P^0}{\Delta t} V_P + \sum_f F_f u_f = -\left(\frac{\partial p}{\partial x}\right)_P V_P + \sum_f \eta_f (\nabla u)_f \cdot \vec{S}_f + f_P^x V_P \tag{6}$$

[1] Actually, the so-called conservative form of (4) is considered, where $\rho Du/Dt$ is replaced by $\frac{\partial(\rho u)}{\partial t} + \nabla \cdot (\rho \vec{v} u)$ on the left hand side.

where the values at the beginning and at the end of the time step are denoted as $()^0$ and written without the superscript, respectively. Fully implicit time integration is used for simplicity, although the following discussion is not limited to this choice of temporal scheme. The subscripts $()_\mathrm{P}$ and $()_\mathrm{f}$ in (6) denote the average values in FV and on FV face, respectively, and the summation $\sum_\mathrm{f}$ comprises all FV faces. Each face is represented by surface vector $\vec{S}_\mathrm{f} = S_\mathrm{f}\,\vec{n}_\mathrm{f}$, where S_f is the face area and $\vec{n}$ the normal vector pointing out of the FV. The velocity has two roles in (6). First, it appears in the mass flux through f-th face $F_\mathrm{f} = \rho_\mathrm{f}\,\vec{v}_\mathrm{f}\cdot\vec{S}_\mathrm{f}$ as *convecting* velocity $\vec{u}_\mathrm{f}$ which should comply with the relation

$$\sum_\mathrm{f} \vec{v}_\mathrm{f}\cdot\vec{S}_\mathrm{f} = 0 \tag{7}$$

following from the assumed incompressibility. Second, the x-component of the velocity, u_f in (6), is also *convected* velocity i.e. the formal unknown of the transport equation (by setting $\phi = u$). To obtain a solvable form of (6), it is written as

$$a_\mathrm{P} u_\mathrm{P} = \sum_{\mathrm{Nb(P)}} a_\mathrm{Nb} u_\mathrm{Nb} + t^0_\mathrm{P} u^0_\mathrm{P} + \left(\frac{\partial p}{\partial x}\right)_\mathrm{P} V_\mathrm{P} + f^x_\mathrm{P}\, V_\mathrm{P}, \tag{8}$$

so all the FV face–averaged values should be expressed in terms of FV-averaged values belonging to neighboring cells. The summation $\sum_{\mathrm{Nb(P)}}$ in (8) then comprises all the neighboring cells; the actual number of cells depends on the type of the approximation used.

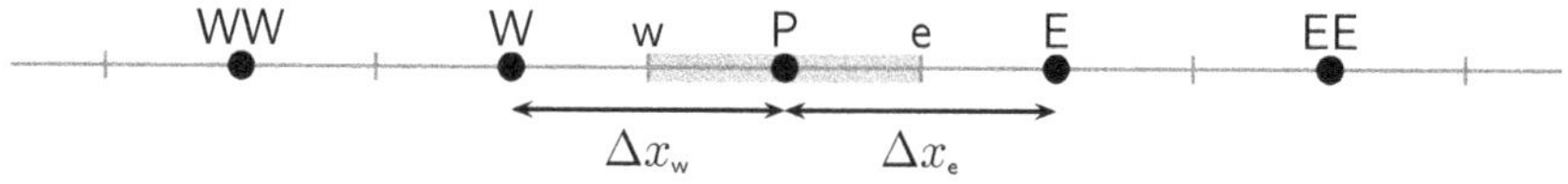

Fig. 2. Schematic example of 1D grid

To concretize and simplify the discussion, we consider a 1-dimensional uniform grid, shown schematically in Fig. 2, so that $S_\mathrm{f} = 1$, $F_\mathrm{e} = \rho_\mathrm{e} u_\mathrm{e}$, $F_\mathrm{w} = -\rho_\mathrm{w} u_\mathrm{w}$, and $\Delta x_\mathrm{e} = \Delta x_\mathrm{w} \equiv \Delta x$. If linear interpolation and central differencing are used to approximate values and derivatives on FV faces, then (8) can be written as

$$a_\mathrm{P} u_\mathrm{P} = a_\mathrm{E} u_\mathrm{E} + a_\mathrm{E} u_\mathrm{E} + t^0_\mathrm{P} u^0_\mathrm{P} - \frac{p_\mathrm{E} - p_\mathrm{W}}{2\Delta x} V_\mathrm{p} + f^x_\mathrm{P}\, V_\mathrm{P} \tag{9}$$

with the coefficients

$$a_\mathrm{E} = \frac{\eta_\mathrm{e}}{\Delta x} + \frac{1}{2}F_\mathrm{e}, \qquad a_\mathrm{W} = \frac{\eta_\mathrm{w}}{\Delta x} + \frac{1}{2}F_\mathrm{w}, \qquad t_\mathrm{P} = \frac{\rho_\mathrm{P} V_\mathrm{P}}{\Delta t}, \qquad a_\mathrm{P} = a_\mathrm{E} + a_\mathrm{W} + t_\mathrm{P}. \tag{10}$$

The solution of (9), written formally as

$$u_\mathrm{P} = \alpha_u \left[\frac{a_\mathrm{E} u_\mathrm{E} + a_\mathrm{W} u_\mathrm{W}}{a_\mathrm{P}} + \frac{t^0_\mathrm{P}}{a_\mathrm{P}} u^0_\mathrm{P} - \frac{p_\mathrm{E} - p_\mathrm{W}}{2\Delta x}\frac{V_\mathrm{p}}{a_\mathrm{P}} + f^x_\mathrm{P}\frac{V_\mathrm{P}}{a_\mathrm{P}}\right] + (1-\alpha_u) u^*_\mathrm{P}, \tag{11}$$

is obtained throughout an iterative process which usually requires under-relaxation of the calculated solution with the one from previous iteration u^*_P, weighted with the under-relaxation factor α_u. We can also notify that u_P, as written above, depends only on p_E and p_W and not on p_P.

To obtain the solution of (9), we need convecting velocities (appearing in the coefficients as factors in FV-mass fluxes) and the values of pressure. The first, u_e for example, can be obtained simply as $\frac{1}{2}(u_P + u_E)$; using (11) for u_P and a matching relation for u_E then yields

$$u_e = \widetilde{u}_e - \frac{1}{2}\left[\frac{V_P}{a_P^\alpha}\frac{p_E - p_W}{2\Delta x} + \frac{V_E}{(a_P^\alpha)_E}\frac{p_{EE} - p_P}{2\Delta x}\right] + \frac{1}{2}\left[\frac{V_P}{a_P^\alpha}f_P^x + \frac{V_E}{(a_P^\alpha)_E}f_E^x\right] \tag{12}$$

where the superscript $()^\alpha$ is used as $a_P^\alpha \equiv a_P/\alpha_u$ and

$$\widetilde{u}_e = \frac{1}{2}\left[\frac{\sum_{Nb(E)} a_{Nb}u_{Nb}}{(a_P)_E} + \frac{\sum_{Nb(P)} a_{Nb}u_{Nb}}{a_P}\right] + \frac{1}{2}\left[\frac{t_P}{a_P^\alpha}u_P^0 + \frac{t_E}{(a_P^\alpha)_E}u_E^0\right] + \frac{1}{2}(1-\alpha_u)[u_P^* + u_E^*]. \tag{13}$$

A corresponding equation can be obtained for u_w.

The pressure field, on the other hand, is calculated from the equation which is obtained from the discrete incompressibility condition (7). On the considered 1D grid it simplifies to

$$u_e - u_w = 0 \tag{14}$$

or, using the above interpolation, to $u_E - u_W = 0$. This means that the conservativeness of mass in cell P, expressed through (7), does not depend on u_P. Furthermore, the pressure field equation

$$\frac{V_E}{(a_P^\alpha)_E}\frac{p_{EE} - p_P}{4\Delta x} - \frac{V_W}{(a_P^\alpha)_W}\frac{p_P - p_{WW}}{4\Delta x} = \widetilde{u}_e - \widetilde{u}_w + \frac{1}{2}\frac{V_E}{(a_P^\alpha)_E}f_E^x - \frac{1}{2}\frac{V_W}{(a_P^\alpha)_W}f_W^x \tag{15}$$

connects the values of p in cells P, EE, and WW. So, the discretized pressure field disintegrates in two mutually independent parts (the other part contains cells E and W), which in practice often results in an unphysical zigzagging pressure pattern. An analogous difficulty appears also on three- and two-dimensional grids. It is manifested in a checkerboarding pattern of the pressure field. One might expect that the problem would disappear on nonuniform grids; unfortunately, this does not happen in practical calculations because the connection between the values of p in neighboring cells remains too weak.

1.3 Staggered versus collocated grids

The origin of the problem described above lies in the mentioned fact that u_P depends on $(\partial p/\partial x)_P$ which is calculated using the values at grid points at two alternate cells (E and W). As the two cells are $2\Delta x$ apart, this means that the gradient is actually obtained from the grid that is twice the coarse than the one actually set. A well known solution to circumvent this unwanted situation is to move the grid belonging to u component for a distance $\frac{1}{2}\Delta x$, so the grids belonging to p and u are staggered to one another.

Staggering the grids belonging to the velocity field components in the direction of the corresponding component first appeared in the paper of Harlow & Welch (1965). It enables more accurate representation of the continuity equation and of the pressure gradient in the Navier-Stokes equation. More importantly, it insures strong pressure–velocity coupling, required to obtain realistic solution of the equation. On the other hand, collocated (i.e. nonstaggered) grids have some obvious advantages over the staggered ones; Perić et al. (1988) describe them as follows:

"(i) all variables share the same location; hence there is only one set of control volumes,

(ii) the convection contribution to the coefficients in the discretized equations is the same for all variables,

(iii) for complex geometries Cartesian velocity components can be used in conjunction with nonorthogonal coordinates, yielding simpler equations than when coordinate oriented velocity components are employed, and

(iv) there are fewer constraints on the numerical grid, since there is no need to evaluate the so-called curvature terms."

In short, the collocated grids offer much simpler CFD code implementation than the staggered counterparts when the domain geometry is complex. This seems to be the main reason why the majority of the popular commercial codes use collocated grids. As the collocated grid arrangement does not inherently insure strong $p - \vec{v}$ coupling which prevents the appearance of a nonphysical checkerboard pressure field, the coupling has to be insured by other means than the grid staggering. The established method for the coupling enhancement is the employment of the momentum interpolation scheme of Rhie & Chow (1983). This scheme together with additional important corrections (Choi, 1999; Gu, 1991; Majumdar, 1988) is described below.

1.4 Corrections of convecting velocity

The interpolated convecting velocity u_e in (12) contains weighted interpolation of the pressure gradient, which leads to the disintegration of pressure field. The idea of the interpolation scheme of Rhie and Chow is to replace the interpolated derivative with the one calculated directly. However, the latter is multiplied with the corresponding interpolated coefficient so that (12) is corrected as

$$u_e := u_e + \frac{1}{2}\left[\frac{V_P}{a_P^\alpha}\frac{p_E - p_W}{\Delta x} + \frac{V_E}{(a_P^\alpha)_E}\frac{p_{EE} - p_P}{\Delta x}\right] - \frac{V_e}{(a_P^\alpha)_e}\frac{p_E - p_P}{\Delta x} \tag{16}$$

where $V_e/(a_P^\alpha)_e \equiv \frac{1}{2}(V_E/(a_P^\alpha)_E + V_P/a_P^\alpha)$ and $a := b$ is read as "a becomes b." Inserting (16) and an equivalent relation for u_w in (14) changes the pressure field equation to

$$\frac{V_e}{(a_P^\alpha)_e}\frac{p_E - p_P}{\Delta x} - \frac{V_w}{(a_P^\alpha)_w}\frac{p_P - p_W}{\Delta x} = \tilde{u}_e - \tilde{u}_w + \frac{1}{2}\frac{V_E}{(a_P^\alpha)_E}f_E^x - \frac{1}{2}\frac{V_W}{(a_P^\alpha)_W}f_W^x \tag{17}$$

which defines more compact computational molecules than (15) and prevents the previously described breakup of the pressure field.

The equation (17) connects the value of p_P with the values in adjacent cells p_E and p_W. Yet, p_P is not directly related with f_P^x, but only with f_E^x and f_W^x. This situation can be resolved with an another important correction, proposed by Gu (1991)

$$u_e := u_e - \frac{1}{2}\left[\frac{V_P}{a_P^\alpha}f_P^x + \frac{V_E}{(a_P^\alpha)_E}f_E^x\right] + \frac{V_e}{(a_P^\alpha)_e}f_e^x \tag{18}$$

so that (17) becomes

$$\frac{V_e}{(a_P^\alpha)_e}\frac{p_E - p_P}{\Delta x} - \frac{V_w}{(a_P^\alpha)_w}\frac{p_P - p_W}{\Delta x} = \tilde{u}_e - \tilde{u}_w + \frac{V_e}{(a_P^\alpha)_e}f_e^x - \frac{V_w}{(a_P^\alpha)_w}f_w^x \tag{19}$$

where f_e^x and f_w^x are the body forces on the corresponding FV faces.

Additional corrections

$$u_e := u_e - \frac{1}{2}(1-\alpha_u)[u_P^* + u_E^*] + (1-\alpha_u)u_e^* \tag{20}$$

and

$$u_e := u_e - \frac{1}{2}\left[\frac{t_P}{a_P^\alpha}u_P^0 + \frac{t_E}{(a_P^\alpha)_E}u_E^0\right] + \frac{1}{2}\left[\frac{t_P}{a_P^\alpha} + \frac{t_E}{(a_P^\alpha)_E}\right]u_e^0 \tag{21}$$

proposed by Majumdar (1988) and Choi (1999), respectively, are obtained in the same spirit. The first one prevents the dependence of u_e on the under-relaxation factor α_u in the converged solution, while the second one diminishes the dependence of u_e on the time-step size. The dependence on Δt is not completely removed because Δt is still contained in a_P^α. The complete removal the dependence on Δt was proposed, for example, by Yu et al. (2002) and recently by Pascau (2011). The latter work shows that this topic is still actual.

2. The problem of nonphysical spikes in velocity field

It turns out that the Rhie-Chow interpolation scheme works well as long as the pressure field is sufficiently smooth, i.e. without any abrupt variations. As explained below, it produces nonphysical spikes in the velocity field such as shown in Fig. 1; the spikes appear near the abrupt variations of the pressure field. The latter are generally a consequence of the abrupt changes in the body force field as can also be understood from the Navier-Stokes equation. Namely, the abrupt variation of $\vec{f}$ in (3) is counter-balanced by such a variation in ∇p. This is more obvious when the fluid is quiescent ($\vec{v} = 0$) so that (3) becomes

$$\nabla p = \vec{f} \tag{22}$$

and the equation for the pressure field is obtained by calculating the divergence of (22)

$$\nabla^2 p = \nabla \cdot \vec{f}. \tag{23}$$

2.1 Example of abrupt body force variation: multiphase flow

In most cases dealing with fluids, the body force field originates from gravity: $\vec{f} = \rho\vec{g}$ where $\vec{g}$ is the gravitational acceleration. Thus, an abrupt variation in ρ results in such a variation in $\vec{f}$ and the former appears often when dealing with multiphase system such as gas-liquid. Multiphase systems present an added difficulty in the flow simulations and require additional modeling. The most widespread approach is the employment of an Eulerian model, which make (3) valid throughout the whole flow domain regardless of the phase; this is typically achieved by using a phase identifying scalar field C, defined as

$$C(\vec{x},t) = \begin{cases} 1, & \vec{x} \text{ occupied by fluid 1 (e.g. gas),} \\ 0, & \vec{x} \text{ occupied by fluid 2 (e.g. liquid).} \end{cases} \tag{24}$$

The finite volume discretization transforms $C(\vec{x},t)$ to volume averaged C_P which represents the volume fraction of fluid 1 in volume P; thus, the cells where $0 < C_P < 1$ contain the interface.

A characteristic representative of Eulerian models is the Volume-of-Fluid (VOF) model (Hirt & Nichols, 1981) in which a two-phase system is treated as a single fluid with material properties defined as a linear combination of the phase specific properties ρ_1, ρ_2, η_1, and η_2

$$\rho = C\rho_1 + (1-C)\rho_2 \qquad \text{and} \qquad \eta = C\eta_1 + (1-C)\eta_2 \tag{25}$$

where C is advected (passively) through the considered domain. This is described by

$$\frac{DC}{Dt} = 0. \tag{26}$$

The solution of (26) is far from trivial and requires special discretization methods which surpass the scope of this chapter; a comprehensive overview of such methods is written, for example, by Scardovelli & Zaleski (1999).

Returning back to the body force, using (25) $\vec{f}$ becomes

$$\vec{f} = (C\rho_1 + (1-C)\rho_2)\,\vec{g} \tag{27}$$

so that the abrupt changes in $\vec{f}$, proportional to the density difference, are present at the phase interface. An another example of the sudden change in the body force field, when employing VOF, is due to the surface tension. The latter is modeled by the continuum surface force (CSF) model (Brackbill et al., 1992) which 'converts' the surface tension to a body force acting in the vicinity of the interface

$$\vec{f} = \sigma\kappa\nabla C \tag{28}$$

where σ and κ are the surface tension coefficient and the curvature of the interface, respectively.

2.2 Examination of the problem in 1D

Clearly, the obtained pressure field equation depends on the interpolation of the velocities on CV boundaries. To investigate the influence of different interpolations in the situations with the presence of the abrupt body force field, we setup a one-dimensional case with $0 < x < 0.1\,\mathrm{m}$ and discretize the defined domain with the uniform grid containing 40 elements. The body force field is defined as $f^x = (C\rho_1 + (1-C)\rho_2)\,g$ where $g = 10.0\,\mathrm{m}^2/\mathrm{s}$. In the considered case, f^x is determined with the phase discrimination function

$$C(\vec{x}) = \begin{cases} 1, & x_1 < x < x_2, \\ 0, & \text{otherwise} \end{cases} \tag{29}$$

where x_1 and x_2 are set to 0.030 m and 0.062 m, respectively. Material properties of air ($\rho_1 = 1.29\,\mathrm{kg/m^3}$, $\eta_1 = 1.8\times10^{-5}\,\mathrm{N\,s/m^2}$) and water ($\rho_2 = 1.0\times10^{3}\,\mathrm{kg/m^3}$, $\eta_2 = 1.0\times10^{-3}\,\mathrm{N\,s/m^2}$) are used.

By setting the velocity at the boundary to zero we expect uniform zero velocity field all over the domain. Also, linear (by parts) pressure field is expected since $\partial p/\partial x$ is counterbalancing the body force field. The resulting pressure field is shown in Fig. 3(a): the pressure field obtained without the Rhie-Chow interpolation scheme (16) exhibits a zigzagging pattern which in this case does not overwhelm the solution. Expectedly, this pattern disappears with the employment of (16). However, Fig. 3(b) shows that this scheme produces unwanted spikes near the discontinuity. Adding Gu's correction does not appear to have a notable effect on the pressure field (Fig. 3(a)) whereas the spikes (Fig. 3(b)) appear even larger in the presented case.

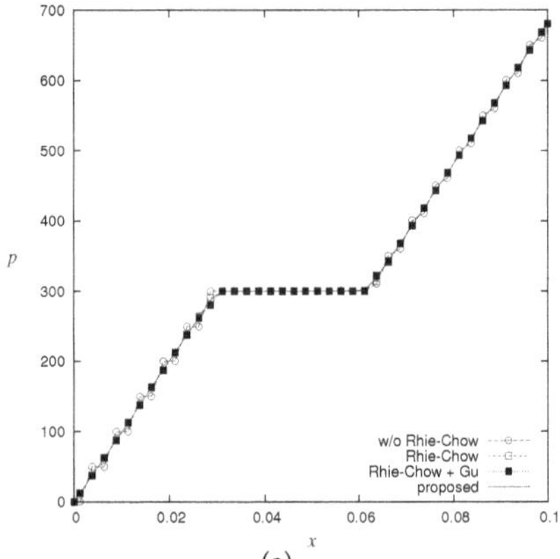

(a)

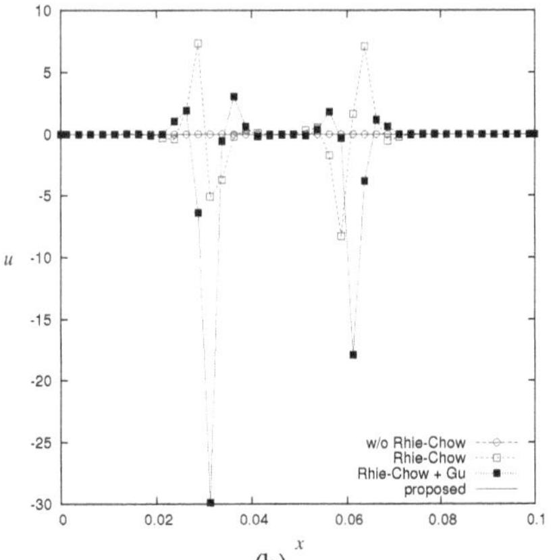

(b)

Fig. 3. The calculated pressure (a) and velocity (b) fields using different interpolations; results are obtained using uniform grid with 40 CVs. Values inside CVs are shown.

2.3 The remedy in 1D

The velocity field in Fig. 3(b) is represented by the values of u inside CVs; the values on CV boundaries are, on the other hand, equal to zero (numerically) as required by (14). To obtain the zero velocity field also inside CVs

$$\frac{p_{\mathsf{E}} - p_{\mathsf{W}}}{2\Delta x} = f_{\mathsf{P}}^{x} \tag{30}$$

must hold. This follows from (9) by setting the velocities to zero. It simply means that the discretized body force and the pressure gradient should be counterbalanced to obtain zero velocity inside FV. At the same time, when both the Rhie-Chow and Gu's scheme are employed, the pressure field is obtained from

$$\frac{V_{\mathsf{e}}}{(a_{\mathsf{P}}^{\alpha})_{\mathsf{e}}} \frac{p_{\mathsf{E}} - p_{\mathsf{P}}}{\Delta x} - \frac{V_{\mathsf{w}}}{(a_{\mathsf{P}}^{\alpha})_{\mathsf{w}}} \frac{p_{\mathsf{P}} - p_{\mathsf{W}}}{\Delta x} = \frac{V_{\mathsf{e}}}{(a_{\mathsf{P}}^{\alpha})_{\mathsf{e}}} f_{\mathsf{e}}^{x} - \frac{V_{\mathsf{w}}}{(a_{\mathsf{P}}^{\alpha})_{\mathsf{w}}} f_{\mathsf{w}}^{x} \tag{31}$$

which follows from (19) for quiescent fluid. The solution of the above equation can be constructed in a rather simple manner: by setting p in a selected starting cell to an arbitrary value and then using the relations, following obviously from (31),

$$p_{\mathsf{E}} = p_{\mathsf{P}} + f_{\mathsf{e}}^{x} \Delta x \qquad \text{and} \qquad p_{\mathsf{W}} = p_{\mathsf{P}} - f_{\mathsf{w}}^{x} \Delta x \tag{32}$$

to calculate the values in the neighboring cells. Inserting (32) into (30) results in

$$f_{\mathsf{P}}^{x} = \frac{1}{2} f_{\mathsf{e}}^{x} + \frac{1}{2} f_{\mathsf{w}}^{x} \tag{33}$$

which is the condition to obtain zero velocity field inside FV when the pressure field is calculated from (31). However, the above equation could also be interpreted as *a rule how to discretize the body force field.* It instructs to construct the body force field inside CVs as a linear combination of the corresponding average values on CV boundaries. The rule eliminates completely (i.e. to round-off error) the unwanted spikes as demonstrated in Fig. 3(b).

The proposed body force discretization rule (33) is obtained for a uniform 1D grid. For a nonuniform grid it becomes

$$f_P^x = \frac{I_e \Delta x_e}{V_P} f_e^x + \frac{I_w \Delta x_w}{V_P} f_w^x \tag{34}$$

where $I_e = (x_e - x_P)/\Delta x_e$ and $I_e = (x_P - x_w)/\Delta x_e$ denote the interpolation factors (with reference to Fig. 2). The above equation is obtained by using

$$\left(\frac{\partial p}{\partial x}\right)_P = \frac{p_e - p_w}{V_P} = \frac{I_e p_E + (1 - I_e) p_P - I_w p_W - (1 - I_w) p_P}{V_P} \tag{35}$$

which follows from the Gauss's theorem (39).

2.4 The generalization of the remedy

Using the same procedure, let us obtain the body force discretization rules for general (i.e. nonorthogonal and/or nonstructured) 2- and 3-dimensional grids. Again, we assume that the solution of the pressure field equation for the quiescent fluid on all CV faces satisfies

$$(\nabla p)_f = \vec{f}_f \tag{36}$$

when both Rhie-Chow and Gu's corrections are used. The dot product of (36) with vector $\vec{\Delta}_f$ results in

$$p_{Nb(f)} - p_P = \vec{f}_f \cdot \vec{\Delta}_f \tag{37}$$

where subscript $()_{Nb(f)}$ denotes the value at neighboring cell Nb(f), sharing face f with cell P, and vector $\vec{\Delta}_f$ points from point P to neighboring point Nb(f) as shown in Fig. 4.

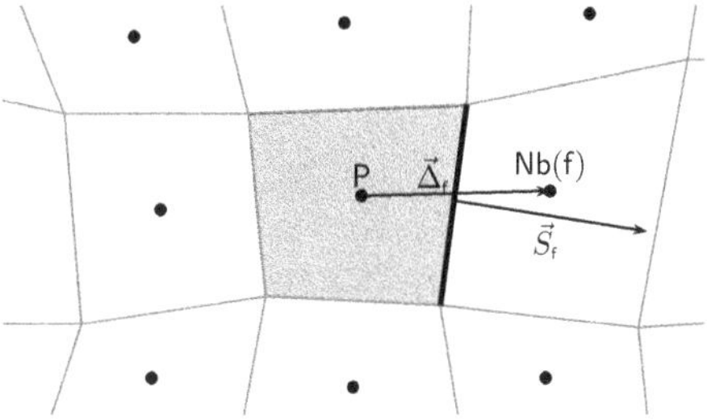

Fig. 4. Schematic example of general two-dimensional FV grid.

When dealing with a 1-dimensional grid, (37) can simply be used to construct the pressure field starting from an arbitrary grid point. On 2- and 3-dimensional grids, however, it has to be noted that condition (36) can be satisfied on all FV faces simultaneously only if the body force field is conservative on the discrete level (Mencinger & Žun, 2007). This is generally not true, and the pressure field can not be constructed simply by using (37). Furthermore, the quiescent solution can not be obtained, i.e. the so-called spurious currents appear. Nevertheless, relation (37) can be considered as a reasonably good approximation and the described methodology still valid.

As the rule is obtained from the requirement for the zero velocity field

$$(\nabla p)_P = \vec{f}_P \tag{38}$$

following from the discretized Navier-Stokes equation, its form depends on the discretization of ∇p used in the equation. For example, $(\nabla p)_{\mathrm{P}}$ can be obtained by employing the Gauss's theorem

$$(\nabla p)_{\mathrm{P}} = \frac{1}{V_{\mathrm{P}}} \sum_{\mathrm{f}} p_{\mathrm{f}} \vec{S}_{\mathrm{f}}. \tag{39}$$

If p_{f} is written in terms of the neighboring cells using the interpolation coefficient I_{f} as

$$p_{\mathrm{f}} = I_{\mathrm{f}} p_{\mathrm{Nb(f)}} + (1 - I_{\mathrm{f}}) p_{\mathrm{P}}, \tag{40}$$

then it can be written using (36) as

$$p_{\mathrm{f}} = p_{\mathrm{P}} + I_{\mathrm{f}} \vec{f}_{\mathrm{f}} \cdot \vec{\Delta}_{\mathrm{f}}. \tag{41}$$

Inserting (41) and (39) in (38) results in

$$\vec{f}_{\mathrm{P}} = \frac{1}{V_{\mathrm{P}}} \sum_{\mathrm{f}} \left[p_{\mathrm{P}} + I_{\mathrm{f}} \vec{f}_{\mathrm{f}} \cdot \vec{\Delta}_{\mathrm{f}} \right] \vec{S}_{\mathrm{f}} = \frac{p_{\mathrm{P}}}{V_{\mathrm{P}}} \sum_{\mathrm{f}} \vec{S}_{\mathrm{f}} + \frac{1}{V_{\mathrm{P}}} \sum_{\mathrm{f}} I_{\mathrm{f}} \left(\vec{f}_{\mathrm{f}} \cdot \vec{\Delta}_{\mathrm{f}} \right) \vec{S}_{\mathrm{f}} \tag{42}$$

where the first summation term on the right hand side equals zero as $\sum_{\mathrm{f}} \vec{S}_{\mathrm{f}} = 0$ must hold for any closed surface. Finally,

$$\vec{f}_{\mathrm{P}} = \frac{1}{V_{\mathrm{P}}} \sum_{\mathrm{f}} I_{\mathrm{f}} \left(\vec{f}_{\mathrm{f}} \cdot \vec{\Delta}_{\mathrm{f}} \right) \vec{S}_{\mathrm{f}} \tag{43}$$

is obtained.

The rule can be generalized even further; if $(\nabla p)_{\mathrm{P}}$ is written in terms of p_{P} and the values of p in the neighboring cells as

$$(\nabla p)_{\mathrm{P}} = \vec{\gamma}_{\mathrm{P}} p_{\mathrm{P}} + \sum_{\mathrm{Nb(f)}} \vec{\gamma}_{\mathrm{Nb(f)}} p_{\mathrm{Nb(f)}} \tag{44}$$

where $\vec{\gamma}_{\mathrm{P}}$ and $\vec{\gamma}_{\mathrm{Nb}}$ are geometrical vector coefficients. The latter can be obtained, for example, with the least squares method. Inserting (44) and (37) in (38) now results in

$$\vec{f}_{\mathrm{P}} = \vec{\gamma}_{\mathrm{P}} p_{\mathrm{P}} + \sum_{\mathrm{Nb(f)}} \vec{\gamma}_{\mathrm{Nb(f)}} \left(p_{\mathrm{P}} + \vec{f}_{\mathrm{f}} \cdot \vec{\Delta}_{\mathrm{f}} \right) = \left[\vec{\gamma}_{\mathrm{P}} + \sum_{\mathrm{Nb(f)}} \vec{\gamma}_{\mathrm{Nb(f)}} \right] p_{\mathrm{P}} + \sum_{\mathrm{Nb(f)}} \vec{\gamma}_{\mathrm{Nb(f)}} \left(\vec{f}_{\mathrm{f}} \cdot \vec{\Delta}_{\mathrm{f}} \right). \tag{45}$$

The term in square brackets in (45) should be zero from the definition (44): the effect of adding a constant field p_0 to p should vanish. Therefore, (45) simplifies to

$$\vec{f}_{\mathrm{P}} = \sum_{\mathrm{Nb(f)}} \vec{\gamma}_{\mathrm{Nb(f)}} \left(\vec{f}_{\mathrm{f}} \cdot \vec{\Delta}_{\mathrm{f}} \right) \tag{46}$$

and presents the discretization rule when (44) is used to obtain $(\nabla p)_{\mathrm{P}}$ in the discretized Navier-Stokes equation. As rule (43), it contains the surface averaged body forces and the geometrical factors.

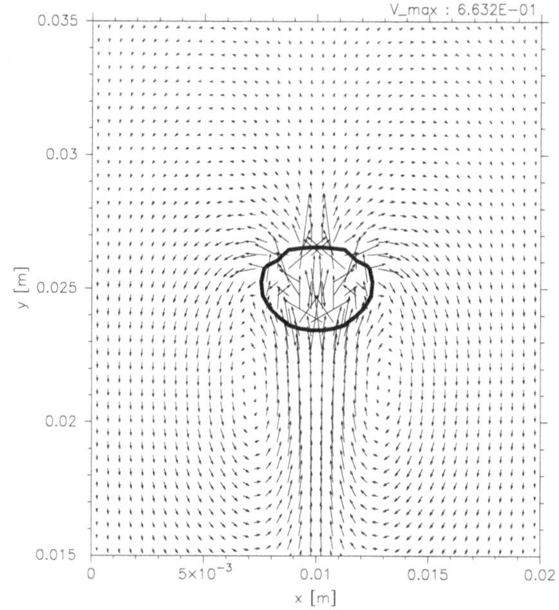

Fig. 5. The velocity field near the rising bubble indicated by the contour; the same calculation as in Fig.1 but with the alternative body force discretization.

3. Testing the alternative discretization

The proposed alternative body force discretization is obtained considering a quiescent fluid which is seldom of research interest. To compare its performance against the standard discretization for a moving fluid we consider two cases: (i) raise of a bubble in a rectangular cavity and (ii) natural convection in a square cavity. The first case demonstrates that the unwanted velocity field spikes are removed or at least largely diminished by using the alternative body force discretization. Whereas an abrupt change of body force field is dealt with in the first case, the body force field is smooth in the second case. Namely, we also want to check the effect of the new discretization in such cases.

3.1 Rising bubble

The case considers a two-dimensional cavity with no-slip boundary condition at all walls. The cavity is rectangular (width w, height h). Initially, it is filled with water and an air bubble of radius R, centered at (x_0, y_0); both the water and the air are quiescent. To follow the rise of the bubble using the VOF model, (26) needs to be solved besides the Navier-Stokes equation. In the latter, both the buoyancy (27) and the surface force (28) are taken into account. Whereas standard central differencing is used to calculate advection and diffusion terms in the momentum equation, the discretization of (26) requires nonstandard differencing scheme in order to reduce numerical diffusion which results in undesirable interface smearing. The low-diffusive CICSAM scheme (Ubbink & Issa, 1999) is used in the presented case. The curvature κ, needed to obtain the forces due to the surface tension by using the CSF model are calculated as suggested by Williams et al. (1998).

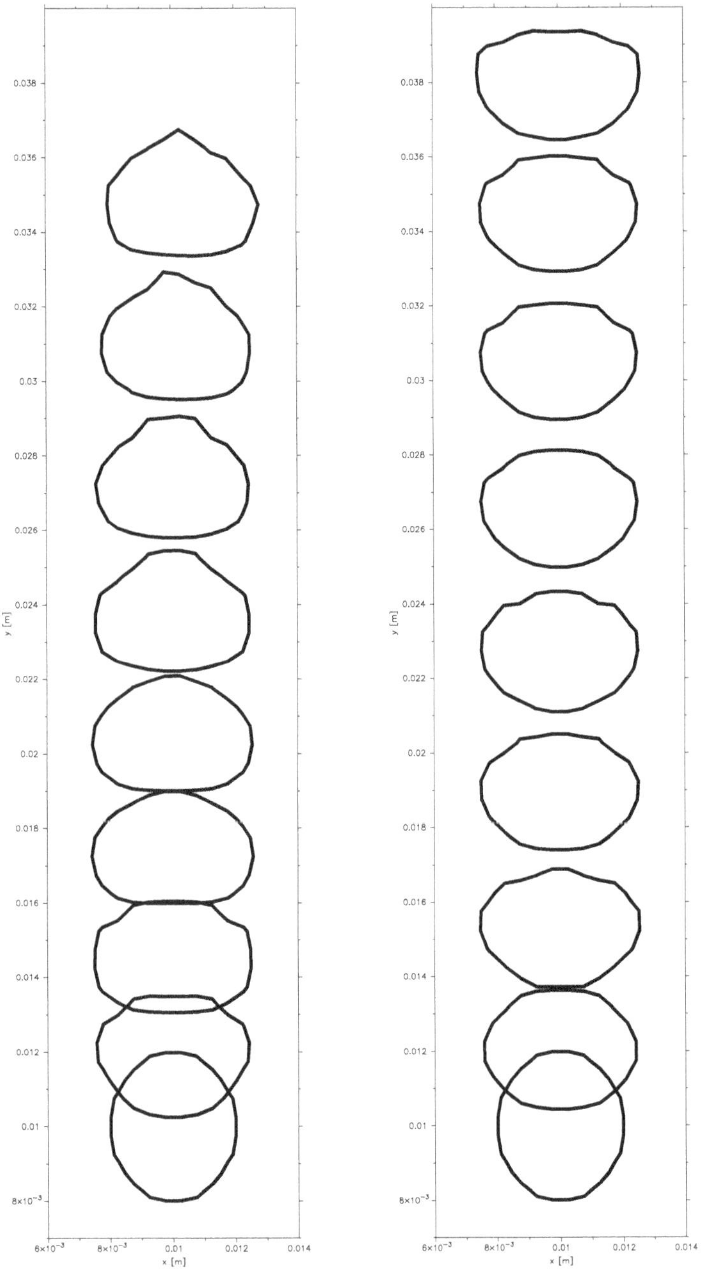

(a) (b)

Fig. 6. Calculated contours of a rising bubble at t=0.0(0.025)0.2 s: standard (a) versus alternative (b) body force field discretization.

In the presented case we set $w = 0.02\,\text{m}$, $h = 0.04\,\text{m}$, $x0 = w/2$, $y0 = h/4$, and $R = 0.002\,\text{m}$. Uniform grid with 40x80 CVs was used. Fig. 5 shows the calculated velocity field around bubble at $t = 0.115\,\text{s}$ when using the alternative discretization; this figure can be compared directly with Fig. 1 which presents the same calculation except for using the standard discretization. Obviously, more realistic results are obtained with the proposed discretization. Interestingly, despite large spikes in the first (i.e. standard) calculation, the bubble contours of the two calculations indicated in Fig. 6 does not differ as much as one would expect. This is perhaps due to the fact that CV face velocities, which appear in the advection terms, are corrected so that they satisfy the continuity equation.

3.2 Natural convection in a square cavity

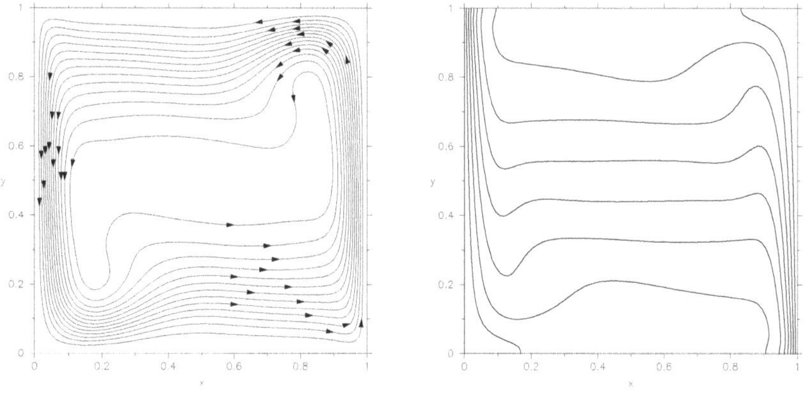

(a) (b)

Fig. 7. Streamlines (a) and isotherms (b) in the stationary state in the natural convection test case with $Pr = 0.71$ and $Ra = 10^6$.

The second case deals with a buoyancy driven flow in a two-dimensional square cavity and presents a classical CFD code benchmark problem (De Vahl Davis, 1983) with well known solutions. The left and the right wall are set to fixed temperatures T_h and T_l ($T_h > T_l$), respectively. Both horizontal walls are thermally insulated. The velocity components vanish at the walls. The velocity field is obtained by solving (3) with Boussinesq approximation

$$\vec{f} = \rho_0 \left(1 - \beta(T - T_0)\right) \vec{g} \tag{47}$$

where β is the thermal expansion coefficient and ρ_0 is the density at reference temperature T_0. The temperature field is determined with the enthalpy transport equation

$$\rho c_p \frac{DT}{Dt} = \nabla \cdot (\lambda \nabla T) \tag{48}$$

where c_p and λ are the thermal conductivity and the specific heat at constant pressure, respectively. Actually, both (3) and (48) are solved in nondimensional form which read

$$\frac{D\vec{v}}{Dt} = -\nabla p + Pr \nabla^2 \vec{v} + Pr\, Ra\, T, \tag{49}$$

$$\frac{DT}{Dt} = \nabla^2 T \tag{50}$$

where, for brevity, the same notation is used for nondimensional and dimensional quantities. Obviously, the problem is determined with the two dimensionless parameters Pr and Ra denoting Prandtl and Rayleigh number, respectively. They are defined as

$$Pr = \frac{c_p \eta}{\lambda}, \qquad Ra = \frac{\beta(T_h - T_l)L^3\rho^2 c_p}{\eta\lambda} \tag{51}$$

where L is the size of the cavity. Following the work of de Vahl Davis, we set $Pr = 0.71$ and consider only the highest value of Ra used in their test: $Ra = 10^6$.

The calculation is performed using relatively coarse uniform grids containing 10x10, 20x20, 40x40, and 80x80 cells. The central discretization scheme is implemented for the interpolation of the values on CV-faces for both (49) and (50). The obtained streamlines and isotherms are shown in Fig. 7.

Fig. 8 shows the variation of the calculated velocity components and the temperature along the horizontal and the vertical centerline of the cavity. Obviously, the difference between the results obtained with the standard (dashed line) and the proposed (solid line) body force discretization vanishes with the increased grid density. The difference on the 80x80 grid is not noticeable on the presented scale and is therefore not drawn in Fig. 8. Both the new and the standard discretization converge to the same solution, thus we can assume that the proposed discretization is *consistent*.

4. Conclusions

Although it was obtained for a quiescent fluid, the proposed discretization works well also for moving fluid as demonstrated in the considered cases. One should keep in mind that the obtained rules are valid when both the Rhie-Chow and Gu's correction are used. Together with the corrections of Choi (1999) and Yu et al. (2002), they approach the calculations on collocated grids to those on staggered grids. It is shown that the proposed discretization of the body force field is more appropriate than the standard one when dealing with abruptly variable body force fields. In fact, it can be used generally as it does not significantly change the calculated solutions when the body force field is smooth.

The proposed discretization does not exactly follow the spirit of the FV method where simply the average body force within a FV is considered. Even so, it is consistent since it converges to the same solution as obtained with the standard discretization. Its form depends on the chosen discretization of ∇p in the Navier-Stokes equation; in the presented calculations, Gauss's divergence theorem was used. Other possibilities such as using the least squares method are also possible, so that the discretization rule changes its form. Nevertheless, the idea remains that FV average value of $\vec{f}$ is to be replaced with the linear combination of the average FV face values.

5. Acknowledgments

The author is grateful to the Slovenian Research Agency (ARRS) for its financial support throughout the P2-0162 project.

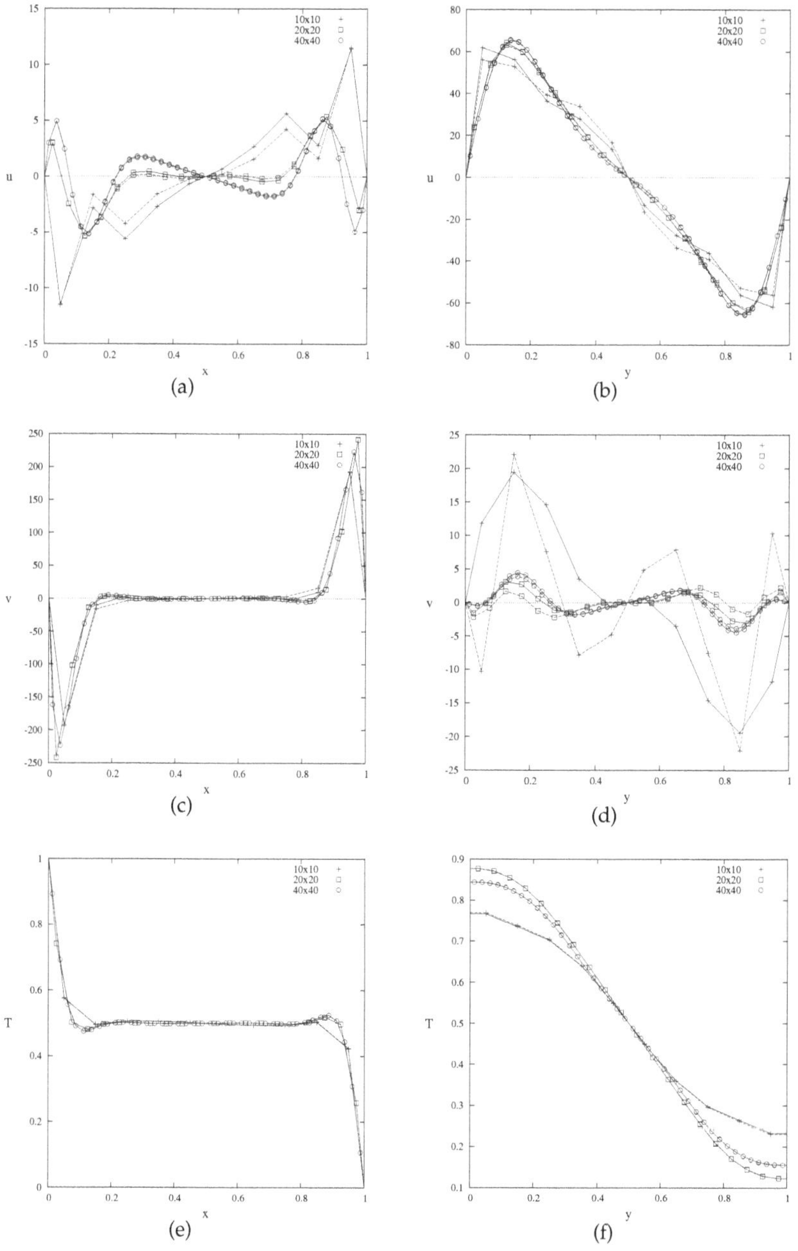

Fig. 8. The variation of u, v and T (first, second and bottom row) along the horizontal (left column) and the vertical (right column) centerline of the cavity, calculated with standard (dashed lines) and proposed (solid lines) discretization of $\vec{f}$.

6. References

Brackbill, J. U., Kothe, D. B. & Zemach, C. (1992). A continuum method for modeling surface tension, *J. Comput. Phys.* 100: 335–354.

Choi, S. K. (1999). Note on the use of momentum interpolation method for unsteady flows, *Numer. Heat Transfer A* 36: 545–550.

De Vahl Davis, G. (1983). Natural convection of air in a square cavity: A bench mark numerical solution, *Int. J. Numer. Meth. Fluids* 3: 249–264.

Ferziger, J. H. & Perić, M. (2002). *Computational Methods for Fluid Dynamics*, Springer.

Gu, C. Y. (1991). Computation of flows with large body forces, *in* C. Taylor & J. H. Chin (eds), *Numerical Methods in Laminar and Turbulent Flow*, Pineridge Press, Swansea, pp. 294–305.

Harlow, F. H. & Welch, J. E. (1965). Numerical calculation of time-dependent viscous flow of fluid with free surface, *The Physics of Fluids* 8: 2182.

Hirt, C. & Nichols, B. (1981). Volume of fluid (VOF) method for the dynamics of free boundaries, *J. Comput. Phys.* 39: 201–225.

Landau, L. D. & Lifshitz, E. M. (1987). *Fluid Mechanics, Second Edition*, Pergamon Press.

Majumdar, S. (1988). Role of underrelaxation in momentum interpolation for calculation of flow with nonstaggered grids, *Numer. Heat Transfer* 13: 125–132.

Mencinger, J. & Žun, I. (2007). On the finite volume discretization of discontinuous body force field on collocated grid: Application to VOF method, *J. Comput. Phys.* 221: 524–538.

Pascau, A. (2011). Cell face velocity alternatives in a structured colocated grid for the unsteady Navier-Stokes equations, *Int. J. Numer. Meth. Fluids* 65: 812–833.

Perić, M., Kessler, R. & Scheuerer, G. (1988). Comparison of finite-volume numerical methods with staggered and colocated grids, *Computers & Fluids* 16: 389–403.

Rhie, C. M. & Chow, W. L. (1983). Numerical study of the turbulent flow past an airfoil with trailing edge separation, *AIAA Journal* 21: 1525–1532.

Scardovelli, R. & Zaleski, S. (1999). Direct numerical simulation of free-surface and interfacial flow, *Annu. Rev. Fluid Mech.* 31: 567–603.

Ubbink, O. & Issa, R. I. (1999). A method for capturing sharp fluid interfaces on arbitrary meshes, *J. Comput. Phys.* 153: 26–50.

Williams, M. W., Kothe, D. B. & Puckett, E. G. (1998). Accuracy and convergence of continuum surface-tension models, *in* W. Shyy & R. Narayanan (eds), *Fluid Dynamics at Interfaces*, Cambridge University Press, Cambridge, pp. 294–305.

Yu, B., Kawaguchi, Y., Tao, W.-Q. & Ozoe, H. (2002). Checkerboard pressure predictions due to the underrelaxation factor and time step size for a nonstaggered grid with momentum interpolation method, *Numer. Heat Transfer B* 41: 85–94.

7

The Finite Volume Method in Computational Rheology

A.M. Afonso[1], M.S.N. Oliveira[1], P.J. Oliveira[2], M.A. Alves[1] and F.T. Pinho[1]
[1]*Transport Phenomena Research Centre, Faculty of Engineering*
University of Porto, Porto
[2]*Department of Electromechanical Engineering, Textile and Paper Materials Unit*
University of Beira Interior, Covilhã
Portugal

1. Introduction

The finite volume method (FVM) is widely used in traditional computational fluid dynamics (CFD), and many commercial CFD codes are based on this technique which is typically less demanding in computational resources than finite element methods (FEM). However, for historical reasons, a large number of Computational Rheology codes are based on FEM.

There is no clear reason why the FVM should not be as successful as finite element based techniques in Computational Rheology and its applications, such as polymer processing or, more recently, microfluidic systems using complex fluids. This chapter describes the major advances on this topic since its inception in the early 1990's, and is organized as follows. In the next section, a review of the major contributions to computational rheology using finite volume techniques is carried out, followed by a detailed explanation of the methodology developed by the authors. This section includes recent developments and methodologies related to the description of the viscoelastic constitutive equations used to alleviate the high-Weissenberg number problem, such as the log-conformation formulation and the recent kernel-conformation technique. At the end, results of numerical calculations are presented for the well-known benchmark flow in a 4:1 planar contraction to ascertain the quality of the predictions by this method.

2. Main contributions

The first contributions to computational rheology in the late nineteen sixties were based on finite difference methods (FDM, Perera and Walters, 1977). In the first major book on computational rheology (Crochet et al, 1984) works using FEM predominate, but the number of contributions using FDM was also significant.

Among the first numerical works to make use of FVM to investigate viscoelastic fluid flows was the study of the benchmark flow around a confined cylinder of Hu and Joseph (1990), who used the simplest differential constitutive equation embodying elastic effects, the upper-convected Maxwell (UCM) model. Velocities were calculated in cylindrical/orthogonal grids staggered relative to the basic mesh for pressure and stresses. The SIMPLER algorithm (Patankar, 1980) was adapted and extended to the calculation of stress tensor components. The inertial terms in the momentum equation were neglected in these low Reynolds number simulations conducted on rather coarse meshes, and convergence was obtained up to Weissenberg numbers of 10.

For creeping flow, the advective terms in the momentum equation can be discarded but the same does not hold for the advective terms in the constitutive equation, which typically originate convergence and accuracy problems. The development of stable and accurate schemes to deal with advection-dominated equations is a fundamental issue which was not addressed in the initial works using FVM. For example, in their sudden contraction calculations Yoo and Na (1991) kept all advective terms, but considered only first order discretization schemes, which are known from classical CFD (Leschziner, 1980) to introduce excessive numerical diffusion, especially when the flow is not aligned with the computational grid (Patankar, 1980).

Staggered meshes, in which different variables are evaluated in different points of the computational mesh (some at the cell centers, others at the cell faces), were used by Yoo and Na (1990), as well as in subsequent works (eg. Gervang and Larsen, 1991; Sasmal, 1995; Xue et al, 1995, 1998 a,b; Mompeam and Deville, 1997; Bevis et al, 1992). Staggered meshes provide an easy way to couple velocities, pressure and stresses, but calculations involving complex geometries become rather difficult and in some cases do not allow for the determination of the shear stress at singular points, such as re-entrant corners. Alternatively, the use of non-orthogonal, or even non-structured meshes, are to be preferred in such cases.

Non-orthogonal meshes have been used in FVM for Newtonian fluids since the mid-nineteen eighties, but its application to finite volume viscoelastic methods happened only in 1995. Initially, the adaptation to computational rheology of some of the techniques previously developed for Newtonian fluids to has been slow, namely on issues like pressure-velocity coupling for collocated meshes (in which all variables are evaluated at the cell centres), time marching algorithms or the use of non-orthogonal meshes. Lately, progress has been quicker on issues of stability for convection dominated flows, as in viscoelastic flows at high Weissenberg numbers and in high speed flows of inviscid Newtonian fluids involving shock waves (Morton and Paisley, 1989; Mackenzie et al, 1993).

Regarding other mesh arrangements, Huang et al (1996) used non-structured methods in a mixed finite element/finite volume formulation by extending the control volume finite element method (CVFEM) of Baliga and Patankar (1983) for the prediction of the journal bearing flow of Phan-Thien and Tanner (PTT) fluids. Nevertheless, the formulation lacked the generality of modern methods in Newtonian fluid calculations on collocated grids (Ferziger and Perić, 2002) and was problematic to extend to higher-order shape functions (usually the convective terms are discretized with some form of upwind). Later, Oliveira et al (1998) developed a general method for solving the full momentum and constitutive equations on collocated non-orthogonal meshes, enabling calculations of complex three-dimensional flows. Their scheme for coupling velocity, pressure and stresses was later improved by Oliveira and Pinho (1999a) and Matos et al (2009). This issue was also addressed in a parallel effort by Missirlis et al (1998), but only for staggered, orthogonal meshes.

As mentioned above, there are hybrid methods, aimed at combining the advantage of finite elements in representing complex geometries and the advantage of finite volumes to ensure conservation of physical quantities; they follow the CVFEM ideas initially proposed by Baliga and Patankar (1983). Within the scope of computational rheology, hybrid methods have been developed especially by Webster and co-workers (Aboubacar and Webster 2001, Aboubacar et al 2002, Wapperom and Webster 1998, 1999), and Sato and Richardson (1994) within the finite-element methodology; and by Phan-Thien and Dou (1999) and Dou and Phan-Thien (1999), within the CVFEM formulation referred to above.

Stability, convergence and accuracy are intimately related, but the early efforts were more concerned with stability and convergence, due to the mixed elliptic/hyperbolic nature of the motion and constitutive equations, than with accuracy. Thus, early developments on the algorithmic side were usually based on first-order discretization methods, such as the classical upwind differencing scheme, leading to lower accuracy (Ferziger and Perić, 2002). Due to computer limitations early works also used rather coarse meshes, but the topic of accuracy started to gain momentum by the mid-nineties, and Sato and Richardson (1994) were among the first to show this concern. Although their approach can be classified as FEM, the constitutive equations were integrated over finite volumes with the advective flux stress terms discretized and stabilized by means of a bounded scheme obeying total variation diminishing (TVD) criteria.

For the pure FVM in computational rheology, there has been a significant effort at developing accurate and stable methods by the authors of this chapter: Oliveira and Pinho (1999b), Alves et al (2000, 2001a, 2003a, 2003b) and Afonso et al (2009, 2012). Oliveira and Pinho (1999b) used second-order interpolation schemes for the advective stress fluxes (either a linear upwind scheme or central differences), but difficulties associated with the intrinsic unboundedness of those schemes led them to the implementation of so-called high-resolution methods, often used in high-speed aerodynamics. These represent important landmark developments, where there was a remarkable improvement both in terms of stability and accuracy (Alves et al, 2000). In fact, high-resolution methods led to solutions having similar accuracy as those obtained with the most advanced FEM (Alves et al, 2001a, 2003b), and also to comparable levels of convergence (measured by the maximum Weissenberg (*Wi*) or Deborah (*De*) numbers above which the methods diverged). For reasons discussed in Fan et al (1999), the lower *De* results showed less discrepancies and FVM could achieve the same accuracy as FEM. Comparisons for the flow in a 4:1 sudden planar contraction are also available in Alves et al (2003b), where the CUBISTA high-resolution scheme especially designed for the treatment of advection in viscoelastic flows is employed (Alves et al, 2003a). Some of the difficulties in iterative convergence of viscoelastic flow calculations of the mid-2000's were solved by such high-resolution schemes for interpolating convective terms in the stress equation as the CUBISTA scheme, which obeys total variation diminishing criteria. These are more restrictive than convection boundedness criteria and the universal limiter of Leonard (1991), as was demonstrated by Alves et al (2003a).

Subsequently, a very relevant development in computational rheology overcame, or at least significantly mitigated, the so-called High-Weissenberg Number Problem, in which calculations breakdown at some critical problem-dependent Weissenberg numbers. In 2004, Fattal and Kupferman proposed a reformulation of the viscoelastic constitutive equations in terms of the matrix logarithm of the conformation tensor to alleviate this problem (Fattal and Kupferman, 2004). This technique, now known as the log-conformation, has been implemented within the framework of FEM (eg. Hulsen et al, 2005) and more recently in the framework of FVM (Afonso et al, 2009, 2011), who maintained the use of the CUBISTA scheme to describe the advection of log-conformation terms for improved accuracy. This technique has been applied to various different flows, including the flow around a cylinder, in which *Wi* on the order of 10 were achieved for the Oldroyd-B model in comparison with previous $Wi \approx 1.2$ attained with the standard version. This approach has been generalized by Afonso et al (2012) considering different functions for the transformation of the tensor evolution equation. This technique, known as kernel-conformation, encompasses the log-conformation approach and assumes particular importance as new phenomena are observed in viscoelastic fluid flows in the context of microfluidics, where elastic effects are enhanced and inertia effects reduced as compared to classical macro-scale fluid flows.

Today it is an undisputable fact that FVM are mature in computational rheology, as indicated by a wide range of computations exhibiting similar or even better performance in terms of accuracy and robustness as other methods (Owens and Phillips, 2002) and presumably at a lower cost, especially in light of the recent developments allowing computations at high *Wi* number.

3. The finite-volume method applied to viscoelastic fluids using collocated meshes

3.1 General methodology

The general finite-volume methodology here described for viscoelastic flow computations is closely patterned along the lines of that previously presented in Oliveira (1992). Numerical calculation of any flow requires solution of two governing equations, for mass conservation and momentum. For a non-Newtonian fluid an additional rheological equation of state is needed. To calculate the pressure it is necessary to solve a thermodynamic equation of state, but since here we are considering incompressible fluid flows only, such equation is used to calculate the fluid density and becomes decoupled from the above mentioned governing equations. Then the flow becomes independent of absolute pressure, and the pressure variations are determined indirectly from the mass conservation equation as discussed later. If temperature variations are important, the energy equation needs also to be considered.

In FVM, described in detail in several textbooks (eg. Patankar, 1980; Ferziger and Perić, 2002; Versteeg and Malalasekera, 2007), the computational domain is divided into contiguous computational cells and within each of these the differential governing equations are volume integrated. Gauss theorem is then invoked to transform the divergence of quantities into surface integral of fluxes in order to guarantee the conservation of the quantities. Next, these surface integrals are represented by summation of fluxes whereas the non-transformed volume integrals are approximated by products of an average value of the integrand and the volume of the computational cells. Finally, the fluxes at the cell faces must be equated as a function of the unknown quantities at the neighbour cell centers. This is achieved differently depending on whether staggered or collocated meshes are used. For the former see Patankar (1980) and for the latter details are given in Ferziger and Perić (2002). The present chapter deals with collocated meshes only.

3.2 Coordinate system

The equations to be solved are written for non-orthogonal coordinate systems aligned with the computational grid for generality in the treatment of complex geometries. This can also be achieved with non-structured meshes (for viscoelastic fluids, see e.g. Huang et al, 1996), but our developments are based on block-structured grids. The equations must obey general principles of invariance, but their discretization in a global mesh composed of six-faced computational cells requires their previous transformation to a non-orthogonal coordinate system (ξ_1, ξ_2, ξ_3), as in Figure 1. It is important to notice that only the coordinates are represented in the non-orthogonal system, whereas velocity and stress components are referred to the original Cartesian system. This means that in the transformation of the conservation equations only the derivatives need to be converted.

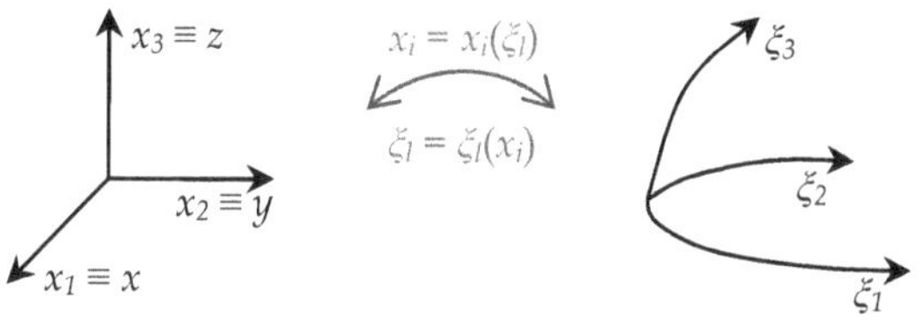

Fig. 1. Schematic representation of the transformation of Cartesian rectangular coordinates to a non-orthogonal system defined by the local orientation of the computational grid.

From the numerical point of view it is advantageous to write the resulting equations in their strong conservative form to help conserve the physical quantities in the final algebraic equations. This is indeed one of the main advantages of FVM: it is essential to maintain conservation of quantities that physically should be conserved, such as mass. The well known transformation rules (see Vinokur, 1989) are given by:

$$\frac{\partial}{\partial t} = \frac{1}{J}\frac{\partial}{\partial t}J$$

$$\frac{\partial}{\partial x_i} = \frac{\partial}{\partial \xi_l}\frac{\partial \xi_l}{\partial x_i} = \frac{1}{J}\frac{\partial}{\partial \xi_l}\beta_{li} = \frac{\beta_{li}}{J}\frac{\partial}{\partial \xi_l} \tag{1}$$

where J is the Jacobian of the transformation $x_i = x_i(\xi_l)$, i.e. $J = \det(\partial x_i / \partial \xi_l)$ and β_{li} are the metric coefficients defined as the cofactor of terms $\partial x_i / \partial \xi_l$ in the Jacobian. These equations are written in terms of indicial notation and the summation convention for repeated indices applies.

3.3 Governing equations

The continuity equation for incompressible fluid flow is

$$\frac{\partial(\rho u_i)}{\partial x_i} = 0 \tag{2}$$

where u_i represents the velocity vector in the Cartesian system and ρ is the density of the fluid, which is retained in Eq. (2) for later convenience. The momentum equation for a generic fluid is given by

$$\frac{\partial \rho u_i}{\partial t}+\frac{\partial \rho u_j u_i}{\partial x_j}=-\frac{\partial p}{\partial x_i}+\rho g_i+\frac{\partial}{\partial x_j}\left(\eta_S\left(\frac{\partial u_i}{\partial x_j}+\frac{\partial u_j}{\partial x_i}\right)\right)+\frac{\partial \tau_{ij}}{\partial x_j} \tag{3}$$

where t represents the time, p the pressure, g_i is the acceleration of gravity, η_S is the solvent viscosity and τ_{ik} is the symmetric extra-stress tensor, which is described by an appropriate rheological constitutive equation. To describe the numerical method, we will adopt the PTT model, which is adequate to explain the variations relative to the method used for Newtonian fluids. Whenever needed the use of a different model will be conveniently mentioned. The extra-stress of the PTT model is given as function of the conformation tensor A_{ij} as

$$\tau_{ij}=\frac{\eta_P}{\lambda}\left(A_{ij}-\delta_{ij}\right) \tag{4}$$

where η_P is the polymer viscosity parameter, λ is the relaxation time and δ_{ij} is the unitary tensor. The conformation tensor is then described by an evolution equation, which for the PTT fluid takes the form

$$\lambda\left(\frac{\partial A_{ij}}{\partial t}+\frac{\partial u_k}{\partial x_k}A_{ij}-\frac{\partial\left(u_i A_{kj}\right)}{\partial x_k}-A_{ik}\frac{\partial u_j}{\partial x_k}\right)=-\mathrm{Y}\left[A_{kk}\right]\left(A_{ij}-\delta_{ij}\right) \tag{5}$$

In its general form function $\mathrm{Y}[A_{kk}]$ for the PTT model is exponential, $Y[A_{kk}]=\exp[\varepsilon(A_{kk}-3)$ (Phan-Thien, 1978), but in this work we will mostly use its linear form, $Y[A_{kk}]=1+\varepsilon(A_{kk}-3)$ (Phan-Thien and Tanner, 1977). When $Y\left[A_{kk}\right]=1$ (i.e. for $\varepsilon=0$) the Oldroyd-B model is recovered. Additionally, if in the momentum equation we set $\eta_S=0$ then the UCM model is obtained. The non-unitary form of $\mathrm{Y}[A_{kk}]$ for the PTT model imparts shear-thinning behaviour to the shear viscosity of the fluid and bounds its steady-state extensional viscosity.

The tensor A_{ij} is a variance–covariance, symmetric positive definite tensor, therefore it can always be diagonalized as $A_{ij}=O_{ik}L_{kl}(O^{T})_{lj}$, where O_{ij} is an orthogonal matrix generated with the eigenvectors of matrix A_{ij} and L_{ij} is a diagonal matrix created with the corresponding three distinct eigenvalues of A_{ij}. This fact provides the possibility of using the log-conformation technique, introduced by Fattal and Kupferman (2004), which has been shown to lead to a significant increase of numerical stability. In this technique a simple tensor-logarithmic transformation is performed on the conformation tensor for differential viscoelastic constitutive equations. This technique can be applied to a wide variety of constitutive laws and in the log-conformation representation the evolution Eq. (5) is replaced by an equivalent evolution equation for the log-conformation tensor, $\mathbf{\Theta}=\log(\mathbf{A})$. The transformation from Eq. (5) to an equation for Θ_{ij} is described by Fattal and Kupferman (2004), and leads to

$$\frac{\partial \Theta_{ij}}{\partial t}+u_k\frac{\partial \Theta_{ij}}{\partial x_k}-\left(R_{ik}\Theta_{kj}-\Theta_{ik}R_{kj}\right)-2E_{ij}=\frac{\mathrm{Y}\left[(\mathrm{e}^{\Theta_{ij}})_{kk}\right]}{\lambda}\left(\mathrm{e}^{-\Theta_{ij}}-\delta_{ij}\right) \tag{6}$$

In Eq. (6) R_{ij} and E_{ij} are a pure rotational tensor and a traceless extensional tensor, respectively, which combine to form the velocity gradient tensor. To recover A_{ij} from Θ_{ij} the inverse transformation, $\mathbf{A}=e^{\Theta}$, is used when necessary. The log-conformation approach is a relevant particular case of the recently proposed general kernel-conformation tensor transformation (Afonso et al 2012), in which several matrix transformations can be applied to the conformation tensor evolution equation.

After application of the transformation rules introduced above, the conservation equations of mass and momentum become (Oliveira, 1992):

$$\frac{\partial(\rho\beta_{li}u_i)}{\partial\xi_l}=0 \tag{7}$$

$$\begin{aligned}&\frac{\partial(J\rho u_i)}{\partial t}+\frac{\partial(\rho\beta_{lj}u_ju_i)}{\partial\xi_l}-\frac{\partial}{\partial\xi_l}\left(\eta\frac{\beta_{l'j}\beta_{l'j}}{J}\frac{\partial u_i}{\partial\xi_l}\right)=\\&-\beta_{li}\frac{\partial p}{\partial\xi_l}+\frac{\partial}{\partial\xi_l}(\beta_{lj}\tau_{ij})+J\rho g_i+\frac{\partial}{\partial\xi_l}\left(\frac{\eta_S\beta_{lj}}{J}\left(\beta_{mj}\frac{\partial u_i}{\partial\xi_m}+\beta_{mi}\frac{\partial u_j}{\partial\xi_m}\right)\right)-\frac{\partial}{\partial\xi_l}\left(\eta\frac{\beta_{l'j}\beta_{l'j}}{J}\frac{\partial u_i}{\partial\xi_l}\right)\end{aligned} \tag{8}$$

with $l' = l$, and no summation over index l'. Note that although the diffusive term of the momentum equation (the term proportional to η_S) involves only normal second derivatives, its transformation to the non-orthogonal system originates mixed second-order derivatives. The artificial diffusion term added in both sides of Eq. (8) has a viscosity coefficient $\eta=\eta_S+\eta_P$ and is especially necessary when $\eta_S=0$.

The rheological constitutive equation becomes

$$\lambda\frac{\partial(J\Theta_{ij})}{\partial t}+\lambda\frac{\partial}{\partial\xi_l}(\beta_{lk}u_k\Theta_{ij})=\lambda J(R_{ik}\Theta_{kj}-\Theta_{ik}R_{kj})+2\lambda JE_{ij}+\Upsilon(A_{kk})J\left(e^{-\Theta_{ij}}-\delta_{ij}\right) \tag{9}$$

together with Eq. (4) and the inverse transformation $\mathbf{A}=e^{\Theta}$.

3.4 Discretization of the equations

The objective of the discretization is to obtain a set of algebraic equations relating centre-of-cell values of the unknown variables to their values at nearby cells. These equations are linearized and the large sets of linear equations are solved sequentially for each variable using well-established iterative methods.

The integration of the governing equations in generalized coordinates is straightforward after an acquaintance with the nomenclature, which is summarized in Figure 2. In the discretization, the usual approximations regarding average unknowns at cell-faces and control volumes apply (for details, see Ferziger and Perić, 2002). For the discretization of the equations in the generalized coordinate system, it suffices to replace the coefficients β_{li} by area components of the surface along direction l, denoted B_{li}, the Jacobian J by the cell volume V, and the derivatives $\partial/\partial\xi_l$ by differences between values along direction l,

$$\frac{\partial\Phi}{\partial\xi_l}=[\Delta\Phi]_l\equiv\Phi_{l^+}-\Phi_{l^-} \tag{10}$$

These differences and the area components can be evaluated at two different locations:

1. at cell centres, here denoted with the superscript P,

$$[\Delta\Phi]_f^{\mathrm{P}} \equiv \Phi_{\mathrm{f}^+} - \Phi_{\mathrm{f}^-} \quad \text{and} \quad B_{fi}^{\mathrm{P}} \tag{11}$$

2. at cell-faces, with superscript f,

$$[\Delta\Phi]_f^{\mathrm{f}} \equiv \Phi_{\mathrm{F}} - \Phi_{\mathrm{P}} \quad \text{and} \quad B_{fi}^{\mathrm{f}} \tag{12}$$

In this notation, index F denotes the centre of the neighbour cell to the generic cell P sharing the same face f (see Figure 2), therefore these two indices, f and F are associated; double characters (FF and ff) refer to the second neighbour and cell face, respectively, along the same direction.

In the discretized equations, variables at a general cell P and at its six neighbours (F = 1 to 6, for W, E, S, N, B and T with compass notation: west, east, south, north, bottom and top, i.e. for $l = \mp 1,\ \mp 2$ and ∓ 3, respectively) are treated implicitly and form the main stencil in the discretization. The six far-away neighbours (FF=1 to 6, for WW, EE, SS, NN, BB and TT) appearing in high order schemes, give rise to contributions which are incorporated into the so-called source term and are treated explicitly being evaluated from known values from the previous iteration/time-step. Thus, the linearized sets of equations for each dependent variable, which need to be solved at every time step, have a well defined block-structured matrix with 7 non-zero diagonals. This is one important difference with the finite-element method, which gives rise to banded matrices with no particular structure inside the band.

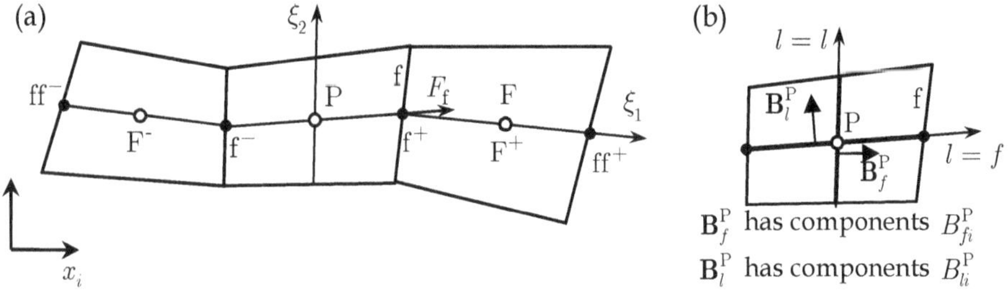

Fig. 2. Nomenclature: (a) general and neighbouring cells; (b) area vectors and components.

3.4.1 Continuity equation

The continuity equation is volume integrated and discretized as follows (sums are explicitly indicated in the discretized equations):

$$\int_{V_{\mathrm{P}}} \frac{\partial}{\partial \xi_l}\left(\rho \beta_{lj} u_j\right) \mathrm{d}V = 0 \rightarrow \sum_{l=1}^{3}\left[\Delta\left(\sum_j \rho B_{lj} u_j\right)\right]_l^{\mathrm{P}} = \sum_{\mathrm{f}=1}^{6}\left(\sum_j \rho B_{fj}^{\mathrm{f}} \tilde{u}_{j,\mathrm{f}}\right) = \sum_{\mathrm{f}=1}^{6} F_{\mathrm{f}} = 0 \tag{13}$$

In this equation, the sum of differences centred at cell centre P has been transformed into a sum of contributions arising from the six cell faces, f. The tilde in $\tilde{u}_{j,\mathrm{f}}$, referring to the cell face velocity component u_j, means that this cannot be computed from simple linear interpolation, in which case no special symbol would be required according to our nomenclature, but need to be evaluated via a kind of Rhie and Chow (1983) interpolation technique, to be explained in Section 3.6. It is this special interpolation that ensures coupling between the pressure and velocity fields in a collocated mesh arrangement. Considering the definition of outgoing mass flow rates ($F_{\rm f}$), the discretized continuity equation expresses the fact that the sum of in-coming mass flow rates (negative) equals the sum of out-going flow rates (positive).

3.4.2 Momentum equation

The integration of each term in Eq. (8), starting from left to right, results in the following algebraic expressions.

Inertial term: This term does not benefit from the application of Gauss' theorem; hence its discretization results in

$$\int_{V_{\rm P}} \frac{\partial}{\partial t}(J\rho u_i)\mathrm{d}V = \frac{\rho V_{\rm P}}{\delta t}\left(u_{i,{\rm P}} - u_{i,{\rm P}}^{(n)}\right) \tag{14}$$

where $u_{i,{\rm P}}^{(n)}$ is the velocity at cell P at the previous time level and $V_{\rm P}$ represents the volume of cell P. The present method is fully implicit meaning that all variables without a time-level superscript are assumed to pertain to the new time-level (n + 1). The superscript (n) denotes a previous time step value. More accurate discretization procedures can be introduced for time-dependent calculations, but at this stage we use the implicit first-order Euler method for simplicity.

Convection term: As in Eq. (13), this term benefits from Gauss' theorem,

$$\int_{V_{\rm P}} \frac{\partial}{\partial \xi_l}\left(\rho\beta_{lj}u_j u_i\right)\mathrm{d}V = \sum_{l=1}^{3}\left[\Delta\left(\sum_j\left(\rho B_{lj}u_j\right)u_i\right)\right]_l^{\rm P} = \sum_{{\rm f}=1}^{6} F_{\rm f}\hat{u}_{i,{\rm f}} \tag{15}$$

with the cell face mass fluxes defined as in Eq. (13) and the convected velocity at face f, $\hat{u}_{i,{\rm f}}$, being given according to the discretization scheme adopted for the convective terms. For the upwind differencing scheme, $\hat{u}_{i,{\rm f}}$ is simply the velocity at the centre of the cell in the upstream direction, which can be written generally by expressing the convection fluxes of momentum as

$$F_{\rm f}\hat{u}_{i,{\rm f}} = F_{\rm f}^{+}u_{i,{\rm P}} + F_{\rm f}^{-}u_{i,{\rm F}} \quad \text{where } F_{\rm f}^{+} \equiv \mathrm{Max}(F_{\rm f},0) \text{ and } F_{\rm f}^{-} \equiv \mathrm{Min}(F_{\rm f},0) \tag{16}$$

Diffusion term: A normal diffusion term is added to both sides of the momentum equation, Eq. (8), in order to obtain a standard convection-diffusion equation when there is no solvent viscosity contribution, $\eta_S = 0$. This choice is akin to the Elastic Viscous Stress Splitting approach (Perera and Walters, 1977; Rajagopalan et al, 1990). The term added to the left hand side of the equation is given by the following expression, and discretized as shown:

$$-\int_{V_{\rm P}} \frac{\partial}{\partial \xi_l}\left(\frac{\eta}{J}\beta_{l'j}\beta_{l'j}\frac{\partial u_i}{\partial \xi_l}\right)\mathrm{d}V = -\sum_{{\rm f}=1}^{6}\frac{\eta_{\rm f}}{V_{\rm f}}B_{\rm f}^2\left[\Delta u_i\right]_f^{\rm f} = -\sum_{{\rm f}=1}^{6}D_{\rm f}\left(u_{i,{\rm F}} - u_{i,{\rm P}}\right) \tag{17}$$

where the surface area of the cell face is $B_{\mathrm{f}} = \sqrt{\sum_j B_{fj}^{\mathrm{f}} B_{fj}^{\mathrm{f}}}$, the volume of a pseudo-cell centred at the face is $V_{\mathrm{f}} = \sum_{j=1}^{3} B_{fj}^{\mathrm{f}} \left[\Delta x_j\right]_f^{\mathrm{f}}$, and $D_{\mathrm{f}} \equiv \eta_{\mathrm{f}} B_{\mathrm{f}}^2 / V_{\mathrm{f}}$ is a diffusion conductance. An identical term is added to the right hand side of the momentum equation, where it is treated explicitly and added to the source term. When iterative convergence is achieved, these two terms cancel out exactly. The solvent viscosity contribution is discretised using a similar approach.

Pressure gradient term: The pressure gradient is centred at P, thus leading to pressure differences across cell-widths. In representing it as S_{u_i} , it is implied that it will become a contribution to the source term of the algebraic equation, and therefore will be calculated explicitly as:

$$-\int_{V_{\mathrm{P}}} \beta_{li} \frac{\partial p}{\partial \xi_l} \mathrm{d}V = -\sum_{l=1}^{3} B_{li}^{\mathrm{P}} \left[\Delta p\right]_l^{\mathrm{P}} \equiv S_{u_i-pressure} \tag{18}$$

Stress-divergence term: Another term benefiting from Gauss' theorem, it becomes

$$\int_{V_{\mathrm{P}}} \frac{\partial}{\partial \xi_l}\left(\beta_{lj}\tau_{ij}\right)\mathrm{d}V = \sum_{\mathrm{f}=1}^{6} B_{fj}^{\mathrm{f}} \tilde{\tau}_{ij,\mathrm{f}} \equiv S_{u_i-stress} \tag{19}$$

where, like with the face velocity in the continuity equation beforehand, the cell-face stress (denoted with tilde) requires a special interpolation method due to the use of the collocated mesh arrangement. The way to do this constitutes one of the contributions of our work and is essential for the applicability of this method described in Section 3.7. This term is also treated explicitly in the context of the momentum equation, i.e. it becomes part of the momentum source term.

Gravity or body-force term: As with the pressure gradient term, this contribution is calculated at the cell centre and is included in the source term of momentum equation,

$$\int_{V_{\mathrm{P}}} J\rho g_i \,\mathrm{d}V = \rho V_{\mathrm{P}} g_i \equiv S_{u_i-gravity} \tag{20}$$

The final discretized form of the momentum equation is obtained at after re-grouping the various terms discussed above, to give:

$$a_{\mathrm{P}} u_{i,\mathrm{P}} - \sum_{\mathrm{F}} a_{\mathrm{F}} u_{i,\mathrm{F}} = S_{u_i} + \frac{\rho V_{\mathrm{P}}}{\delta t} u_{i,\mathrm{P}}^{(n)} \tag{21}$$

where the coefficients a_{F} consist of convection (a_{F}^C , here based on the upwind differencing scheme (UDS)) and diffusion contributions (a_{F}^D):

$$a_{\mathrm{F}} = a_{\mathrm{F}}^D + a_{\mathrm{F}}^C \text{, with } a_{\mathrm{F}}^D = D_{\mathrm{f}} \text{ and}$$

$$a_{\mathrm{F}}^{C} = \begin{cases} -F_{\mathrm{f}}^{-} & \text{(for a negative face, } \mathrm{f}^{-}\text{)} \\ +F_{\mathrm{f}}^{+} & \text{(for a positive face, } \mathrm{f}^{+}\text{)} \end{cases} \tag{22}$$

The central coefficient is:

$$a_{\mathrm{P}} = \frac{\rho V_{\mathrm{P}}}{\delta t} + \sum_{\mathrm{F}} a_{\mathrm{F}} \tag{23}$$

and the total source term is given by the sum

$$S_{u_i} = S_{u_i - pressure} + S_{u_i - gravity} + S_{u_i - stress} + S_{u_i - diffusion} \tag{24}$$

The source term S_{u_i} may contain additional contributions, such as those resulting from the application of boundary conditions, the use of high-resolution schemes for convection, or previous time step values for higher-order time-discretization schemes, amongst others.

3.4.3 Rheological constitutive equation

The two terms on the left hand side of Eq. (9) are discretized as the inertia (Eq. 14) and the convection terms above (Eq. 15), respectively, and do not present any additional difficulty. It should be noted that in all terms the velocity component u_i is replaced by τ_{ij}, and the mass flow rates in the convective fluxes, defined in Eq. (13), should be multiplied by λ/ρ (compare the convective fluxes in Eqs. 8 and 9). Following the same approach as above, the source term in the stress conformation tensor constitutive equation becomes:

$$S_{\Theta_{ij}} = \lambda V_P \left[R_{ik}\Theta_{kj} - \Theta_{ik}R_{kj} \right]_{\mathrm{P}} + 2\lambda V_P E_{ij,\mathrm{P}} + \Upsilon(A_{kk,\mathrm{P}}) V_P \left(e^{-\Theta_{ij,\mathrm{P}}} - \delta_{ij} \right) \tag{25}$$

The final form of the linearized equation is therefore

$$a_{\mathrm{P}}^{\Theta}\Theta_{ij,\mathrm{P}} - \sum_{\mathrm{F}} a_{\mathrm{F}}^{\Theta}\Theta_{ij,\mathrm{F}} = S_{\Theta_{ij}} + \frac{\lambda_{\mathrm{P}} V_{\mathrm{P}}}{\delta t}\Theta_{ij,\mathrm{P}}^{(n)} \tag{26}$$

with the coefficients a_{F}^{Θ} consisting of the convective coefficients in Eq. (22) multiplied by λ/ρ, for the reasons just explained, and the central coefficient is:

$$a_{\mathrm{P}}^{\Theta} = V_{\mathrm{P}} + \sum_{\mathrm{F}} a_{\mathrm{F}}^{\Theta} + \frac{\lambda V_{\mathrm{P}}}{\delta t} - V_{\mathrm{P}} + a_{0}^{\Theta}\Big|_{\mathrm{P}} + \frac{\lambda V_{\mathrm{P}}}{\delta t} \tag{27}$$

Whenever the extra-stress tensor is used in the code, it can be recovered from the conformation tensor using Eq. (4).

3.5 High-resolution schemes

The convective terms of the momentum and constitutive equations contain first derivatives of transported quantities, therefore an interpolation formula for their determination at cell faces is required ($\widehat{u}_{i,\mathrm{f}}$ in Eq. 15 and $\widehat{\Theta}_{ij,\mathrm{f}}$ in the convection term of the constitutive equation). In the previous section the UDS was used for such purpose, in particular for the determination of the convected velocities at cell-faces in Eq. (16). The upwind scheme is the most stable of all the schemes for convection, but has only first order accuracy and therefore gives rise to excessive numerical diffusion. Such problem is aggravated for the hyperbolic-type conformation equations.

Higher order methods for improved calculations have been widely used in CFD, such as second- and third-order upwind schemes (e.g. the QUICK scheme developed by Leonard, 1991). However, these schemes suffer from stability or iterative convergence problems, and are often not limited. To address these difficulties various differencing schemes have been combined in what are called high-resolution schemes (HRS). These methods ensure better convergence and stability properties and are generally bounded to avoid the appearance of spurious oscillations in regions with high gradients of the transported quantity.

The calculation of viscoelastic fluid flows has its own specificities, which are well described in specialized works (e.g. Owens and Phillips, 2002). For instance, HRS with good performance for Newtonian fluids often have problems of convergence and stability with viscoelastic fluids, as discussed by Alves et al (2003a), who developed an HRS particularly adequate for computational rheology, the CUBISTA scheme (Convergent and Universally Bounded Interpolation Scheme for the Treatment of Advection). This HRS is described below as implemented in our viscoelastic flow solver.

In what regards implementation of the HRS we use the so-called deferred correction approach of Khosla and Rubin (1974), where the convective contributions to the coefficients a_{F} and a_{P} are based on the upwind scheme UDS, to ensure positive coefficients for enhanced stability. The difference between the convective fluxes calculated by the HRS and UDS are handled explicitly and are included in the source term. Therefore, the deferred correction provides stability, simplicity of implementation (avoids increasing the computational stencil) and savings in computer memory, since the coefficients a_{P} and a_{F} are the same in the three momentum equations, and a_{P}^{Θ} and a_{F}^{Θ} are also the same in the six stress equations for 3D problems. For time-dependent flows, the use of the deferred correction leads to problems similar to those created by the added diffusive terms of the momentum equation, and the solution is the same: it is necessary to ensure that the added terms cancel each other at each time step by using an iterative procedure within the time-step.

Taking into account the use of the HRS in the scope of the deferred correction, the discretized momentum and constitutive equations can be rewritten as

$$a_{\mathrm{P}} u_{i,\mathrm{P}} - \sum_{\mathrm{F}} a_{\mathrm{F}} u_{i,\mathrm{F}} = S_{u_i} + \frac{\rho V_{\mathrm{P}}}{\delta t} u_{i,\mathrm{P}}^{(n)} + \left(\sum_{\mathrm{f}} F_{\mathrm{f}} \widehat{u}_{i,\mathrm{f}} \right)_{\mathrm{UDS}}^{*} - \left(\sum_{\mathrm{f}} F_{\mathrm{f}} \widehat{u}_{i,\mathrm{f}} \right)_{\mathrm{HRS}}^{*} \tag{28}$$

$$a_{\mathrm{P}}^{\Theta} \Theta_{ij,\mathrm{P}} - \sum_{\mathrm{F}} a_{\mathrm{F}}^{\Theta} \Theta_{ij,\mathrm{F}} = S_{\Theta_{ij}} + \frac{\lambda V_{\mathrm{P}}}{\delta t} \Theta_{ij,\mathrm{P}}^{(n)} + \left(\sum_{\mathrm{f}} \frac{\lambda}{\rho} F_{\mathrm{f}} \widehat{\Theta}_{ij,\mathrm{f}} \right)_{\mathrm{UDS}}^{*} - \left(\sum_{\mathrm{f}} \frac{\lambda}{\rho} F_{\mathrm{f}} \widehat{\Theta}_{ij,\mathrm{f}} \right)_{\mathrm{HRS}}^{*} \tag{29}$$

In both equations the new terms are evaluated at the previous iteration level (indicated by *) and are included in the source term (S_{u-HRS} and $S_{\Theta-HRS}$).

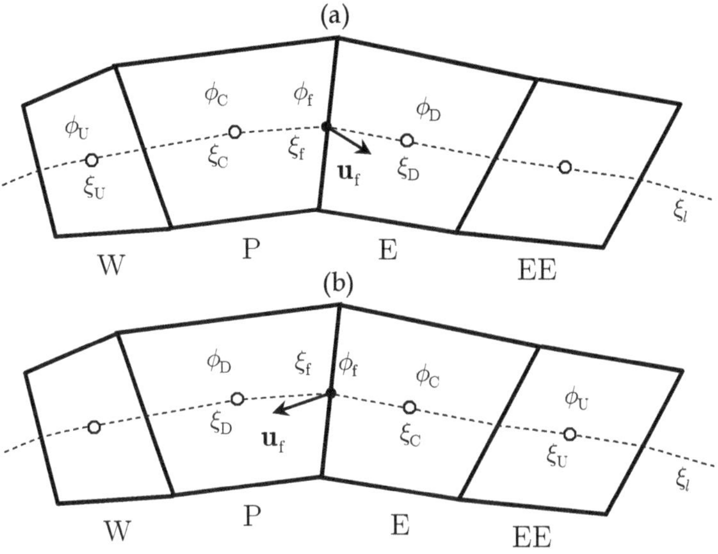

Fig. 3. Definition of local variables and coordinates in the vicinity of face f. (a) Positive velocity along direction ξ_l; (b) negative velocity along direction ξ_l.

The high-resolution schemes are usually written in compact form, using the normalized variable and space formulation of Darwish and Moukalled (1994). In this formulation the transported quantity ϕ (u_i or Θ_{ij}) and the system of general coordinates ξ, shown schematically in Figure 3, are normalized as

$$\widehat{\phi} = \frac{\phi - \phi_U}{\phi_D - \phi_U} \tag{30}$$

$$\widehat{\xi} = \frac{\xi - \xi_U}{\xi_D - \xi_U} \tag{31}$$

where subscripts U and D refer to upwind and downwind cells relative to C, which is immediately upstream of face f. The objective is the calculation of ϕ at cell-face f, via a special interpolation scheme for convection ($\widehat{\phi}_f$).

In order to satisfy the convection boundedness criterion (CBC) of Gaskell and Lau (1988) the functional relationship of an interpolation scheme applied to a cell face f, $\widehat{\phi}_f = fn\left(\widehat{\phi}_C\right)$, must be continuous and bounded from below by $\widehat{\phi}_f = \widehat{\phi}_C$ and from above by unity, in the monotonic range $0 < \widehat{\phi}_C < 1$. However, the CBC is not sufficient to guarantee that a limited scheme has good iterative convergence properties and therefore Alves et al (2003a) also used the "Universal Limiter" of Leonard (1991), which is valid for explicit transient calculations and reduces to Gaskell and Lau's criterion for steady flows, when the Courant number tends to zero. On the other hand, the conditions for an explicit time-dependent method to be Total Variation Diminishing are even more restrictive than the universal limiter and it was based upon these more restrictive conditions that the CUBISTA scheme was formulated to guarantee stability and good iterative convergence properties. The CUBISTA HRS is based on the third-order discretization QUICK scheme, it avoids sudden changes in slope of the functions and ensures limitation of ϕ on the downwind side to preclude ϕ being higher than ϕ_D in its proximity. All these details are extensively discussed in Alves et al (2003a), and give rise to the following function for CUBISTA in non-uniform meshes,

$$\hat{\phi}_f = \begin{cases} \left[1 + \dfrac{\hat{\xi}_f - \hat{\xi}_C}{3(1-\hat{\xi}_C)}\right] \dfrac{\hat{\xi}_f}{\hat{\xi}_C} \hat{\phi}_C & 0 < \hat{\phi}_C < \dfrac{3}{4}\hat{\xi}_C \\ \dfrac{\hat{\xi}_f(1-\hat{\xi}_f)}{\hat{\xi}_C(1-\hat{\xi}_C)} \hat{\phi}_C + \dfrac{\hat{\xi}_f(\hat{\xi}_f - \hat{\xi}_C)}{1-\hat{\xi}_C} & \dfrac{3}{4}\hat{\xi}_C \le \hat{\phi}_C \le \dfrac{1+2(\hat{\xi}_f - \hat{\xi}_C)}{2\hat{\xi}_f - \hat{\xi}_C}\hat{\xi}_C \\ 1 - \dfrac{1-\hat{\xi}_f}{2(1-\hat{\xi}_C)}(1-\hat{\phi}_C) & \dfrac{1+2(\hat{\xi}_f - \hat{\xi}_C)}{2\hat{\xi}_f - \hat{\xi}_C}\hat{\xi}_C < \hat{\phi}_C < 1 \\ \hat{\phi}_C & \hat{\phi}_C < 0 \text{ and } \hat{\phi}_C > 1 \end{cases} \tag{32}$$

where $\hat{\phi}_C$, $\hat{\xi}_C$ and $\hat{\xi}_f$ are defined in Eqs. (30) and (31).

3.6 Formulation of the mass fluxes at cell faces

The mass flow rates (F_f) in coefficients a_F^C and a_P^C have to be calculated with velocities at cell faces ($\tilde{u}_{i,f}$), which must be related to velocities at cell centres. The need to calculate $\tilde{u}_{i,f}$ at a cell face is a consequence of the use of collocated meshes and would not occur if staggered meshes were used. The continuity equation is needed to solve for the pressure field, after a velocity field is calculated from the momentum equation as in the SIMPLE procedure initially developed by Patankar and Spalding (1972) using staggered meshes for the calculation of the velocity and pressure fields. Here, each velocity component data are stored in meshes staggered by half a cell width relative to the original mesh where the scalar quantities are stored and in this way the coupling between velocity and pressure is naturally ensured while momentum and mass is conserved.

By using a single non-orthogonal mesh with the collocated variable arrangement, coupling between the velocity and pressure fields needs a special interpolation scheme to calculate velocities at cell-faces otherwise even-odd oscillations in the pressure or velocity fields may occur. The key idea to solve this decoupling problem was proposed by Rhie and Chow (1983). Oliveira (1992) and Issa and Oliveira (1994) adapted that idea for their time-marching algorithm under a slightly modified form explained hereafter.

The momentum equation (Eq. 21) at node P can be rewritten as

$$a_P u_{i,P} = \sum_F a_F u_{i,F} - \sum_{l=1}^{3} B_{li}^P [\Delta p]_l^P + S'_{u_i} + \left(\frac{\rho V}{\delta t}\right)_P u_{i,P}^{(n)} \tag{33}$$

where the pressure term was extracted from the source term and is written explicitly with a pressure difference evaluated at a cell centre (i.e. $[\Delta p]_l^P = p^{l+} - p^{l-}$, cf. Figure 2).

According to Rhie and Chow's special interpolation method, the cell face velocity $\tilde{u}_f$ is calculated by linear interpolation of the momentum equation, with exception of the pressure gradient which is evaluated as in the original method of Patankar and Spalding for staggered meshes. This idea is applied as described in Issa and Oliveira (1994), by writing

$$\overline{a}_P \tilde{u}_{i,f} = \overline{\sum_F a_F u_{i,F}} + \overline{S_{u_i}^{'}} - B_{fi}^{f}[\Delta p]_f^{f} - \sum_{l \neq f} \overline{B_{li}[\Delta p]_l} + \overline{\left(\frac{\rho V}{\delta t}\right)_P} \tilde{u}_{i,f}^{(n)} \quad (34)$$

where the overbar denotes here an arithmetic mean of quantities pertaining to cells P and F. Notice that the pressure difference along direction $l = f$ is now evaluated at cell-face (i.e. $[\Delta p]_f^{f} = p_F - p_P$), whereas the pressure at cell faces pertaining to directions $l \neq f$ are calculated by linear interpolation of the nodal values of pressure. With this approach the velocity at face f is directly linked to pressures calculated at neighbour cell centres, as in the staggered arrangement, and pressure-velocity decoupling is prevented.

By subtracting Eq. (34) from the averaged momentum equation resulting from averaging all terms of Eq. (33) the following face-velocity equation used to compute F_f is obtained:

$$\tilde{u}_{i,f} = \frac{\overline{a_P u_{i,P}} + \overline{B_{fi}^{P}[\Delta p]_f^{P}} - B_{fi}^{f}[\Delta p]_f^{f} + \overline{\left(\frac{\rho V}{\delta t}\right)_P} \tilde{u}_{i,f}^{(n)} - \overline{\left(\frac{\rho V}{\delta t} u_i^{(n)}\right)_P}}{\overline{a}_P} \quad (35)$$

3.7 Formulation of the cell-face stresses

In the momentum equation it is necessary to compute the stresses at cell faces ($\tilde{\tau}_{ij,f}$ in Eq. 19) from stress values at neighbouring cell centres and there is a stress-velocity coupling problem, akin to the pressure-velocity coupling of the previous subsection. If a linear interpolation of cell centred values of stress is used to compute those face values, a possible lack of connectivity between the stress and velocity fields may result, even with Newtonian fluids, as shown by Oliveira et al (1998). The methodology described here is based on the works of Oliveira et al (1998), Oliveira and Pinho (1999a) and more recently by Matos et al (2009) and constitutes a key ingredient for the success of viscoelastic flow computations with the finite-volume method on general, non-orthogonal, collocated meshes. Following the ideas of Matos et al (2009), the extra-stress at face f is computed as

$$\tilde{\tau}_{ij,f} = \left(\overline{\tau_{ij}}\right)_f + \frac{\eta}{\left(1 + a_0^{\Theta} / V\right)} \left[\frac{1}{V_f}\left(B_{fi}[\Delta u_j]_f + B_{fj}[\Delta u_i]_f\right) - \overline{\frac{1}{V}\left(B_{fi}[\Delta u_j]_f + B_{fj}[\Delta u_i]_f\right)} \right] \quad (36)$$

where the denotes a linear interpolation rather than an arithmetic mean. It is obvious from Eq. (36) that the extra-stress at face f ($\tilde{\tau}_{ik,f}$) is now directly coupled to the nearby cell-centre velocities, through the term in $[\Delta u_i]_f^{f} = u_{i,F} - u_{i,P}$, inhibiting the undesirable decoupling between the stress and velocity fields. In Matos et al (2009) the standard formulation for the constitutive equation was used, based on the extra-stress tensor, and Eq. (36) results directly from the discretization of the extra-stress tensor equation. Here we use the log-conformation methodology, but since the central coefficients of the discretized equations for the extra-stress and for the log-conformation tensors are the same, then Eq. (36) is still applicable.

3.8 Solution algorithm

As in any pressure correction procedure (e.g. Patankar and Spalding, 1972), pressure is calculated indirectly from the restriction imposed by continuity, since the momentum equation, which explicitly contains a pressure gradient term, is used to compute the velocity vector components. The SIMPLEC algorithm of Van Doormal and Raithby (1984) is followed here under a modified form. The original SIMPLEC algorithm was developed for iterative steady flow calculations, but the time-marching version described in Issa and Oliveira (1994) offers some advantages and is used here instead. Time marching allows for the solution of transient flows provided the time step is sufficiently small, with the added advantage that it can be used for steady flows as an alternative to implement under-relaxation.

The incorporation of a rheological constitutive equation produces little changes on the original SIMPLEC method developed for Newtonian fluids, which is mainly concerned with the calculation of pressure from the continuity equation. An overview of the solution algorithm is now given, including the new steps related to the stress calculation:

1. Initially, the conformation tensor A_{ij}, calculated from the extra-stress components τ_{ij} via Eq. (4). At each point the eigenvalues and eigenvectors of A_{ij} are computed and the conformation tensor is diagonalized to calculate Θ_{ij}.
2. The tensors R_{ij} and E_{ij} are calculated, following the procedure described in Fattal and Kupeferman (2004).
3. The discretized form of the evolution equation for Θ_{ij} in Eq. (37) is solved to obtain Θ_{ij} at the new time level,

$$a_{\mathrm{P}}^{\Theta}\Theta_{ij,\mathrm{P}}^{*}-\sum_{\mathrm{F}}a_{\mathrm{F}}^{\Theta}\Theta_{ij,\mathrm{F}}^{*}=S_{\Theta_{ij}} \tag{37}$$

   where the coefficients and source term of these linear equations are based on the previous iteration level variables, and Θ_{ij}^{*} denotes the new time-level of Θ_{ij}.
4. The conformation tensor A_{ij} is recovered and the extra-stress tensor is calculated from the newly computed conformation field using Eq. (4).
5. The momentum equation (38) is solved implicitly for each velocity component, u_i:

$$\left(\sum_{\mathrm{F}}a_{\mathrm{F}}+\frac{\rho V_{\mathrm{P}}}{\delta t}\right)u_{i,\mathrm{P}}^{*}-\sum_{\mathrm{F}}a_{\mathrm{F}}u_{i,\mathrm{F}}^{*}=\sum_{l}^{3}B_{li}\left[\Delta p^{*}\right]_{l}^{\mathrm{P}}+S_{u_i}^{'}+\frac{\rho V_{\mathrm{P}}}{\delta t}u_{i,\mathrm{P}}^{(n)} \tag{38}$$

   where the pressure gradient term is based on previous iteration level pressure field and has been singled out of the remaining source term for later convenience. The stress-related source term (Eq. 19) is based on newly obtained cell-face stress $\tilde{\tau}_{ij}^{*f}$, calculated from Eq. (36), which requires the central coefficient of the log-conformation tensor equation (a_{P}^{Θ}). This is the main reason for solving the constitutive equation before the momentum equation.
6. Starred velocity components (u_i^{*}) do not generally satisfy the continuity equation. The next step of the algorithm involves a correction to u_i^{*}, so that an updated velocity field u_i^{**} will satisfy both the continuity equation and the following split form of the momentum equation:

$$\sum_{\mathrm{F}}a_{\mathrm{F}}u_{i,\mathrm{P}}^{*}+\frac{\rho V_{\mathrm{P}}}{\delta t}u_{i,\mathrm{P}}^{**}=\sum_{\mathrm{F}}a_{\mathrm{F}}u_{i,\mathrm{F}}^{*}-\sum_{l}B_{li}\left[\Delta p^{**}\right]_{l}^{\mathrm{P}}+S_{u_i}^{'}+\frac{\rho V_{\mathrm{P}}}{\delta t}u_{i,\mathrm{P}}^{(n)} \tag{39}$$

It is noted that in Eq. (39) only the time-dependent term is updated to the new iteration level u_i^{**}, a feature of the SIMPLEC algorithm (Issa and Oliveira, 1994). Subtraction of this equation from Eq. (38) and forcing the u_i^{**} field to satisfy continuity ($\sum_f F_f^{**} = 0$, cf. Eq. 13) leads to the pressure correction and velocity correction equations (Eqs. 40 and 41, respectively, where $p' = p^{**} - p^{*}$):

$$a_P^P p_P' = \sum_F a_F^P p_F' - \sum_f F_f^* \quad a_P^P = \sum_F a_F^P ; a_F^P = \frac{\rho B_f^2}{\left(\frac{\rho V}{\delta t}\right)_f} \tag{40}$$

$$\frac{\rho V_P}{\delta t}\left(u_i^{**} - u_i^*\right)_P = -\sum_l B_{li}^P \left[\Delta p'\right]_l^P \tag{41}$$

7. Steps 1-6 are repeated until overall convergence is reached (steady-state calculations), or convergence within a time step (unsteady calculations) followed by advancement until the desired final time is reached.

The various sets of algebraic equations are solved with either a symmetric or a bi-conjugate gradient method for the pressure and the remaining variables, respectively (Meijerink and Van der Vorst, 1977). In both cases the matrices are pre-conditioned by an incomplete LU decomposition.

3.9 Boundary conditions

Appropriate boundary conditions are required for the dependent variables (u_i, p and Θ_{ij}) at the external boundary faces of the flow domain. Four types of boundaries are typically encountered in the applications considered in this work, namely inlets, outlets, symmetry planes and walls. Each one is dealt with briefly below and the interested reader is referred to specific literature for more details.

Inlet: Velocity and stress components are given according to some pre-specified profiles (from theory or measured data), and Θ_{ij} is calculated accordingly. Sometimes the streamwise velocity at inlet is set equal to a uniform value and a null stress field is considered, but most often fully developed flow conditions are assumed for velocity and stress fields. Progress on work dealing with derivation of analytical solutions for viscoelastic models has been made during the past years, and velocity and stress distributions in fully developed duct flows can be found in Oliveira and Pinho (1999c) for the PTT model, Alves et al (2001b) for the full PTT model and Oliveira (2002) for the FENE-P model, amongst other solutions. These are useful not only to prescribe inlet conditions but also to obtain the wall boundary conditions where the convective terms in the equations are null, as for fully developed flow.

Outlet: The outlet planes are located far away from the main region of interest, where the flow can be assumed fully developed. Thus, zero streamwise gradients are prescribed for the velocity, the Θ_{ij} components and the pressure gradient. The latter is equivalent to a linear extrapolation of pressure values from the two internal cells to the outlet boundary face. An additional condition required by the pressure correction equation in incompressible flow is to adjust the velocities at the boundary faces so that overall mass conservation is satisfied.

Symmetry planes: Across a symmetry plane the convective and diffusive fluxes must vanish. These two conditions are applied to all variables using reflection rules in fictitious symmetric cells (Figure 4a) and result in the following procedure to implement these boundary conditions (see Oliveira and Pinho, 1996b for details). At the symmetry plane there is only tangential velocity, i.e. the normal velocity is null. So, at the face coincident with the symmetry plane $u_{i,\mathrm{f}}^{n}=0$ and $u_{i,\mathrm{f}}^{t}=u_{i,\mathrm{f}}$ (superscripts n and t denote normal and tangential components, respectively). Since the fictitious cell P′ is symmetric to cell P, the calculation of the components of the velocity vector u_i at the cell face f ($u_{i,\mathrm{f}}$) is obtained by linear interpolation from the velocities at the adjacent cell nodes leading to

$$u_{i,\mathrm{f}}=u_{i,\mathrm{f}}^{t}=u_{i,\mathrm{P}}-u_{\mathrm{P}}^{n}.n_i \text{ and } u_{\mathrm{P}}^{n}=\sum_j u_{j,\mathrm{P}}n_j \quad (42)$$

where u_{P}^{n} is the component of the velocity vector normal to the symmetry plane and n_i is the i-component of the unit vector normal to the symmetry plane.

For scalar quantities, such as the pressure, the reflexion rule at symmetry planes leads to

$$p_{\mathrm{f}}=p_{\mathrm{P}} \quad (43)$$

Imposition of boundary conditions for the stress is facilitated by recognizing that not all individual stress components are required at the cell face f coincident with the symmetry plane, since the tangential stress vector is zero. Therefore, as seen from Eq. (19), the contribution from face f to the total stress source at cell P is just:

$$\left(S_{u_i-stress}\right)_{\mathrm{f}}=\sum_j B_{fj}^{\mathrm{f}}\tau_{ij,\mathrm{f}}=B_{\mathrm{f}}\sum_j \tau_{ij,\mathrm{f}}n_j \text{ leading to } \left(S_{u_i-stress}\right)_{\mathrm{f}}=T_{n,\mathrm{P}}B_{fi}=T_{n,\mathrm{P}}B_{\mathrm{f}}n_i \quad (44)$$

where the unit normal vector is computed as $n_j=B_{fj}/B_{\mathrm{f}}$. Thus, the boundary condition at the symmetry plane represents only the traction vector normal to face f.

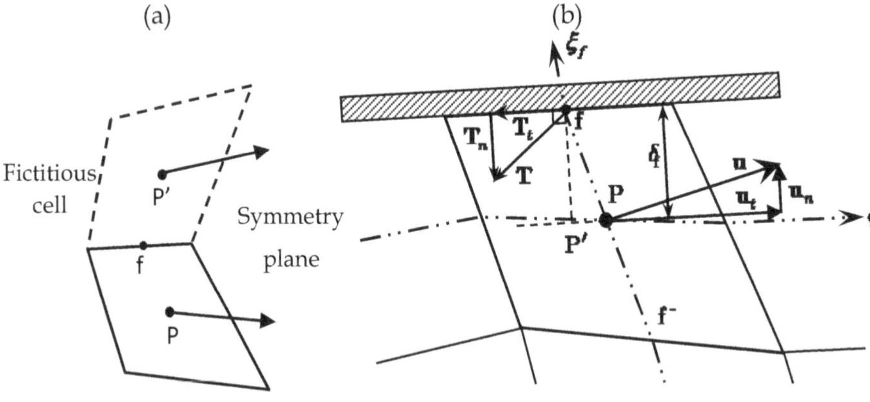

Fig. 4. Cells at boundaries: (a) The fictitious cell adjacent to a symmetry plane; (b) Schematic representation of an internal cell adjoining a wall.

Walls: At walls additional problems in imposing boundary conditions arise, especially for pressure and the stresses. Those problems are more severe when the constitutive equation predicts non null stresses normal to walls, as in the Giesekus or full PTT models. Boundary conditions for the velocity field are easy to impose. For a wall moving at velocity u_w, the no slip condition for the components u_i of the velocity vector are simply

$$u_{i,\mathrm{f}} = u_{i,w} \tag{45}$$

More generally, for a non-porous wall this is mathematically expressed as $u_{i,\mathrm{f}}^n = 0$ and $u_{i,\mathrm{f}}^t = u_{i,\mathrm{f}}$ which are numerically obtained by linear interpolation from velocities at cells adjacent to the wall.

Boundary conditions for stresses are based on the assumption that the flow in the vicinity of a wall is parallel to this boundary, i.e. it is locally a Couette flow. This assumption allows a relatively easy implementation of boundary conditions provided the rheometric material functions of the fluid model are known. A complete explanation of the procedure can be found in Oliveira (2001) and the main points are given here.

Consider Figure 4b which shows the inner cell next to a wall plane. The stress vector near the wall $T_i = \sum_j \tau_{ij} n_j$ has tangential and normal components to the wall ($T_i = T_i^t + T_i^n$). Since the near wall flow is assumed to be a Couette flow, the tangential component of the traction vector is calculated as

$$T_i^t = \mu(\dot{\gamma}) \frac{\partial u_{i,\mathrm{f}}^t}{\partial n} \rightarrow T_i^t = \mu(\dot{\gamma}_\mathrm{f}) \frac{u_{i,\mathrm{f}}^t - u_{i,\mathrm{P}}^t}{\delta_f} \tag{46}$$

where n is the vector normal to the wall and $\mu(\dot{\gamma})$ is the shear viscosity material function of the constitutive model (not to be confused with the parameter η of the constitutive equation), which depends on the invariant $\dot{\gamma}$ of the rate of deformation tensor. This wall shear rate is equal to $\partial u_{i,\mathrm{f}}^t / \partial n$ and is calculated as in Eq. (46), where δ_f is the distance from f to the cell centre P along the normal to the wall (see Figure 4b).

Note that in the finite volume method the discretization of the traction vector is indeed carried out as a component of the momentum equation and appears as the result of the integration and subsequent discretization of the term (19), now applied to a wall. Here

$$\left(S_{u_i - stress}\right)_f = \left(B \sum_j \tau_{ij} n_j\right)_\mathrm{f} = B_\mathrm{f} T_{i,\mathrm{f}} = B_f \left(T_i^t + T_i^n\right)_\mathrm{f} \tag{47}$$

has two contributions: one associated to the tangential stress, given by Eq. (46), and the other due to normal stress at the face which is null for constitutive models with $N_2 = 0$ as those used for the computations in Section 4. For constitutive models with $N_2 \neq 0$, such as the Giesekus or the full PTT, the interested reader is referred to Oliveira (2001) for the determination of T_i^n.

Finally, we must consider the wall boundary condition for pressure. It is usual practice in CFD to extrapolate linearly the pressure to the wall from the two nearest neighbour cells (Ferziger and Perić, 2002) and this practice also works well for some viscoelastic fluids. However, in viscoelastic flow with fluids exhibiting strong normal stresses perpendicular to the wall ($N_2 \neq 0$), pressure extrapolation is not satisfactory and a better formulation can be derived from the momentum equation normal to the wall at the interior point P, as explained in detail in Oliveira (2001), leading to the following corrected extrapolation formula (a_{P} is the central coefficient in the momentum equation)

$$p_{\mathrm{f}} = 2p_{\mathrm{P}} - p_{\mathrm{f}^-} + \frac{a_{\mathrm{P}} u_{n,\mathrm{P}}}{B_{\mathrm{f}}} \tag{48}$$

The two first terms on the right-hand-side of this equation do correspond to linear extrapolation from the two nearest neighbour cells and the last term is a correction which decreases as the mesh is refined close to a wall.

4. Benchmark results in 4:1 planar sudden contraction flows

In this section we assess the capabilities of the finite-volume method described previously by presenting results of simulations for the benchmark flow through a 4:1 planar sudden contraction shown in Figure 5 under conditions of negligible inertia. This is a long standing classic benchmark in computational rheology (Hassager, 1988), where the difficulty lies at the correct prediction of the large stresses and stress gradients in the vicinity of the re-entrant corner (generally all models show the stresses to grow to infinity as the corner is approached) making this flow very sensitive to highly elastic flows. In particular, it is important to know the upstream vortex growth mechanisms due to flow elasticity, and the corresponding large pressure drops and overshoot of the axial velocity along the centreline.

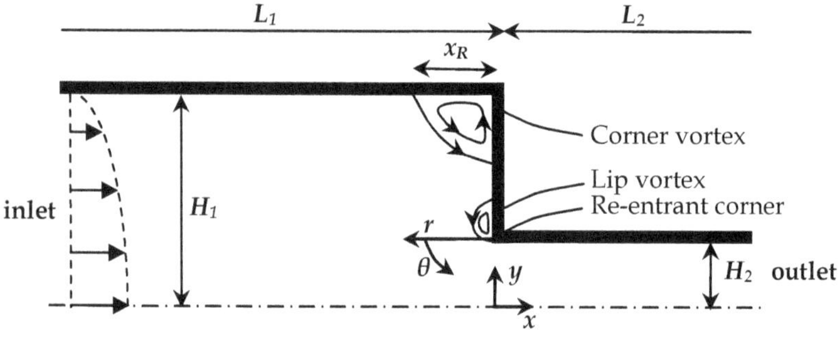

Fig. 5. Schematic representation of a sudden contraction.

4.1 Experimental results

Experiments on sudden contraction flows have been carried out since the 19th century and their characteristics for Newtonian and purely viscous non-Newtonian fluids are well known, especially for the axisymmetric case (cf. the review of Boger, 1987). Contraction flows are very sensitive to fluid properties as well as to geometric characteristics, especially the contraction ratio (*CR*). Therefore, we must distinguish the flow in either planar or axisymmetric contractions, and between elastic fluids having constant viscosity (Boger fluids) and shear-thinning viscosity, as well as fluids having different behaviour in extensional flow (Boger, 1987).

In the circular contraction arrangement, whereas for some fluids there is corner vortex enhancement with fluid elasticity, for some Boger fluids a lip vortex appears first and grows with elasticity while the corner vortex decreases. Then, as elasticity further increases, the lip vortex engulfs the corner vortex, becomes convex and it continues to grow with fluid elasticity until the onset of flow instabilities (Boger et al, 1986). For better understanding, these flow features are illustrated in the sketch of Figure 5.

For the 4:1 planar contraction, the early investigations with Boger fluids by Walters and Webster (1982) did not find any peculiar flow feature, in contrast to their behaviour in a circular 4.4:1 contraction. To help clarify this issue, Evans and Walters (1986, 1989) visualized the flow of shear-thinning elastic fluids in 4:1, 16:1 and 80:1 contractions, and reported elastic vortex growth, even in the smaller *CR*, showing also that an increased *CR* intensified the phenomenon. They also found a lip vortex for the two larger *CR*.

To conclude, for shear-thinning fluids there was elastic vortex growth for both the 4:1 planar and circular contractions whereas for Boger fluids the vortex growth was reported only to occur for the axisymmetric geometry. This was confirmed experimentally by Nigen and Walters (2002), who looked at the behaviour of Boger and Newtonian fluids having the same shear viscosity, in 4:1 and 32:1 sudden contractions: whereas in axisymmetric contractions, elastic vortex growth and increased pressure drop co-existed, the planar contraction flow was Newtonian-like. Lip vortices were reported only for the planar contraction for supercritical flow rates, when the flow was unsteady.

The inexistence of lip vortices in the planar contraction for Boger fluids, and its presence for circular contractions remains to be explained, in spite of existing theoretical work (Binding, 1988; Xue et al, 1998a). Additionally, the experiments and numerical simulations of White and Baird (1986,1988a, 1988b) have established the relevance of extensional stress growth near the contraction plane upon the vortex dynamics for the plane 4:1 and 8:1 contractions. For large circular contractions Rothstein and McKinley (2001) related the dominance of the lip or corner vortices to the competition between shear-induced and extension-induced normal stresses, later confirmed by the simulations of Oliveira et al (2007).

Regardless of the contraction, the growth of elasticity under conditions of negligible inertia inevitably leads to an instability, which may be chaotic at very large *De* or be preceded by periodic unsteady flow. This sequence of events has been seen as early as the late seventies by Cable and Boger (1978a, 1978b, 1979) and Nguyen and Boger (1979) and has been studied in detail by Lawler et al (1986), McKinley et al (1991) and Yesilata et al (1999).

The flows through abrupt contractions have recently been revisited in the context of microfluidics, in which high *De* can be easily attained due to the small characteristic dimensions, even with weakly elastic and viscous fluids as in the experiments of Rodd et al (2005) using dilute and semi-dilute aqueous solutions of poly(ethylene oxide). They observed the onset of divergent streamlines upstream of the recirculation at high *De* in addition to the elastic vortex growth. However, notice that in microfluidics, the flow is often not truly planar (2D) due to the effects of the bounding walls (typically, aspect ratios are of the order of unity in the contraction region), which confer a 3D character to the flow.

4.2 Numerical simulations

Numerical investigations in contraction flows were also initiated in the late seventies, but soon problems of convergence arose leading to the development of robust and accurate numerical methods for predicting steady flows and in particular the elastic vortex growth seen in experiments. These extensive developments are well documented by Owens and Phillips (2002). Here, some of the most accurate and recent results in the 4:1 planar contraction flow for Oldroyd-B and PTT fluids are presented. The PTT fluid used here is the simplified version with N_2=0.

4.2.1 Oldroyd-B fluid

The flow geometry and the notation are shown in Figure 5. The Reynolds number ($Re \equiv \rho U_2 H_2/\eta$) is set to zero by dropping out the convective term in the momentum equation. The Deborah number ($De \equiv \lambda U_2/H_2$) is varied to investigate the effect of elasticity on the flow characteristics, and the solvent viscosity ratio considered was $\beta = \eta_S / \eta = 1/9$, the typical benchmark case.

Results for the Oldroyd-B fluid through the 4:1 sudden contraction were presented by Alves et al (2003b) who used the standard stress formulation and the HRS CUBISTA scheme together with very refined meshes with up to 169 392 computational cells, corresponding to more than one million degrees of freedom. The high accuracy of the results for $De \leq 2.5$, most of which have an uncertainty below 0.3%, indicates these values may be used as benchmark data (Alves et al, 2003b). The new predictions of Afonso et al (2011) for much higher Deborah numbers (up to De = 100), made possible by the log-conformation formulation, are also discussed here.

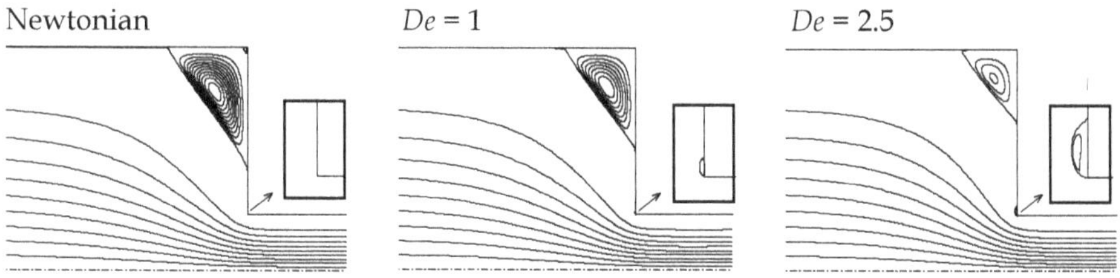

Fig. 6. Streamline plots for the flow of an Oldroyd-B fluid in a 4:1 plane sudden contraction for $De \leq 2.5$. Adapted from Alves et al (2003b).

The evolution of flow patterns with elasticity is shown in the streamline plots of Figure 6. The reduction of the corner vortex length with De for the Oldroyd-B fluid is clear as well as the appearance of a small lip vortex in the re-entrant corner as also happens with the UCM fluid (Alves et al, 2000). At De = 2.5 the lip vortex is still small, but stronger than at lower values of De. Although minute, this lip vortex is not a numerical artefact and has a finite strength. Extrapolation of its size and strength to a zero mesh size from consecutively refined meshes confirms that assertion (see Alves et al, 2003b).

At approximately De = 2.5, local flow unsteadiness is detected near the re-entrant corner. For higher De, a different trend is found, as can be observed in Figure 7. Initially, as the De is increased further, the lip and corner vortex structures merge, as shown for De = 5. Simultaneously, the periodic unsteadiness grows with De leading to a loss of symmetry and eventually, alternate back-shedding of vorticity is observed from the upstream pulsating eddies at higher De. These features are accompanied by a frequency doubling mechanism deteriorating to a complex pattern and eventually to a chaotic regime as shown by the frequency spectra in Afonso et al (2011).

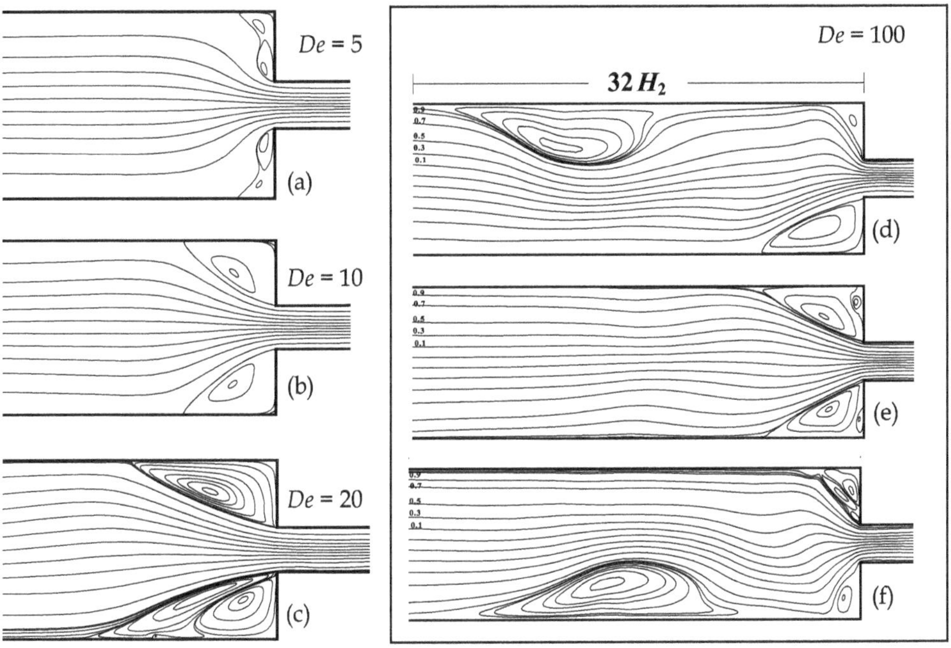

Fig. 7. Streamline plots for the flow of an Oldroyd-B fluid in a 4:1 plane sudden contraction for $De \geq 5$. Adapted from Afonso et al (2011).

The corresponding dimensionless corner vortex length ($X_R \equiv x_R/H_2$) is presented in Figure 8a, showing a non-monotonic variation with De. At low De the vortex size asymptotes to the Newtonian limit, as imposed by continuum mechanics. A semi-analytical investigation of creeping flow of Newtonian fluids by Rogerson and Yeow (1999) estimated the value $X_R = 1.5$ for a 4:1 planar contraction, which coincides with the numerical data of Alves et al (2003b) and Afonso et al (2011) in Figure 8a. Increasing De decreases the vortex size, in agreement with Aboubacar and Webster (2001). For $De \approx 4.5$, a minimum vortex length is attained and for larger values of De the vortex size increases significantly as well as its oscillation amplitude (as indicated by the error bars in Figure 8a).

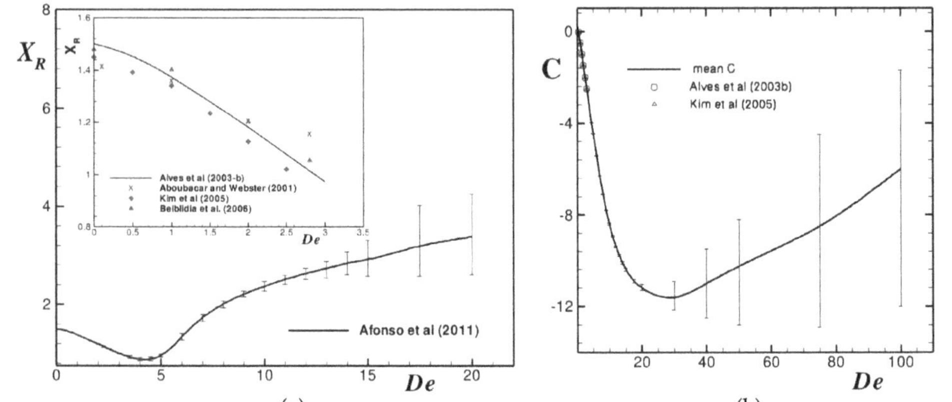

(a) (b)

Fig. 8. Variation with *De* of the dimensionless vortex size (a) and Couette correction (b) for the flow of an Oldroyd-B fluid in a 4:1 planar sudden contraction. The error bars represent the amplitude of the oscillations. Adapted from Alves et al (2003b) and Afonso et al (2011).

Figure 8b plots the variation of the Couette correction (*C*) with *De*. *C* represents the dimensionless localized pressure loss across the contraction, and is defined as

$$C = \frac{\Delta p - (\Delta p_1)_{FD} - (\Delta p_2)_{FD}}{2\tau_w} \tag{49}$$

where $(\Delta p_1)_{FD}$ and $(\Delta p_2)_{FD}$ are the pressure drops associated with fully-developed flows in the inlet and outlet channels, respectively, and τ_w is the wall stress in the downstream channel under fully-developed conditions.

For low *De*, the Couette correction decreases with *De* and becomes negative (elastic pressure recovery), a behaviour which is contrary to the experimental evidence. Only for $De > 20$, an increase in *C* is observed, as seen in numerical studies with the PTT fluid (Alves et al 2003b). Once again, the error bars represent the oscillation amplitude, which is seen to increase significantly with *De* above the minimum *C* value attained.

4.2.2 PTT fluid

Numerical simulations with the PTT fluid model allow us to investigate the combined effects of shear-thinning of the viscometric viscosity and fluid elasticity via the first normal stress difference N_1. From experimental data for contraction flow we know that the behaviour of such fluids is very different from the behaviour of Boger fluids. The results presented here are based on Alves et al (2003b), and were also obtained for creeping flow conditions, as in the previous section for the Oldroyd-B fluid. The PTT model corresponds to the simplified version ($\xi = 0$) with a zero second-normal stress difference ($N_2 = 0$).

Since the PTT fluids have a bounded steady-state extensional viscosity, contrasting with the Oldroyd-B fluid, it was possible to obtain converged solutions up to *De* in excess of 100, even using the standard stress formulation. This PTT model was also combined with a Newtonian solvent to define the same solvent viscosity ratio in the limit of very small rates of shear deformation, corresponding to $\beta = 1/9$. The PTT fluid is shear-thinning both in the shear viscosity as well as in the first-normal stress coefficient. The results presented here correspond to $\varepsilon = 0.25$, a typical value for both concentrated polymer solutions and polymer melts.

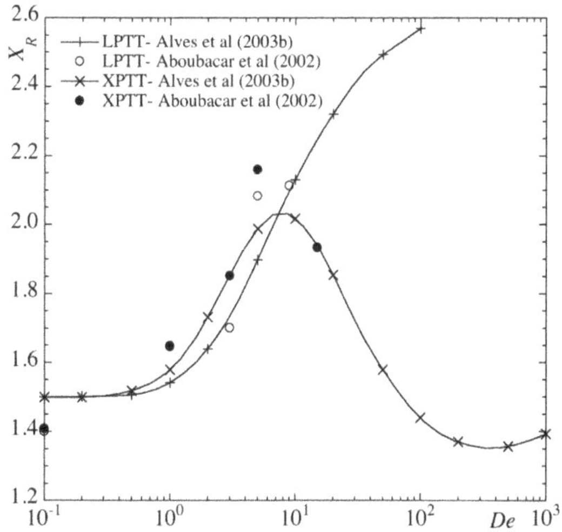

Fig. 9. Influence of elasticity on the recirculation length for the flow of PTT fluids in a 4:1 plane sudden contraction. LPTT: linear PTT; XPTT: exponential PTT. Adapted from Alves et al (2003b).

First, it is noted that the sensitivity of the PTT results to mesh fineness is lower than for the Oldroyd-B model and, in contrast to what happened with the Oldroyd-B fluid, the recirculation length (Figure 9) increases with *De*. This behaviour was expected given the experimental data available, where an intense vortex growth was seen for shear-thinning fluids. The recirculation length tends to stabilize at high *De*, and these predictions do not capture the elastic instabilities observed in experiments, but to assess whether this model is adequate to predict real flow conditions, where the vortex grows and then becomes unstable, simulations must be carried out using 3D meshes and time-dependent approaches with very small time-steps.

The evolution of the streamlines and the growth of the vortex with elasticity can be observed in Figure 10. The comparison with Figures 6-7 emphasises the differences between the behaviour of a shear-thinning fluid and a constant viscosity Boger fluid. At low *De*, the corner vortex grows towards the re-entrant corner while its size increases, but no lip vortex is observed. When the corner vortex occupies the whole contraction plane (*De* = 2) its shape changes from concave to convex and, as elasticity is further increased from $De \sim 2$ to $De \sim 10$, the vortex grows upstream even further. Simultaneously the eddy centre moves towards the re-entrant corner with the growth of elasticity. At high Deborah numbers the increase in X_R and Ψ_R becomes progressively less intense.

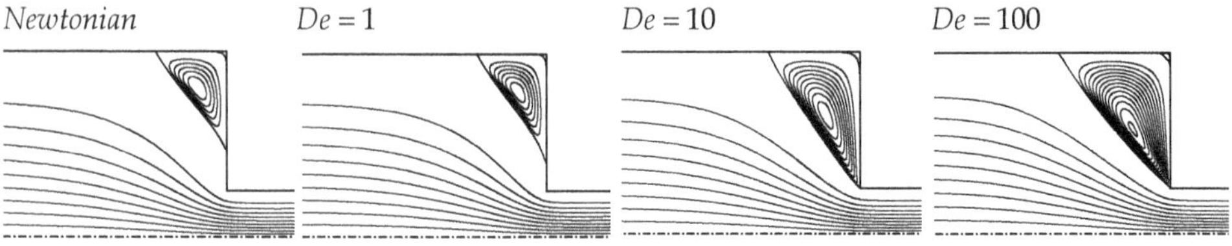

Fig. 10. Evolution of the flow patterns with *De* for a linear PTT fluid (ε = 0.25) in a 4:1 plane sudden contraction. Adapted from Alves et al (2003b).

The variation of the Couette correction for the linear PTT fluid is shown in Figure 11. *C* decreases with *De* until a minimum negative value is reached at $De \approx 20$, and then increases. Hence, the growth of *C* takes place only for $De > 20$ and therefore it is possible again to question the usefulness of this model to predict the enhanced pressure losses observed experimentally in sudden contraction flows. However, we should note the measurements of Nigen and Walters (2002) in a plane sudden contraction, which are not fully conclusive: in their Figure 13 there are negative values of *C* (C = -1.37) for a flow rate of 40 g/s with their Boger fluid 2 and shear-thinning syrup 2. It is possible that such negative values are due to the large experimental uncertainty, since their measurements in an axisymmetric sudden contraction with a short outlet pipe have shown a continuous drop in excess pressure drop ($C > 0$). More details of the predictions can be found in Alves et al (2003b).

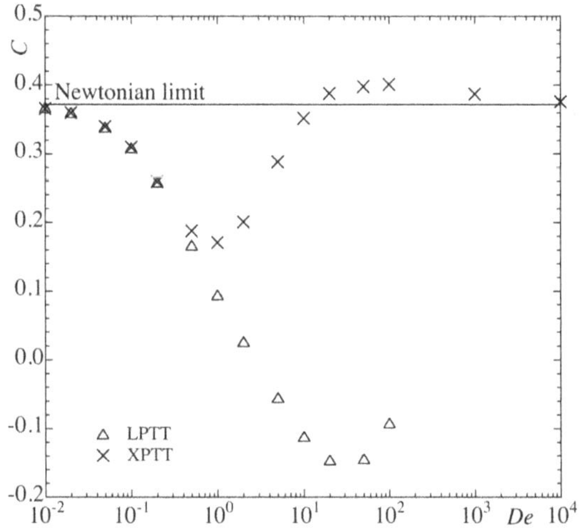

Fig. 11. Variation of the Couette correction with *De* for PTT fluids (ε = 0.25) in a 4:1 plane contraction flow. LPTT: linear PTT; XPTT: exponential PTT. Adapted from Alves et al (2003b).

The exponential stress coefficient in the PTT model (XPTT) substitutes the high strain plateau of the extensional viscosity by strain-thinning after the peak extensional viscosity. This affects the fluid dynamical behaviour of the PTT model since the fluid rheology tends to Newtonian-like at high shear and extensional deformation rates, as shown in Figures 9 and 11 which include results of predictions for the XPTT model. Our results for this fluid agree with the literature for X_R, Ψ_R and *C*, tending to the Newtonian values at very high *De*, as expected. For this model we could obtain iterative convergence at extremely high *De* ≈ 10 000, even using the standard stress formulation. Although more in line with experimental observations in terms of the variation of *C* with *De*, the computed values of *C* are still quite lower than those measured.

As a final comment, we remark that although part of the increase in *C* does correspond to a real increase in pressure drop, the major effect here is related to the normalization employed to define *C* where the pressure drop is scaled with the wall shear stress under fully-developed conditions in the downstream channel. Since this wall stress decreases significantly because of the shear-thinning behaviour of the PTT fluid (see Oliveira and Pinho, 1999c), the coefficient increases. This, and previous comments, pinpoint a crucial issue associated with constitutive equation modelling, as the current models are unable to predict correctly the enhanced entry pressure loss measured when elastic liquids flow through contractions. Most likely, models with increased internal energy dissipation are required for such purpose.

5. Acknowledgments

We are indebted to Fundação para a Ciência e a Tecnologia and FEDER for funding support over the years.

6. References

Aboubacar M and Webster MF (2001). *J. Non-Newt. Fluid Mech.*, Vol. 98, pp. 83-106.

Aboubacar M, Matallah H and Webster MF (2002). *J. Non-Newt. Fluid Mech.*, Vol. 103, pp. 65-103.

Afonso AM, Oliveira PJ, Pinho FT and Alves MA (2009). *J. Non-Newt. Fluid Mech.*, Vol. 157, pp. 55-65.

Afonso AM, Oliveira PJ, Pinho FT and Alves MA (2011). *J. Fluid Mech.*, Vol. 677, 272-304.

Afonso AM, Pinho FT and Alves MA (2012). *J. Non-Newt. Fluid Mech*, Vol. 167-168, pp. 30-37.

Alves MA, Pinho FT and Oliveira PJ (2000). *J. Non-Newt. Fluid Mech.*, Vol. 93, pp. 287-314.

Alves MA, Pinho FT and Oliveira PJ (2001a). *J. Non-Newt. Fluid Mech.*, Vol. 97, pp. 205-230.

Alves MA, Pinho FT and Oliveira PJ (2001b). *J. Non-Newt. Fluid Mech.*, Vol. 101, pp. 55-76.

Alves MA, Oliveira PJ and Pinho FT (2003a). *Int. J. Numer. Meth. Fluids*, Vol. 41, pp. 47-75.

Alves MA, Oliveira PJ and Pinho FT (2003b). *J. Non-Newt. Fluid Mech*, Vol. 110, pp. 45-75.

Baliga RB and Patankar SV (1983). *Num. Heat Trans.*, Vol. 6, pp. 245-261.

Bevis MJ, Darwish MS and Whiteman JR (1992). *J. Non-Newt. Fluid Mech.*, Vol. 45, pp. 311-337.

Binding DM (1988). *J. Non-Newt. Fluid Mech.*, Vol. 27, 173-189.

Bird RB, Armstrong RC and Hassager O (1987). *Dynamics of Polymeric Liquids, Volume 1: Fluid Mechanics*, John Wiley, New York.

Boger DV, Hur DU and Binnington RJ (1986). *J. Non-Newt. Fluid Mech.*, Vol. 20, pp. 31-49.

Boger DV (1987). *Annual Rev. of Fluid Mech*, Vol. 19, 157-182.

Cable PJ and Boger DV (1978a). *AIChE J*, Vol. 24, pp. 868-879.

Cable PJ and Boger DV (1978b). *AIChE J*, Vol. 24, pp. 992-999.

Cable PJ and Boger DV (1979). *AIChE J*, Vol. 25, pp. 152-159.

Coates PJ, Armstrong RC and Brown RA (1992). *J. Non-Newt. Fluid Mech.*, Vol. 42, pp. 141-188.

Crochet MJ, Davies AR and Walters K (1984). *Numerical Simulation of Non-Newtonian Flow*, Elsevier, Amsterdam.

Darwish MS and Moukalled F (1994). *Num. Heat Transfer*, Part B, Vol. 26, pp. 79-96.

Dean WR and Montagnon PE (1949). *Proc. Cambridge Philo. Soc.*, Vol. 45, pp. 389-394.

Debbaut B, Marchal JM and Crochet MJ (1988). *J. Non-Newt. Fluid Mech.*, Vol. 29, pp. 119-146.

Dou HS and Phan-Thien N (1999). *J. Non-Newt. Fluid Mech.*, Vol. 87, pp. 47-73.

Evans RE and Walters K (1986). *J. Non-Newt. Fluid Mech.*, Vol. 20, pp. 11-29.

Evans RE and Walters K (1989). *J. Non-Newt. Fluid Mech.*, Vol. 32, pp. 95-105.

Fan Y, Tanner RI and Phan-Thien N (1999). *J. Non-Newt. Fluid Mech.*, Vol. 84, pp. 233-256.

Fattal R and Kupferman R, (2004). *J. Non-Newt. Fluid Mech.* Vol. 123, pp. 281-285.

Ferziger JH and Perić (2002). *Computational Methods for Fluid Dynamics*. Springer Verlag, 3rd edition, Berlin.

Gaskell PH and Lau AKC (1988). *Int. J. Num. Meth. Fluids*, Vol. 8, pp. 617-641.

Gervang B and Larsen PS (1991). *J. Non-Newt. Fluid Mech.*, Vol. 39, pp. 217-237.

Hassager O (1988). *J. Non-Newt. Fluid Mech.*, Vol. 29, pp. 2-55.

Hinch EJ (1993). *J. Non-Newt. Fluid Mech.*, Vol. 50, pp. 161-171.

Hu HH and Joseph DD (1990). *J. Non-Newt. Fluid Mech.*, Vol. 37, pp. 347-377.

Huang X, Phan-Thien N and Tanner RI (1996). *J. Non-Newt. Fluid Mech.*, Vol. 64, 71-92.

Issa RI and Oliveira PJ (1994). *Computers and Fluids*, Vol. 23, pp. 347-372.

Khosla PK and Rubin SG (1974). *Computers and Fluids*, Vol. 2, pp. 207-209.

Lawler JV, Muller SJ, Brown RA and Armstrong RC (1986). *J. Non-Newt. Fluid Mech.*, Vol. 20, pp. 51-92.

Leschziner MA (1980). *Comp. Methods Appl. Mech. Eng.*, Vol. 23, pp. 293-312.

Leonard BP (1991). *Comp. Meth. Appl. Mech. Eng.*, Vol. 88, pp. 17-74.

Mackenzie JA, Crumpton PI and Morton KW (1993). *J. Comput. Phys.*, Vol. 109, 1-15.

Matallah H, Townsend P and Webster MF (1998). *J. Non-Newt. Fluid Mech.*, Vol. 75, pp. 139-166.

Matos HM, Alves MA and Oliveira PJ (2009). *Num. Heat Transfer, Part B*, Vol. 56, pp. 351-371.

McKinley G, Raiford WP, Brown RA and Amstrong RC (1991). *J. Fluid Mech.*, Vol. 223, pp. 411-456.

Meijerink JA and Van Der Vorst HA (1977). *Math. of Comp.*, Vol. 31, pp. 148-162.

Missirlis KA, Assimacopoulos D and Mitsoulis E (1998). *J. Non-Newt. Fluid Mech.*, Vol. 78, pp. 91-118.

Moffatt HK (1964). *J. Fluid Mech.*, Vol. 18, pp. 1-18.

Mompean G and Deville M (1997). *J. Non-Newt. Fluid Mech.*, Vol. 72, pp. 253-279.

Morton KW and Paisley MF (1989). *J. Comput. Physics*, Vol. 80, pp. 168-203.

Nguyen H and Boger DV (1979). *J. Non-Newt. Fluid Mech.*, Vol. 5, pp. 353-368.

Nigen S and Walters K (2002). *J. Non-Newt. Fluid Mech.*, Vol. 102, pp. 343-359.

Oliveira PJ (1992). Computer modelling of multidimensional multiphase flow and applications to T- junctions. PhD thesis, Imperial College, Univ. of London, UK.
Oliveira PJ, Pinho FT and Pinto GA (1998). *J. Non-Newt. Fluid Mech.*, Vol. 79, pp. 1-43.
Oliveira PJ and Pinho FT (1999a). *Num. Heat Transfer, Part B*, Vol. 35, pp. 295-315.
Oliveira PJ and Pinho FT (1999b). *J. Non-Newt. Fluid Mech.*, Vol. 88, pp. 63-88.
Oliveira PJ and Pinho FT (1999c). *J. Fluid Mech.*, Vol. 387, pp. 271-280.
Oliveira PJ (2001). *Num. Heat Transfer*, Part B, 40, pp. 283-301.
Oliveira PJ (2002). *Acta Mechanica*, Vol. 158, pp. 157-167.
Oliveira MSN, Oliveira PJ, Pinho FT and Alves MA (2007). *J. Non-Newt. Fluid Mech.*, Vol. 147, pp. 92-108.
Owens RG, Chauvière C and Phillips TN (2002). *J. Non-Newt. Fluid Mech.*, Vol. 108, pp. 49-71.
Owens RG and Phillips TN (2002). *Computational Rheology*. Imperial College Press, London, UK.
Patankar SV (1980). *Numerical Heat Transfer and Fluid Flow*. McGraw-Hill, New York.
Patankar SV and Spalding DB (1972). *Int. J. Heat Mass Transfer*, Vol. 15, pp. 1787-1806.
Perera MGN and Walters K (1977). *J. Non-Newt. Fluid Mech.*, Vol. 2, pp. 49-81.
Phan-Thien N (1978). *J. Rheol.*, Vol. 22, pp. 259–283.
Phan-Thien N and Dou H-S (1999). *Comput. Methods Appl. Mech. Eng.*, Vol. 180, pp. 243-266.
Phan-Thien N and Tanner RI (1977). *J. Non-Newt. Fluid Mech.*, Vol. 2, pp. 353–365.
Phillips TN and Williams AJ (1999). *J. Non-Newt. Fluid Mech.*, Vol. 86, pp. 215-246.
Rajagopalan D, Armstrong RC and Brown RA (1990). *J. Non-Newt. Fluid Mech.*, Vol. 36, pp. 159-192.
Rhie CM and Chow WL (1983). *AIAA J*, Vol. 21, pp. 1525-1532.
Rodd LE, Scott TP, Boger DV, Cooper-White JJ, McKinley GH (2005). *J. Non-Newt Fluid Mech.* Vol. 129, pp. 1–22
Rogerson MA and Yeow YL (1999). *J. Appl. Mech.- Trans ASME*, Vol. 66, pp. 940-944.
Rothstein JP and McKinley GH (2001). *J. Non-Newt. Fluid Mech.*, Vol. 98, pp. 33-63.
Sasmal, GP (1995). *J.Non-Newt. Fluid Mech.*, Vol. 56, pp. 15-47.
Sato T and Richardson S (1994). *J. Non-Newt. Fluid Mech.*, Vol. 51, pp. 249- 275.
Walters K and Webster M F (1982). *Phil. Trans. R. Soc. London A*, Vol. 308, pp. 199-218.
Wapperom P and Webster MF (1998). *J. Non-Newt. Fluid Mech.*, Vol. 79, pp. 405-431.
Wapperom P and Webster MF (1999). *Comput. Methods Appl. Mech. Eng.*, Vol. 180, pp. 281-304.
White SA and Baird DG (1986). *J. Non-Newt. Fluid Mech.*, Vol. 20, pp. 93-101.
White SA and Baird DG (1988a). *J. Non-Newt. Fluid Mech.*, Vol. 29, pp. 245-267.
White SA and Baird DG (1988b). *J. Non-Newt. Fluid Mech.*, Vol. 30, pp. 47-71.
Williamson CHK (1996). *Ann. Rev. Fluid Mech.*, Vol. 28, pp. 477-539.
Van Doormal JP and Raithby GD (1984). *Num. Heat Transfer*, Vol. 7, pp. 147-163.
Versteeg HK and Malalasekera W (2007). *An Introduction to Computational Fluid Dynamics: The Finite Volume Method*, 2nd Ed., Pearson Education Limited, Edinburgh, UK.
Vinokur M (1989). *J. Comp. Physics*, Vol. 81, pp. 1-52.
Xue S-C, Phan- Thien N and Tanner RI (1995). *J. Non-Newt. Fluid Mech.*, Vol. 59, pp. 191-213.

Xue S-C, Phan- Thien N and Tanner RI (1998a). *Rheol. Acta.*, Vol. 37, 158-169.
Xue S-C, Phan- Thien N and Tanner RI (1998b). *J. Non-Newt. Fluid Mech.*, Vol. 74, pp. 195-245.
Yesilata B, Öztekin A and Neti S (1999). *J. Non-Newt. Fluid Mech.*, Vol. 85, pp. 35-62.
Yoo JY and Na Y (1991). *J. Non-Newt. Fluid Mech.*, Vol. 39, pp. 89-106.
Zana E, Tiefenbruck F and Leal LG (1975). *Rheol. Acta*, Vol. 14, pp. 891-898.

Part 2

Studies of Particular Problems Through FVM, Development of New Ways for Their Solution

8

Volume-of-Fluid (VOF) Simulations of Marangoni Bubbles Motion in Zero Gravity

Yousuf Alhendal and Ali Turan
The University of Manchester
United Kingdom

1. Introduction

This chapter deals with two-phase flows, i.e. systems of different fluid phases such as gas and liquid. A typical example of a two-phase flow is a motion of a particle (bubble or droplet) in a stagnant fluid (liquid or gas). In many branches of engineering it is important to be able to describe the motion of gas bubbles in a liquid (Krishna & Baten, 1999). In multiphase flow, the simultaneous flow strongly depends on the gravity force. However, in zero gravity conditions, buoyancy effects are negligible and as an alternative, three different methods were found to make the bubbles or drops move in zero gravity. They are electrocapillary, solutalcapillary and thermocapillary motion.

When a temperature gradient exists on the interface, the surface tension varies along the interface, resulting in bulk fluid motion, called thermocapillary (Marangoni) flow. In normal gravity this thermocapillary flow tends to be weighed down by buoyancy driven flow. However, for small geometry and/or zero gravity environments, this is not the case and thermocapillary is dominant and it could become an important driving force.

Bubbles suspended in a fluid with a temperature gradient will move toward the hot region due to thermocapillary forces. Surface tension generally decreases with increasing temperature and the non-uniform surface tension at the fluid interface leads to shear stresses that act on the outer fluid by viscous forces, thus inducing a motion of the fluid particle (a bubble or a drop) in the direction of the thermal gradient. In space, where buoyancy forces are negligible, thermocapillary forces can be dominant and can lead to both desirable and undesirable motion of fluid particles.

Particle dynamics has become a very important study area for fundamental research and applications in a zero gravity environment, such as space material science, chemical engineering, space-based containerless processing of materials e.g., glass is believed to have the potential of producing very pure materials, (Uhlmann, 1982), and thermocapillary migration may provide mechanisms to remove bubbles from the melt. Control of vapor bubbles forming in both the fuel systems of liquid-rockets (Ostrach, 1982) and the cooling system of space habitats may be achievable using thermocapillary migration. Thermocapillary migration may also lead to accumulation of gas bubbles on the hot surface of heat exchangers and reducing their efficiency. Ostrach (1982) studied various types of fluid flows that could occur under low-gravity conditions and pointed out that Marangoni convection is one of the important flows. In practical applications, it is frequently necessary to deal with a large number of bubbles or drops and their collective behavior may differ from what one might expect based on results for a single particle.

1.1 The need for CFD in zero gravity investigation

An understanding of the behaviour of two phase fluids flow in zero gravity conditions is important for designing useful experiments for the space shuttle and the international space station. In addition, such understanding is important for the future design of thermo-fluid systems and machinery that might be employed in comparable environments. However, it is quite expensive and time-consuming to design and fabricate a space experiment and very little is known of the microgravity behavior of fluids due to the relative difficulty of obtaining experimental data under such conditions.

The available experimental results are usually supposed to be the major source of information on the behaviour of the physical process of two phase flow at zero gravity. However, because of the difficulties in obtaining experimental results, the numerical methods modelling turns out to be the ideal tool allowing to investigate the behaviour of two phase flow and capture the flow physics in plenty of time and of course less cost. The available literature of the zero gravity particle behavior is limited to low Re and Ma because of the difficulties in obtaining experimental results in microgravity (Kang et. al., 2008), and computer simulation can help understand the basic fluid physics, as well as help design the experiments or systems for the microgravity condition. Experiments under normal and microgravity conditions are too costly as well as being complicated, (Bozzano, 2009), and hence, numerical simulations become an important tool in research studies of two-phase flows under a microgravity environment. wolk et al, (2000) referred to the important of prediction of the flow pattern in a gas/liquid flow in two phase flow and especially concerning the particle dynamics in zero gravity and the need to predict accurately the existing flow patterns. A number of experiments have been conducted in drop towers, sounding rockets and aboard space shuttles; see the extensive review of (Subramanian et al., 2002). These experiments have noted complicated transients and time-dependent behavior in regimes where the flow has finite viscous and thermal inertia (Treuner et al., 1996; Hadland et al., 1999; Wozniak et al., 2001). Those experimental studies noted that there are no theoretical or numerical results with which to compare their experiments. It is also uneasy to get complete information about the behavior of bubble in space and a CFD study has been undertaken by many researchers to compare and analyse their experimental results, (Treuner et al., 1996). The shape and the area of the varying interface are very complex and a simulation study is required, (Subramanian et al., 2009). Two phase flow experiments generally require continuous observation of moving fluid during a test, which makes the experiment complicated. It is also a challenge for space researchers to design a space experiment to accommodate most of their objectives. For the above reasons and more it is necessary to carry out appropriate numerical simulations for the behavior of bubble/drop in microgravity and numerical simulation can help understand the basic fluid physics, as well as assist design the experiments or systems for the zero gravity environment.

In this chapter, we are specifically targeting the use of computation in order to simulate thermocapillary (Marangoni) fluid flow in a zero gravity condition to better understand the physical processes behind many of the observed physical phenomena of zero gravity environments. It also allows sensitivity and feasible studies to be carried out for different parameters and design.

1.2 Description of the chapter and objectives

In this chapter we will use the ANSYS-Fluent (Release13.0, 2011) code to analyze and design bubble flow system in a zero gravity environment and investigate sensitivity studies to various parameters.

The literature review of this chapter examines the challenges facing fluid experiments aboard orbiting spacecraft. This illustrates the importance of the CFD simulations and examines the effect of several external and internal forces on the bubble flow in zero gravity. The finite volume method (FVM) with a fixed non uniform spatial grid was used to computationally model 2-D axis-symmetric and 3-D domains. The model's solution algorithms, boundary conditions, source terms, and fluid properties used in the simulation are described in details in this chapter with sample calculation.

This chapter provides an interesting opportunity to test the capability of finite volume method (FVM) simulation and the Volume of fluid (VOF) method to accurately represent thermocapillary flow of single and multi bubbles. In addition, several published articles concerning the use of the Volume of Fluid (VOF) method in bubble/drops simulation are reviewed. Lastly, the simulation was able to examine scenarios not covered experimentally in order to verify the theoretical prediction of the effect of temperature different, column-particle aspect ratio, and effect of rotational on the coalescence of multi bubbles under the effect of both fluid rotation and surface tension. Though a huge amount of publications (textbooks, conference proceedings and journal articles) concern two-phase, publications on two phase flow in microgravity is a very little studied field and information about bubble behavior in a rotating column in particular is not so complete comparing with other physical phenomena in normal gravity. These are the main reasons for carrying out simulation research in microgravity.

2. Fluid dynamics

In order to study two-phase fluids in microgravity we should understand concepts and notions of fluid dynamics used in later sections and to remind the reader of some fundamental dimensionless quantities which will be frequently encountered.

2.1 Compressible and incompressible fluid

Fluids are compressible if changes in pressure or temperature will result in changes in density. However, in many situations the changes in pressure and temperature are sufficiently small that the changes in density are negligible. In this case the flow can be modeled as an incompressible flow.

2.2 Newtonian flow

Sir Isaac Newton showed how stress and the rate of strain are related for many common fluids. The so-called Newtonian fluids are described by a coefficient called viscosity, which depends on the specific fluid.

2.3 Viscosity

Viscosity describes fluids resistance to flow and may be thought as a measure of fluid friction. In viscous problems fluid friction has significant effects on the fluid motion. Real fluids are fluids which have a resistance to shear stress. Fluids which do not have any resistance are called ideal fluid. Viscosity is divided in two types: dynamic (or absolute) and kinematic viscosity. Dynamic viscosity is the ratio between the pressure exerted on the surface of a fluid and the velocity gradient. The SI physical unit of dynamic viscosity is Pa.s. In many situations, we are concerned with the ratio of the viscous force to the inertial force, the latter characterized by the fluid density ρ. This ratio is characterized by the kinematic viscosity, defined as follows:

$$\nu = \frac{\mu}{\rho} \tag{1}$$

Where v is the kinematic viscosity in m^2/s, ρ is the density and μ is the dynamic viscosity.

2.4 Reynolds number

The Reynolds number is a measure of the ratio of inertial to viscous forces and quantifies the importance of these two types of forces for given flow conditions. It is the most important dimensionless number in fluid dynamics and it is commonly used to provide a criterion for determining dynamic similitude. Reynolds number is defined as follows:

$$\mathrm{Re} = \frac{Vo\ R}{\mathrm{v}} \tag{2}$$

Where R is the particle radius in m, V_o is the velocity in m/s.

The velocity V_0derived from the tangential stress balance at the free surface is used for scaling the migration velocity in Eq. (2) and (4):

$$V_0 = \frac{\frac{d\sigma}{dt}\frac{dT}{dx}}{\mu} \tag{3}$$

Where, the constant $\frac{d\sigma}{dT}$ (or σ_T) is the rate of change of interfacial tension and $\frac{dT}{dX}$, the temperature gradient imposed in the continuous phase fluid.

2.5 Marangoni number

In zero gravity a bubble or a drop in an immiscible fluid will move toward the warmer side when subjected to a temperature gradient. Such a phenomenon is known as Marangoni flow or the thermocapillary migration. Marangoni flow is induced by surface tension gradients asa result of temperature and/or concentration gradients.

$$Ma = \frac{RV_0}{\alpha} = \mathrm{Re.Pr} \tag{4}$$

Here, α is the temperature diffusivity

2.6 Prandtl number (Pr)

Pr is the ratio of kinematic viscosity to thermal diffusivity. Very high Pr (Pr >> 1) fluids are usually very viscous, while highly thermally conducting fluids, usually liquid metals, have low Pr (Pr << 1).

$$\Pr = \frac{\nu}{\alpha} \tag{5}$$

In 1959, Young et al. (1959) first investigated the thermocapillary migration of bubbles and drops with their linear model, which is suitable for small Reynolds number (Re) and small Marangoni number (Ma). This model is now called the YGB model. The Reynolds number is the ratio of inertial forces to viscous forces. Small Re means that inertial effects are negligible. The Marangoni number is the ratio of convective transport of energy to heat conduction. Small Ma means that convective heat transfer is negligible compared to heat conduction. YGB model velocity V_{YGB} (Young et al. 1959) is expressed as follows:

$$V_{YGB} = \frac{2|\sigma_T| R\nu\Delta T_\infty}{(2\mu + 3\mu')(2k + k')} \tag{6}$$

where R is the radius of the bubble; $\Delta T\infty$ is the temperature gradient; μ and μ', k and k' are the dynamic viscosity, thermal conductivity of gas and continuous phase, respectively.

2.7 Microgravity

Gravity is a force that governs motion throughout the Universe. It holds us to the ground, keeps the Moon in orbit around the Earth, and the Earth in orbit around the Sun. The nature of gravity was first described more than 300 years ago. Gravity is the attraction between two masses. Bigger the mass, most apparent the attraction is. The acceleration of an object caused only by gravity, near the surface of the Earth, is called normal gravity, or 1g. The condition of microgravity comes about whenever an object is in "free fall": that is, it falls faster and faster, accelerating with exactly the acceleration due to gravity (1g). Objects in a state of free-fall or orbit are said to be "weightless."

3. Computational procedure

3.1 VOF model

The CFD software offer several models to incorporate multiphase flows; every model is developed for its own specific flow type. The governing continuum conservation equations for two phase flow were solved using the commercial software package (ANSYS, 2011), and the Volume of Fluid (VOF) method (Hirt et al. 1981) was used to track the liquid/gas interface.

Volume of Fluid (VOF) model is designed for two or more immiscible fluids, where the position of the interface between the fluids is of interest (ANSYS, 2011). Applications of the VOF model include the prediction of jet break-up, the motion of large bubbles in a liquid, the motion of liquid after a dam break, or the steady or transient tracking of any liquid-gas interface and If the bubble is so large that it extends across several control volumes, the VOF formulation is appropriate to track its boundary. Most ANSYS-FLUENT models are available in combination with the VOF model. For example, the sliding-mesh model can be used to predict the shape of the surface of a liquid in a mixing tank. The deforming mesh capability is also compatible with the VOF model. The porous media model can be used to track the motion of the interface between two fluids through a packed bed or other porous region. The effects of surface tension may be included (for one specified phase only), and in combination with this model, you can specify the wall adhesion angle. Heat transfer from walls to each of the phases can be modeled, as can heat transfer between phases. The basic idea of the volume of fluid (VOF) method is to consider a colour function, defined as the volume fraction of one of the fluids within each cell, to capture the interface. This function will be one if the cell is filled with the gas phase, zero if the cell is filled with the liquid phase, and between zero and one in the cells where there is an interface. The VOF method belongs to the so called "one" fluid method, where a single set of conservation equations is solved for the whole domain. The VOF method codes have been used extensively to calculate the hydrodynamics of bubbles rising in liquid by Kawaji et al.,(1997). They noted a new result which was not observed experimentally when they compared the numerical simulation results from a two dimensional simulation of a Taylor bubble rising in a stagnant liquid filled tube to experimental analysis. The work of Tomiyama et al., (1993) illustrated the capability of VOF to accurately simulate bubble shape, and the simulated shapes were shown to agree with experimental published data. The researchers showed their results of predicted bubble shapes were in good agreement with those of Bhaga & Weber (1981). The VOF simulations with gas-liquid systems could be used as an investigative tool for studying bubble rise and bubble-bubble interactions in gas-liquid bubble columns, (Alhendal et al.,2010). The Volume of Fluid (VOF) method is made for flows with completely separated phases; the phases do not diffuse into each other (ANSYS, 2011). "geo-reconstructed-VOF" method in Fluent is chosen for this investigation. Geo-reconstruction is added to the VOF scheme to define the free surface more accurately, (Hirt et al., 1981). Applications of the VOF model include stratified flows, free-surface flows, the steady or transient tracking of any liquid gas interface (ANSYS, 2011).

3.2 Formulation of the problem and the solution strategy

The movement of the gas-liquid interface is tracked based on the distribution of α_G, the volume fraction of gas in a computational cell, where $\alpha_G = 0$ in the liquid phase and $\alpha_G=1$ in the gas phase. Therefore, the gas-liquid interface exists in the cell where α_G lies between 0 and 1. The geometric reconstruction scheme that is based on the piece linear interface calculation (PLIC) method of Youngs (1982) is applied to reconstruct the bubble free surface. A single momentum equation, which is solved throughout the domain and shared by all the phases, is given by:

$$\frac{\partial}{\partial t}(\rho\vec{v}) + \nabla.(\rho\vec{v}\vec{v}) = -\nabla p + \nabla.[\mu(\nabla\vec{v} + \nabla\vec{v})^T] + \rho + \vec{F} \tag{7}$$

where v is treated as the mass-averaged variable.

$$v = \frac{\alpha_G\rho_G v_G + \alpha_L\rho_L v_L}{\rho} \tag{8}$$

In this bubble simulation, $\vec{F}$ represents the volumetric forces at the interface resulting from the surface tension force per unit volume. The CSF model of Brackbill et al., (1992) was used to compute the surface tension force for the cells containing the gas-liquid interface:

$$\vec{F} = \sigma\frac{\rho k n}{\frac{1}{2}(\rho_L + \rho_G)} \tag{9}$$

Where σ is the coefficient of surface tension, n is the surface normal which is estimated from the gradient of volume fraction, κ is the local surface curvature calculated as follows:

$$k = -(\nabla \hat{n}) = \frac{1}{n}\left[\frac{n}{|n|}\nabla|n| - (\nabla . n)\right] \tag{10}$$

The tracking of the interface between the gas and liquid is accomplished by the solution of a continuity equation for the volume fraction of gas, which is:

$$\frac{\partial}{\partial t}(\alpha_G \rho_G) + \nabla \cdot (\alpha_G \rho_G \vec{v}_G) = 0 \tag{11}$$

The volume fraction equation is not solved for the liquid; the liquid volume fraction is computed based on the following constraint:

$$\alpha_G + \alpha_L = 1 \tag{12}$$

where α_G and α_L is the volume fraction of gas and liquid phase respectively.

The properties appearing in the transport equations are determined by the presence of the component phases in each control volume and are calculated as volume-averaged values. The density and viscosity in each cell at interface were computed by the application of following equations:

$$\rho = \alpha_G \rho_G + (1 - \alpha_G)\rho_L \tag{13}$$

$$\mu = \alpha_G \mu_G + (1 - \alpha_G)\mu_L \tag{14}$$

where ρ_G, ρ_L, μ_G and μ_L is density and viscosity of gas and liquid phase respectively, while α_G is the volume fraction of gas. The energy equation is also shared among the phases:

$$\frac{\partial}{\partial t}(\rho E) + \nabla . [\vec{v}(\rho E) + p)] = \nabla . (\kappa_{eff} \nabla T) \tag{15}$$

The VOF model treats energy, E, and temperature, T, as mass-averaged variables:

$$E = \frac{\sum_{q=1}^{n} \alpha_q \rho_q E_q}{\sum_{q=1}^{n} \alpha_q \rho_q} \tag{16}$$

where E_q for each phase is based on the specific heat of that phase and the shared temperature. The effective thermal conductivity k_{eff} is also shared by the phases.

3.3 Using Fluent - Grid generation and independence

The bubble was initially placed at the centre of the cylindrical domain by using the region adaptation option of ANSYS-Fluent. To do that, a spherical patch or domain was selected at

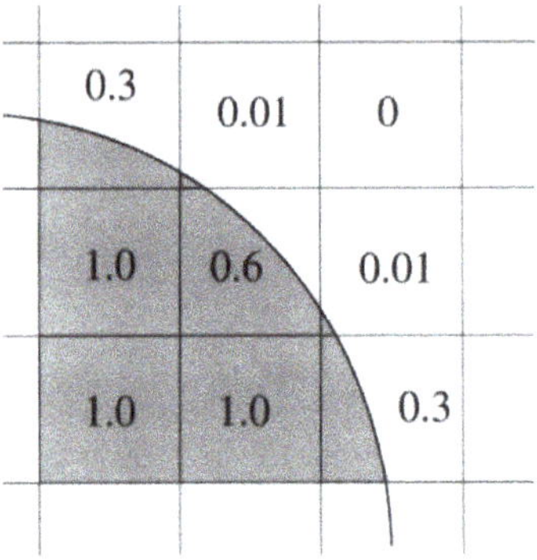

Fig. 1. Volume of fluid (VOF) interface reconstruction

Fig. 2. Typical mesh used for Marangoni cases

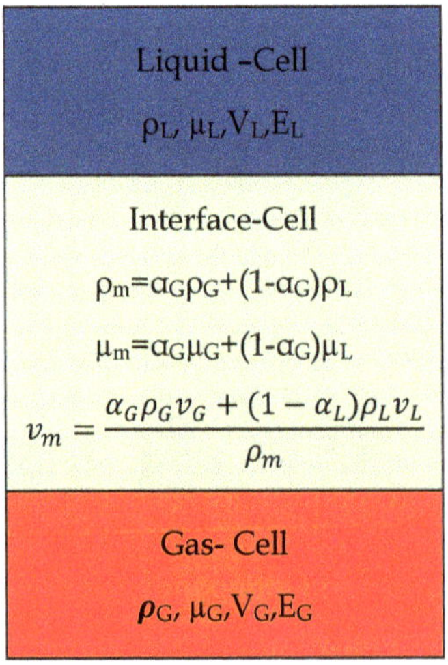

Fig. 3. Volume fraction and properties in each cell in the bubble

the cylinder with void fraction of 1. The rest of the cylindrical domain was specified as liquid, i.e., with a void fraction of zero. Figure 4 depicts the bubble inside the computational domain at the initiation of a simulation. The geometry of the two-dimensional model is somewhat similar to the one used by Thompson et al. (1980), and extension to fully 3-dimensional model is followed by the 2-dimensional study. The initial rise velocity for the bubble was zero. The upper surface of the fluid is a hooter than the bottom surface boundary, which is assumed to be flat with no-slip walls. The grid details are shown in fig. 2. After doing the required sensitivity grid tests, a non-uniform grid of 7200 cells was used throughout the 2-D axis simulations with the grid lines clustered towards the centre, and uniform grid of 372000 cells for the 3-D cases.

Fig. 4. Initial condition for the bubble inside the 2-D axis.

3.3.1 Grid independence

For the 2-d axis, the domain was split into two areas to create non-uniform mesh clustered in the area of interest (centre), see fig. 2 for domain mesh technique. Another important point is that, when using the axisymmetric solver, you create a mesh only for the half of your domain, thus reducing drastically the number of cells you use, and consequently the time of calculation. Fluent will accept the axis in the direction of the positive X-axis only. The model was verified for grid independence and geometry-related such as 2-d axis and rotational periodic flow as shown in figs. 5 & 6. Grid independence was examined by using three grid systems with 20x20x80 (96000 cells), 30x30x120 (324000 cells) and 40x40x160 (768000 cells). The three simulations showed nearly bubble migration for the simulations with the maximum difference of less than 1% between the cases. For calculation efficiency, it was found that the 30x30x120 mesh produces a grid-independent solution and was used in this calculation with 0.001 time step. For some cases when using the periodic flow solver, you create a mesh only for the quarter, half, three quarters of the domain, thus reducing the number of cells you use, and consequently the time of calculation. Periodic boundary condition is recommended for the bubble located at the centre of the domain.

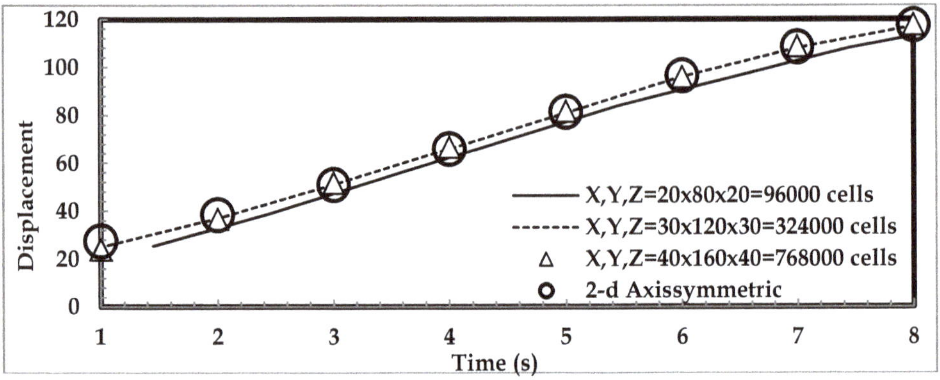

Fig. 5. Grid independence shows identical bubble migration for 2-d axis & 3-D simulations

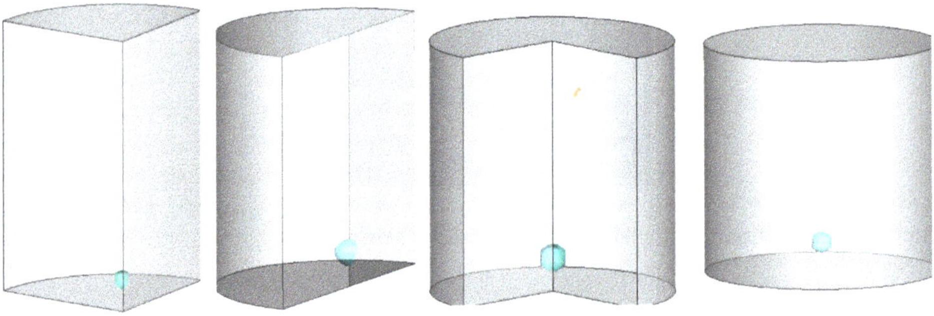

Fig. 6. Rotational periodic/full geometry used to calculate bubbles/drops migration

4. Numerical procedure and simulation

In this first test case, a bubble with a diameter of 6 mm was placed 10 mm from the bottom wall (cold). Hence, the size of the computational wall bounded domain was chosen as 120 x60 mm with no "inflow or overflow" from the sides. For the purposes of these simulations, ethanol properties were taken to be same as the properties given in table (1) from Thompson et al. (1980). The unsteady 2-D axisymmetric and 3-D models were formulated using the commercial software package (ANSYS-FLUENT® v.13, 2011) in modeling the rise of a bubble in a column of liquid in zero gravity (Marangoni flow). Surface tension and its temperature coefficient used in the simulations for the ethanol and N_2 are (σ =27.5 dynes/cm and σ_T =-0.09 dynes/cm°C), (Kuhlmann, 1999). A numerical prescription for the surface tension vs. temperature behavior is provided via a user defined function (UDF) which can be dynamically linked with the FLUENT solver.

	Unit	Ethanol	Nitrogen (N_2)
Density (ρ)	kg/m^3	790	1.138
Specific Heat (Cp)	j/kg-k	2470	1040.7
Thermal Conductivity (k)	w/m-k	0.182	0.0242
Viscosity (μ)	kg/m-s	0.0012	1.66e-5
Surface tension coefficient (σ_T)	w/m.k	0.00009	----
Velocity V_O Eq. (3)	(m/s)	0.075	
Velocity V_{YGB} Eq. (6)	(m/s)	0.034	
Velocity -V (CFD)	(m/s)	0.014	
Scaled velocity V/ V_{YGB}	-	0.41	
Prandtl Number (Pr)	Eq. (5)	16.28	
Reynolds Number(Re)	Eq. (2)	197.5	
Marangoni Number(Ma)	Eq. (4)	3216.4	

Table 1. Physical properties of the liquids employed in the simulation at 300K and sample results for bubble diameter= 9 mm for (Pr=16.28).

4.1 Description of test cases and results

Predicted simulations have been compared with the experimental work of Thompson et al., (1980) as shown in fig. 8 and agreement obtained. In a non-uniform temperature gradient, the fluid on the bottom is cold, and therefore has greater surface tension. The fluid on the top is hot, therefore possesses weaker surface tension. The tendency of the fluid with greater surface tension is going to pull the fluid with less surface tension towards it. This motion would be from top to bottom. "Whenever surface is created, heat is absorbed, and whenever surface is destroyed heat is given off. Therefore a swimming bubble absorbs heat at its hot end and rejects heat at its cold end" (Nas & Tryggvason, 1993) as seen in temperature contour of fig 9.

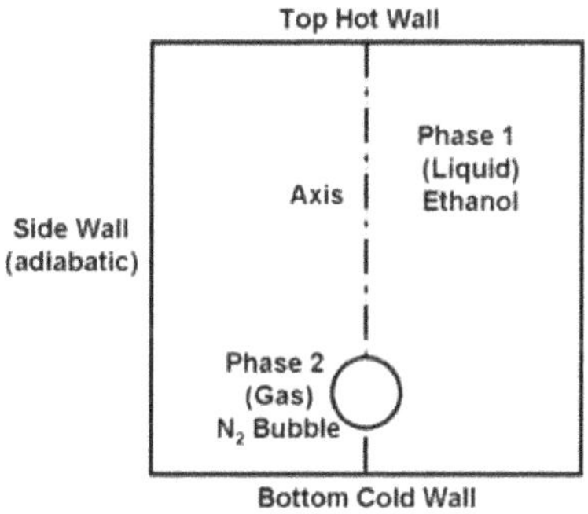

Fig. 7. Schematic of bubble migration in a uniform temperature gradient.

The ANSYS-FLUENT code allows for the use of different gravity values. With this option, we ran several cases for different geometry conditions. The cases that are presented in this chapter were conducted in a zero gravity environment so that the flow would be driven by surface tension instead of buoyancy. On earth-based, where gravity presents, the flow is mainly driven by buoyancy, and most of its effects have already been studied. On the other hand, surface tension driven flows have not been studied to such a high level and this was an opportunity to do so.

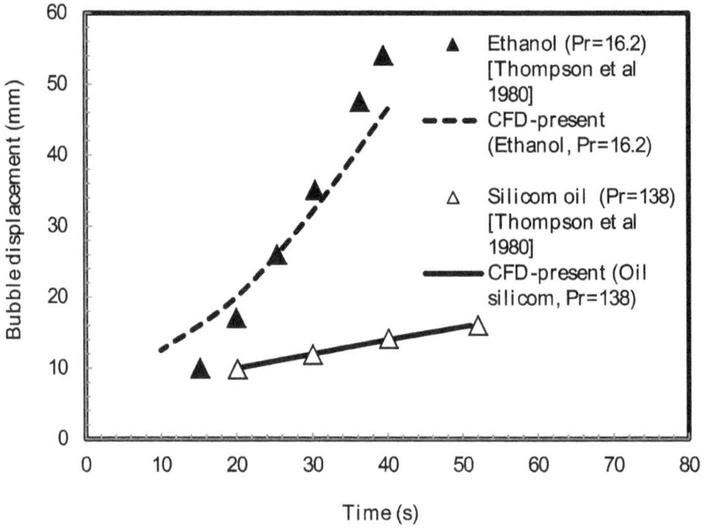

Fig. 8. Result validation of present-CFD calculation with (Thompson et al., 1980) for N2 = 6mm diameter

4.2 Effect of liquid temperature upon bubble migration

The second test case in this chapter is the most representative and interesting case. For this test case the linear temperature distribution was used between the walls. As we ran and visualized these cases we interpreted how they affected the bubble migration speed toward the hotter side. Fig. (9) shows that the different temperature differences between the walls made the bubble move faster for greater temperature differences, or slower for small temperature differences. For the design fluid, Ethanol in this case, hotter side temperature above 330K will not have significant effect on the bubble speed and less than 320K possibly will not move the bubble. With these concepts in mind we selected a temperature range vary from (320-335) K for the top wall and 300K was fixed for the bottom wall. With these design concepts we started to set boundary conditions and to slightly change some initial conditions. All these cases gave us insight on the behavior of such a temperature differences. In these test cases simulations of bubble thermocapillary migration, Re and Ma range from 228.6 to 274.3 and from 3722.7 to 4467.3, respectively. Sensitivity tests results for different bulk liquid temperatures in figures 9 and 10 show the capability of ANSYS-Fluent to simulate two phase flow in microgravity environment using VOF model.

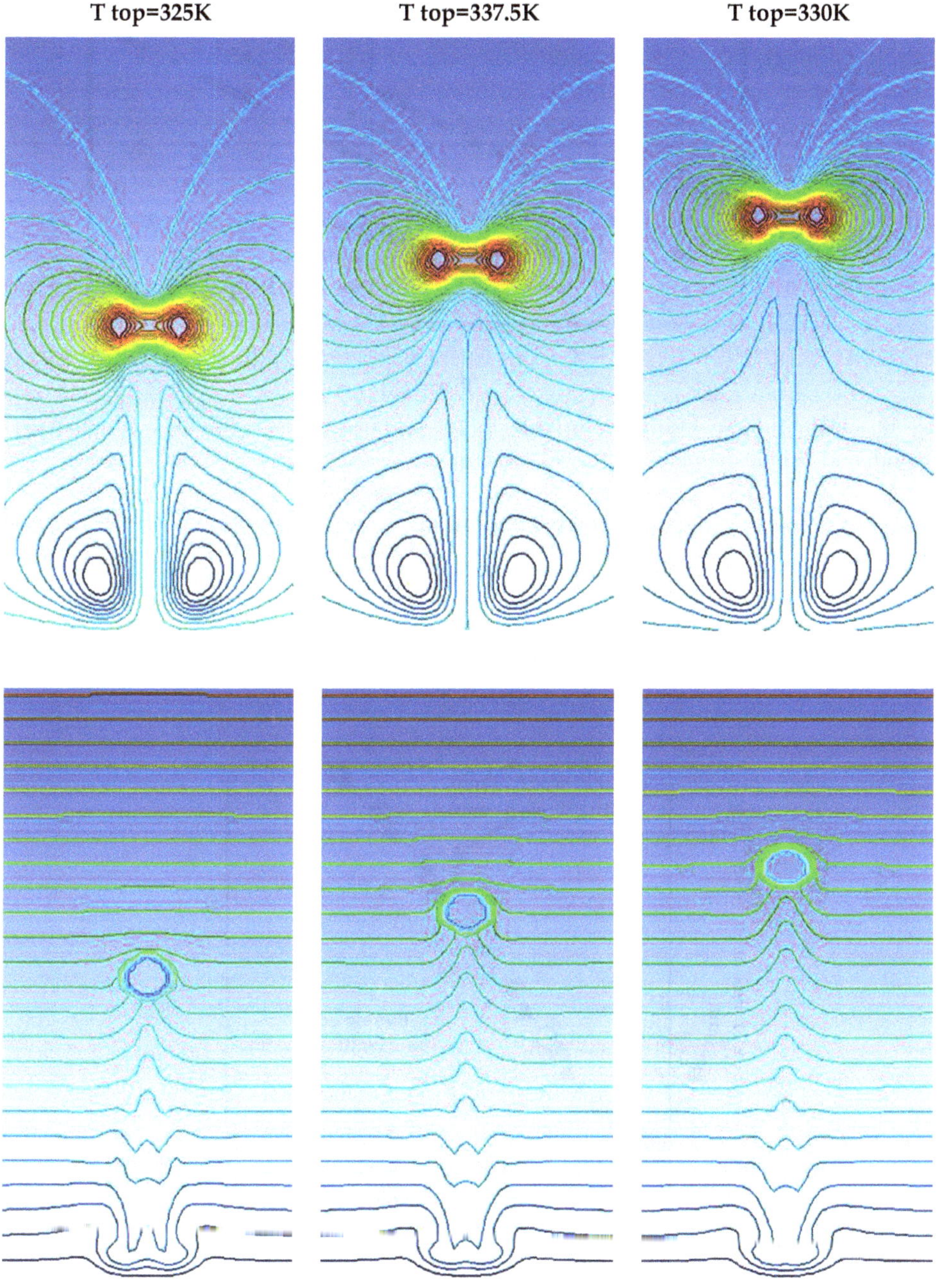

Fig. 9. Temperature contours (bottom) and streamlines (top) for the single bubble (d=10 mm) at t=5 s.

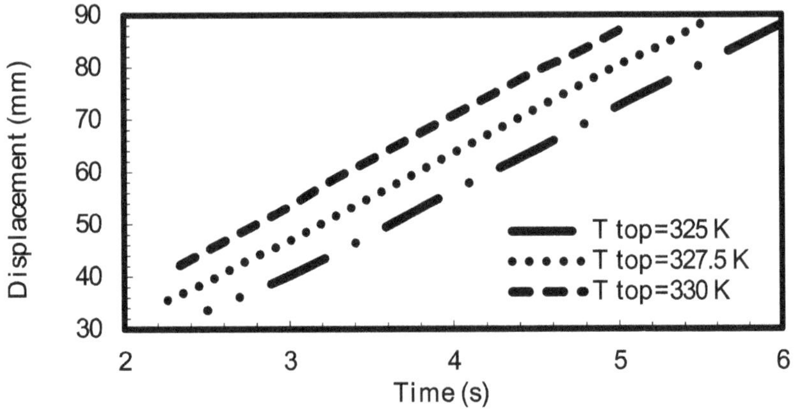

Fig. 10. Sensitivity tests results for different bulk liquid temperatures (Pr=16.3) at time (t)=5 second, and bubble diameter (d) for N2 =10 mm

4.3 Effect of aspect ratio upon bubble migration

We report the results of an extensive numerical investigation on the speed of rise of gas bubbles diameters in the size range of 4-10 mm in stagnant Ethanol liquid in a column with diameters of 20, 40, 60, 80 mm. The column diameter was found to have a significant effect on the rise velocity of the bubbles as shown in fig. 11 below.

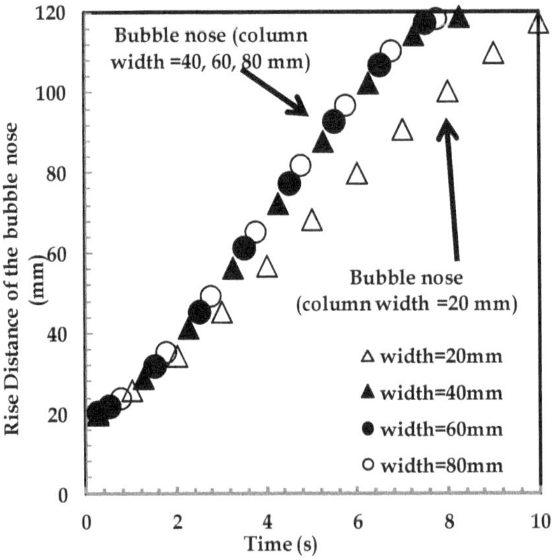

Fig. 11. Compares the x-coordinates of the nose of 10 mm bubbles rising in columns of 20, 40, 60, and 80 mm diameters

Figure 11 shows the effect of the column width on the raise speed of the bubble. The liquid phase stream lines profiles for three simulations is illustrated in figure 12. Figure 13 shows when the Aspect Ratio (AR) of the bubble diameter to the column diameter, is smaller than 0.355 the influence of the column diameter on the rise velocity is negligible. With increasing AR there is a significant reduction of the rise velocity.

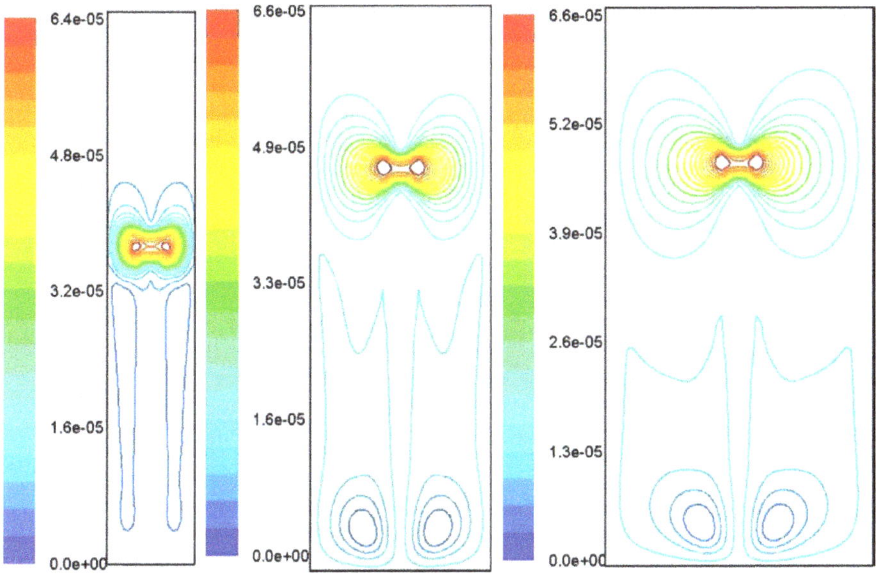

Fig. 12. CFD simulation for studies on rise velocity of single bubble in side a column AR=(0.5, 0.25, 0.1667) from the left to the right.

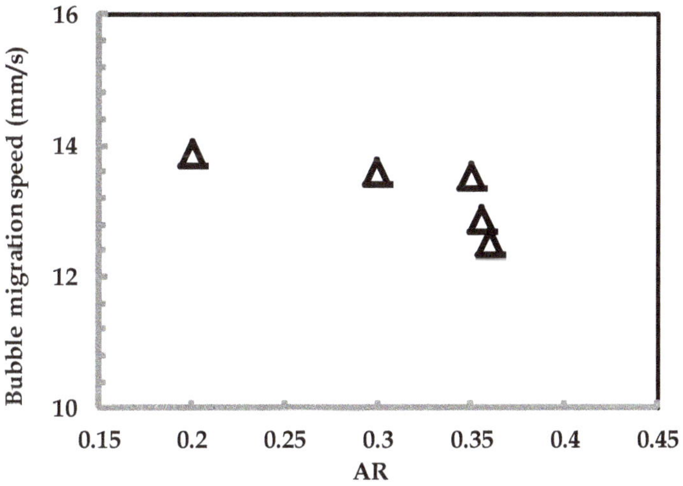

Fig. 13. Shows the effect of (AR) on the rise speed of single bubble

4.4 Effect of rotating cylinder upon bubble coalescence

In a rotating field, fluid particles which are less dense than the surrounding media migrate inward toward the axis of rotation (Annamalai et al. 1982). The results of three dimension rotating cylinder can complement the previous study in providing techniques for causing small bubble to move or bringing several unmovable small bubbles within a cylinder to coalesce into a large bubble which might subsequently be extracted or otherwise manipulated within the cylinder. The same technique can be used in glass melts processing. Under such conditions, rotation of the melt, followed by thermocapillary migration of the coalesced bubbles results in a "centrifugal fining" operation for bubble removal (Annamalai et al. 1982).

The effect of both cylindrical rotation and surface tension on the trajectory of the single bubble is illustrated in figure 12. In this figures, it can be seen that the angular velocity (ω) pulled the bubble towards the centre of rotation, and at the same time Marangoni force moved the bubble towards the hotter side.

As the cylinder begins to rotate from (5-50 rpm), the linear movement of the bubble changes and the bubble starts moving further towards the axis of rotation (centre). Results show that by adjusting the rotational speed, it is possible to change the gas bubble behavior in a thermocapillary flow. The results can help determine the new migration time and speed in the rotating cylinder. Figs 15-16 show that the transient development of the radial migration of the three bubbles will move from the rest towards the centre of the cylinder and towards the top side (hotter side). This movement depends on the angular speed (ω) and thermocapillary flow. These figures illustrate that bubble breakage and agglomeration can be controlled

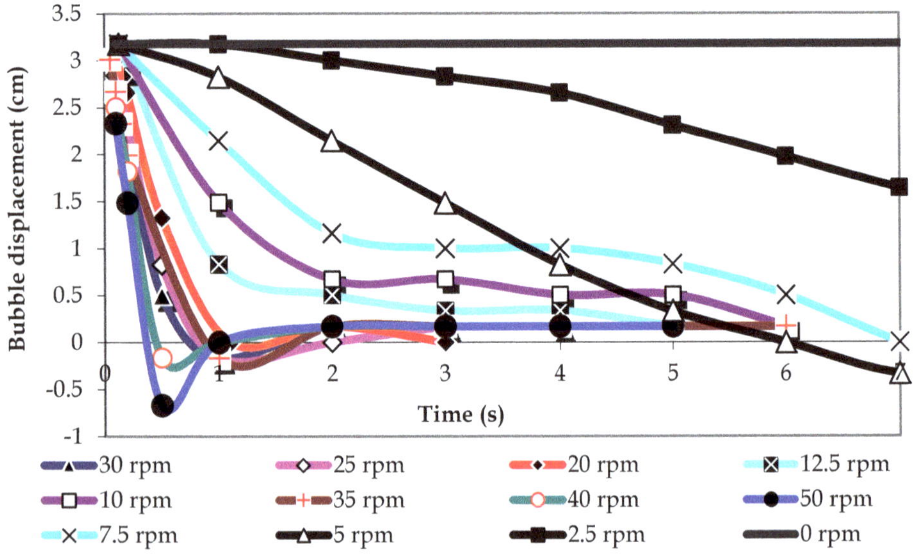

Fig. 14. Bubble displacement (cm) from the releasing position (3 cm from the axis of rotation) toward the axis of rotation and the hotter side (Marangoni)

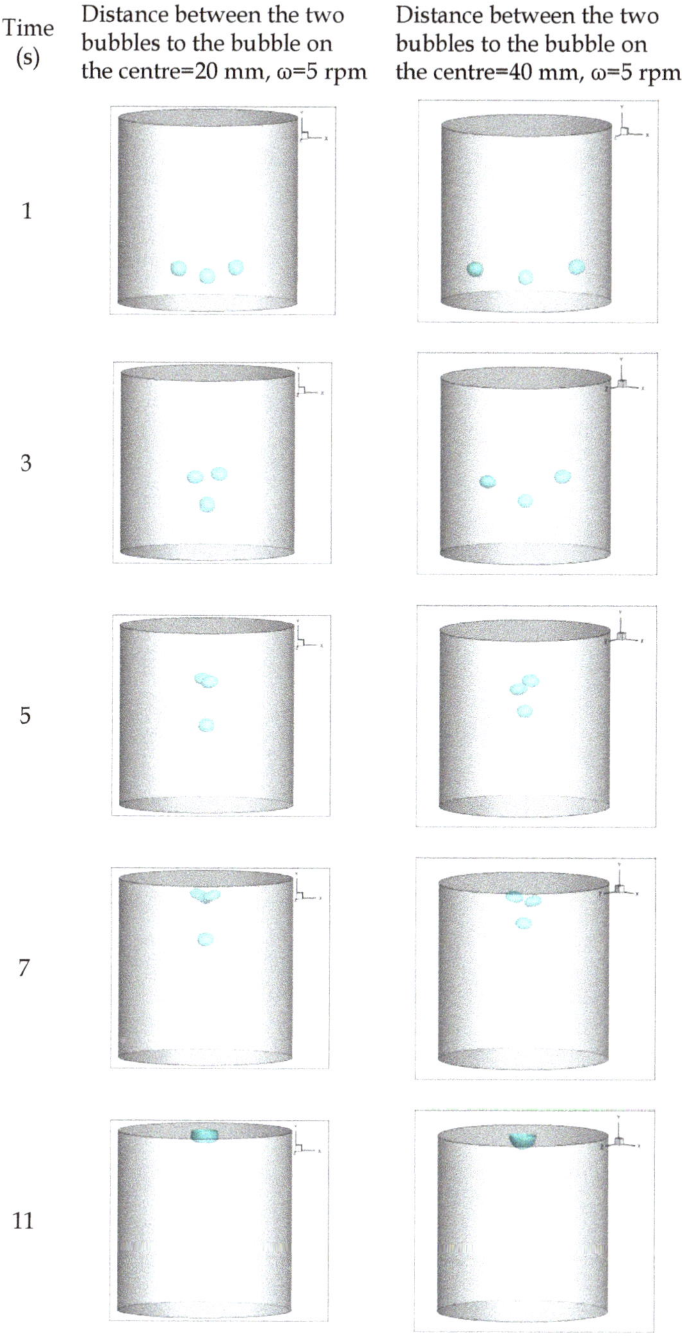

Fig. 15. Shows the coalescence sequence of 10 mm diameter bubbles

Time (s)	Distance between the two bubbles to the bubble on the centre=40 mm, ω=5 rpm	Distance between the two bubbles to the bubble on the centre=50 mm, ω=5 rpm
Initial		
2		
4		
6		
8		
10		

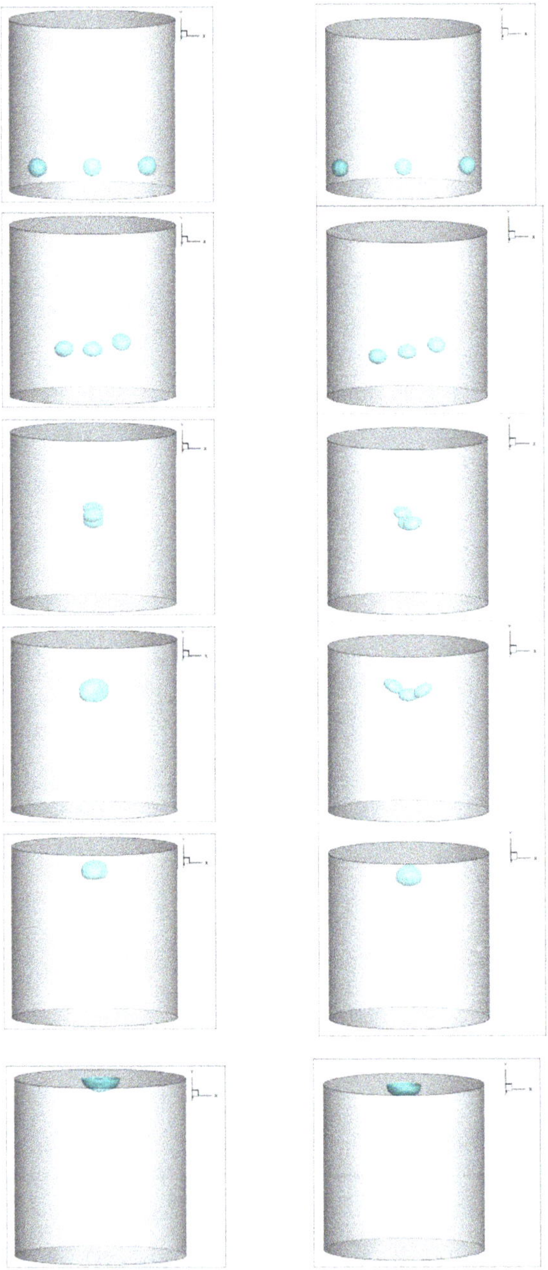

Fig. 16. Shows the coalescence sequence of 12 mm diameter bubbles

5. Conclusions and future work

In This chapter, it was shown by figures the conclusive existence of Marangoni bubble flow phenomena in a zero gravity environment. We discovered the effect of temperature, wall aspect ratio, and rotating cylinder on the bubble behavior at zero gravity using ANSYS 13. It has been proven that VOF is a robust numerical method for the simulation of gas-liquid two-phase flows, and the ability to simulate surface tension as a function of temperature (thermocapillary flow) using a UDF is possible for routine design and development engineering activities

A constant angular speed (ω) was applied to the walls of the cylinder which imparts an extra radial forced vortex motion to the adjacent fluid layer, the effect of which translates as a velocity towards the axis. The 3D simulation was complemented to the previous axissymmetry cases in providing techniques for bringing several unmovable small bubbles within a cylinder to coalesce into a large bubble or to its centre which might subsequently be extracted or otherwise manipulated within the cylinder. These results can help in determine the new migration speed and behavior of gas bubbles by adjusting the angular speed.

Most experimental in microgravity are limited to the case of shortage in time and some other space limitations. Computer simulations, on the other hand, are not restricted to such circumstance, and any arbitrary geometry can be simulated in addition to rotating system. Thanks to the increasing computer power, it is nowadays possible to solve more complex multiphase models. With that the computer solution proves to be a valuable tool to study the complex problem under the conditions of zero and reduced gravity, and from the results of this chapter we can see the ability of ANSYS-Fluent code to simulate Marangoni 3-d cases.

The behavior of the compound rotational and surface tension driven motion, shapes, and trajectories of bubbles is a new area of study, and it is planned to help support research area based on space applications. According to our knowledge, the present study 3-D-VOF-based method for simulating the influence of both rotational and thermocapillary on single and multi bubbles located off centre is first numerical study case in this field and no comparable investigation has been published in the open literature due to difficulty for researchers to access to microgravity facilities. In the future, more detail discussions for different breakage and coalescence times for a group of bubbles will be investigated and used to make kernels function which can be used in population balance equations.

6. References

Alhendal, Y.; Turan A. & Wael A. (2010). VOF Simulation of Marangoni Flow of Gas Bubbles in 2D-axisymmetric Column. Lecture Notes in Computer Science. Vol. 1. Issue 1. 673-680. May. DOI: 10.1016/j.procs.2010.04.072

Annamalai P., Shankar N., Cole R., & Subramanian R.S. (1982). Bubble Migration Inside a Liquid Drop in a Space Laboratory, Appl. Sci. Res., 38, 179-186

ANSYS-Fluent (2011). Users Guide.

Bhaga, T. & Webber, M. (1981). Bubbles in viscous liquids: shapes, wakes and velocities. J. Fluid Mech. 105, 61–85

Bozzano, G. & Dente M., (2009). Single bubble and drop motion modelling. AIDIC Conference Series, 09, 53-60 D01:10.3303/ACOS0909007

Hadland, P. H., Balasubramaniam, R., Wozniak, G. & Subramanian, R. S. (1999). Thermocapillary migration of bubbles and drops at moderate to large Marangoni number and moderate Reynolds number in reduced gravity. Expt. Fluids 26,240-248.

Hirt, C. & Nichols, B., (1981). Volume of Fluid (VOF) Method for the Dynamics of Free Boundaries. J. of Comp. Physics, 39, 201-225.

Kang, Q., Cui H. L., Hu L., & Duan L., (2008). On-board Experimental Study of Bubble Thermocapillary Migration in a Recoverable Satellite, Microgravity Sci. Technol (2008) 20:67 - 71-DOI 10.1007/s12217-008-9007-6

Kawaji, M., DeJesus, J.M., & Tudose, G., (1997). Investigation of Flow Structures in Vertical

Krishna, R. & J. M. van Baten, (1999). Rise Characteristics of Gas Bubbles in a 2D Rectangular Column: VOF simulations vs Experiments. International Communications in Heat and Mass Transfer 26, 965-974

Kuhlmann H.C., (1999). Thermocapillary Convection in Models of Crystal Growth, 152, Springer, Berlin, Heidelberg Springer Tracts in Modern Physics.

Nas, S. & Tryggvason, G. (2003). Thermocapillary interaction of two bubbles or drops.Int. J. Multiphase Flow 29, 1117-1135.

Ostrach, (January 1982). Low-Gravity Fluid Flows Annual Review of Fluid Mechanics,Vol. 14: 313-345-DOI: 10.1146/annurev.fl.14.010182.001525
Slug Flow", Nuclear Engineering and Design, Vol. 1 75, pp. 37-48

Subramanian, K., Paschke S., Repke JU, & Wozny G., (2009). Drag force modelling in CFD simulation to gain insight of packed columns, AIDIC Conference Series, 09, 299-308 D01:10.3303/ACOS0909035

Subramanian, R. S., Balasubramaniam, R. & Wozniak, G. (2002). Fluid mechanics of bubbles and drops. In Physics of Fluids in Microgravity (ed. R. Monti), pp. 149-177. London: Taylor and Francis.

Thompson, R.L., DeWITT, K.J. & Labus, T.L., (1980). Marangoni Bubble Motion Phenomenon in Zero Gravity,Chemical Engineering Communications 5 299-314.

Treuner, M., Galindo, V., Gerbeth, G., Langbein, D. & Rath, H. J. (1996). Thermocapillary bubble migration at high Reynolds and Marangoni numbers under low gravity. J. Colloid Interface Sci. 179, 114-127.

Uhlmann, D.R., (1982). Glass Processing in a Microgravity Environment. Materials Processing in the Reduced Gravity Environment of Space, Edited by Rindone, G.E., Elsevier, NY, USA, 269—278.

Wolk, G., M. Dreyer & H.J. Rath, (2000). Flow Pattern in Small Diameter Vertical Non-circular Channels. International Journal of Multiphase Flow, Vol. 26, pp. 1037-1061

Wozniak, G., Balasubramaniam, R., Hadland, P. H. & Subramanian, R. S. (2001). Temperature fields in a liquid due to the thermocapillary motion of bubbles and drops. Expt. Fluids 31, 84-89.

Young, N. O., Goldstein, J. S. & Block, M. J. (1959). The motion of bubbles in a vertical temperature gradient. J. Fluid Mech. 6, 350-356.

Youngs, D.L., (1982). Time-dependent multi-material flow with large fluid distortion. Numerical Methods for Fluid Dynamics, eds. K.W. Morton & M.J. Baines, Academic Press, pp. 273–285, Brackbill, J. U., Kothe, D. B. and Zemach, C., 1992, A continuum method for modeling surface tension, J.Comp. Physics 100, 335.

9

Rayleigh–Bénard Convection in a Near-Critical Fluid Using 3D Direct Numerical Simulation

Accary Gilbert
Holy-Spirit University of Kaslik
Faculty of Engineering
Lebanon

1. Introduction

Convection in a fluid close to its gas-liquid critical point (CP) has been a subject of growing interest since the exhibition of the piston-effect (PE), this thermo-acoustic effect responsible for the fast thermal equilibrium observed in such a fluid in the absence of convection. In 1987, under microgravity conditions, Nitsche and Straub observed a fast and homogeneous increase of the temperature inside a spherical cell containing SF_6 slightly above the CP when it was subjected to a heating impulse. This phenomenon was then explained theoretically (Zappoli et al., 1990; Onuki et al., 1990; Boukari et al., 1990) by the well-known critical anomalies, more precisely by the divergence of the thermal expansion coefficient and the vanishing of its thermal diffusivity when approaching the CP. Indeed, the heating of a cell containing a supercritical fluid (SCF) induces along the heated wall a thin thermal boundary layer in which density shows large variations because of the divergence of the thermal expansion coefficient; this thermal layer expands compressing adiabatically the rest of the fluid leading by thermo-acoustic effects (the so-called PE) to a fast and homogeneous heating of the bulk of the cell. Several experiments were carried out subsequently, mainly in microgravity (Guenoun et al., 1993; Straub et al., 1995; Garrabos et al., 1998) but also on Earth (Kogan & Meyer, 1998), and confirmed the existence of the PE.

Since 1996, many experimental and numerical studies were devoted to the interaction between the PE and natural convection. The Rayleigh-Bénard configuration (bottom heating) received a particular attention (Kogan et al., 1999; Amiroudine et al., 2001; Furukawa & Onuki, 2002) because the hydrodynamic stability of the SCF in that case is governed by an interesting and non-common criterion. Owing to the PE, the thermal field exhibits a very specific structure in the vertical direction. A very thin hot thermal boundary layer is formed at the bottom, then a homogeneously heated bulk settles in the core at a lower temperature, and at the top, a cooler boundary layer is formed in order to continuously match the bulk temperature with the colder temperature of the upper wall. The linear analysis, carried out by Gitterman and Steinberg in 1970, showed that the hydrodynamic stability of these thermal boundary layers, when subjected to a gravity field, depends on the interaction between two stability criteria which, for a normally compressible fluid, are separately available at very different space scales: on one hand, the classical Rayleigh criterion, derived from the Boussinesq approximation, hence available at small space scales, and on the other hand, the Schwarzschild criterion, usually encountered in atmospheric science, where the stabilizing effect of the hydrostatic pressure becomes appreciable. Indeed, because of the divergence of the isothermal compressibility of a SCF, the Schwarzschild criterion becomes available at small space scales; this was proven theoretically (Gitterman & Steinberg, 1970b; Carlès & Ugurtas, 1999), experimentally (Kogan & Meyer, 2001), and numerically (Amiroudine et al., 2001). Taking advantage of the interaction between those two stability criteria, a numerical study (Accary et al., 2005a) showed that, in spite of convection onset in the thermal boundary layers according to the classical Rayleigh criterion, a reverse transition to stability through the Schwarzschild line is possible without any external intervention. The hydrodynamic stability of the thermal boundary layers developed in this configuration has been exhaustively investigated (Accary et al., 2005b), and recently a numerical study (Accary et al., 2009) focused on the convective regime of the flow. Because of the particular physical properties of the fluid in the vicinity of the CP, the convective regime of the Rayleigh-Bénard problem is turbulent for unusually low intensities of heating (~*mK*). In this last study, 3D direct numerical simulations are carried out for Rayleigh numbers varying from 2.68×10^6 up to 160×10^6. For a perfect gas (PG), this range of Rayleigh numbers corresponds to the transition between the soft and the hard turbulence; however, this is not always the case for the SCF because of its strong stratification induced by its high isothermal compressibility.

In § 2, the problem under consideration is presented. In § 3, the mathematical model is described together with the acoustic filtering of the Navier-Stokes equations. The details of the numerical method are presented in § 4 and the simulation conditions are mentioned. In § 5, several aspects of the Rayleigh-Bénard convection in a near-critical fluid are reported: the hydrodynamic stability of the thermal boundary layers, the convection onset and the beginning of the convective regime, the steady-state turbulent regime, details of the temperature and the dynamic fields, and the global thermal balance of the cavity. In § 6, a comparison with the case of the PG is presented at equal Rayleigh number. Finally, the chapter is concluded in § 7.

2. The problem under consideration

We consider a SCF in a cube-shaped cavity (of height $H' = 10\ mm$) subjected to the earth gravity field $g' = 9.81\ m.s^{-2}$ (Fig. 1). The horizontal walls are isothermal while the sidewalls are insulated, and no-slip conditions are applied to all the walls. Initially the fluid is at rest, in thermodynamic equilibrium at a constant temperature T'_i slightly above the critical temperature T'_c, such that $T'_i = (1+\varepsilon)T'_c$, where $\varepsilon << 1$ defines the non-dimensional proximity to the CP. Under the effect of its own weight, the fluid is stratified in density and in pressure, with a mean density equal to its critical value ρ'_c. While maintaining the top wall at its initial temperature T'_i, the temperature of the bottom wall is gradually increased (during one second) by $\Delta T'$ (a few *mK*).

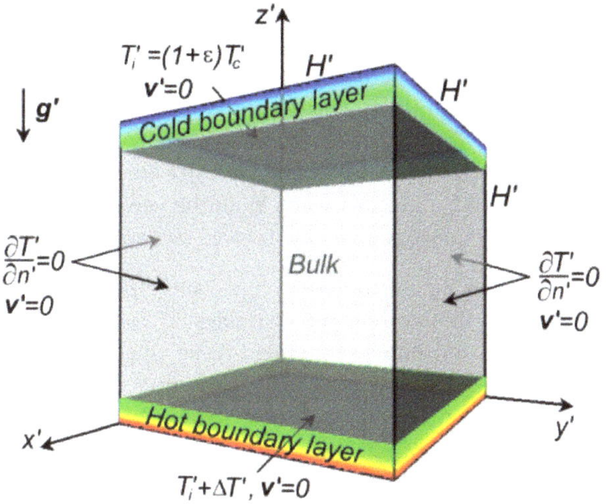

Fig. 1. Geometry of the cube-shaped cavity and the velocity and temperature conditions applied to the boundaries. The vertical axis z' is co-linear with the acceleration due to the earth gravity g'. Since the first seconds of heating, the temperature field is vertically stratified, divided in three distinct zones: two thermal boundary layers and the bulk of the cavity.

3. The mathematical model

3.1 Equations governing near-critical fluid buoyant flows

The mathematical model for a SCF flow (Zappoli, 1992) is described by the Navier-Stokes and energy equations written for a Newtonian and highly conducting van der Waals fluid.

$$\text{Continuity: } \frac{\partial \rho'}{\partial t'} + \nabla.(\rho'.\mathbf{v}') = 0 \tag{1}$$

$$\text{Momentum: } \frac{\partial(\rho'.\mathbf{v}')}{\partial t'} + \nabla.(\rho'.\mathbf{v}'.\mathbf{v}') = -\nabla P' + \mu'\left[\nabla^2\mathbf{v}' + \frac{1}{3}\nabla(\nabla.\mathbf{v}')\right] + \rho'\mathbf{g}' \tag{2}$$

$$\text{Energy: } \frac{\partial(\rho'.e')}{\partial t'} + \nabla.(\rho' e' \mathbf{v}') = -P'(\nabla.\mathbf{v}') + \nabla.(\lambda'\nabla T') + \Phi' \tag{3}$$

$$\text{Van der Waals: } P' + a'\rho'^2 = \frac{\rho' r' T'}{1 - b'\rho'} \tag{4}$$

Where P' is the pressure, T' is the temperature, and ρ' is the density. $v'(u',v',w')$ is the velocity, $g' = (0,0,-g')$, e' is the internal energy, and Φ' is the viscous energy dissipation. μ' is the dynamic viscosity, λ' is the thermal conductivity, r' is the PG constant, a' and b' are respectively the energy parameter and the co-volume related to the critical coordinates T'_c and ρ'_c by: $b' = 1/(3\rho'_c)$ and $a' = 9r'T'_c/(8\rho'_c)$.

In spite of its simplicity, the van der Waals' equation of state gathers the required conditions for the existence of the CP and yields to a critical divergence[1] as $(T'/T'_c\text{-}1)^{-1}$ of the thermal expansion coefficient β'_P, of the isothermal compressibility χ'_T, and of the heat capacity at constant pressure C'_P. The critical divergence of the thermal conductivity is described by: $\lambda' = \lambda'_0(1+\Lambda\varepsilon^{-1/2})$ where $\Lambda = 0.75$ and λ'_0 is the thermal conductivity for a PG. The heat capacity at constant volume C'_V and the dynamic viscosity are supposed to be constant and equal to those of a PG, C'_{V0} and μ'_0 respectively. With the van der Waals' equation of state, the expression of the internal energy is given by: $\delta e' = C'_V\delta T' - a'.\delta\rho'$.

In order to make the variables dimensionless, T'_c, ρ'_c, and $r'\rho'_cT'_c$ are used respectively as representative scales of the thermodynamic variables T', ρ', and P'. The independent variables of length $\mathbf{x}'(x',y',z')$ and time t' are scaled respectively by the height of the cavity H' and the PE time-scale[2] given by $t'_{PE} = \Psi^{-1}\rho'_c H'^2/\mu'_0$ with $\Psi = \varepsilon^{-1}(\Lambda^{-1}+\varepsilon^{-0.5})$ (Zappoli, 1992; Zappoli *et al.*, 1999). Hence, the representative scale of velocity is $V'_{PE} = H'/t'_{PE}$. This scaling introduces the Reynolds numbers $Re = \Psi$, the Froude number $Fr = (V'_{PE})^2/(g'H')$, the Prandtl number based on the properties of the PG assumption ($Pr_0 = \mu'_0C'_{P0}/\lambda'_0$), and the Mach number $Ma = V'_{PE}/c'_0$ where $c'_0 = (\gamma_0 r'T'_c)^{1/2}$ is the speed of sound for a PG (with $\gamma_0 = C'_{P0}/C'_{V0}$). Note that the PE time-scale obtained by (Onuki *et al.*, 1990) is given by $t'_1 = H'^2/D'_T(\gamma-1)^2$, where D'_T is the thermal diffusivity and $\gamma = C'_P/C'_V$ is the specific-heat ratio. Adapted for a van der Waals' gas [3] and in the assumption that $\varepsilon \ll 1$, the PE time-scale $t'_1 \approx Pr/\Lambda\gamma_0(\gamma\text{-}1)\times t'_{PE}$.

3.2 The acoustic filtering of the governing equations

Despite its high isothermal compressibility, the sound speed c' in a SCF, defined by $c'^2 = C'_P/C'_V\times\chi'_T{}^{-1}$, does not vanish at the CP according to the van der Waals' equation of state, indeed C'_P/C'_V and χ'_T diverge with the same critical exponent of -1, which allows the acoustic filtering of the equations. In the basic assumption that $Ma \ll 1$, all the primary dimensionless variables of the problem $\boldsymbol{\phi} = {}^t(v, T, P, \rho)$ can be expanded in series of Ma^2 (Paolucci, 1982) as follows: $\boldsymbol{\phi} = \boldsymbol{\phi}^{(0)} + \gamma_0 Ma^2\boldsymbol{\phi}^{(1)} + o(Ma^2)$ where $\boldsymbol{\phi}^{(0)}$ and $\boldsymbol{\phi}^{(1)}$ are $O(1)$. The $O(1)$ and $O(Ma^2)$ parts of the governing equations resulting from this expansion that need to be solved are given by:

$$\text{O(1) continuity: } \frac{\partial\rho}{\partial t} + \nabla.(\rho.\mathbf{v}) = 0 \tag{5}$$

$$\text{O(1) momentum: } \nabla P_{th} = 0 \tag{6}$$

$$\text{O(Ma}^2\text{) momentum: } \frac{\partial(\rho\mathbf{v})}{\partial t} + \nabla.(\rho\mathbf{v}\mathbf{v}) = -\nabla P^{(1)} + \frac{1}{\text{Re}}\left[\nabla^2\mathbf{v} + \frac{1}{3}\nabla(\nabla.\mathbf{v})\right] + \frac{\mathbf{e}_g}{Fr}(\rho - \rho_i) \tag{7}$$

[1] The real critical exponents (which are the same for all fluids) differ from those obtained from the van der Waals' equation of state that remains a good approximation to carry out qualitative studies.
[2] The PE time-scale is the time necessary the PE to homogenize the temperature in the core of the cavity, this time-scale is between the acoustic time scale (H'/c') and the thermal diffusion one (H'^2/D'_T) where c' is the sound velocity in the SCF and D'_T is the thermal diffusivity.
[3] For a van der Waals' gas, $\gamma/\gamma_0 = C_P/C_{P0} = 1+(1-1/\gamma_0)\varepsilon^{-1}$.

$$\text{O(1) energy: } \frac{\partial(\rho T)}{\partial t}+\nabla.(\rho \mathbf{v}T)=-(\gamma_0-1)(P_{th}+a\rho^2)(\nabla.\mathbf{v})+\frac{\gamma_0}{\mathrm{Re}\,\mathrm{Pr}_0}\nabla.\left[\left(1+\Lambda(T-1)^{-0,5}\right)\nabla T\right] \quad (8)$$

$$\text{O(1) van der Waals: } P_{th}+P_{hyd}=\frac{\rho T}{1-b\rho}-a\rho^2 \quad (9)$$

In these equations, $v(u,v,w)$, T, P, and ρ refer to the $O(1)$ of the dimensionless variables, the superscript (0) has been omitted for conciseness. P_{th} and P_{hyd} are respectively the thermodynamic pressure (homogeneous in space but time varying according to the $O(1)$ momentum equation) and the time independent hydrostatic pressure ($P^{(0)} = P_{th} + P_{hyd}$). $a = 9/8$ and $b = 1/3$ are the dimensionless parameters of the van der Waals' equation of state, $\mathbf{e}_g = (0,0,-1)$. Before the heating begins, the initial dimensionless density distribution ρ_i, the initial thermodynamic pressure P_{thi}, and the hydrostatic pressure P_{hyd} are obtained from the initial thermodynamic and static equilibrium with the constraint of a dimensionless mean density equal to 1. This results in:

$$T_i = 1+\varepsilon\,,\ \rho_i=\frac{K}{e^K-1}e^{Kz}\,,\ P_{thi}=\frac{1+\varepsilon}{1-b}-a\,,\ P_{hyd}=\chi_T^{-1}(\rho_i-1) \quad (10)$$

where $\chi_T^{-1}=\frac{1+\varepsilon}{(1-b)^2}-2a$ and $K=\chi_T\frac{\gamma_0 Ma^2}{Fr}$

This low Mach number approximation (adapted to SCF buoyant flows) differs from the classical one where $\rho_i = 1$ and consequently $P_{hyd} = 0$. Indeed, owing to the divergence of the isothermal compressibility of the SCF, the hydrostatic pressure induces density variations ($\rho_i - 1$) comparable to those resulting from a weak heating. This has been done by keeping the buoyancy term $(\gamma_0 Ma^2/Fr)\rho^{(0)}\mathbf{e_g}$ in the leading order $O(Ma^2)$ of the momentum equation (Eq. 7) while in the classical low Mach number approximation, this term is shifted to the $O(Ma^4)$ order. It has been shown (see § 5.2) that this modification is essential for a correct prediction of the convection onset in the thermal boundary layers (Accary et al., 2005c).

We consider the carbon dioxide critical coordinates ($T'_c = 304.13$ K, $\rho'_c = 467.8$ Kg.m^{-3}) and physical properties ($r' = 188$ J.Kg^{-1}.K^{-1}, $\mu'_0 = 3.44\times10^{-5}$ Pa.s, $C'_{V0} = 658$ J.Kg^{-1}.K^{-1}, $\mathrm{Pr}_0 = 2.274$, $\lambda'_0 = 0.01$ W.m^{-1}.K^{-1},). The simulations were carried out for $T'_i - T'_c = 1$K ($\varepsilon = 3.29\times10^{-3}$); in this case, $t'_{PE} = 0.256$ s, $V'_{PE} = 3.9$ cm.s^{-1}, Re = 5710, Fr = 1.55×10^{-2}, $K = (4/9\varepsilon)(\gamma_0 Ma^2/(Fr)) = 2.32\times10^{-4}$, and the effective Prandtl number $\mathrm{Pr} = \mathrm{Pr}_0.\varepsilon^{-1/2} = 39.6$.

4. Numerical method

In describing the numerical method used in this analysis, it is assumed that the reader is familiar with the basis of the standard finite volume method and with the velocity-pressure coupling algorithms extensively reported in (Patankar, 1980). In this section, we will draw the outlines of the method the space and the time accuracies, the velocity-pressure coupling, the linear systems solvers, and the solver performance.

4.1 Space and time discretization

The computational domain is subdivided into a number of cells using a wall refined mesh for a better description of the solution in the boundary layers, the mesh is refined in the vicinity of the walls; as one moves away from the wall, the control volume size increases according to a geometric progression. If N is the number of cells in a direction (x for example), the dimensionless positions of the cells interfaces between the wall and the center of the computation domain (i.e. for $0 < x(i) \leq 0.5$) would be:

$$x(i) = \frac{1}{2}\left(\frac{2i}{N}\right)^q \quad \text{for } i = 1,\ldots,N/2 \tag{11}$$

For $0.5 \leq x(i) < 1$, cells interfaces position are obtained by symmetry with respect to x = 0.5. Depending on the value of q ($q \geq 1$), this mesh refinement is termed 'power q law' type; $q=1$ provides a uniform mesh, the simulations were carried out with q = 2. The variables location is staggered: the scalar variables are stored at the cells centers while the velocity components are defined at the midpoints of the cells faces perpendicular to the velocity direction. This staggering practice avoids the high-frequency noise in the solution resulting from the well-known problem of the zigzag pressure filed which would be made up of arbitrary values of pressure arranged in a checkerboard pattern (Patankar, 1980). In return, this staggering has no adverse consequences in the simple rectilinear domain considered here.

The convection-diffusion transport equation of a variable ϕ to be solved in computational fluid dynamics can be written as:

$$\frac{\partial \rho\phi}{\partial t} + \nabla.(\rho \mathbf{v} \phi) = \nabla.(\Gamma_\phi \nabla \phi) + S_\phi \tag{12}$$

Where Γ_ϕ is the diffusion coefficient and S_ϕ is a source term. While integrating Eq. 12 over the control volume (CV) of a discrete variable ϕ_p and over a time step, the value of a variable ϕ and its gradients are assumed to be constant on its CV faces; therefore the space accuracy of the method is already limited to the second order, depending on the interpolation scheme in the direction perpendicular to the considered face used to approach the value of ϕ and its derivative.

The time integration is fully implicit providing the method a non-conditional stability as far as the time step is concerned; and to allow large time scale simulations, the unsteady terms are approached by a standard Euler time scheme with four time levels, leading to a third order truncation error in time. The time integration of Eq. 12 at the instant t^n and for a uniform time step δt is done as follows:

$$\frac{1}{\delta t}\left[\frac{11}{6}(\rho\phi)^n - 3(\rho\phi)^{n-1} + \frac{3}{2}(\rho\phi)^{n-2} - \frac{1}{3}(\rho\phi)^{n-3}\right] + \nabla.(\rho \mathbf{v} \phi)^n = \nabla.(\Gamma_\phi \nabla \phi)^n + S_\phi^n \tag{13}$$

Careful space discretization and integration of the transport equation is needed to reach the second order space accuracy ceiling of the method particularly for a non-uniform mesh. This concerns the integration of the source and the unsteady terms, and the evaluation of the variable ϕ and its gradients at the faces of a CV. A linear interpolation is used to evaluate the density at the faces of a CV, while a harmonic mean is considered for the thermal conductivity as recommended in (Patankar, 1980). Figure 2 shows the grid structure and the notations in one dimension.

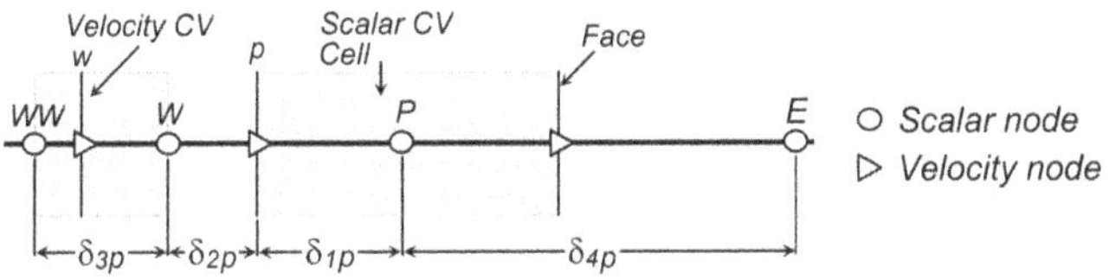

Fig. 2. Mesh structure and notations in one-direction.

For a second order space integration, the integrand should be localized in the centre of the CV; it is the case for the scalar variables (see Fig. 2), while for the velocity components a neighbor contribution must be considered when the grid is not uniform. For example, if ϕ refers to a velocity component, its value ϕ^*_w at the centre of the velocity CV (Fig. 2) can be written as:

$$\phi^*_w = \phi_w + \frac{1}{2}\left(1 - \frac{\delta_{3p}}{2\delta_{2p}}\right)(\phi_p - \phi_w) \tag{14}$$

$$\phi^*_w = \phi_w \text{ if } \delta_{3p} = 2\,\delta_{2p} \text{ (uniform mesh)}$$

A central difference (CD) scheme approaches the diffusion terms. For the velocity components, since CV faces are localized at midway between two consecutive velocity nodes (see Fig. 2), the standard two-point formulation provides a second order approximation of the velocity gradients at the faces of a CV. It is not the case for a scalar variable on non-uniform mesh, where a three-point scheme is necessary to reach the second order accuracy. For example, the gradient of a scalar ϕ at the face p can be written as:

$$\left.\frac{\partial\phi}{\partial n}\right|_p = \frac{\phi_P - \phi_W}{\delta_{1p} + \delta_{2p}} + (\delta_{2p} - \delta_{1p}) \times \left[\frac{\phi_W}{(\delta_{1p} + \delta_{2p} + \delta_{4p})(\delta_{1p} + \delta_{2p})} - \frac{\phi_P}{\delta_{4w}(\delta_{1p} + \delta_{2p})} + \frac{\phi_E}{\delta_{4w}(\delta_{1p} + \delta_{2p} + \delta_{4p})}\right] \tag{15}$$

In this scheme, a correction is added to the standard CD formulation, a correction depending on $(\delta_{2p} - \delta_{1p})$, hence that vanishes for a uniform mesh.

A second order hybrid scheme (SHYBRID) is used for the convection terms (Li & Rudman, 1995). It uses a four-point formulation to interpolate the values of a variable at the faces of its CV, and combines the QUICK, the second order upwind (SOU), and the CD schemes whose respective weights in the formulation depend on the local Peclet number which is the ratio of the convection flux to the diffusion one. The value ϕ_p of a variable ϕ at a face p is determined as follows:

$$\begin{aligned} &\text{if } v_p > 0,\ \phi_p = (1-\alpha)\phi_W + \alpha\phi_P - q_p^-(\phi_P - (1+\beta^-)\phi_W + \beta^-\phi_{WW}) \\ &\text{if } v_p < 0,\ \phi_p = (1-\alpha)\phi_W + \alpha\phi_P - q_p^+(\phi_W - (1+\beta^+)\phi_P + \beta^+\phi_E) \end{aligned} \tag{16}$$

where, $\alpha = \dfrac{\delta_{2p}}{\delta_{1p} + \delta_{2p}}$, $\beta^- = \dfrac{\delta_{1p} + \delta_{2p}}{\delta_{3p}}$, $\beta^+ = \dfrac{\delta_{1p} + \delta_{2p}}{\delta_{4p}}$

The CD, SOU, and QUICK schemes fall into this formulation for appropriate choices of q^-_p and q^+_p; for the CD scheme $q^+_p = q^-_p = 0$, for the SOU scheme $q^-_p = \alpha$ and $q^+_p = 1-\alpha$, and for the QUICK one $q^-_p = \alpha\delta_{1p}/(\delta_{1p}+\delta_{2p}+\delta_{3p})$ and $q^+_p = (1-\alpha)\delta_{2p}/(\delta_{1p}+\delta_{2p}+\delta_{4p})$. The values of q^-_p and q^+_p are automatically adjusted during the simulation according to the local Peclet number (Pe_p) in order to minimize the potential oscillations by minimizing the remote-nodes contributions while maintaining positive neighbor-nodes coefficients (see Patankar, 1980, 2nd basic rule). This minimization (Li & Rudman, 1995) results in,

$$\begin{aligned} &if\ Pe_p = 0,\ q_p^- = q_p^+ = 0 \\ &if\ Pe_p \neq 0,\ q_p^- = max\left(0,(1-\alpha)-\frac{1}{|Pe_p|}\right)\ and\ q_p^+ = max\left(0,\alpha-\frac{1}{|Pe_p|}\right) \end{aligned} \tag{17}$$

A quick inspection of these choices shows that when the transport is diffusion-dominated (Pe_p very small) SHYBRID scheme becomes the CD, when the transport is convection-dominated (Pe_p very large or infinity) SHYBRID approaches the SOU scheme, and between these two limits it may go through the QUICK scheme for certain values of Pe_p. Thus, SHYBRID is a stable second order accurate scheme for a wide range of the Peclet number. In order to suppress non-physical oscillations when predicting solutions with sharp gradients (Li & Rudman, 1995), a flux-correction transport was necessary; it was however useless for the gradient magnitudes encountered in our study.

Integrating Eq. 12 for ϕ_p, the discrete transport equation may be written after some manipulation (Patankar, 1980), as:

$$a_P\phi_P = \sum_{nb=1}^{6} a_{nb}\phi_{nb} + S_c \tag{18}$$

where a_P and a_{nb} are discretization coefficients, S_c is the discrete source term, and the subscript nb designates the six direct neighbors of the node P, any remote-node contribution resulting from second order space discretization is included in S_c. The linear system (Eq. 18) is solved using iterative methods (Barrett et al., 1994); the pressure symmetric equation is solved using the Conjugate Gradient method with Jacobi preconditioning, while the Bi-Conjugate Gradient Stabilized with the same preconditioner is used for the other non-symmetric transport equations. In addition, the pressure equation is preconditioned using a SCF equivalent of the artificial compressibility method proposed by (Chorin, 1997). In spite of the computational optimization of these solvers, most of the computation effort was spent on solving the linear systems and especially the pressure one.

4.2 The coupling algorithm

The pressure splitting in the low Mach number results into two more variables in the governing equations, the hydrostatic pressure P_{hyd} and P_{th} the thermodynamic one. P_{hyd} is time-independent and is given by the initial stratification (Eq. 10), while P_{th} (constant in space) can be determined at any moment using the conservation of the total mass whose dimensionless value is equal to 1.

$$\int_\Omega \rho.d\mathbf{x} = \int_\Omega \rho_i.dV = 1 \tag{19}$$

In order to determine P_{th}, one must provide the function $\rho = \rho(P_{th}, T)$; thus at each time step and at each iteration k of the velocity-pressure coupling algorithm, after computing the temperature field, the density is linearized using the van der Waals' equation as follows:

$$\rho^k = \frac{P_{th}^k + P_{hyd}}{\dfrac{T^k}{1 - b\rho^{k-1}} - a\rho^{k-1}} = \frac{P_{th}^k + P_{hyd}}{f(T^k, \rho^{k-1})} \tag{20}$$

The thermodynamic pressure at iteration k is computed from the conservation of the total mass of the fluid, since:

$$\int_\Omega \rho^k dV = 1 = P_{th}^k \int_\Omega \frac{dV}{f(T^k, \rho^{k-1})} + \int_\Omega \frac{P_{hud} dV}{f(T^k, \rho^{k-1})} \tag{21}$$

Thus for each time step, a global iterative process (Fig. 3) consists first in solving the energy equation (since the flow is temperature driven), updating the thermodynamic pressure (Eq. 21) and the density (Eq. 20) using the linearized equation of state, and then solving the dynamic field. The global convergence is assumed obtained when the L_∞-norm of all governing equations residuals (momentum, energy, and state) reach an imposed stopping criterion. The velocity-pressure coupling is treated by a PISO algorithm (Jang et al., 1986); at each sweep, a velocity prediction, a pressure prediction, and a velocity correction are performed; at this stage, a second correction of the dynamic field is useless since the density will be severely perturbed in the next iteration. PISO algorithm determines the pressure field using a pressure equation and requires no pressure correction that introduces instability into the convergence process of unsteady solutions. More details about the numerical method can be found in (Accary et al., 2006) where the code has been thoroughly validated using an artificial analytical solution and on several benchmark problems of natural convection.

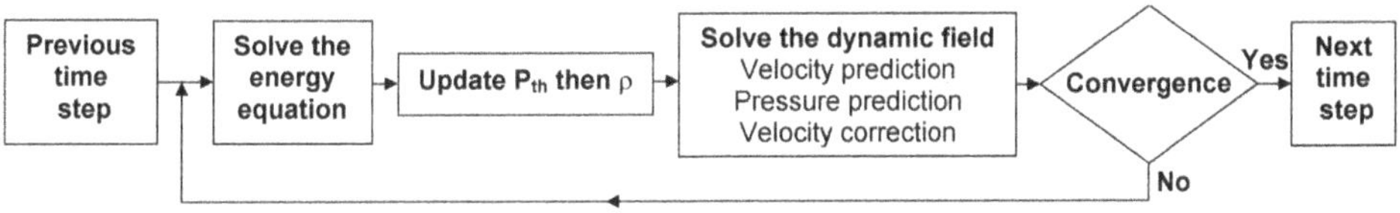

Fig. 3. Global iterative procedure at each time step.

The dimensionless computational domain is a cube of unit length, $\Omega = [0,1]^3$. For the momentum equation, Dirichlet conditions ($v = 0$) are applied on all boundaries. For the energy equation, homogeneous Neumann conditions are applied on the vertical boundaries and Dirichlet conditions on the horizontal ones: $T(z=1) = T_i$ and after one second of simulation $T(z=0) = T_i + \Delta T$, $\Delta T = \Delta T'/T'_c$ being the dimensionless intensity of heating. The mesh size and time step depend on the heating applied to the bottom plate; the mesh size varies between 100^3 and 200^3 computation points and the dimensionless time step varies between 0.01 and 0.1. At each time step, the converged solution is supposed to be obtained when the residuals of all transport equations reach 10^{-9} in non-dimensional form.

5. Rayleigh-Bénard convection in a near critical fluid

5.1 Hydrodynamic stability of the thermal boundary layers

As mentioned earlier, because of the PE, the temperature field is stratified vertically with three distinct zones since the first seconds of heating: the hot boundary layer, the cold boundary layer and the bulk of the cavity. Regardless the considered heating, as long as the flow is dominated by the diffusion and by the PE, the thermal boundary layers grow as $(D'_T.t')^{1/2}$ with $D'_T = 5.18\times10^{-5}\ cm^2.s^{-1}$. For $\Delta T' = 1\ mK$, figure 4(a) shows the fast and homogeneous increase of the temperature in the bulk of the cavity by the PE and the growth of the thermal boundary layers. Figure 4(b) shows the corresponding density profiles; we notice that the density variations induced by the heating are comparable to those resulting from the hydrostatic pressure, which justifies the adaptation of the low Mach number approximation by including the fluid stratification in the model.

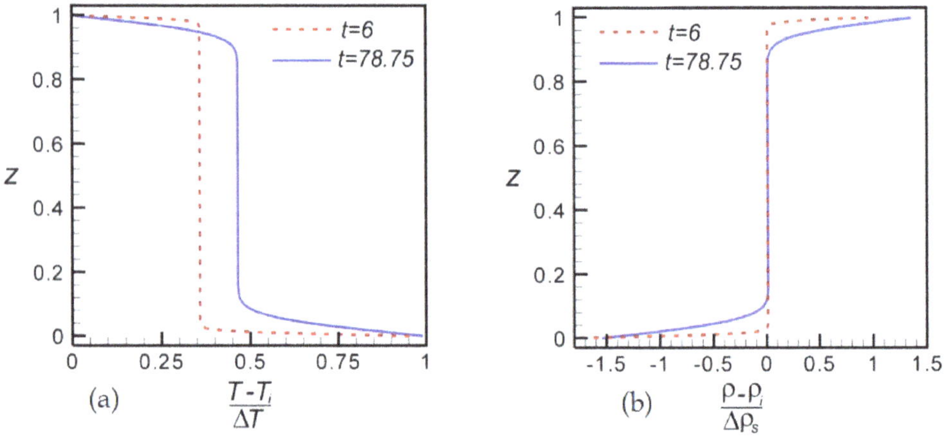

Fig. 4. (a) Temperature profiles for $\Delta T' = 1\ mK$ showing the action of the PE and the growth of the thermal boundary layers before the convection onset. (b) The corresponding density profiles scaled by the density variation due to stratification in the dimensionless form $\Delta\rho_s = K = (4/9\varepsilon)(\gamma_0 Ma^2/(Fr)$.

The thickness h' of the hot boundary layer was defined as the average distance from bottom wall where the local vertical temperature gradient becomes equal to the global one between the horizontal plates, $\Delta T'/H'$. The total temperature variation inside the hot boundary layer is denoted $\delta T'$. The normalized variables $h = h'/H'$ and $\delta T = \delta T'/T'_c$ are also defined. For $\Delta T' = 1\ mK$, figure 5(a) shows the time evolution of h and of δT until the beginning of the convective regime. δT increases to reach a maximum after one second of heating, and then it decreases progressively according to the function $e^t \times erfc(t^{1/2})$ (Zappoli & Durand-Daubin, 1994) as a result of the PE action that increases the temperature of the core. For a SCF diffusing-layer, the local Rayleigh number based on h and δT is given by (Gitterman & Steinberg, 1970b; Carlès & Ugurtas, 1999):

$$Ra^{corr}(h,\delta T) = \frac{g'\rho_c'^2\beta_P'C_P'h'^4}{\lambda'\mu'}\left(\frac{\delta T'}{h'} - \frac{\Delta T_a'}{H'}\right) \tag{22}$$

To account for the high compressibility of the SCF, the classical expression of the Rayleigh number is modified in Eq. 22 by the adiabatic temperature gradient ($\Delta T'_a/H' = g'\beta'_P T'_i/C'_P$) obtained by moving a fluid particle along the hydrostatic pressure gradient. This term, that can be neglected for a normally compressible fluid, represents the stabilizing contribution of the Schwarzschild criterion commonly encountered for large air columns, and according to which the fluid layer is stable if:

$$\frac{\delta T'}{h'} < \frac{\Delta T_a'}{H'} = \frac{g'\beta_P'T_i'}{C_P'} \tag{23}$$

In the considered model, the adiabatic temperature gradient $\Delta T'_a/H' = 0.34\ mK/cm$ and does not depend on the proximity ε to the CP since β'_P and C'_P have the same critical exponent of *-1*. To better estimate the interaction between natural convection and stratification, the normalized intensity of heating of the bottom wall ΔT is henceforth expressed in terms of $\Delta T_a = \Delta T'_a/T'_c$.

Figure 5(b) shows the time evolution of $Ra^{corr}(h,\delta T)$ for $\Delta T = 3\Delta T_a$. According to Eq. 22, $Ra^{corr}(h,\delta T)$ behaves as $h^3 \times \delta T \sim t^{3/2} \times e^t \times erfc(t^{1/2})$; in fact, $Ra^{corr}(h,\delta T)$ can be very well fitted in Fig. 5(b) by the curve $180 \times t^{3/2} \times e^t \times erfc(t^{1/2})$, and we can easily prove at long time scales that $erfc(t^{1/2}) \sim e^{-t} \times t^{-1/2}$, which explains the linear time evolution of $Ra^{corr}(h,\delta T)$. When the local Rayleigh number exceeds the critical value of about *1100* (Chandrasekar, 1961), the hot boundary layer becomes unstable[4]. Convective cells start to get organized along the bottom plate; figure 6(b) enables the visualization of these vortical structures using the *Q*-criterion[5] (Dubief & Delcayre, 2000) along the bottom plate. Then, the intensity of these vortices rises exponentially with time; this can be easily seen on the time evolution of the mean enstrophy in the hot boundary layer shown in Fig. 5(b) and defined by:

$$Ens(h) = \frac{1}{2h}\int_0^h \left[\left(\frac{\partial u}{\partial y} - \frac{\partial v}{\partial x}\right)^2 + \left(\frac{\partial u}{\partial z} - \frac{\partial w}{\partial x}\right)^2 + \left(\frac{\partial v}{\partial z} - \frac{\partial w}{\partial y}\right)^2\right].dz \tag{24}$$

The intensity of these convective cells keeps rising until producing enough amount of convective transport to deform the isotherms causing the collapse of the thermal boundary layers. For $\Delta T = 3\Delta T_a$, the collapse of the hot boundary layer occurs around $t = 120$ and corresponds to the symbol (□) in Figs. 5(a) and 5(b), afterwards the convective regime starts and the definition of the hot boundary layer holds no more. The cold boundary layer developed along the top plate is governed by the same mechanisms and its hydrodynamic

---

[4] The considered critical Rayleigh number is that of fluid layer with mixed (solid-free) boundary conditions. However, because the hot boundary layer is connected to the bulk of the cavity, its upper boundary is not sharply defined, the real critical Rayleigh number should slightly defer from *1100*; but this value, even though not very precise, remains the most suitable theoretical value for the considered configuration.

[5] $Q = \frac{1}{2}(\Omega_{ij}\Omega_{ji} - S_{ij}S_{ji})$ where S_{ij} and Ω_{ij} denote respectively the symmetric and anti-symmetric parts of ∇v.

stability depends on the same criterion (Accary et al., 2005b); the same scenario occurs for the cold boundary layer.

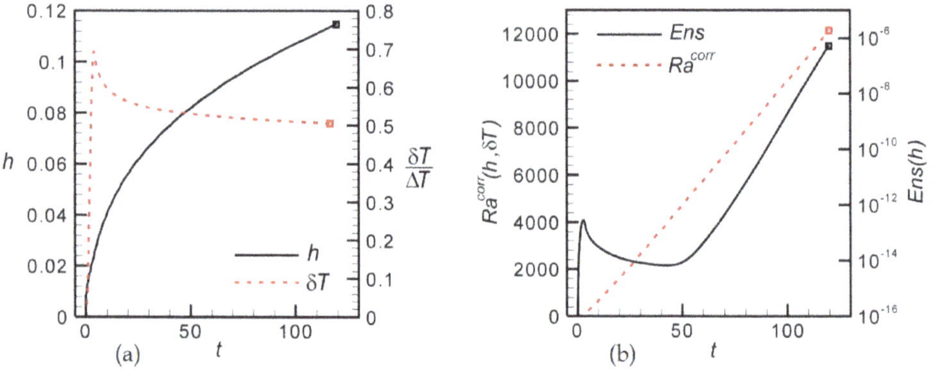

Fig. 5. (a) Time evolution of the hot boundary layer thickness h and of the temperature difference inside it δT for $\Delta T' = 1mK$. The symbol (□) indicates the beginning of the convective regime. (b) Time evolution of the local Rayleigh number $Ra^{corr}(h, \delta T)$ related to the hot boundary layer by Eq. 22 and of the mean enstrophy in the hot boundary layer (Eq. 24) showing the exponential increase of the intensity of convection.

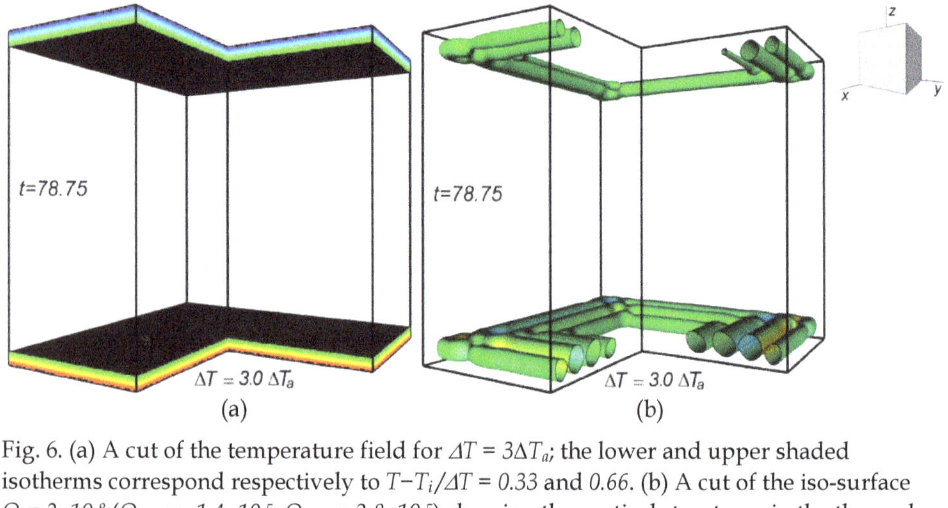

Fig. 6. (a) A cut of the temperature field for $\Delta T = 3\Delta T_a$; the lower and upper shaded isotherms correspond respectively to $T-T_i/\Delta T = 0.33$ and 0.66. (b) A cut of the iso-surface $Q = 2\times10^{-8}$ ($Q_{min} = -1.4\times10^{-5}$, $Q_{max} = 2.8\times10^{-5}$) showing the vortical structures in the thermal boundary layers shown in subfigure (a).

In Fig. 7, the critical value of δT for the convection onset in the hot boundary layer is derived from Eq. 22 and plotted versus h (the thick solid line) defining the unstable zone. This neutral stability curve consists of two lines representing the limits of the convection-onset criterion depending on h. For small values of h, the fluid compressibility can be neglected and the stability of the hot boundary layer is governed by the classical Rayleigh criterion, obtained from Eq. 22 by dropping the term $g'\beta'_P T'_i/C'_P$, while for larger values of h, viscosity and thermal diffusion are neglected, and the stability depends on the Schwarzschild criterion obtained from Eq. 23. For several intensities of heating ΔT, figure 7 shows the evolution curves $\delta T(h)$ for the hot boundary layer (the solid blue lines) until the beginning of the convective regime, which corresponds to the symbol (□). Figure 7 shows also results obtained in a 2D approximation with periodic vertical boundaries (the dashed red lines) (Accary et al., 2005b). The boundary effects induced by the presence of the lateral walls in the 3D case accelerates the development of convection and, at equal intensity of heating, the convective regime is reached earlier in comparison with the 2D approximation with periodic vertical boundaries. For low intensities of heating, practically for $\Delta T \leq 0.72\Delta T_a$, once the hot boundary layer has become unstable, the intensity of the convective cells rises exponentially with time until deforming the isotherms. However, this deformation is not large enough to induce the collapse of the hot boundary layer that keeps growing and the curve $\delta T(h)$ crosses the Schwarzschild line back into the stable zone again and a reverse transition to stability obtained without any external intervention (Accary et al., 2005a). This phenomenon requires that the thermal boundary layers grow enough without reaching the centre of the cavity (in order to avoid their interaction); a height of $1.5H'$ at least is needed in this case.

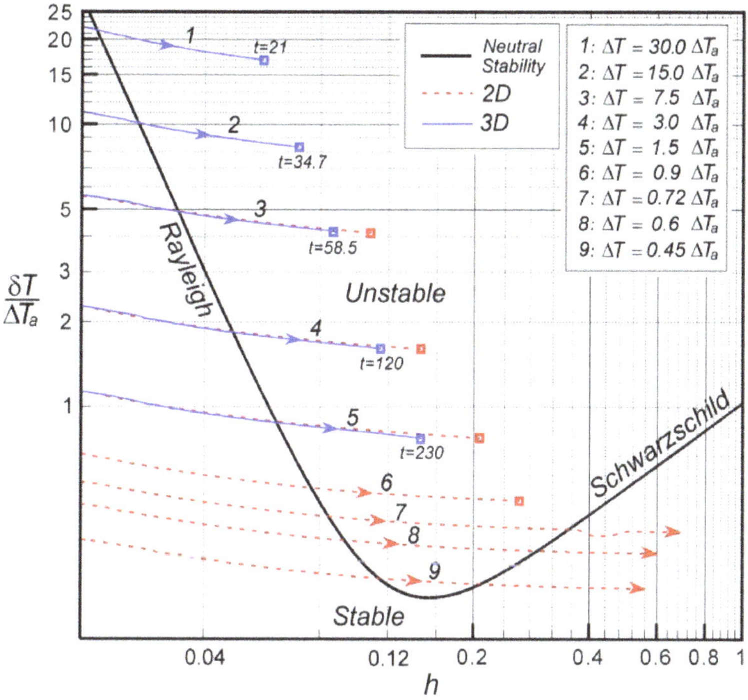

Fig. 7. Evolution of the temperature difference δT across the hot boundary layer as a function of its thickness h. Time evolves in the arrows' direction and the symbols (□) correspond to the beginning of the convective regime. The neutral stability line was derived from Eq. 20. The 2D results were obtained using a cavity of height $1.5H'$ with periodic vertical boundaries; for $\Delta T \leq 0.72\Delta T_a$, a reverse transition to stability is obtained though the Schwarzschild line.

5.2 Effect of the adapted low Mach number approximation on convection onset

Figure 8 shows a comparison for the evolution curves $\delta T(h)$ between the low Mach number approximation adapted to the SCF flow (ALMN) described in § 3.2 and the classical one (CLMN) obtained from the model by setting $\rho_i = 1$ in Eq. 10. The simulations are carried-out in a 2D approximation but with periodic vertical boundaries in order to suppress the disturbance resulting from the side walls. The straight dashed line with a slop of (-3) represents the classical Rayleigh criterion obtained from Eq. 22 by removing the adiabatic temperature gradient term. We notice that for a relatively strong heating ($\Delta T > 3\Delta T_a$) the flows predicted with both models are similar but not identical and the collapse of the thermal boundary layers occur at about the same time; this is due to the fact that a strong heating covers the effects of stratification, but the weight of this latter becomes more significant as the heating decreases.

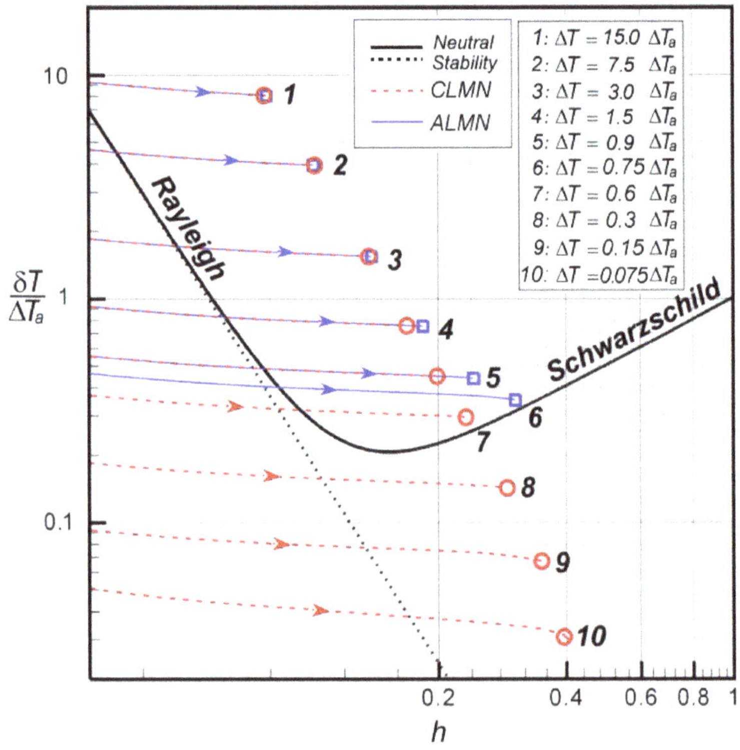

Fig. 8. Comparison between the adapted low Mach number model (ALMN) and the classical one (CLMN) for the evolution of the temperature difference δT across the hot boundary layer as a function of its thickness h. The symbols (□) and (o) correspond to the beginning of the convective regime and the neutral stability line was derived from Eq. 22. The simulations are carried-out in a 2D approximation with periodic vertical boundaries.

If we consider for example the case $\Delta T=0.9\Delta T_a$, the collapse of the hot boundary layer is observed with about $90{\times}t'_{PE}$ of time gap between the two models. For $\Delta T=0.9\Delta T_a$, figure 9 shows the temperature fields obtained $100 \times t'_{PE}$ after the collapse of the hot boundary layer. Compared to Fig. 9(a) (ALMN), the thermal plumes are much more developed in Fig. 9(b) (CLMN) since their motion is not hindered by the stratification which, when taken into account, prevents the free growth of the plumes. However for the considered heating intensity ($\Delta T=0.9\Delta T_a$), the thermal plumes manage to deform the temperature field giving rise the slowly moving structures. As the heating gets weaker ($\Delta T < 0.6\Delta T_a$ for example), the buoyant force being not strong enough for pulling the fluid particles through the hydrostatic pressure gradient, the hot boundary layer predicted with the ALMN approximation remains stable. In return, not including the stratification, the CLMN model is unable to take account of this stabilizing effect and persists in predicting a convective instability (according to the classical Rayleigh criterion) provided that the height H' of the cavity allows enough growth of the thermal boundary layers.

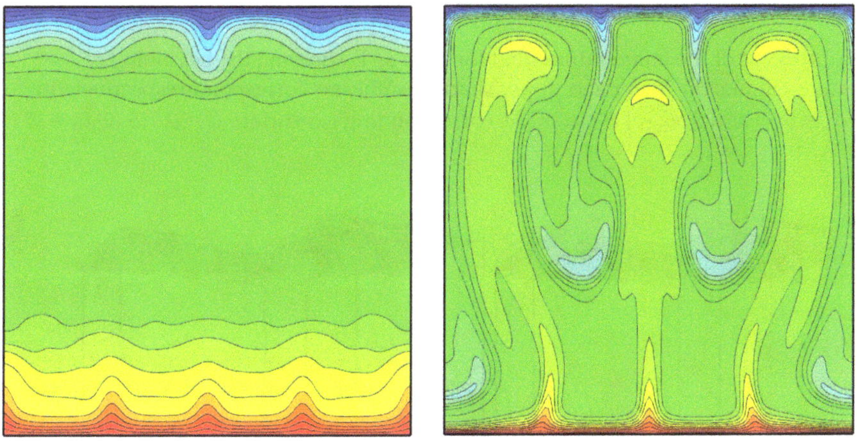

(a) (b)

Fig. 9. Comparison of the temperature fields obtained in a 2D approximation (with periodic vertical boundaries) for $\Delta T=0.9\Delta T_a$, $100 \times t'_{PE}$ after the collapse of the hot boundary layer. (a) Adapted low Mach number model (ALMN), (b) Classical low Mach number model (CLMN).

5.3 The beginning of the convective regime

The convective regime starts with several plumes rising from within the thermal boundary layers as shown in Fig. 10(a). These plumes are encircled by donut-shaped structures shown by the Q-criterion in Fig. 10(b). Convection improves the heat transfer between the isothermal walls and the bulk of the cavity, resulting into a faster thermal balance in the whole fluid volume. For all the heating cases that we considered, the hot boundary layer has always become unstable before the cold one. As the heating increases, convection is triggered earlier since the instability criterion ($Ra^{corr}(h,\delta T) > 1100$) is satisfied earlier; consequently, the thickness of the thermal boundary layer is smaller when the convection arises and the size of the convective structures decreases as shown in Fig. 11. A detailed study of the size of the convective structures has been done in a 2D approximation in (Accary et al., 2005b).

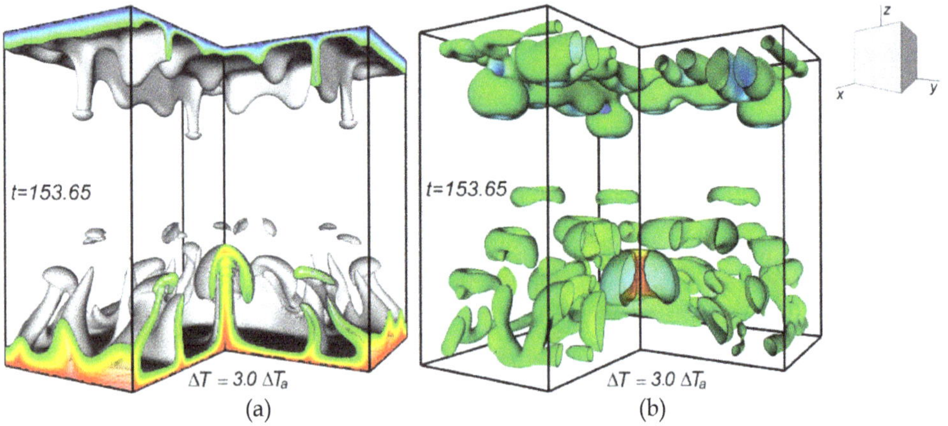

Fig. 10. (a) A cut of the temperature field for $\Delta T = 3\Delta T_a$ showing the beginning of the convective regime; the lower and upper shaded isotherms correspond respectively to $T-T_i/\Delta T = 0.33$ and 0.66. (b) A cut of the corresponding iso-surface $Q = 0.015$ ($Q_{min} = -0.15$, $Q_{max} = 0.15$).

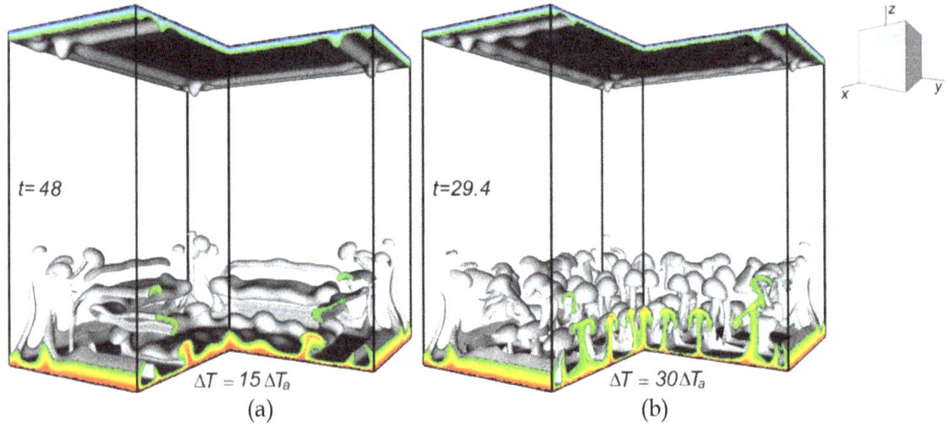

Fig. 11. Cuts of temperature fields for (a) $\Delta T = 15\Delta T_a$ and (b) $\Delta T = 30\Delta T_a$ showing the effect of the intensity of heating on the temperature field at the beginning of the convective regime. The lower and upper shaded isotherms correspond respectively to $T-T_i/\Delta T = 0.33$ and 0.66.

5.4 Transition to turbulence

In the convective regime of the flow that follows the convection onset, the Rayleigh number, based on the total height H' of the cavity and on the temperature difference $\Delta T'$ between the isothermal walls (Eq. 25), becomes a better indicator of the regime of the flow.

$$Ra^{corr} = \frac{g'\rho_c'^2\beta_P'C_P'H'^4}{\lambda'\mu'}\left(\frac{\Delta T'}{H'} - \frac{\Delta T_a'}{H'}\right) \tag{25}$$

For $\Delta T < \Delta T_a$, the Rayleigh number obtained from Eq. 25 is negative; this, however, does not prevent convection to arise in the thermal boundary layers when the local Rayleigh number (Eq. 22) exceeds *1100*. But for $\Delta T > \Delta T_a$, for example for $\Delta T = 1.5\Delta T_a$, the term in front of the parentheses in Eq. 25, which diverges as $\varepsilon^{-1.5}$, is very large and results in a Rayleigh number of 2.68×10^6, while for a PG, the Rayleigh number is directly proportional to ΔT.

The turbulent Rayleigh-Bénard convection is characterized by a statistically steady state of heat transfer. In the considered configuration, the settlement of the turbulent regime may be identified on the time evolution of the mean Nusselt numbers on the isothermal walls given by:

$$Nu = -\frac{1}{\Delta T}\frac{\partial T}{\partial z} \tag{26}$$

For $\Delta T = 7.5\Delta T_a$ which corresponds to $Ra^{corr} = 80\times10^6$ (Eq. 25), figure 12 shows the time evolution of the mean Nusselt numbers on the bottom wall and of the top one. The convection onset is easily identified by the improvement of the heat transfer corresponding to the increase in the mean Nusselt numbers that stabilize afterwards around almost the same value, which indicates the settlement of the turbulent flow. Figure 13(a) shows the temperature field obtained in the turbulent regime. We notice first the appearance of crest-like patterns defining on the isothermal walls flat regions where the temperature is almost homogeneous in the (x,y) plan, we notice also the spreading of the isotherms along the adiabatic walls. Figure 13(b) shows the chaotic flow that takes place in the turbulent regime. The vortical structures have no particular shape; the tubular and toroïdal structures obtained at the beginning of the convective flow have completely disappeared.

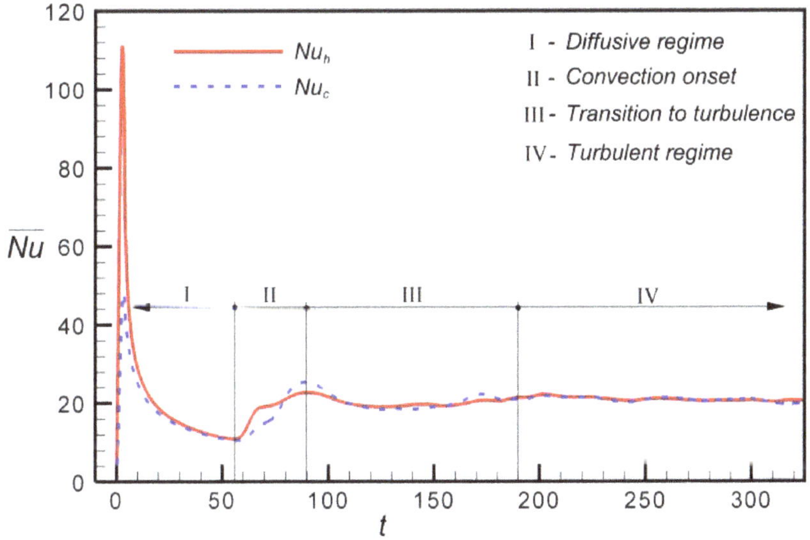

Fig. 12. Time evolution of the mean Nusselt numbers (Eq. 26) on the bottom wall (Nu_h, *h* for hot) and the top one (Nu_c, *c* for cold) for $\Delta T = 7.5\Delta T_a$ ($Ra^{corr} = 80\times10^6$).

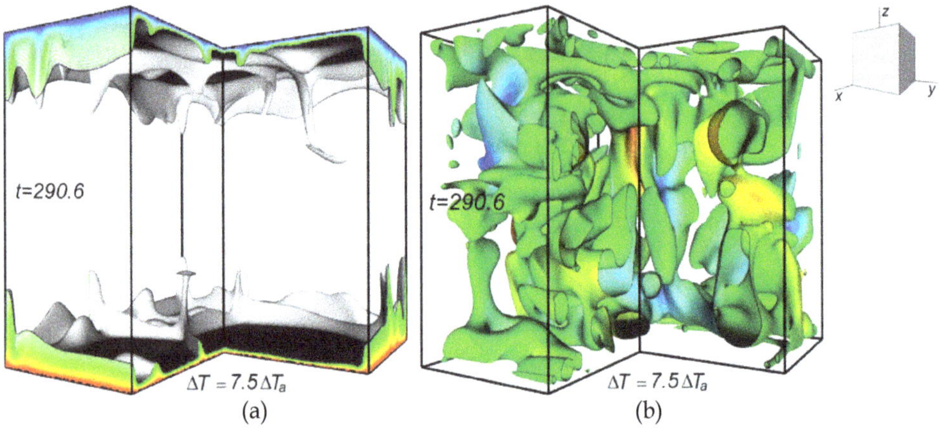

Fig. 13. (a) A cut of the temperature field for $\Delta T = 7.5\Delta T_a$; the lower and upper shaded isotherms correspond respectively to $T-T_i/\Delta T = 0.33$ and 0.66. (b) A cut of the corresponding iso-surface $Q = 0.015$ ($Q_{min} = -0.37$, $Q_{max} = 0.67$).

In order to better estimate the size of the vortical structures and its time evolution, a discrete Fourier transformation[6] of the vertical velocity component w has been carried out in both x and y directions. Along the line ($y = y_0$, $z = z_0$) and for a wavelength H'/k associated to the mode k, the Fourier coefficient of $w(x,y_0,z_0)$ is given by:

$$W_x(k,y_0,z_0) = \int_0^1 \left[w(x,y_0,z_0) - (-1)^k w(1-x,y_0,z_0)\right] sin(2\pi kx).dx \quad (27)$$

Once the coefficients $W_x(k, y_0, z_0)$ are computed for all (y_0, z_0), the mean contribution of the mode k to the field of w is determined by:

$$\overline{W}_x(k) = \int_0^1\int_0^1 \left|W_x(k,y,z)\right|.dy.dz \quad (28)$$

Figure 14 shows the contribution of the different modes to the spectrum of the component w in the x and y directions at the beginning of the convective regime and in the turbulent one. We notice an important contribution of small wavelengths ranging between $H'/11$ and $H'/4$ at the beginning of convection (at $t = 89.25$, see Fig. 12). But as time goes by, the spectra of w show a much higher contribution of large wavelengths exceeding sometimes half the of the cavity width. Similar results were obtained for the horizontal velocity components, u and v. Thus, the turbulent flow consists mainly of large vortical structures.

Figure 15 shows cuts of the temperature field in the vertical median plans of the cavity with the corresponding velocity fields that confirm the presence of large convective structures in the steady-state turbulent regime. We notice that the temperature field consists mainly of two unstable thermal boundary layers exchanging heat and mass with the bulk of the cavity in

[6] The operation required the fictive assumption of a periodic and odd distribution of w in the horizontal directions with a period of $2H'$.

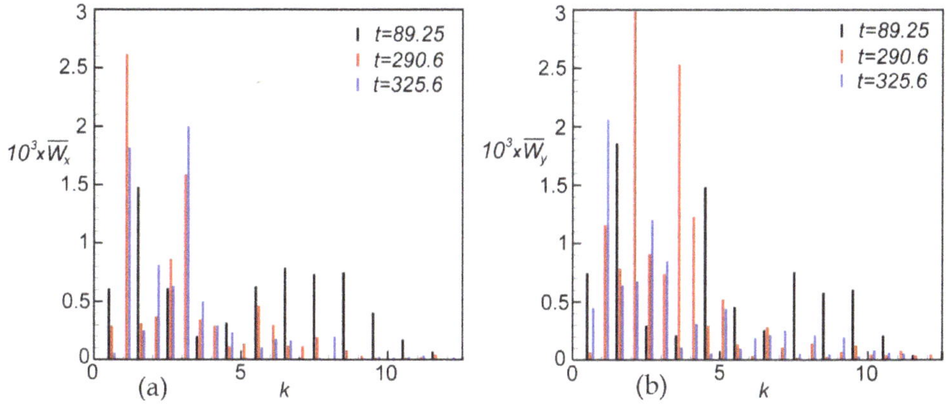

Fig. 14. The weights of the different wave-vectors k in the spectrum of the vertical velocity component w obtained in the directions x (a) and y (b) for $\Delta T = 7.5\Delta T_a$. In average, the convective structures are clearly larger in the steady-state turbulent regime ($t = 290.6$ and 325.6) than at the beginning of convection ($t = 89.25$).

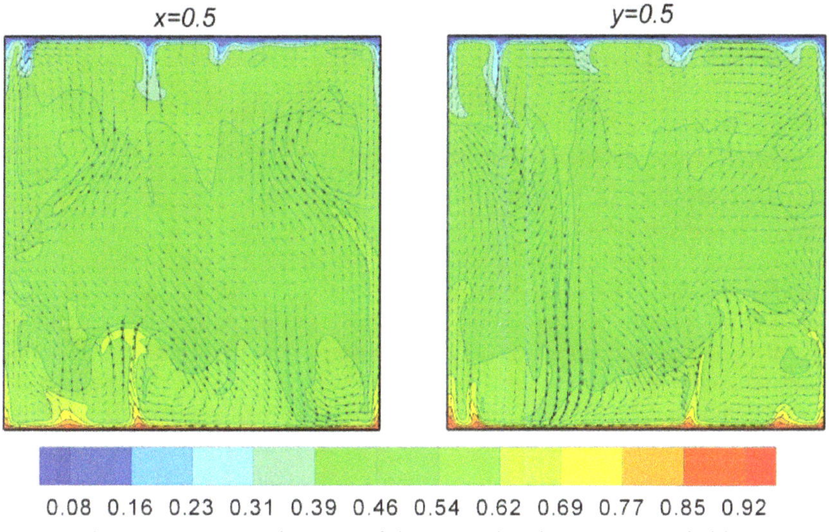

Fig. 15. Vertical cuts at $x = 0.5$ and $y = 0.5$ of the normalized temperature field $T-T_i/\Delta T$ shown in Fig. 13(a) ($\Delta T = 7.5\Delta T_a$, $t = 290.6$), with the corresponding velocity fields.

which the convective activity induces a quasi-homogeneous temperature. Figure 16 shows the time evolution, along the vertical axis of the cavity ($x = y = 0.5$), of the velocity magnitude and of the temperature at the free boundaries of the thermal layers ($z = 0.05$ and $z = 0.95$)[7] and at the centre of the cavity; the velocity components have the same order of magnitude. Figure

[7] Despite convection, the thickness of the thermal boundary layers may be computed at each point of the horizontal walls using the same definition of section 5.1; the normalized values of the thermal boundary layers' thicknesses (that were averaged in space and in time) are around 0.05.

16(a) underlines the chaotic convection that takes place in the whole fluid volume; the velocity has been monitored at 25 different points of the cavity and confirms that chaotic behavior. In the steady-state turbulent regime, figure 16(b) shows a slight difference of the time-averaged temperature between positions $z = 0.05$ et $z = 0.95$, which reveals the existence of a temperature gradient in the bulk of the cavity that will be investigated in § 5.5.

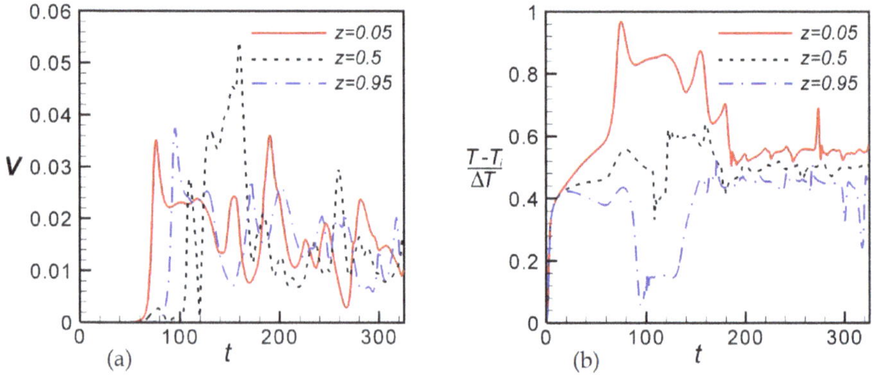

Fig. 16. Time evolution of the local velocity magnitude (scaled by $V'_{PE} = H'/t'_{PE} = 3.9\ cm.s^{-1}$) (a) and the local normalized temperature (b) at three positions ($z = 0.05, 0.5,$ and 0.95) along the line $x = y = 0.5$, for $\Delta T = 7.5\Delta T_a$.

5.5 The global thermal balance of the cavity

The steadiness of the mean Nusselt numbers on the isothermal walls (Fig. 12, turbulent regime) reflects the settlement of a statistically steady-state heat transfer across the cavity. However, figure 17 reveals the strong non-uniformity of the Nusselt numbers distributions on the isothermal walls. These patterns are directly related to those of the temperature field: the Nusselt number's minima are reached under the crest-like patterns shown in Fig. 13(a), while the maxima are obtained inside the cells determined by those patterns. These cells are thus characterized by very thin thermal boundary layers; for the temperature field shown in Fig. 13(a), the minimal normalized thicknesses of the thermal boundary layer were about 0.014 for the hot boundary layer and 0.012 for the cold one and were obtained where the distributions of the Nusselt numbers reach their maxima. Despite the strong non-uniform distributions of the Nusselt numbers, in the steady-state turbulent regime, the mean Nusselt numbers on both isothermal walls fluctuate around the same value.

For different intensities of heating and hence Rayleigh numbers, figure 18(a) reports the mean Nusselt numbers (the filled circles). For a PG, the experimental results (Poche et al., 2004) and those issued from a scaling theory (Siggia, 1994) show that Nusselt number behaves as $Ra^{2/7}$. This behavior can be observed for the of the Nusselt number corrected by the adiabatic temperature gradient (Kogan & Meyer, 2001), given by:

$$Nu^{corr} = \frac{-\dfrac{\partial T}{\partial z} - \Delta T_a}{\Delta T - \Delta T_a} = \frac{\Delta T}{\Delta T - \Delta T_a}\left(Nu - \frac{\Delta T_a}{\Delta T}\right) \tag{29}$$

For $\Delta T >> \Delta T_a$, $Nu^{corr} \rightarrow Nu$; but as the intensity of heating decreases, the corrected expression of the Nusselt number (the filled squares in Fig. 18(a)) enables the retrieval of the $Ra^{2/7}$ law. However, it should be reminded that the effective heat transfer is described by the classical expression of the Nusselt number given by Eq. 26, not by the corrected one.

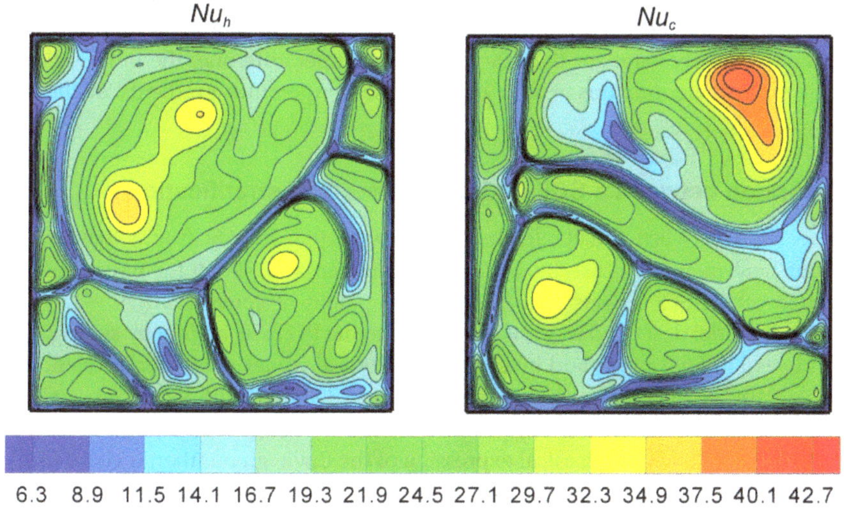

Fig. 17. Distributions of the Nusselt number on the bottom (Nu_h) and the top (Nu_c) isothermal walls, corresponding to temperature field shown in Fig. 13(a) ($\Delta T = 7.5\Delta T_a$, $t = 290.6$).

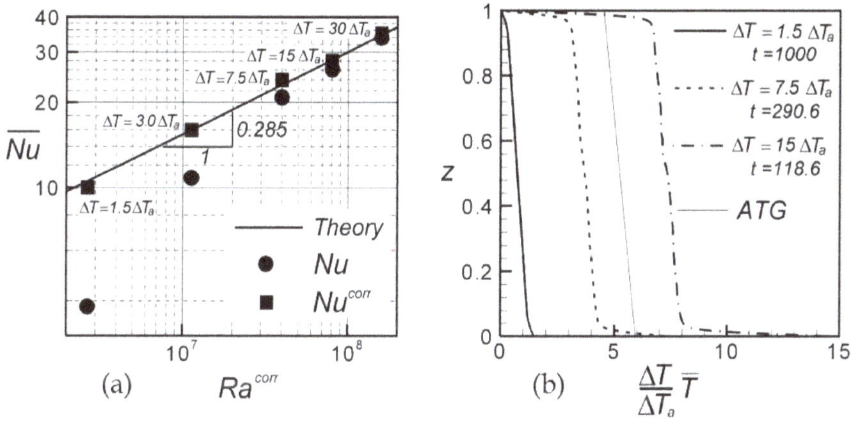

Fig. 18. In the steady state regime of the turbulent flow: (a) the classical (Eq. 26) and the corrected (Eq. 29) mean Nusselt numbers versus the corrected Rayleigh number (Eq. 25); (b) temperature profiles (averaged in the xy-plan) for different intensities of heating ($\overline{T}$ varies between 0 and 1), ATG stands for 'adiabatic temperature gradient'.

At the global thermal balance of the cavity and for all intensities of heating, figure 18(b) reveals the existence of a mean temperature gradient in the bulk of the cavity equal to the adiabatic temperature one. This is a natural structure of the mean temperature field that ensures the minimal temperature gradients in the thermal boundary layers with the constraint of a globally stable bulk of the cavity. Indeed, if the mean temperature gradient in the bulk of the cavity were larger than the adiabatic temperature one, the bulk of the cavity would lose its hydrodynamic stability. In return, if the mean temperature gradient in the bulk of the cavity were smaller than the adiabatic temperature one, the bulk of the cavity would be 'too' stable, but this would increase the temperature gradients in the thermal boundary layers.

6. Comparison between a SCF and a PG, effects of stratification

The comparison between the Rayleigh-Bénard convection in a SCF and that in a PG is carried out here in the 3D case for a Rayleigh number of 2.68×10^6 for which the density stratification of the SCF affects clearly the development of convection ($\Delta T = 1.5\Delta T_a$). The mathematical model described in § 3 was adapted to the PG case, mainly by setting $a = b = 0$, $\Lambda = 0$, and $\varepsilon = 0$ in Eqs. 5 to 9, and by choosing reference values of temperature and density compatible with the PG assumption, these were set to $300\ K$ and $1.8\ Kg.m^{-3}$ respectively. An intensity of heating $\Delta T' = 5\ K$ was applied to bottom wall in the PG case and the height H' of the cavity, deduced from the classical expression of the Rayleigh number, is equal to $13.8\ cm$. A mesh of 100^3 and a time step of $0.125s$ have been used.

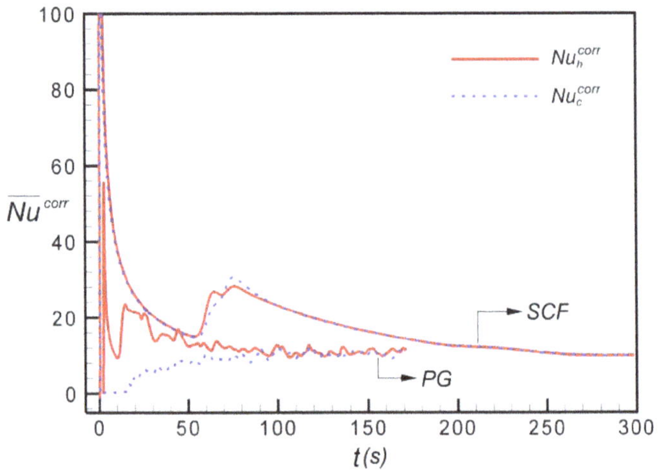

Fig. 19. Time evolution of the mean Nusselt numbers on the isothermal walls obtained in the 3D case for a Rayleigh number of 2.68×10^6 for a SCF ($\Delta T = 1.5\Delta T_a$) and for a PG. The curves were shifted by $2s$ for the PG to show prominently the first peak. For the SCF, the first peak of the mean Nusselt number on the bottom wall reaches the value of 370, and the beginning of convection at about $60\ s$ is consistent with the result shown in Fig. 7, where the convective regime starts at $t' = 120\times t_{PE} = 58.9\ s$.

Figure 19 shows the time[8] evolution of the mean Nusselt numbers for the SCF and for the PG[9]. The large temperature gradients obtained at the very first seconds of heating in the case of the SCF are responsible for the very high peak of the Nusselt number. For the SCF, figure 19 reports very similar evolutions of the mean Nusselt numbers on both isothermal walls. By contrast for the PG, while the mean Nusselt number on the bottom wall shows a similar behavior to that of the SCF during the diffusive regime, no heat transfer is detected on the top wall ($Nu_c = 0$) until the beginning of convection. Because the PE is practically inexistent for the PG, the heat transfer is only activated on the top wall when the thermal plumes rising from the hot boundary layer reach it. Even though the Prandtl number is about 18 times smaller[10] (Verzicco & Camussi, 1999), convection in the PG is much more developed than in the SCF at the same Rayleigh number, as shown by Fig. 20. The fluctuating time evolution of the mean Nusselt numbers for the PG results from this intense convective activity. By contrast, the trace of the diffusion-dominated temperature field (Fig. 20(a)) obtained for SCF due to its strong stratification is visible on the time evolution of the mean Nusselt numbers after the convection onset. Under these conditions, the global thermal balance of the cavity is mainly achieved by diffusion at long time scales because of the critical vanishing (as $\varepsilon^{1/2}$) of the thermal diffusivity of the SCF. We notice finally that even though the temperature field of the SCF is diffusion dominated while it is convection-dominated for the PG, the corrected mean Nusselt number at the global thermal balance of the cavity is the same in both cases.

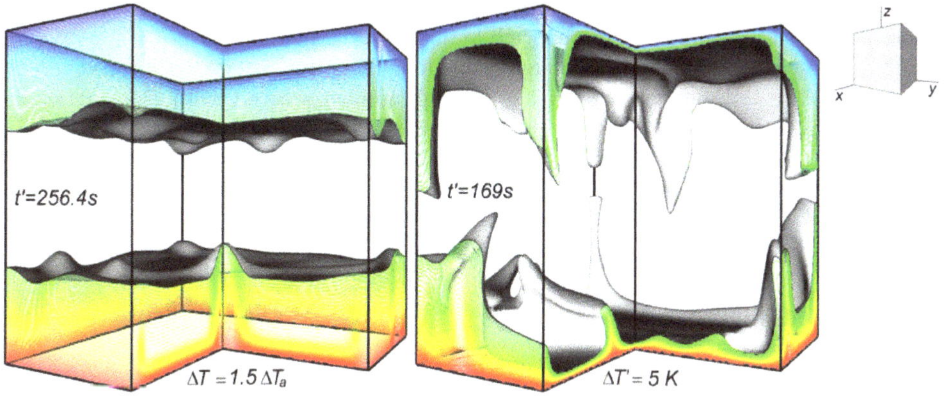

(a) (b)

Fig. 20. Cuts of temperature fields for a Rayleigh number of 2.68×10^6 (a) for a SCF ($\Delta T = 1.5\Delta T_a$) and (b) for a PG, showing how the strong stratification of the SCF holds back the development of convection. The lower and upper shaded isotherms correspond respectively to $T-T_i/\Delta T = 0.33$ and 0.66.

[8] Time is not scaled in this case because the PE does not exist for the PG.

[9] The PG adiabatic temperature gradient is very small compared to $\Delta T'/H'$, hence: $Nu^{corr} \rightarrow Nu$.

[10] According to the model, the PG Prandtl number is about 2.27 against 39.6 for the SCF.

7. Conclusions

In this chapter, the mathematical model for SCF buoyant flows with the appropriate acoustic filtering has been recalled, then a description of the different stages of the SCF flow in a cube-shaped cavity heated from below were reported from the first seconds of heating until the settlement of a statistically steady-state of heat transfer, and this for Rayleigh numbers ranged from 2.68×10^6 up to 160×10^6. While the scenarios of the convection onset and disappearance (reverse transition to stability) can be observed in a 2D approximation, the convective regime and the transition to turbulence requires 3D simulations. At the beginning of convection, tubular convective structures appear inside the thermal boundary layers while the thermal plumes are encircled by toroïdal vortical structures; the size of these structures decreases as the intensity of heating increases. In the turbulent regime, the convective structures grow until their size exceeds half of the cavity, and create on the isothermal walls several cells where an intense heat transfer takes place. Despite the non-homogeneous heat transfer on the isothermal walls, the steadiness of the mean Nusselt numbers around the same value reflects the global thermal balance of the cavity. The relation between that equilibrium Nusselt number and the Rayleigh number obtained for a PG ($Nu \sim Ra^{2/7}$) is applicable to the SCF, provided that the adiabatic temperature gradient is taken into account in the expressions of both numbers. In the turbulent regime, the temperature field consists mainly of two unstable thermal boundary layers and a bulk characterized by a mean temperature gradient equal to the adiabatic temperature one. For relatively high intensities of heating ($\Delta T >> \Delta T_a$), the global thermal balance of the cavity is achieved by a chaotic convection invading in the whole fluid volume. By contrast for weak intensities of heating ($\Delta T \gtrsim \Delta T_a$), the strong density stratification, due to the high isothermal compressibility of the fluid, prevents the free development of convection whose penetrability is dramatically reduced; in this case, the thermal balance of the cavity is mainly achieved by diffusion and therefore on long time scales. Finally, the comparison between the SCF and the PG for the same Rayleigh number showed two major differences. The first, related to the PE, is the absence of heat transfer on the top wall for the PG until the beginning of convection; while for the SCF, the time evolutions of the mean Nusselt numbers on both isothermal walls are similar. The second, related to the stratification of the SCF and thus only encountered for $\Delta T \gtrsim \Delta T_a$, is the diffusion-dominated thermal balance of the cavity for the SCF, while it is convection-dominated for PG.

8. Acknowledgments

The author is much indebted to Dr. Patrick Bontoux, Dr Bernad Zappoli, and Dr Horst Meyer for fruitful and illuminating discussions on supercritical fluids. Thanks are also due to the CNES (Centre National d'Etudes Spatiales, France) for the financial support and to the IDRIS (Institut du Developpement et des Ressources en Informatique Scientifique, France) for providing the computation resources. The reported numerical simulations were carried out at the M2P2 laboratory (UMR 6181 CNRS, France).

9. References

Accary, G.; Raspo, I.; Bontoux, P. & Zappoli, B. (2005a). *Reverse transition to hydrodynamic stability through the Schwarzschild line in a supercritical fluid layer*, Phys. Rev. E 72, 035301

Accary, G.; Raspo, I.; Bontoux, P. & Zappoli, B. (2005b). *Stability of a supercritical fluid diffusing layer with mixed boundary conditions*, Phys. of Fluids 17, 104105

Accary, G.; Raspo, I.; Bontoux, P. & Zappoli, B. (2005c). *An adaptation of the low Mach number approximation for supercritical fluid buoyant flows*, C. R. Méc. 333, pp. 397

Accary, G. & Raspo, I. (2006). *A 3D finite volume method for the prediction of a supercritical fluid buoyant flow in a differentially heated cavity*, Comp. & Fluids 35(10), pp. 1316

Accary, G.; Bontoux, P. & Zappoli, B. (2009). *Turbulent Rayleigh–Bénard convection in a near-critical fluid by three-dimensional direct numerical simulation*. J. Fluid Mech., 619, pp. 127

Amiroudine, S.; Bontoux, P.; Larroudé, P.; Gilly, B. & Zappoli, B. (2001). *Direct numerical simulation of instabilities in a two-dimensional near-critical fluid layer heated from below*, J. Fluid Mech. 442, pp. 119

Barrett, R.; Berry, M.; Chan, T.F.; Demmel, J.; Donato, J.M.; Dongarra, J.; Eijkhout, V.; Pozo, R.; Romine, C. & Van der Vorst, H. (1994). *Templates for the solution of linear systems: building blocks for iterative methods*, Siam, Philadelphia

Boukari, H.; Schaumeyer, J.N.; Briggs, M.E. & Gammon, R.W. (1990). *Critical speeding up in pure fluids*, Phys. Rev. A 41, pp. 2260

Carlès, P. & Ugurtas, B. (1999). *The onset of free convection near the liquid-vapour critical point. Part I: Stationary initial state*, Physica D 162, pp. 69

Chandrasekar, S. (1961). *Hydrodynamic and hydromagnetic stability*, Clarendon Press, Oxford.

Chorin, A.J. (1997). *A numerical method for solving the incompressible and low speed compressible equations*, Journal of Computational Physics,137(2), pp. 118

Dubief, Y. & Delcayre, F. (2000). *On coherent-vortex identification in turbulence*, J. of Turbulence 1, 011

Furukawa, A. & Onuki A. (2002). *Convective heat transport in compressible fluids*, Phys. Rev. E 66, 016302

Garrabos, Y.; Bonetti, M.; Beysens, D.; Perrot, F.; Fröhlich, T.; Carlès, P. & Zappoli B. (1998). *Relaxation of a supercritical fluid after a heat pulse in the absence of gravity effects: Theory and experiments*, Phys. Rev. E 57, pp. 5665

Gitterman, M. & Steinberg V.A. (1970b). *Criteria for the commencement of convection in a liquid close to the critical point*, High Temperature USSR 8(4), pp. 754

Guenoun, P.; Khalil, B.; Beysens, D.; Garrabos, Y.; Kammoun, F.; Le Neindre, B. & Zappoli, B. (1993). *Thermal cycle around the critical point of carbon dioxide under reduced gravity*, Phys. Rev. E 47, pp. 1531

Jang, D.S.; Jetli, R. & Acharya, S. (1986). *Comparison of the Piso, Simpler, and Simplec algorithms fort he treatment of the pressure-velocity coupling in steady flow problems*, Numerical Heat Transfer,10, pp. 209

Kogan, A.B. & Meyer, H. (1998). *Density Response and Equilibration in a Pure Fluid Near the Critical Point: ^{3}He*, J. Low Temp. Phys. 112, pp. 417

Kogan, A.B. & Meyer, H. (2001). Heat transfer and convection onset in a compressible fluid: ^{3}He near the critical point, Phys. Rev. E 63, 056310

Kogan, A.B.; Murphy, D. & Meyer, H. (1999). *Onset of Rayleigh Bénard convection in a very compressible fluid: ^{3}He, near T_c*, Phys. Rev. Lett. 82, pp. 4635

Li, Y. & Rudman, M. (1995). *Assessment of higher-order upwind schemes incorporating FCT for convection-dominated problems*, Numerical Heat Transfer B, 27, pp. 1

Nitsche, K. & Straub J. (1987). *The critical "hump" of C_v under microgravity, results from D-spacelab experiment "Wärmekapazität"*, Proceedings of the 6th European Symp. on Material Sc. under Microgravity Conditions. ESA SP-256, pp. 109

Onuki, A.; Hao, H. & Ferrell, R.A. (1990). *Fast adiabatic equilibration in a single-component fluid near the liquid-vapor critical point*, Phys. Rev. A 41, pp. 2256

Paolucci, S. (1982). *On the filtering of sound from the Navier-Stokes equations*, Sandia National Lab. Report, SAND 82, 8257

Patankar, S.V. (1980). *Numerical Heat Transfer and Fluid Flow*, Hemisphere, New York

Poche, P.E.; Castaing, B.; Chabaud, B. & Hébral, B. (2004). *Heat transfer in turbulent Rayleigh-Bénard convection below the ultimate regime*, J. Low Temp. Phys. 134(5/6), 1011

Siggia, E.D. (1994). *High Rayleigh number convection*, Ann. Rev. Fluid. Mech. 26, 137

Straub, J.; Eicher, L. & Haupt, A. (1995). *Dynamic temperature propagation in a pure fluid near its critical point observed under microgravity during the German Spacelab Mission D-2*, Phys. Rev. E 51, pp. 5556

Verzicco, R. & Camussi, R. (1999). Prandtl number effects in convective turbulence. J. Fluid Mech. 383, pp. 55

Zappoli, B.; Bailly, D.; Garrabos, Y.; Le Neindre, B.; Guenoun, P. & Beysens, D. (1990). *Anomalous heat transport by the piston effect in supercritical fluids under zero gravity*, Phys. Rev. A 41, pp. 2264

Zappoli, B. (1992). The *response of a nearly supercritical pure fluid to a thermal disturbance*, Phys. Of Fluids A 4, pp. 1040

Zappoli, B. & Duran-Daubin, A. (1994). *Heat and mass transport in a near supercritical fluid*, Phys. Of Fluids 6(5), pp. 1929

Zappoli, B.; Jounet, A.; Amiroudine, S. & Mojtabi, K. (1999). Thermoacoustic heating and cooling in hypercompressible fluids in the presence of a thermal plume. J. Fluid Mech. 388, pp. 389

10

A Concept of Discretization Error Indicator for Simulating Thermal Radiation by Finite Volume Method Based on an Entropy Generation Approach

H. C. Zhang, Y. Y. Guo, H. P. Tan and Y. Li
Harbin Institute of Technology
P. R. China

1. Introduction

Heat transfer is frequently dominated by Thermal Radiation (TR) in many scientific and engineering applications, especially at high temperature (Howell et al., 2010). Usually, three main fundamental approaches are supplemented to investigate TR problem, including analytical, experimental and numerical methods (Modest, 2003), however, among those TR problems, only quite a few of them can be analytically or experimentally solved. Recently, because of a rapid growth of computer and information techniques, numerical approximation has been eventually become the major simulating tool towards TR problems. The general equation to describe TR transport is the Radiative Transfer Equation (RTE), and several computational algorithms were proposed for solution of the RTE, which have achieved great advancement (Howell et al., 2010; Modest, 2003; Shih et al., 2010).

The Finite Volume Method (FVM) has validated to be an efficient algorithm with satisfactory precision (Raithby & Chui, 1990), which has been applied to various problems. Besides, much innovation to improve its performance is also proposed. FTn method is used to predict TR characteristics for a 3D complex industrial boiler with non-gray media (Borjini et al., 2007). FVM is applied with Lattice Boltzmann method in a transient 2D coupled conduction-radiation problem by an inverse analysis (Das et al., 2008). Combined mixed convection-radiation heat transfer is dealt with by a FVM (Farzad & Shahini, 2009). Transient radiative heating characteristics of slabs in a walking beam type reheating furnace is predicted by FVM (Han et al., 2009). A complex axisymmetric enclosure with participating medium is investigated by using FVM with an implementation of the unstructured polygonal meshes (Kim et al., 2010). A particular procedure as a first-order skew, positive coefficient, upwind scheme was presented (Daniel & Fatmir, 2011), which is incorporated in FVM.

Essentially, FVM can be categorized as a numerical method applied to investigate radiative heat transfer problems. Because algebraic equations for the FVM are determined through discretization of the RTE over user-selected control volumes and specific control solid angles, it will inevitably encounter various errors, which is an important and integral part in connection with the solution procedures. The most common discretization errors occurring in the FVM are called the ray effect and the false scattering, which were initially identified by Chai et al. (Chai et al., 1993). Only the error caused by spatial discretization is discussed in this paper, and it is also referred to as numerical scattering or numerical smearing (Zhang & Tan, 2009), which is analogous to false diffusion in the context of computational fluid dynamics caused by discretization of spatial coordinates (Patankar, 1980). It has been shown that many factors can cause false scattering (Tan et al., 2004) influencing solution accuracy, including grid quality (Kallinderis & Kontzialis, 2009), spatial discretization schemes (Coelho, 2008), radiative properties and volumetric heat sources (Kamel et al., 2006). However, there are few effective routines for evaluating the spatial discretization error, and it is necessary to formulate an innovative framework to explore parameters or define indicator to analyze its uncertainty and accuracy.

The concept and theory of entropy, based on the second law of thermodynamics, has been an innovative and effective approach to study computational errors within the fields of fluid flow and heat transfer (Naterer & Camberos, 2003). The entropy production is used to predict numerical errors for viscous compressible flow (Camberos, 2000). The concept of information entropy (Cover & Thomas, 2003) has been shown to be an appropriate method and has been widely applied to error analysis for Euler's equations and the stability of numerical solution (Camberos, 2007). Although some work has been done based on radiation entropy generation (Caldas & Semiao, 2005; Liu & Chu, 2007), much work has been focused on error analysis in computational fluid dynamics, heat conduction and heat convection, instead of error analysis for TR. In the previous work, an entropy formula based on information theory is proposed to investigate uncertainty in FVM towards artificial benchmarks (Zhang et al. 2011), which show its adaptability in field of TR.

In this chapter, an artificial benchmark model of central laser incidence on a two-dimensional (2D) rectangle containing a semi-transparent medium is used as a framework to investigate the numerical scattering, using reference data from the Monte Carlo method (MCM), which has been proven to generate no false scattering (Tan et al., 2004). Based on the local entropy generation approach (Herwig & Kock, 2004) derived from the second law of thermodynamics, which is considered a very effective method to analyze the process of energy transfer, a discretization error indicator is defined. Within the framework of the current model, grid independence is first validated. The effects of the spatial differential scheme, the spatial grid number and the absorption coefficient deviation of the medium on numerical scattering in the FVM are presented.

2. Mathematical model and artificial benchmark

In an emitting, absorption and scattering medium, the RTE can be written as:

$$\frac{\mathrm{d}I_\lambda(\vec{s},\vec{\Omega})}{\mathrm{d}s} = -\kappa_{a\lambda} I_\lambda(\vec{s},\vec{\Omega}) - \kappa_{s\lambda} I_\lambda(\vec{s},\vec{\Omega}) + \kappa_{a\lambda} I_{b\lambda}(\vec{s}) + \frac{\kappa_{s\lambda}}{4\pi}\int_{4\pi} I_\lambda(\vec{s},\vec{\Omega}')\Phi_\lambda(\vec{\Omega}',\vec{\Omega})\mathrm{d}\Omega' \quad (1)$$

For an opaque, diffuse emitting and reflective boundary wall, the corresponding boundary condition can be written as:

$$I_\lambda(\vec{\Omega}) = \varepsilon_\lambda I_{b\lambda} + \frac{1-\varepsilon_\lambda}{\pi}\int_{\vec{n}_b\cdot\vec{\Omega}'<0} I_\lambda(\vec{\Omega}')\,|\,\vec{n}_b\cdot\vec{\Omega}'\,|\,d\vec{\Omega}' \quad (2)$$

When FVM is used to solve the RTE, hemispherical space of 4π steradians is divided to a solid angular grid, i.e., a limited number of directions. Along a specific angular direction Ω^l, a relationship is necessary to correlate the radiative intensities at the face of the control volume to the intensity at the centre of control volume. This yields a spatial differential scheme, which in general can be represented as:

$$I_p^{\Omega^l} = \sum f_\alpha I_\beta^{\Omega^l} \tag{3}$$

where f_α denotes different values for f_x , f_y and f_z corresponding different types of differential schemes, and I_β means radiative intensities at different interfaces of a control volume, including I_e, I_w, I_n, I_s, I_t and I_b.

Frequently, the following three kinds of differencing schemes are selected, namely, the step scheme with $f_\alpha = 1.0$, the diamond scheme with $f_\alpha = 0.5$ and the exponential scheme with:

$$f_\alpha = \left[1 - \exp\left(-\tau_\alpha^{\Omega^l}\right)\right]^{-1} - \left(\tau_\alpha^{\Omega^l}\right)^{-1} \tag{4}$$

where τ is the optical thickness.

2.1 Numerical scattering

It is generally considered that numerical scattering is a multi-dimensional problem caused by spatial coordinates discretization. When the first term of the truncation error is the second-order space derivative, its error is dissipative. The derivative term of radiative intensity is used as the first-order difference scheme in the solution of RTE and truncation error is the second-order space derivative. As a result, numerical scattering is caused. Its premise is that: intensity at certain grids is related to other grids by space deferential schemes, and these grids are not in the transfer direction. In a multi-dimensional problem, if a derivative is substituted by first-order difference, if the profile of radiative intensity is assumed by spatial differencing schemes and if the direction of transfer is intersected with grid and non-negative intensity gratitude exists in the direction perpendicular with transportation direction, numerical scattering is still generated (Tan et al., 2004).

Because numerical scattering is a multi-dimensional problem, a 2D case is taken into account in the current study, without considering the scattering in the medium.

The accurate solution for the spectral radiative intensity in a specific angular direction Ω^l of the FVM is denoted by $I_{(\vec{s})}^{\Omega^l}$, and the symbol $L(I)_{\vec{s},\Omega^l}$ is defined as an operator for the differential process of $I_{(\vec{s})}^{\Omega^l}$ at a control volume of $\left(\vec{s},\Omega^l\right)$, where vector $\vec{s}$ denotes the spatial coordinates of the point, as shown in Eq. (5):

$$L(I)_{\vec{s},\Omega^l} = \left[\xi\frac{\partial I}{\partial x} + \eta\frac{\partial I}{\partial y} - \kappa_{\alpha\lambda} I - S\right]_{\vec{s},\Omega^l} \tag{5}$$

If the symbol $L_{\Delta\vec{s}}\left[I_{(\vec{s})}^{\Omega^l}\right]$ is used for the difference operator for $I_{(\vec{s})}^{\Omega^l}$, we then have:

$$L_{\Delta\vec{s}}\left[I_{(\vec{s})}^{\Omega^l}\right]=\xi\frac{I_{i+1}^{\Omega^l}-I_i^{\Omega^l}}{\Delta x}+\eta\frac{I_{j+1}^{\Omega^l}-I_j^{\Omega^l}}{\Delta y}-\kappa_{\alpha\lambda}I^{\Omega^l}-S_s^{\Omega^l} \tag{6}$$

In Eq. (6), $L_{\Delta\vec{s}}\left[I_{(\vec{s})}^{\Omega^l}\right]$ denotes the step scheme for a 2D differential equation. In this way, the cut-off for the discrete differential equations denotes the difference between the difference operators and the corresponding differential operators, which can be expressed using *TE*, i.e.

$$TE=L_{\Delta\vec{s}}\left[I_{(\vec{s})}^{\Omega^l}\right]-L(I)_{\vec{s},\Omega^l} \tag{7}$$

Eq. (7) can be deduced using the Taylor expansion of the difference equation. For the case above, the Taylor expansions of $I_{(\vec{s})}^{\Omega^l}$ and $I_{(\vec{s}+d\vec{s})}^{\Omega^l}$ in the space position $\vec{s}=\vec{s}(i,j)$ can be substituted into the difference equation and reorganized, and we then obtain the correlation function shown in Eq. (8):

$$TE=O\left(\Delta x^2,\Delta y^2\right)+\left(S_s^{\Omega^l}-S\Big|_{s,\Omega^l}\right) \tag{8}$$

For a 2D problem, if the derivative is substituted for a first-order difference, if the radiative intensity profile is assumed by the space differential schemes, and if the transfer direction intersects with the grid and a non-negative radiative intensity gradient exists in the direction perpendicular to the transportation direction, numerical scattering is still generated.

Generally, a relationship should be derived to correlate the radiative intensities in the face of the control volume and the intensities at the centre of the control volume, which constitute a kind of spatial differential scheme. In view of this, the numerical scattering is also related to the spatial differential factor, which is shown in Eq. (9):

$$TE=\left|\frac{\partial I^{\Omega^l}}{\partial x}-\frac{I_e^{\Omega^l}-I_w^{\Omega^l}}{\Delta x}\right|=\left|\frac{(1-2f_x)\Delta x}{2!}\left(\frac{\partial^2 I^{\Omega^l}}{\partial x^2}\right)\right| \tag{9}$$

In a multi-dimensional problem, the numerical scattering acts in a similar manner to the way it acted in one dimension.

Also, the fact that the spatial discretization error can be reduced only by increasing the grid numbers is not a simple problem. From Eq. (5) to Eq. (9), the radiative intensity I varies with the wavelength λ, based on the assumption of spectral band consistency, which affects the radiative property, i.e., $\kappa_{\alpha\lambda}$. Therefore, errors caused by the radiative property are also included in the numerical scattering within the scope. For the approximation complexity of the spectral absorption coefficient, a ratio X is applied to denote its deviation, which is shown in Eq. (10):

$$\kappa_{\alpha\lambda}^{*}=\chi\kappa_{\alpha\lambda} \tag{10}$$

To summarize, the factors that affect *TE* can be implemented in a correlation, which is shown in Eq. (11), which can be similarly extended to a three-dimensional problem.

$$TE = TE\left[(\Delta x, \Delta y), f_{\alpha,\beta}, \kappa_{\alpha\lambda}^{*}\right] \tag{11}$$

For brevity, the detailed explanation of the reason towards its generation can be referred in the previous works (Zhang & Tan, 2009; Tan et al., 2004; Zhang et al. 2011).

2.2 Artificial benchmark model

Consider a 2D rectangle containing an absorbing-emitting grey medium without scattering; its refractive index is uniform, and is equal to that of the surroundings. In real cases, the medium may be gaseous, solid or liquid; however, in the current work, a generalized version of the participating medium is used.

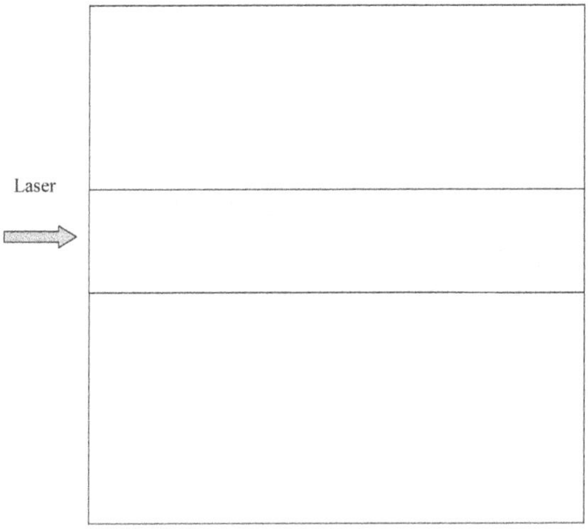

Fig. 1. Artificial benchmark model for central laser incidence to a two-dimensional rectangle

The four interfaces of the medium are all diffusely reflective, opaque and grey. Its east and west interfaces are coated with a sheet of thin film, individually. At certain wavelength, the west interface shows to be a semitransparent and specularly reflective surface. While, at other wavelengths, it is shown to be a diffusely reflective, opaque and grey surface. The thin film in its east interface is opaque, which is a diffusely reflective surface, except for the wavelength 10.6 μm. A laser beam of wavelength 10.6 μm is projected to the centre the west interface (the shadowed area), which is shown in Figure 1.

Detailed information on the current model selected can be referred in the previous works [12].

2.3 Discretization error indicator by local entropy generation rate

In case of steady state, the temperature can be determined by the following correlation, in which

$$\nabla \cdot \vec{\mathbf{q}}^{r} = \sum_{k=1}^{M_b} \kappa_k \left(4\pi I_{bk} - H_k\right) = \sum_{k=1}^{M_b} \kappa_k \left[4B_{k,T_p}\sigma T_P^4 - \sum_{\Omega^l \in 4\pi} I_k^{\Omega^l}\,\Omega^{\Omega^l}\right] \tag{12}$$

According to the expression of the local radiative entropy generation rate in a participating medium (Caldas & Semiao, 2005; Liu & Chu, 2007), an error indicator can be defined as:

$$\dot{S}'''_{gen} = \int_0^{\infty}\int_{4\pi}\left[-\left(\kappa_{a\lambda}+\kappa_{s\lambda}\right)\frac{I_\lambda(\mathbf{r},\mathbf{s})}{T_\lambda(\mathbf{r},\mathbf{s})}+\kappa_{a\lambda}\frac{I_{b,\lambda}(\mathbf{r})}{T_\lambda(\mathbf{r},\mathbf{s})}+\frac{\kappa_{s\lambda}}{4\pi}\int_{4\pi}\frac{I_\lambda(\mathbf{r},\mathbf{s}')}{T_\lambda(\mathbf{r},\mathbf{s})}\Phi(\mathbf{s}',\mathbf{s})\mathrm{d}\Omega'\right]\mathrm{d}\Omega\mathrm{d}\lambda$$
$$+\int_0^{\infty}\int_{4\pi}\kappa_{a\lambda}\frac{\left[I_\lambda(\mathbf{r},\mathbf{s})-I_{b,\lambda}(\mathbf{r})\right]}{T(\mathbf{r})}\mathrm{d}\Omega\mathrm{d}\lambda \tag{13}$$

Since there is no scattering in the medium, and the refractive index is uniform, there is no refraction of the laser beam. Laser is projected from the normal direction of the west interface and when it arrives at the east interface, it is then specularly reflected to the inverse direction without angle variation. In other words, there is no scattering in the process of laser propagation through the semi-transparent in view of actual physical process; within finite time, temperature will increase only in the region where laser irradiated. When this process is simulated by numerical method and if scattering phenomenon happens, i.e., the entropy generation increasing in non-central region where the region is not irradiated by laser, numerical scattering is deemed to appear, and vise versa.

In addition, to obtain local entropy generation rate, the RTE is first solved by FVM, then it can be derived. The radiative heat transfer process in the artificial model above and the detailed derivation of the governing equation can be found in the reference (Zhang & Tan, 2009).

3. Simulation result and analysis

Radiative properties and computing parameters are: geometry of the computing domain $L_x = L_y = 0.25\,\mathrm{m}$; refractive index of medium $n = 1$, spectral absorption coefficient $\kappa_{a\lambda} = 1\,\mathrm{m}^{-1}$, spectral scattering coefficient $\kappa_{s\lambda} = 0$, and therefore, optical thickness along x and y coordinates are $\tau_x = \tau_y = 0.25$. Emissaries of the four interfaces $\varepsilon_{k,e}$, $\varepsilon_{k,w}$, $\varepsilon_{k,s}$ and $\varepsilon_{k,n}$ are uniformly specified as 0.8 and reflectivity of the four interfaces $\rho_{k,e}$, $\rho_{k,w}$, $\rho_{k,s}$ and $\rho_{k,s}$ are uniformly specified as 0.2; when $\lambda = 10.6\,\mu\mathrm{m}$, $\varepsilon_{k,w} = 0$ and $\gamma_{k,w} = 0.8$. Surrounding temperatures T_e, T_w, T_s and T_n are uniformly specified as 1000 K; initial temperature of the rectangle medium is set as T_0 = 1000 K. Moreover, the incident wavelength of the laser is set as $\lambda_{la} = 10.6\,\mu\mathrm{m}$, and the power flux density of the incident laser is specified as q_{la} = 2 MW. Also, the thermal conductivity of this medium is specified to be extremely small to ensure that thermal radiation is the dominant heat transfer method.

3.1 Verification of computation code

To validate reliability and compare result of the algorithm, the following expression for temperature increment is defined as:

$$\Delta T_{i,j(or\ m,n)} = T_{i,j(or\ m,n)} - T_0 \tag{14}$$

An error indicator $EI_{\mathrm{La,T}}$ is defined as the following, which is the maximum temperature increment where node is without laser incidence region to minimum temperature increment where node is within laser incidence region, i.e.,

$$EI_{\mathrm{La,T}} = \frac{\max\left(\Delta T_{i,j}\right)(i,j \notin \text{Laser Incidence})}{\min\left(\Delta T_{m,n}\right)(m,n \in \text{Laser Incidence})} \tag{15}$$

Based on local entropy generation rate expression, as in shown Eq. (8), an error indicator $EI_{\mathrm{La,Sgen}}$ can be accordingly defined as:

$$EI_{\mathrm{La,\,Sgen}} = \frac{\max[\dot{S}'''_{gen}(i,j)][(i,j) \subset \text{Laser Incidence}]}{\min[\dot{S}'''_{gen}(m,n)][(m,n) \not\subset \text{Laser Incidence}]} \tag{16}$$

Both $EI_{\mathrm{La,T}}$ and $EI_{\mathrm{La,Sgen}}$ are used to evaluate numerical scattering for different situations. Since $EI_{\mathrm{La,Sgen}}$ is an absolute value for every control volume, it supplies the information of energy dissipation of the solution process, i.e., numerical dissipation, not a physically real process.

Although statistical error exists in MCM, the numerical scattering does not exist in the MCM, and its results can be used as benchmark solution to test accuracy. In MCM, the most important factor which affects its simulation accuracy is the random bundle number *NM*, and sensitivity of MCM with different random bundle numbers is tested, the result is shown in Tab.1, in which spatial grid number is set to $NX \times NY = 10 \times 10$.

NM	$\mathrm{EI_{La,T}}$ (%)	$\mathrm{EI_{La,Sgen}}$ (%)
10^2	2.13	0.11
10^3	1.78	0.08
10^4	0.17	0.01
10^5	0.16	0.0
10^6	0.14	0.0

Table 1. Sensitivity of MCM with different random bundle numbers

From Tab.1, it can be seen that when NM is larger than 10^4, the results of MCM is stable and less accurate, so in the following calculation, NM = 10^6 is used in all simulations. Because numerical scattering does not exist in the MCM, the results of MCM can be used as a benchmark solution to test FVM accuracy. Furthermore, the advantage of $EI_{\mathrm{La,Sgen}}$ over $EI_{\mathrm{La,T}}$ is that, as error indicators, the value of $EI_{\mathrm{La,T}}$ is affected by statistical error in MCM. Meanwhile, the value of $EI_{\mathrm{La,Sgen}}$ is independent from the statistical error in MCM. Therefore, it is shown to be a better error indicator in the current framework.

The next step is to validate the effect of solid angle discretization in FVM with several angular schemes presented, and the results are shown in Tab.2, in which spatial grid number is specified as $NX \times NY = 10 \times 10$.

In Tab.2, it can be seen that, when $N\theta \times N\varphi > 16 \times 20$, both results of $EI_{\mathrm{La,T}}$ and $EI_{\mathrm{La,Sgen}}$ is shown to be stable, and it denotes that the results of error indicator is independent from numbers of solid angle discretization grids. Therefore, in the following calculation, solid angle discretization number is set to $N\theta \times N\varphi = 24 \times 36$, which is also used in all the following simulations.

Finally, it is necessary to test the grid independence of spatial discretization number in FVM to show the uncertainty of two categories of error indicator free from spatial grid numbers. The step scheme for FVM is used, and the results are shown in Tab. 3.

$N\theta \times N\varphi$	$EI_{\mathrm{La,T}}$ (%)	$EI_{\mathrm{La,Sgen}}$ (%)
4×8	23.54	15.69
10×12	8.96	7.94
12×16	6.24	5.27
16×20	5.72	2.73
20×28	5.17	2.75
24×36	5.18	2.76

Table 2. Independence of solid angle discretization number's test

$NX \times NY$	$EI_{\mathrm{La,T}}$ (%)	$EI_{\mathrm{La,Sgen}}$ (%)
5×5	5.19	2.79
10×10	5.18	2.76
20×20	5.14	2.75

Table 3. Independence of spatial discretization numbers for FVM (step scheme)

In Tab.3, it can be seen that, when $NX \times NY > 5 \times 5$, both results of $EI_{\mathrm{La,T}}$ and $EI_{\mathrm{La,Sgen}}$ is shown to be stable, which denotes that the results of error indicator is independent from numbers of spatial grids.

3.2 Numerical scattering simulation and error indicator distribution

For the case shown in Fig.1, to make the effect of numerical scattering more clear, the contours of temperature profile computed by MCM, FVM by step scheme (FVM1), FVM by diamond scheme (FVM2) and FVM by exponential scheme (FVM3) are shown in Fig.2-Fig.4.

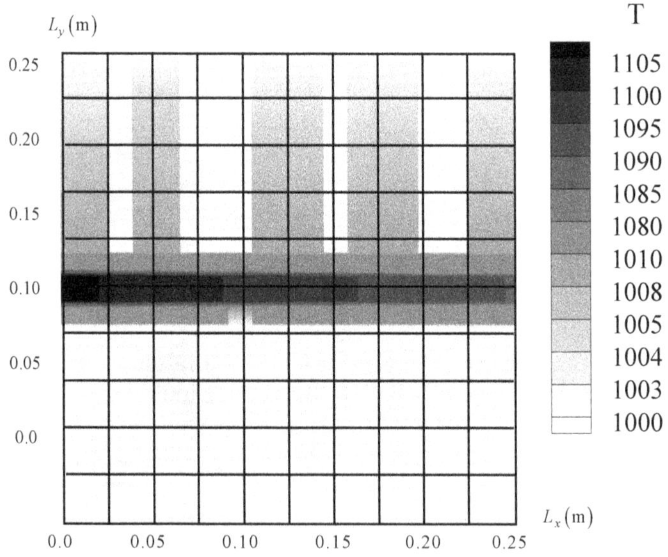

Fig. 2. Temperature contour by MCM with grid number $NX \times NY = 20 \times 20$

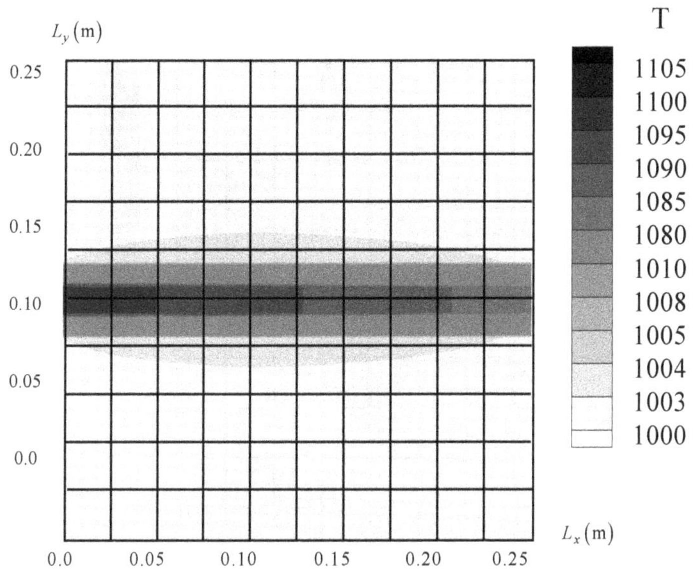

Fig. 3. Temperature contour by FVM1 with grid number $NX \times NY = 20 \times 20$

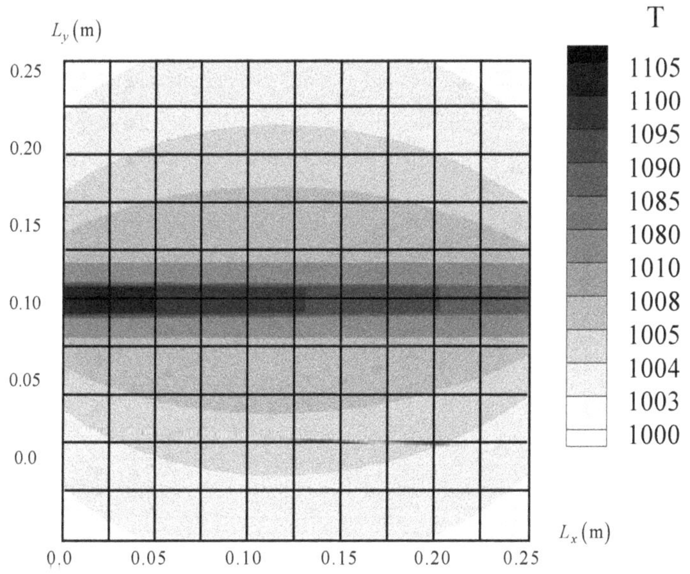

Fig. 4. Temperature contour by FVM2 with grid number $NX \times NY = 20 \times 20$

In those cases, grid number $NX \times NY = 20 \times 20$ is adopted, in which grid number of unit optical thickness is $\zeta = 80$.

The temperature distribution of each scheme in the region of laser incidence is shown in Fig.6.

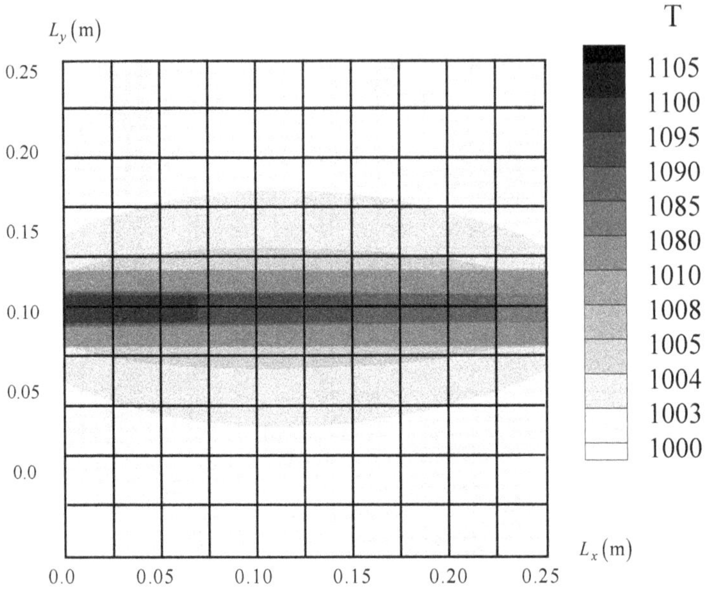

Fig. 5. Temperature contour by FVM3 with grid number $NX \times NY = 20 \times 20$

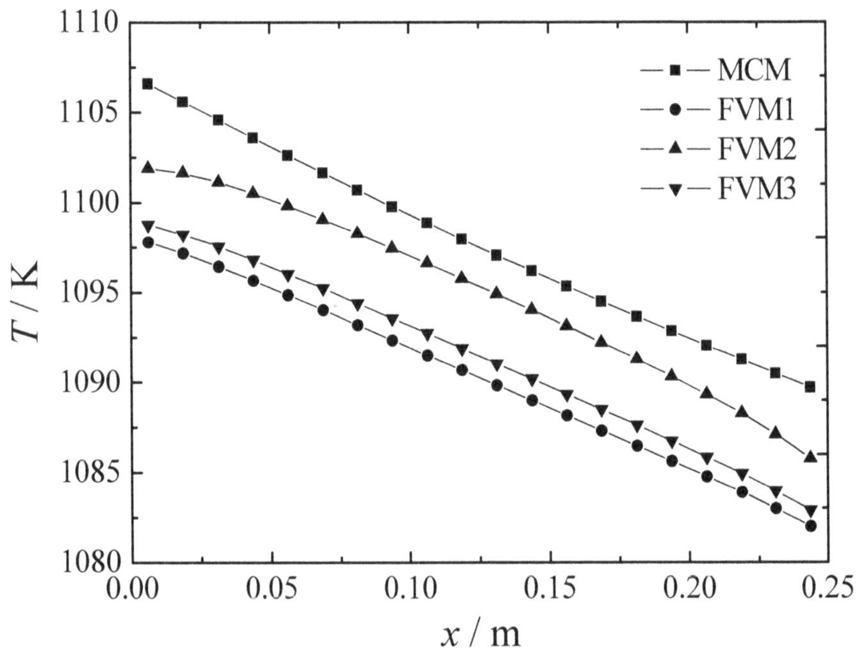

Fig. 6. Temperature profile of central laser incidence by MCM and FVM with different spatial differential schemes, $NX \times NY = 20 \times 20$

It can be seen that: among them, the accuracy of the diamond scheme is the highest, and the exponential scheme is a bit lower, the lowest accuracy of the three schemes is the step scheme.

$EI_{La,T}$ for central laser incidence by MCM and FVM with different spatial differential schemes and different gird numbers is tabulated in Tab.4 .

	FVM1 (%)	FVM2 (%)	FVM3 (%)	MCM(%)
$NX \times NY = 5 \times 5$ $Er_{La,T}$	5.19	9.76	5.82	0.15
$NX \times NY = 10 \times 10$ $Er_{La,T}$	5.18	11.21	8.09	0.06
$NX \times NY = 20 \times 20$ $Er_{La,T}$	5.14	8.75	5.67	0.00

Table 4. $EI_{La,T}$ for central laser incidence by MCM and FVM with different spatial differential schemes and different gird numbers

$EI_{La,Sgen}$ for central laser incidence by MCM and FVM with different spatial differential schemes and different gird numbers is tabulated in Tab.5.

	FVM1 (%)	FVM2 (%)	FVM3 (%)	MCM(%)
$NX \times NY = 5 \times 5$ $Er_{La,Sgen}$	2.79	5.25	3.08	0.00
$NX \times NY = 10 \times 10$ $Er_{La,Sgen}$	2.76	5.24	3.07	0.00
$NX \times NY = 20 \times 20$ $Er_{La,Sgen}$	2.75	5.22	3.05	0.00

Table 5. $EI_{La,Sgen}$ for central laser incidence by MCM and FVM with different spatial differential schemes and different gird numbers

It is also interesting to see the distribution of numerical scattering. Choosing the x-axis position where the maximum temperature increment without laser incidence happens, in different height, the distribution of numerical scattering of different spatial differential schemes with grid numbers is shown in Fig. 7.

From the Fig. 7, it is shown that, if we set the direction of laser incidence as central axis, it can be seen that numerical scattering distributed symmetry along the axis, which can be called as symmetrical cross-scattering. All of the three schemes show symmetrical cross-scattering.

It can be seen from the above tables and figures that, for grid number, when its number is increasing, numerical scattering will be reduced. This is the same tendency as in all other fields. However, on one aspect, the accuracy of FVM will also be affected by the spatial differential scheme and among them, the diamond scheme has the highest, and exponential scheme has less accuracy, while step scheme has the least accuracy of the three schemes. On the other aspect, the degree of numerical scattering is reverse, i.e., the step scheme produces minimum numerical scattering, and exponential scheme produces more, while the diamond scheme produces maximum among three methods.

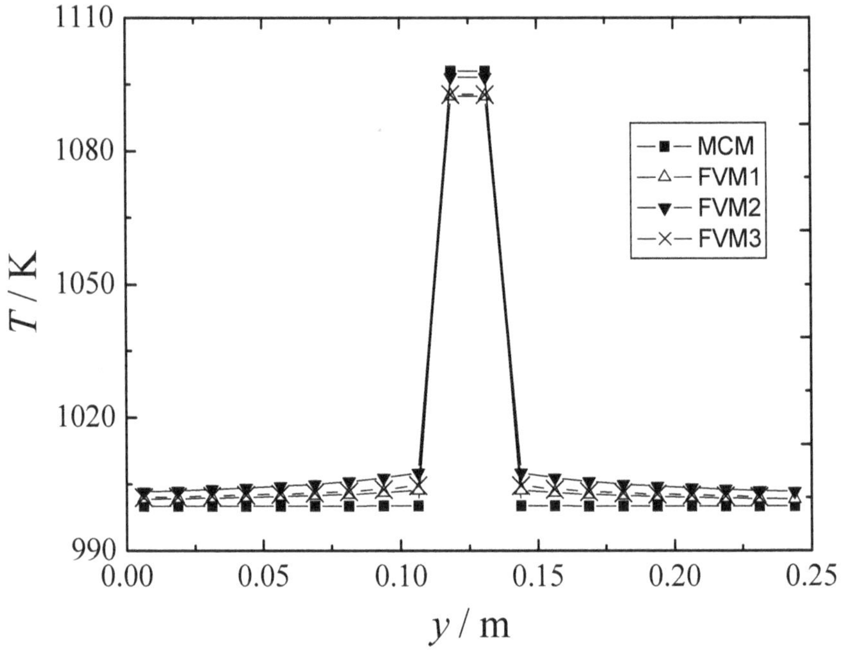

Fig. 7. Distribution of numerical scattering of FVM with different spatial differential schemes, x=0.1063, $NX \times NY = 20 \times 20$

3.3 Effect of absorption coefficients deviation

The purpose to discuss numerical scattering is to examine how to constitute better differential scheme of the intensity to obtain solution with good accuracy without less oscillation. For this reason, the hypothesis of uniform property is also included in the false scattering. By considering the individual absorption coefficients $\kappa^{*}_{\alpha\lambda}$ of 0.1, 1.0, 2.0 and 10.0, and the corresponding optical thicknesses τ = 0.025, 0.25, 0.5, and 2.5 individually. The numerical test results towards $EI_{La,T}$ and $EI_{La,Sgen}$ of grid numbers $NX \times NY = 5 \times 5$ and $NX \times NY = 20 \times 20$ for the MCMs and FVM1s are shown in Tab. 6 and Tab. 7.

$NX \times NY$	$\kappa_{\alpha\lambda}$	$Er_{La,T}$ / MCM (%)	$Er_{La,T}$ / FVM1 (%)
5×5	0.1	0.0	0.50
	1.0	0.15	5.17
	2.0	0.58	10.73
	10.0	23.37	65.56
10×10	0.1	0.0	0.35
	1.0	0.11	4.41
	2.0	0.49	9.51
	10.0	41.38	67.33

Table 6. $EI_{La,T}$ for MCM and FVM of different absorption coefficients in central laser incidence

$NX \times NY$	$\kappa_{a\lambda}$	$Er_{\text{La, Sgen}}$ / MCM (%)	$Er_{\text{La, Sgen}}$ / FVM1 (%)
5×5	0.1	0.0	0.02
	1.0	0.05	0.03
	2.0	0.26	9.63
	10.0	12.37	40.34
10×10	0.1	0.0	0.04
	1.0	0.03	0.12
	2.0	0.19	12.51
	10.0	9.27	59.52

Table 7. $EI_{\text{La,Sgen}}$ for MCM and FVM of different absorption coefficients in central laser incidence

It can be seen that when the absorption coefficient deviation is high, the numerical scattering cannot be eliminated, even with higher grid numbers.

4. Conclusion

Based on the theory of local entropy generation rate used in fluid flow and heat transfer, an error indicator is defined to evaluate and compare discretization errors caused by different factors in FVM for solving the RTE, which is proven to be an effective approach. In addition, since the discretization error is a quality generated in the solution process, while the theory of local entropy generation is focused on process evaluation, therefore, it is shown to be better, comparing with the former error indicator defined by temperature increasing.

An artificial benchmark model of central laser incidence on a 2D rectangle containing a semi-transparent medium is proposed to investigate the numerical scattering in the FVM, along with the use of reference data from the MCM, which has been proven to generate no false scattering. Meanwhile, the value of new error indicator is independent from the statistical error in MCM.

Within the framework of the current model, it is shown that numerical scattering for the FVM is affected by the spatial grid numbers and is also affected by the different spatial discretization schemes to a large degree, with the diamond scheme being best, then the exponential scheme and finally the step scheme, in ranked order. Numerical scattering also varies with the amount of absorption coefficient deviation. When the absorption deviation is large, the numerical scattering cannot be eliminated solely by increasing the grid number. Also, numerical scattering is distributed symmetrically along the laser incidence direction, and all of the schemes show symmetrical cross-scattering.

5. Acknowledgment

The work described herein is supported by the National Natural Science Foundation of China (nos. 51006026, 90916020), the Development Program for Outstanding Young Teachers in Harbin Institute of Technology (no. HITQNJS. 2009. 022) and the Fundamental Research Funds for the Central Universities (Grant No. HIT.NSRIF. 2012072), to whom grateful acknowledgment is expressed.

6. Nomenclature

f = spatial differencing factor [-]
H = information entropy indicator [bit]
I_λ = spectral radiative intensity $\left[\mathrm{W}/\left(\mathrm{m}^2\text{-sr-}\mu\mathrm{m}\right)\right]$
$I_{b\lambda}$ = blackbody spectral radiative intensity, $\left[\mathrm{W}/\left(\mathrm{m}^2\text{-sr-}\mu\mathrm{m}\right)\right]$
i, j = index of nodal point in the region without laser incidence [-]
NX = spatial discretization grid number along x axis [-]
NY = spatial discretization grid number along y axis [-]
$N\theta$ = angular discretization grid number along θ direction [-]
$N\varphi$ = angular discretization grid number along φ direction [-]
$\vec{n}_b$ = normal vector of the boundary [-]
p = probability of temperature increasing due to numerical scattering [-]
$\vec{s}$ = spatial position vector [m]
T = temperature [K]

Greek

γ = transmittance [-]
ε = emissivity [-]
$\kappa_{a\lambda}$ = spectral absorption coefficient of medium [m^{-1}]
$\kappa_{s\lambda}$ = spectral scattering coefficient of medium [m^{-1}]
λ = wavelength [μm]
τ = optical thickness [-]
Φ = scattering-phase function [-]
$\vec{\Omega}$ = solid angle ordinate direction [-]
$\vec{\Omega}'$ = solid angle ordinate for scattering direction [-]

Subscripts

b = bottom boundary of control volume p
e = east boundary of control volume p
i, j = index of nodal point in the region without laser incidence
n = north boundary of control volume p
p = control volume p
s = south boundary of control volume p
t = top boundary of control volume p
w = west boundary of control volume p
x, y, z = coordinates directions
λ = spectrum (wavelength)
0 = initial value

Superscripts

Ω^l = a certain selected angular direction

7. References

Howell, J.R.; Siegel, R. & Menguc, M. P. (2010). *Thermal Radiation Heat Transfer* (5th Edition), CRC Press, ISBN 978-143-9805-33-6, New York, USA

Modest, M.F. (2003). *Radiative Heat Transfer* (2nd Edition), Academic Press, ISBN 978-012-5031-63-9, San Diego, USA

Shih, T. M.; Thamire, C., Sung, C. H. & Ren, A. L. (2010). Literature Survey of Numerical Heat Transfer (2000-2009): Part I. *Numerical Heat Transfer, Part A: Applications,* Vol. 57, No. 3, pp. 159-296, ISSN 1040-7782

Raithby, G. D. & Chui, E. H. (1990). A Finite-Volume Method for Predicting a Radiant Heat Transfer in Enclosures with Participating Media, *ASME Transactions Journal of Heat Transfer,* Vol. 112, pp. 415-423, ISSN 0022-1481

Borjini, M. N.; Guedri, K. & Said R. (2007). Modeling of Radiative Heat Transfer in 3D Complex Boiler with Non-gray Sooting Media, *Journal of Quantitative Spectroscopy & Radiative Transfer,* Vol. 105, No. 2, pp. 167-179, ISSN 0022-4073

Das, R.; Mishra, S. C.; Ajith M.; et al. (2008). An inverse analysis of a transient 2-D conduction-radiation problem using the Lattice Boltzmann method and the finite volume method coupled with the genetic algorithm, *Journal of Quantitative Spectroscopy & Radiative Transfer,*Vol.109, No.11, pp. 2060-2077, ISSN 0022-4073

Farzad, B.T. & Shahini, M. (2009). Combined Mixed Convection-Radiation Heat Transfer within a Vertical Channel: Investigation of Flow Reversal, *Numerical Heat Transfer Part A: Applications,* Vol. 55, No. 3, pp. 289-307, ISSN 1040-7782

Han, S. H.; Baek, S. W. & Kim, M. Y. (2009). Transient Radiative Heating Characteristics of Slabs in a Walking Beam Type Reheating Furnace, *International Journal of Heat and Mass Transfer,* Vol. 52, No. 3, pp. 1005-1011, ISSN 0017-9310

Kim,C.; Kim, M.Y.; Yu M.J. & Mishra S.C. (2010). Unstructured Polygonal Finite-Volume Solutions of Radiative Heat Transfer in a Complex Axisymmetric Enclosure, *Numerical Heat Transfer Part B: Fundamentals,* Vol. 57, No. 3, 2010, pp. 227-239, ISSN 1040-7790

Daniel, R. R. & Fatmir, A. (2011). A consistent interpolation function for the solution of radiative transfer on triangular meshes. I-comprehensive formulation, Numerical *Numerical Heat Transfer Part B: Fundamentals,* Vol. 59, No. 2, pp. 97-115, ISSN 1040-7790

Chai, J. C.; Lee, H. S. & Patankar, S. V. (1993). Ray effect and false scattering in the discrete ordinates method, *Numerical Heat Transfer Part B: Fundamentals,* Vol. 24, No. 4, pp. 373-389, ISSN 1040-7790

Zhang, H. C. & Tan, H. P. (2009). Evaluation of Numerical Scattering in Finite Volume Method for Solving Radiative Transfer Equation by a Central Laser Incidence Model, *Journal of Quantitative Spectroscopy & Radiative Transfer,* Vol. 110, No. 18, pp. 1965-1977, ISSN 0022-4073

Patankar, S. V. (1980). *Numerical Heat Transfer and Fluid Flow,* Hemisphere Pub, ISBN 978-089-1165-22-4, Washington, DC, USA

Tan, H. P.; Zhang, H.C. & Zhen, B. (2004). Estimation of Ray Effect and False Scattering in Approximate Solution Method for Thermal Radiative Transfer Equation, *Numerical Heat Transfer Part A: Applications,* Vol. 46, No. 8, pp. 807-829, ISSN 1040-7782

Kallinderis, Y. & Kontzialis, C. (2009). A Priori Mesh Quality Estimation via Direct Relation between Truncation Error and Mesh Distortion, *Journal of Computational Physics*, Vol. 228, No. 3, pp. 881-902, ISSN 0021-9991

Coelho, P. J. (2008). A Comparison of Spatial Discretization Schemes for Differential Solution Methods of the Radiative Transfer Equation, *Journal of Quantitative Spectroscopy & Radiative Transfer*, Vol. 109, No. 2, pp. 189-200, ISSN 0022-4073

Kamel, G.; Naceur, B. M.; Rachid M. & Rachid S. (2006). Formulation and Testing of the FTn Finite Volume Method for Radiation in 3-D Complex Inhomogeneous Participating Media, *Journal of Quantitative Spectroscopy & Radiative Transfer*, Vol. 98, No. 3, pp. 425-445, ISSN 0022-4073

Naterer, G. F. & Camberos J. A. (2003). Entropy and the Second Law Fluid Flow and Heat Transfer Simulation, *AIAA Journal of Thermophysics and Heat Transfer*, Vol. 17, No. 3, 2003, pp. 360-371, ISSN 0887-8722

Camberos, J. A. (2000). The Production of Entropy in Relation to Numerical Error in Compressible Viscous Flow, *AIAA Paper*, No. 2000-2333, AIAA Fluid 2000 Symposium, Denver, CO, June, 2000

Cover, T. M. & Thomas, J. A. (2003). *Elements of Information Theory*, Tshinghua University Press, ISBN 730207285X, Beijing, China

Camberos, J. A. (2007). A Review of Numerical Methods in Light of the Second Law of Thermodynamics, 39th AIAA Thermophysics Conference, 25-28, June 2007, Miami, FL: 410-444.

Caldas, F. & Semiao, V. (2005). Entropy Generation through Radiative Transfer in Participating Media: Analysis and Numerical Computation, *Journal of Quantitative Spectroscopy & Radiative Transfer*, Vol. 96, No.3-4, pp. 423-437, ISSN 0022-4073

Liu L. H. & Chu S. X. (2007). Verification of Numerical Simulation Method for Entropy Generation of Radiation Heat Transfer in Semitransparent Medium, *Journal of Quantitative Spectroscopy & Radiative Transfer*, Vol. 103, No. 1, pp. 43-56, ISSN 0022-4073

Zhang, H. C.; Tan H. P. & Li Y. (2011). Numerical Uncertainty for Radiative Transfer Equation by an Information Entropy Approach, *AIAA Journal of Thermophysics and Heat Transfer*, Vol. 25, No. 4, pp. 635-638, ISSN 0887-8722

Herwig, H. & Kock, F. (2004). Local entropy production in turbulent shear flows: A high-Reynolds number model with wall functions, *International Journal of Heat and Mass Transfer*, Vol. 47, No. 10-11, pp. 2205-2215, ISSN 0017-9310

On FVM Transport Phenomena Prediction in Porous Media with Chemical/Biological Reactions or Solid-Liquid Phase Change

Nelson O. Moraga[1] and Carlos E. Zambra[2]
[1]*Universidad de La Serena/Departamento de Ingeniería Mecánica, La Serena*
[2]*Centro de Investigación Avanzada en Recursos Hídricos y Sistemas Acuosos (CIDERH)*
CONICYT-REGIONAL GORE-TARAPACÁ, Iquique
Chile

1. Introduction

The aim of the chapter is to show the FVM capabilities in the accurate and efficient prediction of transport phenomena in porous media including either biological, chemical reactions or liquid-solid phase transformations. Four applied technological problems are solved with the FVM. Problem 1 is related to heat and mass diffusion in a saturated porous media where chemical and biological reactions occur. Such a situation can be found in compost piles resulting from contaminated water treatments. Field experimental data are used to assess the quality of FVM calculations for temperature and oxygen concentration distribution time variations along the prediction of thermal explosions (Moraga, 2009) and effect of the moisture in compost piles self-heating (Zambra, 2011). Problem 2 deals with the improvement of thermal energy efficiency and pollution reduction in hydrocarbons combustion. Methane combustion with air in a cylindrical porous burner is investigated by solving 2D unsteady continuity, linear momentum, energy, and chemical species governing equations with the FVM. Sensibility studies performed via numerical tests allowed to obtain numerical results for unsteady velocity and temperature distributions, along to the displacement of the combustion zone. The effects of inlet reactants velocity (methane and air) in the range 0.3-0.6 m/s; excess air ratios between 3 and 6 and porosities of 0.3 up to 0.6 in the fluid dynamics, forced convection heat transfer and combustion process are described (Moraga, 2008). Problem 3 is devoted to characterize 3D natural convection and heat conduction with solidification inside a cavity filled with a porous media, in which a Darcy-Brinkman-Forchheimer flow model is used. This infiltration technique can be applied to produce new materials with enhanced physical properties. A fixed grid method is used along the FVM to solve, with a temperature dependent liquid fraction, the moving boundary problem by using a power law model for binary non Newtonian alloys (Moraga, 2010; Moraga, 2010). Problem 4 describes 3D turbulent convective heat transfer in a bioreactor, including the self-heating of the porous material due to chemical and biological reactions. FVM simulation based on a coupled heat mass transfer external forced convection model is used to assess the effects of reactor geometry, self- heating parameters, air flow and temperature in the bioreactor performance. The mathematical model includes the convective turbulent flow of momentum, energy and oxygen concentration, with the κ-ε turbulence model, and the diffusion of energy and oxygen concentration in the saturated porous medium. Numerical results for the dependent variables are successfully validated with experimental data. Issues such as the use of dynamic time steps, under-relation of dependent variables and local refined meshes are discussed in each one of the problems solved. The pressure-velocity-temperature-chemical species coupling is discussed and a novel PSIMPLER method for the FVM is presented (Moraga, 2010). The stability, rate of convergence and efficiency of the PSIMPLER method is determined by solving natural, forced and mixed heat convection inside cavities by comparison with the solution obtained by using the standard SIMPLE algorithm. Improvements achieved in convergence rates by modifying the predictor-corrector schemes used to solve the discretized fluid mechanics, heat and mass transfer equations are discussed in some of the numerical experiments presented.

2. General scheme for the classic Finite Volume Method (Patankar, 1980)

Convective fluid dynamics/heat and mass transfer, for either laminar or turbulent flows of Newtonian or non-Newtonian fluids, with phase change in porous media is described by partial differential equations. Systems of nonlinear second order partial differential equations can be efficiently solved numerically using the finite volume method. Each governing equation is treated in the generalized form for a transport equation, with unsteady, convection, diffusion and linearized source terms:

$$\frac{\partial(\rho\phi)}{\partial t} + div(\rho\vec{u}\phi) = div(\Gamma\ grad\phi) + S_p\phi + S_c \tag{1}$$

where ϕ is the dependent variable, ρ density, t time, $\vec{u}$ velocity vector, Γ diffusion coefficient, $S_c$independent source term and $S_p$dependent source term. No convection is considered inside the volume occupied by the MWM, since the flow velocities are zero in that region, and diffusion becomes the only transport mechanism inside this porous medium. This "blocked" region is implemented by setting a very large numerical viscosity (10^{30}) for the control volumes enclosed in such region, which renders the velocity essentially zero ($\sim 10^{-30}$) inside the material.

The time integration is performed with an explicit Euler scheme:

$$\frac{\partial\phi}{\partial t} = \frac{\phi^{t+\Delta t} - \phi^{t}}{\Delta t} \tag{2}$$

At each time step, the system of discretized nodal equations for each main dependent variable (velocity components, temperature and mass fraction) is solved iteratively by internal iterations, with a combination of the alternating tri-diagonal matrix algorithm (TDMA) and Gauss-Seidel method. The sequential coupling of these main variables is accomplished by the SIMPLE method in the external iterations. Under-relaxation is always applied to the dependent variables during these external iterations.

3. Self-heating in compost pile solved with FVM

3.1 General mathematical models for porous media applied to compost pile

Richard equation (RE) (Richards, 1931) is a standard, frequently used approach for modeling and describing flow in variably saturated porous media. When do not consider gravitational and the source term effects, and is introduced a new term $D_{(\theta)}$, the follow equation can be used:

$$\frac{\partial \theta}{\partial t} = \nabla D_{(\theta)} \nabla \theta \tag{3}$$

The volumetric water content θ is the quotient between water volume and the total sample volume, so it is has not unit and its values are between 0 and 1.

The effects of the porosity and type of soil should be introduced by the $D_{(\theta)}$ parameter. A non-linear equation for this parameter is reported for Serrano (Serrano, 2004).

$$D_{(\theta)} = \upsilon_1 e^{\lambda(\theta)^{\alpha}} - \upsilon_2 \tag{4}$$

The constants $\upsilon_1, \upsilon_2, \lambda$ and α, may be obtained by experimental field test. The Eqs. (3) and (4) are used when the specific hydraulic properties of the compost pile are not available. Oxidation and microorganism activity inside the pile are incorporated in the model by volumetric heat generation. For simplicity, local thermal equilibrium is assumed, which is a common assumption for porous medium and packed particle beds (Nield & Bejan, 1992). The equations for the temperature and the oxygen concentration are (Zambra et al., 2011):

$$\frac{\partial \left(\rho C_{p,T}\right)_{eff} T}{\partial t} = \nabla K_{eff} \nabla T + Q_c \varepsilon_s A_c^* \rho_c C_{ox} e^{\left(\frac{-E_c}{RT}\right)} + Q_b \varepsilon_s \rho_b \rho_c \frac{A_1^* e^{\left(\frac{-E_1}{RT}\right)}}{1 + A_2 e^{\left(\frac{-E_2}{RT}\right)}} - L_v \rho_{va} q(\theta) X_v \tag{5}$$

$$\varepsilon_{air} \frac{\partial C_{ox}}{\partial t} = \nabla D_{eff} \nabla C_{ox} + \varepsilon_s A_c^* \rho_c C_{ox} e^{\left(\frac{-E_c}{RT}\right)} \tag{6}$$

In Eqs. (5) and (6) $A_1^* = A_1 * \upsilon_1$ and $A_c^* = A_c * \upsilon_c$, where A_1 and A_c are the pre-exponential factor for the oxidation of the cellulose and bio-mass growth, respectively. The coefficients υ_1 and υ_c are parameters that allows relations between the variables T and C_{ox} which are function of the moisture and oxygen concentration at time *t*. The constants E_c, E_1, E_2, are the activation energy for the cellulose, bio-mass growth and inhibition of biomass growth, respectively. The effects of the vaporization of water in the internal energyare calculated with the third term of the right hand side of Eq (5), where L_v is the vaporization enthalpy, ρ_{va} is the water vapor density, $q(\theta)$ is the mass water flux and X_v is the vapor quality. The total porosity (ε_{fl}) is calculated in terms of the apparent density and real density, ε_w and ε_{air} are the fraction of water and air into the pore respectively.

$$\varepsilon_{fl} = \frac{\rho_{real} - \rho_{app}}{\rho_{app}}; \qquad \varepsilon_{fl} = \varepsilon_w + \varepsilon_{air}; \qquad 1 - \varepsilon_{fl} = \varepsilon_s \tag{7}$$

Thermodynamics equilibrium and ideal mixture between oxygen and water are assumed in the porous medium,

$$K_{eff} = \varepsilon_{air} k_{air} + \varepsilon_w k_w + \varepsilon_s k_c \tag{8}$$

$$\left(\rho C_{p,T}\right)_{eff} = \varepsilon_{air} \rho_{air} C_{p,air} + \varepsilon_w \rho_w C_{p,w} + \varepsilon_s \rho_c C_{p,c} \tag{9}$$

$$D_{eff} = \varepsilon_{air} D_{air,c} \tag{10}$$

where K_{eff} and D_{eff} are the effective properties which are considered dependent of temperature, and $C_{p,air}$, $C_{p,w}$ and $C_{p,c}$ are specific heat capacity of the air, water and cellulose. Oxygen concentrations variations are affected by cellulosic oxidation. This assumption is incorporated in the second term of right hand side Eq. (6).

3.2 Diffusion of temperature and oxygen concentration

3.2.1 Mathematical model

Cellulosic oxidation and micro-organism activity inside the compost pile are taken into account by the model in the form of a volumetric heat generation source. When is considered one fluid phase (air), the heat transfer eq. (5) take the form.

$$\left(\rho C_{p,T}\right)_{eff} \frac{\partial T}{\partial t} = K_{eff} \nabla^2 T + Q_c (1-\varepsilon) A_c \rho_c C_{ox} e^{\left(\frac{-E_c}{RT}\right)} + Q_b (1-\varepsilon) \rho_b \rho_c \frac{A_1 e^{\left(\frac{-E_1}{RT}\right)}}{1 + A_2 e^{\left(\frac{-E_2}{RT}\right)}} \tag{11}$$

while the oxygen concentration within the pile is described by

$$\varepsilon_{air} \frac{\partial C_{ox}}{\partial t} = D_{eff} \nabla^2 C_{ox} + (1-\varepsilon) A_c \rho_c C_{ox} e^{\left(\frac{-E_c}{RT}\right)} \tag{12}$$

In equations (3) and (4), A_c is the pre-exponential factor for the oxidation rate of the cellulose, and. Heat and mass transfer properties in the porous medium are defined in terms of the pile porosity ε as

$$K_{eff} = \varepsilon_{fl} k_{air} + (1-\varepsilon_{fl}) k_c\,; \;\; \left(\rho C_{p,T}\right)_{eff} = \varepsilon_{fl} \rho_{air} C_{p,air} + (1-\varepsilon_{fl}) \rho_c C_{p,c}\,; \;\; D_{eff} = \varepsilon_{fl} D_{air,c} \tag{13}$$

where K_{eff} and D_{eff} are respectively the effective thermal conductivity and diffusion coefficient, which are considered independent of the temperature and concentration. In these expressions the subscript c refers to the cellulose. Details for the formulation of the term representing the heat generated by the biomass have been given by Chen and Mitchell (Chen & Mitchell, 1996). The parameter values used in the mathematical model for the processes that take place inside the porous block of waste material were obtained from reference (Sidhu et al., 2007) and are presented in Table 1.

Parameter	Value	Parameter	Value
A_C	$1.8x10^4$	$C_{ox,0}$	0.272
A_1	$2x10^6$	Q_b	$7.66x10^6$
A_2	$6.86x10^{32}$	Q_C	$5.5x10^9$
C_{air}	1005	k_{air}	0.026
C_C	3320	k_C	0.18
$D_{o,air}$	$2.4x10^{-7}$	ε_{fl}	0.3
E_C	$1.1x10^5$	ρ_{air}	1.17
E_1	$1x10^5$	ρ_b	575
E_2	$2x10^5$	ρ_c	1150

Table 1.Constant values used in the mathematical model for the internal processes in the compost pile.

3.2.2 Numerical results of temperature time evolution and grid study

A 2D case of self-heating in a rectangular porous pile, with 2.5 m height (H) and 5 m length (L), was investigated using three grids, with 100x100, 200x200 and 300x300 nodes, and three time steps: 300s, 600s and 3600s.The temperature time evolution was calculated in the three positions: f) H/4, L/4; g) H/2, L/2 and h) 3H/4, 3L/4.Results of the time needed to cause auto-ignition, in days, are shown in Table 2, for the three positions. The use of a time step of 600s and a grid with 300x300 nodes allows to calculate a time for auto-ignition: 247, 246 and 250 days, at the three vertical positions, respectively, independently of the time a space discretizations, as is showed in the table 2.

Grids	Δt (s)	f) H/4,L/4	g) H/2,L/2	h) 3H/4, 3L/4
100x100	3600	247	245	249
	600	248	248	250
	300	244	241	246
200x200	3600	248	246	251
	600	250	249	253
	300	246	244	249
300x300	3600	248	247	252
	600	247	246	250
	300	247	246	250

Table 2.Days before the self-ignition in positions f), g) and h) within the compost pile.

Fig. 1 shows the time evolution results of temperature in the f) position: H/4, L/4, calculated with a grid of 300x300 nodes using two time steps: 300s and 3600s. A typical heating curve for the temperature is observed in the Figure 1a. Auto-ignition at H/4, L/4 occurs after 247 days. The temperature increased suddenly in one day from 370 to 515K. A complex system of solid, liquid and gaseous fuels as a final result of the cellulosic oxidation originated a volumetric heat generation causing the self-ignition process. Temperature decreasing in time, characterized the last stage, in which the fuel reserves in the location are exhausting. Figure 1b, a zoom view of Fig. 3a for the time interval between days 240 and 255, shows that a time step reduction from 3600s to 300s allows to determine a more accurate prediction (within 1 day) for the time needed to insatiate the self-ignition in a 2.5m height compost pile. Due to the previous analysis, a mesh with 300x300 nodes and a dynamic time step with 300s during the auto-ignition and 3600s in other states may be used in the calculation of the processes of self-heating with thermal explosion.

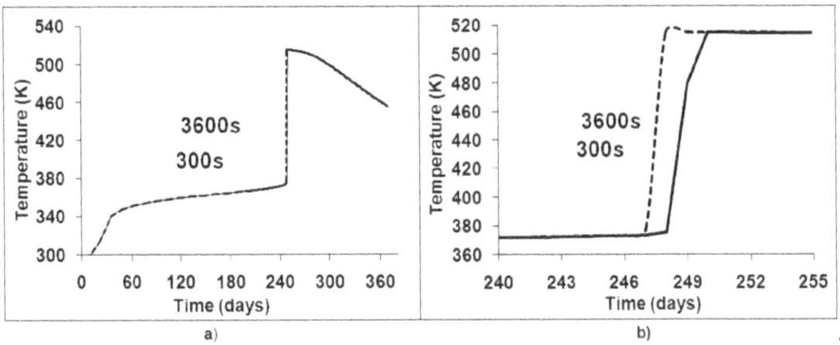

Fig. 1. Temperature evolution calculated with two time steps, 300x300 mesh, at position H/4, L/4, for a 2.5 m high pile, a) full time scale, b) during thermal explosion.

3.2.3 Comparison between experimental and numerical results.

The experimental data and numerical results in 2D obtained with the FVM for a 2.5 m high pile and trapezoidal form, are compared in Figure 2. Data obtained from numerical calculations were plotted considering daily output at 12:00 AM. The experimental and predicted data follow the same general trends. Near the surface (0.35 m depth) during the third week the main differences found are not larger than 3 °C. The best description of the experimental data was obtained at a depth of 2.1 m.

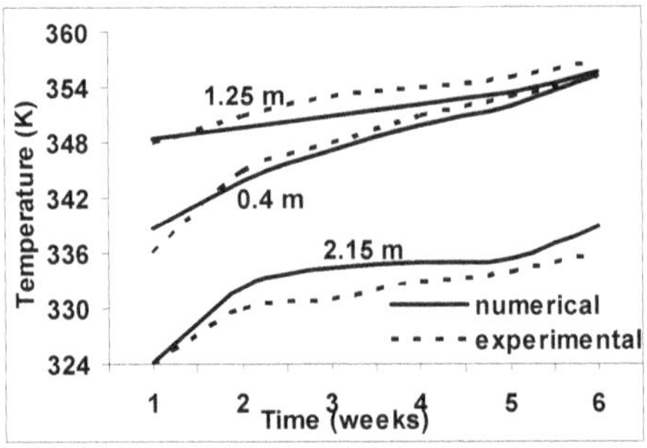

Fig. 2. Comparison between experimental and numerical temperature values during six weeks.

3.3 Inhibition of the self-ignition in the sewage sludge waste water treatment

The coupled heat and mass diffusion equations system of partial differential equation (11)-(15), is solved in 3D by the finite volume method. In previous works Moraga et al. (Moraga et al., 2009) established that statics compost pile with high less than 1.5 m do not have self-ignition. Other statics compost pile higher than 1.5 m may have self-ignition. This restriction decreases the possibilities of the storage in field of this material. Moraga and Zambra (Moraga and Zambra, 2008) proposed a novel method for the inhibition of the self-ignition in large piles. A uniform mesh with 94x32x62 nodes in x, y and z directions was found to be adequate by comparison of results between plane 2D and the central plane 3D (x and y directions) in a trapezoidal pile. Figure 3 compares the numerical results 2D and 3D in three positions inside the pile. The small differences of the values are produced for the third dimension incorporated in the simulations. The main differences in the temperature values occur when the self-ignition reached the tested position.

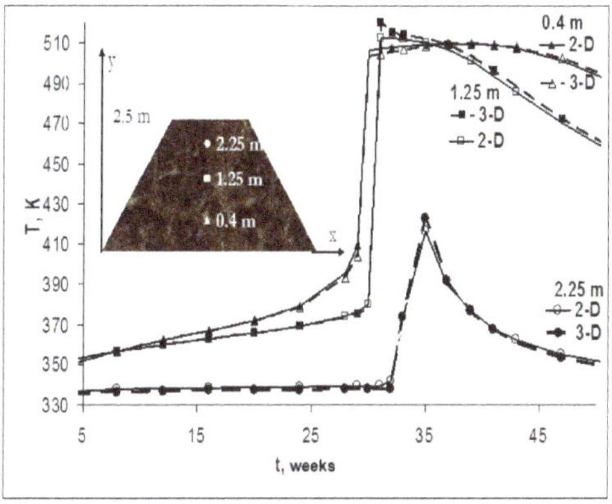

Fig. 3. Comparison of numerical results of temperature for a trapezoidal plane 2D and central plane 3D.

In figure 4 the physical situation used for the inhibition of the ignition is shown.

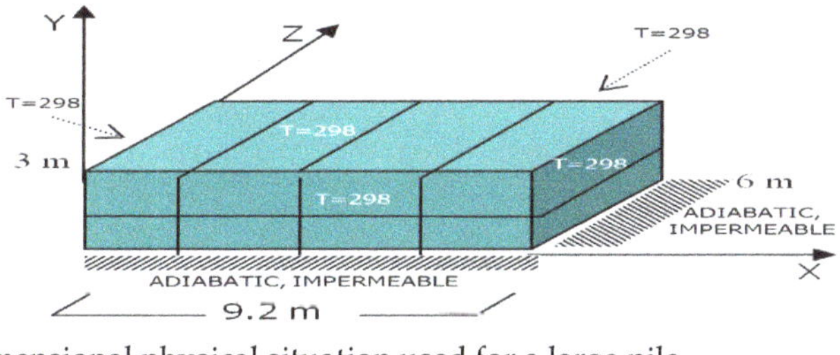

Fig. 4. Three-dimensional physical situation used for a large pile.

A parallelepiped of 9.2 m, 3 m and 6 m in x, y and z directions respectively, was used. Insulation and impermeable walls was introduced to regular interval in vertical and horizontal directions within the pile. The pile base is adiabatic and impermeable to the oxygen diffusion. A temperature of 298 K, was imposed at the border. Initially, the temperature and oxygen concentration inside the pile is constant. Perpendicular steel walls with thick 0.1 m and each 2.225 m to separate the pile in 8 sections to the long of the x coordinate were used. The table 1 shows the parameters used in the mathematical model. Numerical results of maximum values for the temperatures and oxygen concentrations are presented in figure 5. Clearly the self-ignition not occurs. The steel walls allow the inhibition of the thermal explosion because the oxygen diffusion within the pile is restricted. The maximum temperatures and minimum oxygen concentration reaches the 355 K and 0.262 kg/m^3 , respectively.

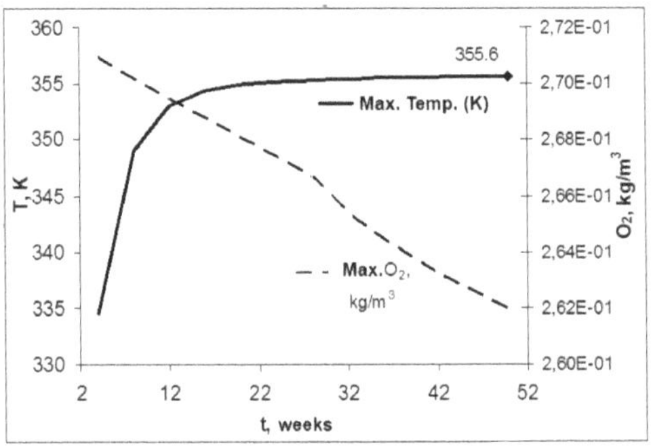

Fig. 5. Maximum values of the temperatures and oxygen concentrations for the sectioned pile.

Distribution of the temperature and the oxygen concentrations in a central plane of the sectioned pile are presented in figure 6. The walls do not allows the oxygen diffusion and the below sections have similar behavior to the 1.5 m high single pile. The temperature isconduced trough the walls but do not have influence in the self-ignition of the neighbor section.

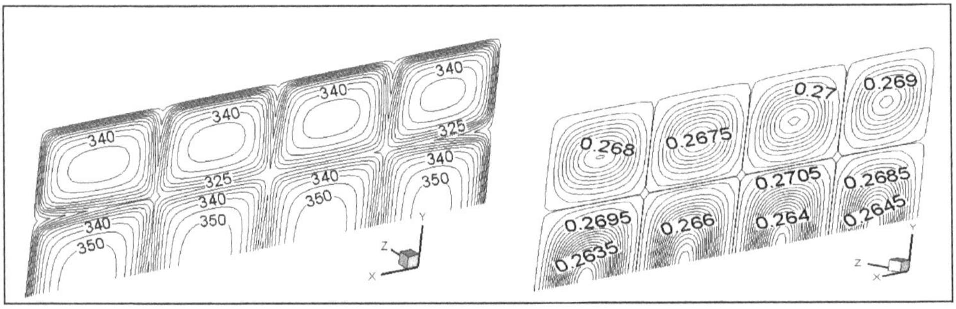

Fig. 6. Temperature and oxygen distribution within of the sectioned pile.

4. Combustion and convective heat transfer in porous medium combustors

4.1 Numerical simulation of a cylindrical porous medium burner

A porous media combustor, built on base of alumina spheres placed inside of an axi-symmetric cylindrical quartz tube of 0.52 m in length and 0.076 m in diameter, is shown in Fig. 7. Methane and air mixture enter to the combustor at ambient temperature with uniform velocity. To start the combustion a temperature profile of one step type, with a maximum temperature of 1150 K and a thickness of 4 cm is assumed to simulate the ignition by means of an external energy source. In the combustion zone, the products: CO_2, H_2O, O_2 and N_2 are generated. Air, gas and products are assumed to behave as ideal gases and hence density is calculated in terms of temperature from the ideal gas state equation. Burners based on this technology has been investigated and tested for many industrial applications (Foutko et al., 1996, Zhdanok et al., 1995).

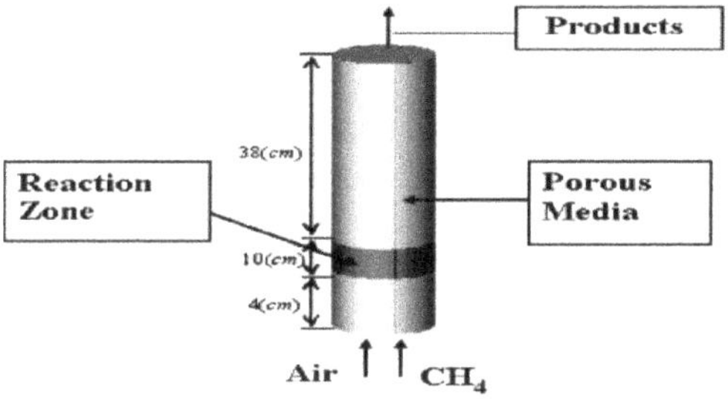

Fig. 7. Porous media combustor.

The assumptions used to build the mathematical model include: single-step chemical reaction, laminar 2D flow of Newtonian fluid of ideal gases. The mathematical model includes the porosity terms in both the energy equations for the solid as well as for the gas; similarly, also the continuity, linear momentum and fuel mass fraction equations are included. All physical properties are variable with temperature, and it is postulated that density varies according to the ideal gases state equation. The chemical reaction for methane is considered to be in a single step, with excess air included.

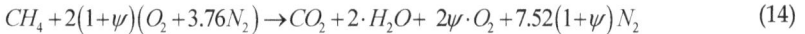

$$CH_4+2(1+\psi)(O_2+3.76N_2)\rightarrow CO_2+2\cdot H_2O+2\psi\cdot O_2+7.52(1+\psi)N_2 \tag{14}$$

Continuity and ideal gas equations

$$\frac{\partial(\rho\cdot v_r)}{\partial r}+\frac{\rho\cdot v_r}{r}+\frac{\partial(\rho\cdot v_z)}{\partial z}=0; \qquad \rho=\frac{\rho_0\cdot T_0}{T} \tag{15}$$

Linear momentum in radial r and axial z directions:

$$\frac{\partial(\rho\cdot v_r)}{\partial t}+v_r\frac{\partial(\rho\cdot v_r)}{\partial r}+v_z\frac{\partial(\rho\cdot v_r)}{\partial z}=-\frac{\partial p}{\partial r}+\frac{\partial^2(\mu\cdot v_r)}{\partial r^2}+\frac{1}{r}\frac{\partial(\mu\cdot v_r)}{\partial r}-\frac{\mu\cdot v}{r^2}+\frac{\partial^2(\mu\cdot v_r)}{\partial z^2} \tag{16}$$

$$\frac{\partial(\rho\cdot v_z)}{\partial t}+v_r\frac{\partial(\rho\cdot v_z)}{\partial r}+v_z\frac{\partial(\rho\cdot v_z)}{\partial z}=-\frac{\partial p}{\partial z}+\frac{\partial^2(\mu\cdot v_z)}{\partial r^2}+\frac{1}{r}\frac{\partial(\mu\cdot v_z)}{\partial r}+\frac{\partial^2(\mu\cdot v_z)}{\partial z^2} \tag{17}$$

Fuel mass conservation equation

$$\frac{\partial(\rho\cdot w)}{\partial t}+v_r\frac{\partial(\rho\cdot w)}{\partial r}+v_z\frac{\partial(\rho\cdot w)}{\partial z}=\frac{\partial}{\partial r}\left(D^M\cdot\rho\cdot\frac{\partial w}{\partial r}\right)+\frac{\partial}{\partial z}\left(D^M\cdot\rho\cdot\frac{\partial w}{\partial z}\right)-\rho\cdot K\cdot w\cdot e^{\frac{-Ea}{Ro\cdot T}} \tag{18}$$

Two energy equations were used for gas and solid in the porous medium

$$\varepsilon\cdot\left(\frac{\partial(\rho\cdot Cp\cdot T)}{\partial t}+v_r\frac{\partial(\rho\cdot Cp\cdot T)}{\partial r}+v_z\frac{\partial(\rho\cdot Cp\cdot T)}{\partial z}\right)=-\alpha\left(T_G-T_S\right)-\varepsilon\cdot\rho\cdot\Delta h\cdot K\cdot w\cdot e^{\frac{-Ea}{Ro\cdot T}} \tag{19}$$

$$(1-\varepsilon)\cdot\left(\frac{\partial(\rho_S\cdot Cp_S\cdot T_S)}{\partial t}\right)=\frac{1}{r}\frac{\partial}{\partial r}\left(r\cdot\lambda_{eff}\frac{\partial T_S}{\partial r}\right)+\frac{\partial}{\partial z}\left(\lambda_{eff}\frac{\partial T_S}{\partial z}\right)+\alpha\left(T_G-T_S\right) \quad (20)$$

where the chemical reaction speed and the effective conductivity of the solid including radiation are:

$$r_f=w_i\cdot\rho_G\cdot K\cdot\exp\left(-\frac{Ea}{Ro\cdot T}\right)\ \lambda_{eff}=\left(1-\varepsilon\right)\lambda_S+\frac{32\cdot\sigma\cdot\varepsilon\cdot dp\cdot T_S^3}{9\cdot\left(1-\varepsilon\right)} \quad (21)$$

The coefficient of convective heat transfer between solid and gas is calculated as follows

$$\alpha=\frac{6\left(1-\varepsilon\right)}{dp}\cdot\frac{\lambda_G}{dp}\cdot Nu;\quad Nu=2.0+1.1\cdot\left(\Pr^{1/3}\cdot\mathrm{Re}^{0.6}\right);\ \mathrm{Re}=\frac{\varepsilon\cdot u\cdot dp\cdot\rho}{\mu} \quad (22)$$

Conjugate boundary conditions between the porous burner section and the annular one are used in the internal and external areas of the inner tube:

$$T_i(r_i,z,t)=T_t(r_i,z,t);\ -\lambda_{eff}\frac{\partial T_i}{\partial r}=-\lambda_t\frac{\partial T_t}{\partial r};\ T_t(r_e,z,t)=T_e(r_e,z,t);\ -\lambda_t\frac{\partial T_t}{\partial r}=-\lambda_a\frac{\partial T_a}{\partial r} \quad (23)$$

The parameters used in the simulation included the Stephan–Boltzmann constant (σ), combustion enthalpy (Δh_{COMB}), frequency factor (K), activation energy (E_a), and the universal gas constant (Ro), whose respective values are: $\sigma=5.67\cdot10^8 W/m^2K^4$; $\Delta h_{COMB}=50.15\cdot10^6 J/kg$; $K=2.6\cdot10^8(1/s)$ and $Ea/Ro=15643.8K$. The mass diffusion coefficient was found by assuming that the Lewis number was equal to 1,

$$D^M=\frac{\lambda_G}{(\rho\cdot Cp)_G};\quad (Le=1);\quad \lambda_G=\frac{\mu\cdot Cp}{\Pr};\ D^M=\frac{\mu}{\rho\cdot\Pr} \quad (24)$$

Inlet and outlet boundary conditions of the heat exchanger are:

$$\begin{aligned}
&Z=0.0,\quad when\begin{cases}0\le r\le0.038 & \Rightarrow U_0=0.43\wedge T=300K\\ 0.038\le r\le0.077 & \Rightarrow\frac{\partial u}{\partial z}=\frac{\partial T}{\partial z}=0\end{cases}\\
&Z=1.5,\quad when\begin{cases}0\le r\le0.038 & \Rightarrow U_0=0.43\wedge T=300K\\ 0.038\le r\le0.077 & \Rightarrow U_0=\frac{\mathrm{Re}\cdot v}{D_h}\wedge T=283K\end{cases}
\end{aligned} \quad (25)$$

4.1.1 Solution procedure

The coupled, strongly non-linear system of partial differential equations was solved numerically using the FVM, with the SIMPLE algorithm (Patankar, 1980). A fifth power law was used to calculate the convective terms while the diffusion terms were determinated by linear interpolation functions for the dependent variables between the nodes. Each one of the governing equations was written in the general form of the transport equation, with unsteady, convective, diffusion and linearized source terms:

$$\frac{\partial(\rho\cdot\phi)}{\partial t}+div(\rho\cdot\vec{v}\cdot\phi)=div(\Gamma\cdot grad\phi)+Sc+Sp\cdot\phi \quad (26)$$

The convergence criteria used for gas and solid temperature, fuel mass fraction and for the two velocity components were

$$\left|\phi_{i,j}^{k}-\phi_{i,j}^{k-1}\right| \leq 10^{-3}\text{, for } T_G, T_S, w_i \text{; and } \left|\phi_{i,j}^{k}-\phi_{i,j}^{k-1}\right| \leq 10^{-4}\text{, for } u \text{ and } v \tag{27}$$

A non-uniform grid with 622 x 15 nodes, in axial and radial directions, respectively, was found by a trial and error procedure to be efficient to solve the discretized model with accuracy and a reasonable computation time (Moraga et al., 2008). The iterative solution procedure was based on the use of the line by line method,that combines the TDMA algorithm with the under-relaxed Gauss-Seidel algorithm. The under-relaxation coefficients used were equal to 0.1, for velocity components, and 0.5, for gas and solid temperatures and for the fuel ratio. A strategy based on a dynamic time step was implemented to calculate the unsteady terms. The initial time step used in the calculation procedure was equal to 0.00001s until t = 0.001s, and then was increased by and order of magnitude as time increased one order of magnitude, to end up with a time step equal to 0.1s when time was over 0.1s.

4.1.2 Results and discussion

Numerical experiments were performed to assess the effect of porosity, inlet gas velocity and excess air coefficient on the fluid dynamics and heat transfer in the porous media burner. The effect of increasing porosity from 0.3 to 0.6 on the axial velocity at four axial locations is shown in figure 8a, while the influence of the inlet velocity is depicted in figure 8b. The increments of porosity causes a reduction in axial velocity and in the velocity gradients near the walls. Axial velocity increases with time and a maximum value is reached for $\varepsilon = 0.3$, when t = 600s, at 0.15m from the inlet. Secondary flows are observed near the wall (z = 0.21m) when $\varepsilon = 0.3$. The axial velocity profile increases with time, when the initial velocity Uo = 0.3m/s, reaching a maximum at 900s, at z= 0.15m and up to 1500s for z = 0.21m.

Superadiabatic combustion in the porous media combustor causes temperature increaments from 300 K, near the inlet and close to the oulet, to 1600 K in the flame region, as depicted in figure 9. A displacement of the combustion front toward the middle of the combuster is achieved by increasing the air excess ratio ψ from 3.0 to 6.0. Similar effects in the temperature distribution can be noticed when porosity increased from $\varepsilon = 0.3$ to $\varepsilon = 0.6$ and when the inlet velocity increased from Uo = 0.3 m/s to Uo = 0.3 m/s. The displacement of the combustion zone, typical for porous combustor with uniform porosity and cross section, in the direction of the burner exit has been shown to be strongly influenced by increments in the values of: excess air ratio, porosity and inlet reactants velocity. The porous combustor design requires a reduced combustion front displacement. The results obtained (table 2) show that the displacement velocity of the combustion front decreases when the porosity and inlet reactants are reduced, and when the excess air ratio increases.Table 3 shows that a change in the mathematical model, from 1D to 2D, caused increments of about 5% in the gas temperature, 7% in the solid temperature and up to 3% in the combustion front velocity.

Figure 10 shows that main changes caused by the radial diffision of heat, captured with the 2D model, are larger after the combustion zone, where higher temperature are obtained.

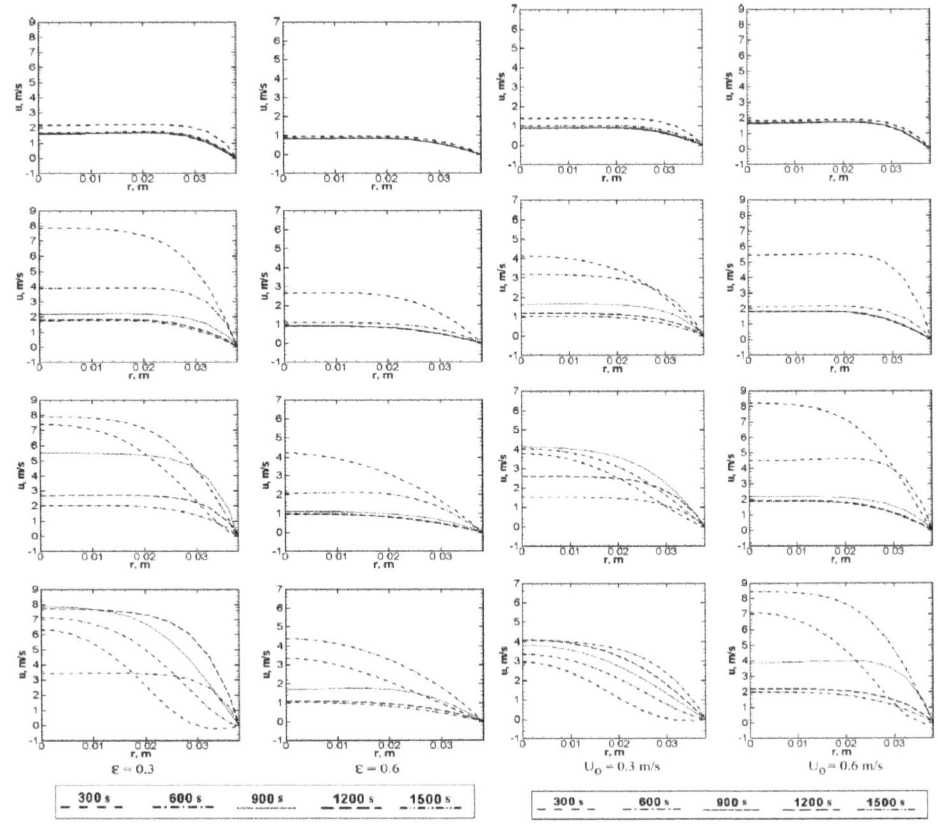

a) Effects of porosity. *Uo*= 0.43 m/s,*ψ*= 4.88 b) Effects of inlet velocity. *ψ*= 4.88, *ε*= 0.4

Fig. 8. Effects of porosity and inlet velocity on axial velocity along the burner. Axial locations (from top to bottom): 0.05m, 0.10m, 0.15m and 0.21m.

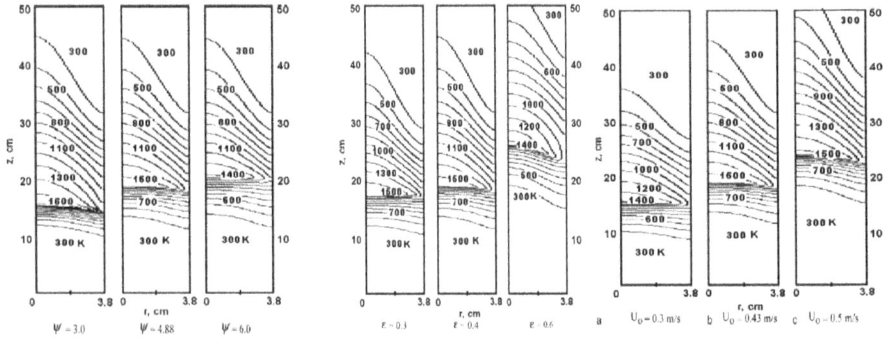

Fig. 9. Effects of excess air (left), porosity (center) and inlet velocity wright) on gas temperature at time t = 15 min.

Excess of air	Porosity	v_{in}(m/s)	Model	Gas T_{max} (K)	Solid T_{max} (K)	Combustion front velocity (m/s)
3.00	0.4	0.43	1D	1569.87(4.50%)	1417.39(7.06%)	9.538E-05(1.74%)
			2D	1640.45	1517.47	9.417E-05
4.88	0.4	0.43	1D	1404.44(4.99%)	1323.31(6.85%)	1.250E-04(0.67%)
			2D	1474.57	1413.95	1.258E-04
4.88	0.3	0.43	1D	1424.87(4.99%)	1354.25(6.98%)	1.108E-04(1.50%)
			2D	1496.03	1448.81	1.092E-04
4.88	0.6	0.43	1D	1368.515.03%)	1258.09(6.29%)	1.858E-04(2.69%)
			2D	1437.38	1337.18	1.908E-04
4.88	0.4	0.30	1D	1333.79(5.82%)	1246.03(7.32%)	9.167E-05(2.73%)
			2D	1411.39	1337.21	8.917E-05
4.88	0.4	0.60	1D	1465.13(3.53%)	1394.07(1.98%)	1.750E-04(0.48%)
			2D	1513.48	1421.52	1.758E-04

Table 3. Comparison between results of 1D and 2D models for porous media burner.

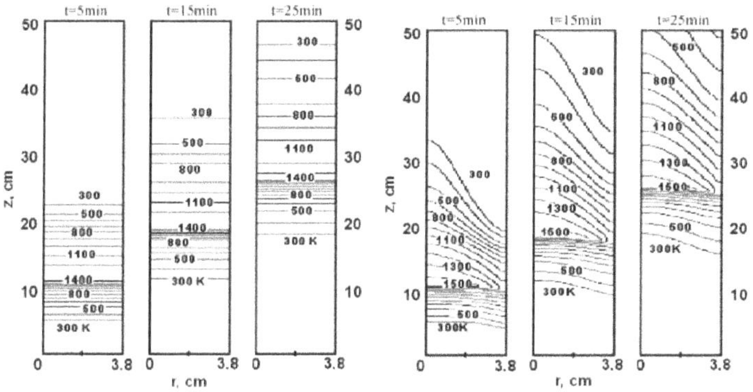

Fig. 10. Temperature distribution calculated with a 1D model (left side) and with a 2D model (right side)

4.2 Wood stove with porous medium post-combustor

4.2.1 Physical situation and mathematical model

The design of a porous post-combustor for a wood stove is investigated with FVM in order to reduce emissions. Figure 11 shows the post-combustor location in the gas exhaust tube along with some of the boundary conditions used in the analyis.Primary air inlet is located in the lower section of the stove front, secondary air enters in the upper section of the lateral walls through a three ways valve that allows high, medium or low secondary air injection and a terciary air inlet located in the upper section on the back, that allows for pre-heating of the auxiliary air.

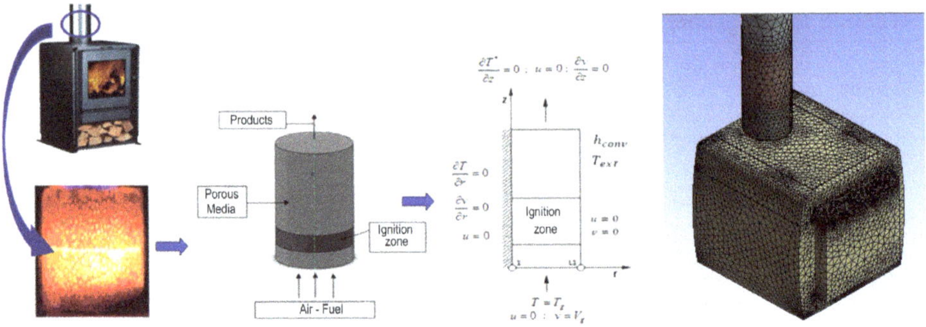

Fig. 11. Wood stove with a porous post combustor and mesh used in the FVM simulation.

The alumina porous combustor with a length of 300mm and a diameter of 150mm, was designed with two sections with different porous diameter of 1.52mm (length equal to 100mm) and 5.6 mm (with a length of 200mmm). An ignition temperature of 1150K was assumed in the simulations.

4.2.2 Solution procedure

A two stages procedure was used in the numerical study. In the first stage, a 3D turbulent k-e model was used to describe, with ANSYS/Fluent, the fluid mechanics and the convective heat transfer in the stove. Temperature and velocity distributions at the entrance of the exhaust stove pipe were found from the 3D model. Then, in the second stage, a secondary combustion process was described from a 2D laminar forced convection model in the porous post-combustor.

a. 3D k-e model for the turbulent gas flow inside the stove: Continuity, ideal gas, linear momentum, energy, species transport, turbulent kinetic energy and rate of dissipation equations:

$$\frac{\partial(\rho v_j)}{\partial x_j}=0; \quad \rho=\frac{\rho_0 T_0}{T} \tag{28}$$

$$\frac{\partial}{\partial x_j}(\rho v_j v_i)=-\frac{\partial p}{\partial x_i}+\frac{\partial}{\partial x_j}\left((\mu+\mu_t)\frac{\partial v_i}{\partial x_j}\right)+\frac{\partial}{\partial x_j}\left((\mu+\mu_t)\frac{\partial v_j}{\partial x_i}\right)+\rho g_i \tag{29}$$

$$\frac{\partial}{\partial x_j}(\rho v_j T)=\frac{\partial}{\partial x_j}\left(\lambda_{eff}\frac{\partial T}{\partial x_j}\right)+S_C \quad ; \quad \lambda_{eff}=\lambda+\frac{c_P\mu_t}{\Pr_t}; \quad \Pr_t=0.85 \tag{30}$$

$$\frac{\partial}{\partial x_j}(\rho v_j Y_{m'})=\frac{\partial}{\partial x_j}\left(\left(D_{m'}^M\rho+\frac{\mu_t}{Sc_t}\right)\frac{\partial Y_{m'}}{\partial x_j}\right)+S_{m'}; \quad Sc_t=0.7 \tag{31}$$

$$\frac{\partial}{\partial x_i}(\rho k v_i)=\frac{\partial}{\partial x_j}\left[\left(\mu+\frac{\mu_t}{\sigma_k}\right)\frac{\partial k}{\partial x_j}\right]+G_k+G_b-\rho\varepsilon \quad ; \quad \mu_t=\rho C_\mu\frac{k^2}{\varepsilon} \tag{32}$$

$$\frac{\partial}{\partial x_i}(\rho \varepsilon v_i) = \frac{\partial}{\partial x_j}\left[\left(\mu + \frac{\mu_t}{\sigma_\varepsilon}\right)\frac{\partial \varepsilon}{\partial x_j}\right] + C_1 \frac{\varepsilon}{k}(G_k + C_3 G_b) - C_2 \rho \frac{\varepsilon^2}{k} \tag{33}$$

$$G_k = \mu_t \left[\frac{\partial v_i}{\partial x_j} + \frac{\partial v_j}{\partial x_i}\right]\frac{\partial v_i}{\partial x_j} \quad ; \quad G_b = \beta g_i \frac{\mu_t}{\Pr_t}\frac{\partial T}{\partial x_i}; \quad \beta = -\frac{1}{\rho}\frac{\partial \rho}{\partial T} \tag{34}$$

The values for the five constants used in the k-e turbulence model are those suggested by Launder and Spalding, 1974.

$$C_1 = 1.44; \quad C_2 = 1.92; \quad C_3 = 1.0; \quad C_\mu = 0.09; \quad \sigma_k = 1.0; \quad \sigma_\varepsilon = 1.3 \tag{35}$$

b. 2D Darcy-Brikman-Forcheimer model for the laminar gas flow in the porous media post-combustor: Continuity, ideal gas, linear momentum, energy and species transport equations

$$\frac{\partial(\rho v_j)}{\partial x_j} = 0 \quad ; \quad \rho = \frac{\rho_0 T_0}{T} \tag{36}$$

$$\frac{\partial}{\partial x_j}(\rho v_j v_i) = -\frac{\partial p}{\partial x_i} + \frac{\partial}{\partial x_j}\left(\mu \frac{\partial v_i}{\partial x_j}\right) - \rho g_i - \left(\frac{\mu}{\alpha} + \frac{C\left|\vec{V}\right|}{\sqrt{\alpha}}\right)v_i; \quad C = 0.55 \tag{37}$$

$$\frac{\partial}{\partial x_j}(\rho v_j T) = \frac{\partial}{\partial x_j}\left(\lambda_{ef}\frac{\partial T}{\partial x_j}\right) \quad ; \quad \lambda_{ef} = \gamma \lambda_f + (1-\gamma)\lambda_s \tag{38}$$

$$C_P \frac{\partial}{\partial x_j}(\rho v_j Y_{m'}) = \frac{\partial}{\partial x_j}\left(D_{m'}^M \rho \frac{\partial Y_{m'}}{\partial x_j}\right) \tag{39}$$

The air flow for the three operational modes inside the stove combustion chamber is shown in figure 12.

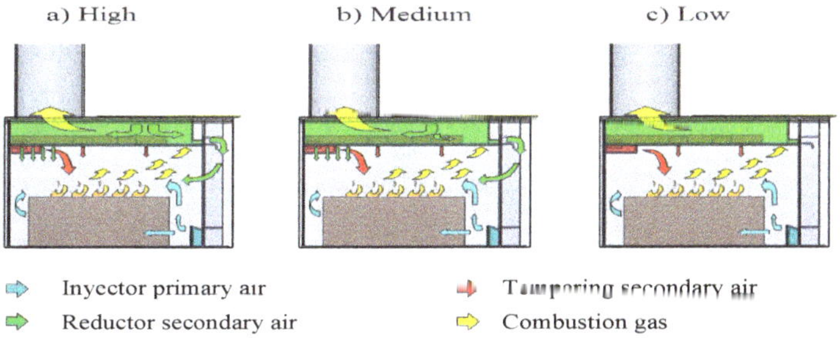

Fig. 12. Air flow inside the primary combustor chamber for the three operacional modes.

The 3D model for the combustion zone of the stove was discretised with 736,767 tethahedral, 18,032 wedge and 280 pyramidal elements. Under-relaxation coefficients for pressure were: 0.2 for pressure, 0.3 for the three velocity components, 0.7 for the kinetic energy and for the dissipation rate of the kinetic energy and 0.9 for temperature. Axial discretization for the 2D porous postcombustor included 40 nodes in the pre-heating section, from z = 0 to z = 2cm, 450 nodes in the flame region, up to z= 24cm, 100 nodes in the post-secondary combustion zone (from z=24 to z=30cm) and 30 nodes in the last zone (from z= 30cm to z= 60cm in z direction). The convergence was assumed when the maximum deviation for each dependent variable, at each control volume and for all time steps, $\Phi^{k}_{i,j}$-$\Phi^{k-1}_{i,j}$ was smaller than 0.0001 for the velocity components and smaller than 0.001 for solid and gas temperature. Under-relaxation factors were equal to 0.1 for the velocity components, 0.5 for gas and solid temperature and equal to 0.3 for pressure.

4.2.3 Results and discussion

Velocity and temperature distributions in the central plane of the wood stove are described in Figure 13, for three air operation modes. Higher velocities, in the order of 2m/s are found in the left lower section and in the central part of the stove. A 30% velocity increment is obtained in the low mode. Temperature in the primary combustor is the range between 800K and 1365K, with higher temperature reached in the low mode.

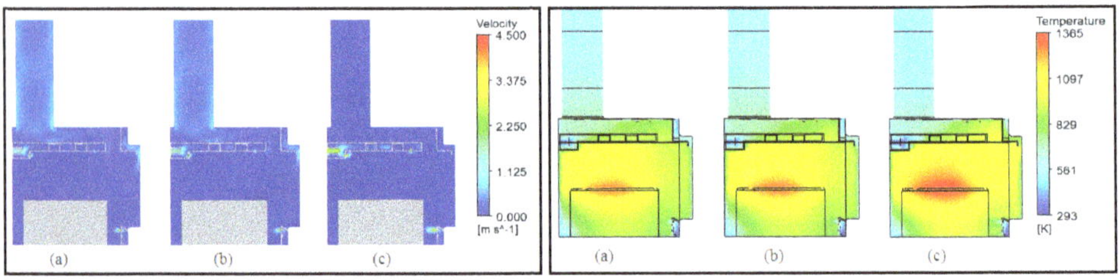

Fig. 13. Velocity and temperature distributions inside the wood stove.

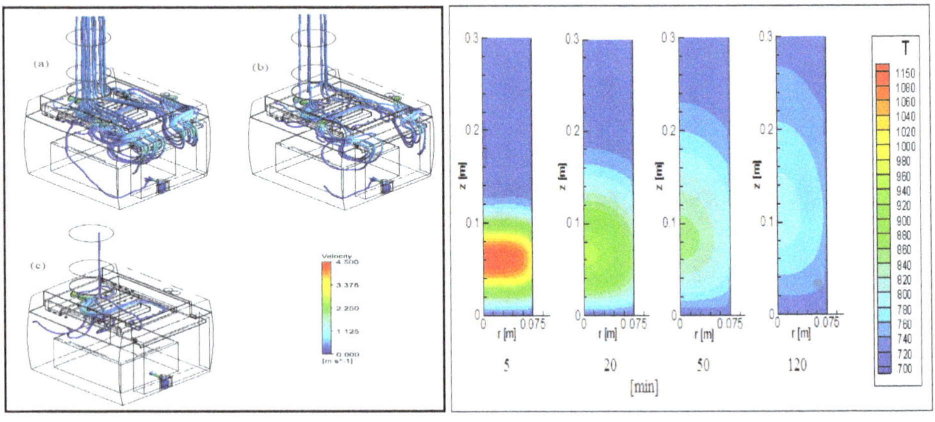

Fig. 14. Air trajectories inside the stove (at left side) and time evolution of temperature inside the porous post-combustor.

Figure 14, in the left side, depicts the air and gas trajectories in the primary combustor, where higher velocities in the range of 2.3m/s are found for the three operation modes. Time evolution for the temperature distribution in the porous post-combustor, is shown in the right hand side of Figure 13. A hot region, with temperatures in the range between 800K and 1150K can be observed to last for 5 minutes and a zone with temperatures higher than 850K can be noticed during the first 50 minutes of operation.

5. Alloy solidification and natural non-Newtonian convection predicted with a porous model

Liquid to solid phase change is a relevant process in many industrial applications, such as: polymer casting moulding, pure metals and alloys solidification, solar energy storage and food freezing and thawing. The pourpose of this section is to describe numerical solutions obtained with a porous media model and the FVM that have been applied to solidification of pure metals and alloys. The sequential solution of the discretized system of fluid mechanics and convective heat transfer is accomplished by the PSIMPLER algorithm (Moraga et al., 2010).

5.1 Physical situation and mathematical model

The physical situation related to each one of the three cases studied is shown schematically in Figure 15. The first case corresponds to the solidification of pure aluminum, and the second is the solidification of aluminum alloys with 1.7% Si. In all cases the liquid to solid phase change occurs inside a square cavity with the right vertical wall and the horizontal ones being adiabatic. The left-side wall is subjected to a convective condition, of the Robin type, for case 1, while in case 2 an imposed temperature condition is assumed, of the Dirichlet type. The fluid mechanics in the mushy zone and in the liquid phase is based on laminar flow, with a power law non-Newtonian model (n=0.5). Density is assumed to vary linearly with temperature, according to the approximation of Boussinesq. A porous medium model, with a Darcy numberDa=1.37×10^{-5}[26], for an average pore diameter dm = 1.2x10^{-5} is proposed to describe the fluid motion in the mushy zone, along with the general Darcy-Brinkman-Forchheimer porous flow model.

Case 1 (Al); Bi = 3.3

Case 2: (Al- 1.7%Si) and $\qquad h \cdot (I\,|_{x=0} - T_{\infty}) = h \left. \dfrac{\partial T}{\partial x}\right|_{x=0}$

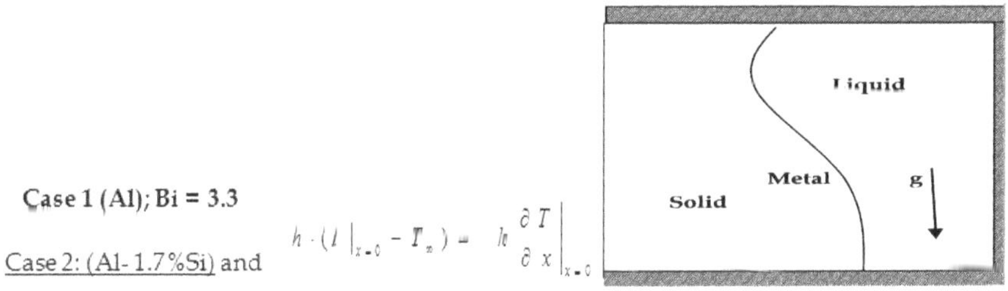

Fig. 15. Solidification of a pure metal, a binary and a ternary alloy in a square cavity.

The mathematical model includes continuity, linear momentum and energy equations in porous media

$$\frac{\partial u}{\partial x}+\frac{\partial v}{\partial y}=0 \tag{40}$$

$$\frac{\rho}{\varepsilon}\frac{\partial u}{\partial t}+\frac{\rho}{\varepsilon^2}\left[u\frac{\partial u}{\partial x}+v\frac{\partial u}{\partial y}\right]=-\frac{\partial p}{\partial x}+\frac{\eta}{\varepsilon}\left[\frac{\partial^2 u}{\partial x^2}+\frac{\partial^2 u}{\partial y^2}\right]-\left[\frac{\eta}{K}+\frac{\rho\cdot C}{\sqrt{K}}\left|\vec{V}\right|\right]u \tag{41}$$

$$\frac{\rho}{\varepsilon}\frac{\partial v}{\partial t}+\frac{\rho}{\varepsilon^2}\left[u\frac{\partial v}{\partial x}+v\frac{\partial v}{\partial y}\right]=-\frac{\partial p}{\partial y}+\rho\cdot g\cdot\beta\cdot(T-T_C)+\frac{\eta}{\varepsilon}\left[\frac{\partial^2 v}{\partial x^2}+\frac{\partial^2 v}{\partial y^2}\right]-\left[\frac{\eta}{K}+\frac{\rho\cdot C}{\sqrt{K}}\left|\vec{V}\right|\right]v \tag{42}$$

$$\left[\rho\cdot\left[1+\varepsilon\frac{h_{fs}}{C_p}\cdot\frac{\partial f_{PC}}{\partial T}\right]\cdot\frac{\partial T}{\partial t}+\rho\cdot\left(u\frac{\partial T}{\partial x}+v\frac{\partial T}{\partial y}\right)\right]=\frac{k}{C_p}\left[\frac{\partial^2 T}{\partial x^2}+\frac{\partial^2 T}{\partial y^2}\right] \tag{43}$$

where the liquid phase change fraction, being equal to 0 when $T<T_S$ and equal to 1 when $T > T_L$ is calculated from

$$f_{pc}= [(T- T_S)/(T_L - T_S)]^m \quad \text{for } T_S < T < T_L \tag{44}$$

The dynamic apparent viscosity is $\eta = \eta_l / f_{pc}$, for Newtonian and non-Newtonian fluids, with η_l and the deformation rate for the power law model defined as follows

$$\eta_l=\mu\cdot\dot{\gamma}^{n-1} \quad \dot{\gamma}=\left\{2\cdot\left[\left(\frac{\partial u}{\partial x}\right)^2+\left(\frac{\partial v}{\partial y}\right)^2\right]+\left[\frac{\partial u}{\partial y}+\frac{\partial v}{\partial x}\right]^2\right\}^{1/2} \tag{45}$$

5.2 Computational implementation

The mathematical models presented in the previous equations were solved with the finite volume method and the PSIMPLER algorithm developed and programmed in FORTRAN. The PSIMPLER algorithm is a mixture of two algorithms, SIMPLER and PISO (Moraga et al., 2010). In all the cases, the grid used was of the overlapping type, and 40x40, 60x60 and 80x80 grid sizes were evaluated. The results are presented for a 60x60 grid, which was efficient in time and accurate enough. The values for the under-relaxation factors used for the two velocity components, temperature and pressure were different for pure metals and for binary and ternary alloys

$$a_u = a_v = 0.2\,;\ a_T=0.1\,;\ a_P= 0.7\ \text{(pure metal) and } a_u= a_v = 0.5\,;\ a_T = 0.3\,;\ a_P= 0.9\ \text{(alloys)} \tag{46}$$

The iterative procedure was finished when the difference between $\Phi^k_{i,j}$ - $\Phi^k_{i,j}$ at two successive iterations was smaller or equal to ξ, for all control volumes and at each time step, with$\xi=10^{-6}$ for velocity and $\xi=10^{-3}$ for temperature.

5.3 Results and discussion

The first case describes the solidification of pure aluminum metal, with Ra = 10^5, by using the proposed Darcy-Brinkman-Forchheimer porous model for the mushy zone and three alternative temperature dependent liquid phase fractions, defined by changing the exponent

m in Eq. (44). Figure 16 shows that the time evolution for the temperature distributions obtained with the FVM, for m = 0.5, 1.0 and 2.0, in dashed lines is in agreement with the values calculated by the finite element method and the classical mathematical model, shown in continuos lines (Cruchaga et al., 2000).

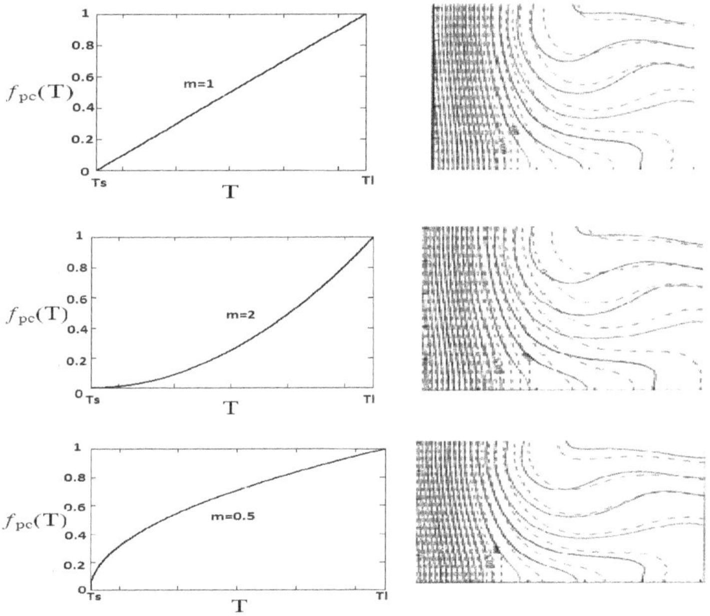

Fig. 16. Isotherms time evolution for Al solidification with Ra=10^5, present porous model and Cruchaga et al., 2000 results.

The time evolution of the solidification front for the aluminum, calculated with the general porous media model is shown in figure 17 to be in agreement with the results obtained with the classical model.

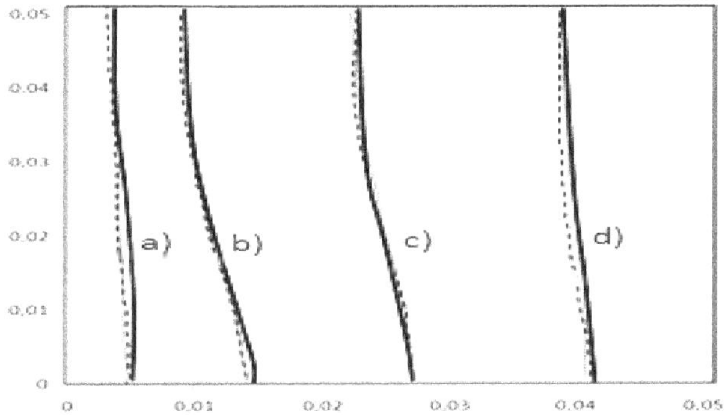

Fig. 17. Time evolution of the Al solidification front.

In case 2, the melt characterization by a non-Newtonian power law model, with a power index n = 0.5, is investigated along the use of the DBF porous model for the mushy zone, to describe the solidification of the binary Al-1.7wt%Si, when Ra=2.5x10^5. Figure 18 describes the evolution in time of the isotherms (at the left side) and of the steamlines in the liquid phase and in the mushy zone (at the rigth side), calculated with FVM by assuming either a Newtonian or a non-Newtonian power law fluid models. The pseudoplastic fluid assumption (n=0.5) originates a slighty faster convection, requiring lower time to complete the solid to liquid phase transformation.

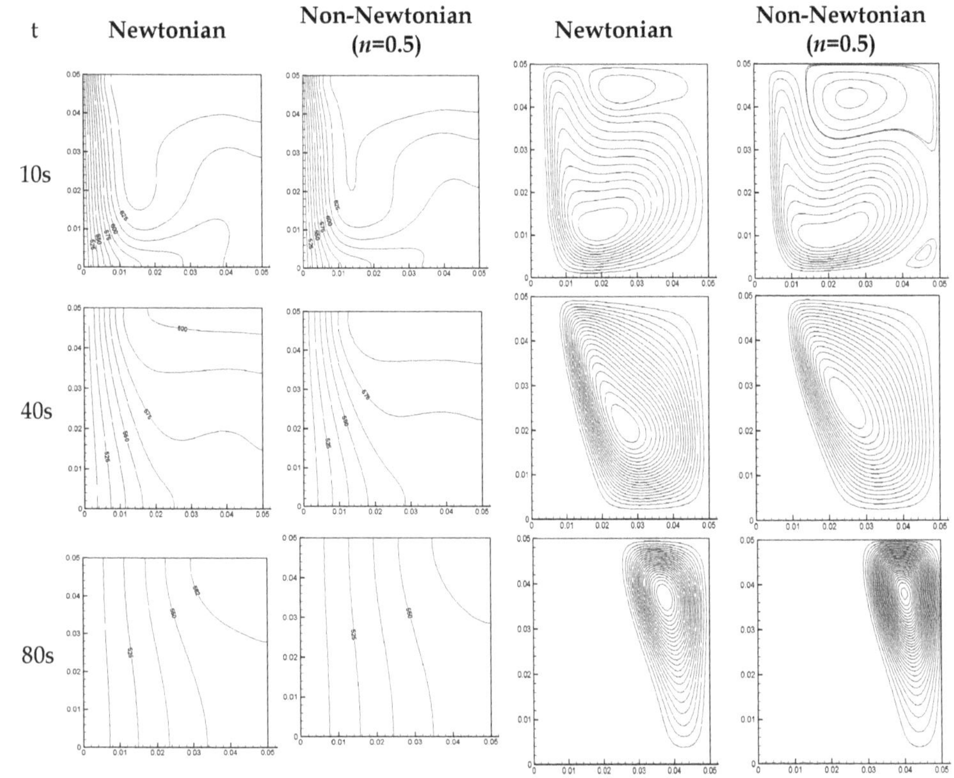

Fig. 18. Isotherms and stream function for Al-1.7%Si solidification calculated for Ra = 2.5 x 10^5.

6. Mixed turbulent convection and diffusion in a bioreactor

In the last decades, the municipal waste materials (MWM) outputs in many countries have increased significantly due to the large increase in population and industries. These wastes contain organic matter that can be recovered, and through recycling, MWM can be returned to the environment. One option in that direction, for example, is to use them as fertilizers. Another interesting possibility is to use them as fuels. In this way, in addition to solving the problem of MWM disposal, the recycling of these materials becomes a useful source of energy. For this, the MWM require usually a pre-treatment to eliminate water, in order to have good combustion. This process is essentially a method for humidity control of the waste, which has a great impact in combustion efficiency.

The numerical modeling of heat and mass transfer involved in the drying process without self-heating has motivated several studies using finite difference methods (Kaya et al., 2006; Kaya et al., 2008, Chandra & Talukdar, 2010; Sheng et al., 2009). These studies have provided useful insight into the phenomenon, but the air flow regime in a typical drying bioreactor is nevertheless turbulent. Therefore there is interest in the analysis and characterization of this process in the fully turbulent flow regime. In this case also the self-heating of the material due to chemical and biological reactions is described.

6.1 Physical situation of an experimental bioreactor

In order to compare the model results with experimental data, we perform the computations for a case with the dimensions of the experimental bioreactor: 1 m × 1 m × 1 m, in the x, y and z coordinate directions respectively (see Fig. 19). The walls are assumed adiabatic and impermeable. Air at ambient temperature is forced into the reactor through an inflow in the plane z = 0. This inflow has a section of 0.05 m × 0.05 m and is located at the center of the lower wall. The air entering through this opening flows past the block of MWM inside the reactor fulfilling the drying of this material. The air outflow is located at the center of the top wall (z = 1 m) and has the same dimensions of the inflow located below.

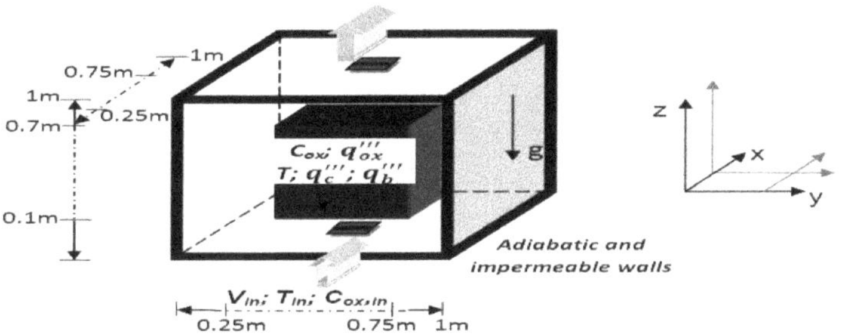

Fig. 19. 3D physical situation used for an experimental bioreactor.

The physical situation presented in this work is the same that is studied experimentally in reference (Rada et al., 2007), which provides a good source for validation of the general mathematical modeling.

6.2 Mathematical model for the bioreactor

The κ-ε turbulence model, equations (28) to (35), is used for the modeling of air flow within the reactor (Launder & Spalding, 1974), coupled with a diffusion model proposed by Sidhu et al. (Sidhu et al., 2007), and presented in equations (11) to (15).

The air flow inside the reactor is governed by continuity and by the Navier-Stokes equations.

The initial velocity is taken as zero: $\vec{\upsilon}(x,y,z,0)=0$. The prescribed value for the inlet velocity is 1m/s in the z coordinate:

$$u_{in}(x,y,0,t)=0\,;\; v_{in}(x,y,0,t)=0\,;\; w_{in}(x,y,0,t)=1m/s \tag{47}$$

At the outlet, zero-gradient outflow boundary conditions are applied. The walls are considered adiabatic and impermeable. The temperature of the air at the inlet is constant and an initial linear distribution in the domain was assumed:

$$T(x,y,0,t)=283\,K\,; \qquad T(x,y,z,0)=283-\frac{z}{0.2} \tag{48}$$

Inside the MWM the process is considered essentially as diffusion in porous medium, and therefore the velocities are zero in the region occupied by the block of material. Constant oxygen concentration is imposed around the MWM volume,

$$C_{ox}(x,y,z,t)=0.272\,kg/m3\left[for(x,y,z)\text{ outside the MWM}\right] \tag{49}$$

The initial temperature and oxygen concentration within the MWM are considered homogeneous,

$$T(x_{wm},y_{wm},z_{wm},0)=283\,K\,;\; C_{ox}(x_{wm},y_{wm},z_{wm},0)=0.272\,kg/m^3 \tag{50}$$

6.3 Numerical simulation with finite volume method

The sequential coupling of these main variables (external iterations) is done with the SIMPLE method. Under-relaxation is applied during these external iterations, with a relaxation coefficient of 0.5 for all the variables, except for the pressure correction for which a value of 0.8 was used instead.

The bioreactor was discretized with a non-uniform mesh of 32×62×82 finite volumes in x, y and z directions, respectively. Additionally, a buffer zone extending for 1 m beyond the exit of the reactor is considered, in order to ensure a zero-gradient condition at the outlet of the computational domain. The waste material occupies a volume inside the reactor equivalent to a parallelepiped with dimensions of 0.5×0.5×0.6 m^3. The discretization of this sub-region of the domain comprises 16×40×40 finite volumes in the x, y and z directions, respectively. The simulation was carried out for a total time of 696 hr. The time step was 3600 s for most of this period. During the first hour a dynamic time step was used (starting with a step of 0.001 s) in order to achieve convergence at each time step during the initial fast transient period starting from the prescribed initial condition.

6.4 Comparison with experimental data

In the experimental bio-drying reactor analyzed in reference (Rada et al, 2007), three thermocouples were installed inside the MWM, along the z direction, at equal intervals of 0.2 m in z direction inside of MWM, to measure the internal temperature in the material. Figure 20 presents the numerical results of temperature evolution obtained with the mathematical model and the numerical simulation described in previous sections. The numerical results in Fig. 20 were obtained in the following three (x, y, z) positions: P1(0.49,0.44,0.19), P2(0.49,0.44,0.39) and P3(0.49,0.44,0.59). The values for all coordinates are in meters. The results for the time evolution of the temperature in Fig. 20 are consistent with the behavior showed by Rada et al. (2007). The self-heating, product of the biological and chemical heat generation is clearly observable. Initially, a sudden increase of temperature occurs in the three sampled positions until 90 h. For all times, the temperature at point P2 is greater than the others. This can be attributed to its more inner position, making more difficult for the locally generated heat to diffuse to the surface, where is removed by the flowing air. Points P1 and P3, being located nearer to the surface, can release more easily the generated heat to the surrounding air. The asymmetry in the temperature at P1 and P3 is due to the location of P1 on the side subjected directly to the impinging flow of cold air.

After about 96 h, the temperatures at the three points moderate their rate of increase, and the maximum temperatures occur at about t = 168 h for all these positions. The maximum temperature reached inside the MWM is close to 340 K, at the central position P2. The existence of these maxima can be associated with an intrinsic self-moderation of the rate of heat generation, given by the last term in Eq. (11). That term, describing the heat released by the biological activity of micro-organisms initially increases with the temperature, as their population grows, but after exceeding 318 K approximately this source of heat starts to decrease, because of the progressive inhibition in micro-organism growth as the temperature continues to increase. After 168 h, the temperature decreases in all positions as the internal heat source has become weaker and at the same time a heat diffusion pattern inside the material has established, allowing and effective removal of the generated heat towards the surrounding air.

The table 4 shows a comparison between computed temperatures at the sampled points and experimental data from Ref. (Rada et al., 2007) at corresponding positions. The temperatures calculated in the present study approximate very well the experimental values in the three compared positions. A maximum difference of 5 K can be observed at 360 h in the position P3.

	Temperature, K					
Time, h	Thermocouple 2, Ref. [3].	P1	Thermocouple 3, Ref. [3].	P2	Thermocouple 4, Ref. [3].	P3
96	317	315	335	337	330	329
168	323	325	335	338	335	334
360	307	307	331	329	312	317
696	290	291	298	299	292	293

Table 4. Comparison of temperatures between experimental data and calculated values.

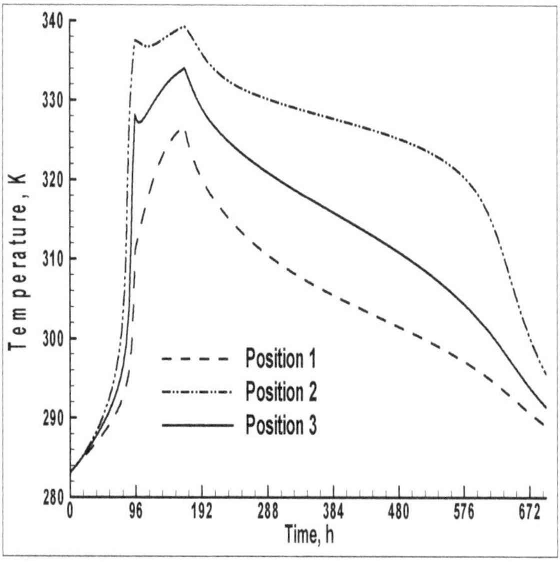

Fig. 20. Temperature evolution in three positions of the MWM.

7. Conclusions

The FVM capabilities to produce efficient and accurate prediction of fluid mechanics, heat and mass transfer in porous media, including biological and chemical reactions or liquid to solid phase transformations, have been shown by solving four practical examples. In the first case, auto-ignition of compost piles was studied and field experimental data were used to assess the quality of FVM results for the evolution of temperature and oxygen concentration distributions. The conclusion found for this case was that a pile heigth equal to 1.7m is a critical value to produce self-ignition. Combustion in porous media, predicted by FVM, allows to conclude in the second case, that improvements on thermal energy efficiency and pollution reduction in a methane porous combustor and in a wood stove, can be achieved. The use of the FVM and a generalized DBF flow porous media model for the mushy zone was found to describe convective cooling during solidification of non-Newtonian melted binary alloys in the third case. A conclusion found for this type of phase change processes was that the CPU time requiered for the numerical simulation can be reduced one order of magnitude by using a new improved predictor-corrector sequential algorithm, PSIMPLER, for the pressure-velocity-temperature-concentration calculation procedure. In the fourth case, the FVM along the k-ε turbulence model, were used to describe 3D turbulent convective heat transfer, with self heating of the porous media due to chemical and biological reactions, in a bioreactor. The conclusions found in this case were that the FVM simulations for the dependent variables were successfully validated with experimental values and that the effects of reactor geometry, self-heating parameters, air flow and temperature in the bioreactor perfomance can be evaluated. Finally, it is concluded that the use of FVM and adequate mathematical models along to experimental physical results can be used to investigate physical, biological and chemical coupled problems in order to achieve improved thermal efficiency, adequate use of energy resources and pollution reduction in these processes.

8. Acknowledgements

This work was conducted with support of CONICYT-Chile to projects Fondecyt 1111067, Fondecyt 11110097 and Fondef DO8I1204.

9. References

Chandra, V.P. & Talukdar, P. (2010). Three dimensional numerical modeling of simultaneous heat and moisture transfer in a moist object subjected to convective drying. *International Journal of Heat and Mass Transfer*, Vol.53, pp.4638–4650, ISSN 0017-9310.

Chen, X.D. & Mitchell D.A. (1996). Star-up strategize of self-heating and efficient growth in stirred bioreactor for solid state bioreactors, *Proceedings of the 24th Annual Australian and New Zealand Chemical Engineering Conference (CHEMECA 96)*, pp.111-116, ISSN 0858256584.

Cruchaga, M. & Celentano, D. (2000). A Finite Element Thermally Coupled Flow Formulation for Phase-Change Problems, *Int. J. Numer. Meth. Fluids*, Vol. 34, pp. 279-305.

Ferziger, J. H. & Peric, M. (2002). Computational Methods for Fluid Dynamics, Springer, Berlin, ISBN 3540594345.

Foutko, S.I.; Shanbunya, S.I. & Zhdnadok, S.A. (1996). Superadiabatic combustion wave in a diluted methane-air mixture under filtration in a packed bed, in: *Proceedings of the 26th International Symposium on Combustion/The Combustion Institute*, Mink, Belarus.

Jin-Sheng Leu; Jiin-Yuh Jang & Wen-Cheng Chou. (2009). Convection heat and mass transfer along a vertical heated plate with film evaporation in a non-Darcian porous medium. *International Journal of Heat and Mass Transfer*, Vol.52, Vol.5447–5450, ISSN 0017-9310.

Kaya, A.; Aydın, O. & Dincer, I. (2006). Numerical modeling of heat and mass transfer during forced convection drying of rectangular moist objects, *International Journal of Heat and Mass Transfer*, Vol.49, pp.3094–3103, ISSN 0017-9310.

Kaya, A.; Aydin, O. & Dincer, I. (2008). Heat and mass transfer modeling of recirculating flows during air drying of moist objects for various dryer configurations. *Numerical Heat Transfer; Part A: Applications*, Vol.53, No.1, pp. 18-34, ISSN 1521-0634.

Launder, B.E. & Spalding, D.B. (1974). The numerical computation of turbulence flow. *Computer Methods in Applied Mechanics and Engineering*, Vol.3, pp.269-289, ISSN 0045-7825.

Moraga, N., Zambra, C. (2008). Autoigniciόn 3D en depόsitos de lodos provenientes de tratamientos de aguas residuales. INGENIARE, Revista Chilena de Ingeniería, Vol.9, No.3, pp. 352-357, ISSN: 0718-3291.

Moraga, N.O.; Andrade, M.A. & Vasco, D. (2010). Unsteady mixed convection phase change of a power law non-Newtonian fluid in a square cavity. *International Journal of Heat and Mass Transfer*, Vol.53, pp.3308-3318, ISSN 0017-9310.

Moraga, N.O.; Corvalán, F.; Escudey, M.; Arias, A. & Zambra, C.E. (2009). Unsteady 2D Coupled Heat and Mass Transfer in Porous Media With Biological and Chemical Heat Generations. *International Journal of Heat and Mass Transfer*, Vol.52, pp.25-26, ISSN 0017-9310.

Moraga, N.O.; Ramírez, S.C. & Godoy, M.. (2010). Study of convective Non-Newtonian alloy solidification in moulds by the PSIMPLER/ Finite Volume Method. *Numerical Heat Transfer,* Vol.57, pp.936-953, ISSN 1521-0634.

Moraga, N.O.; Rosas, C.E.; Bubnovich, V.I. & Tobar J. (2008). On predicting two-dimensional heat transfer in a cylindrical porous combustor. *International Journal of Heat and Mass Transfer*, Vol.51, pp.302-311, ISSN 0017-9310.

Moraga, N.O.; Sánchez, G.C. & Riquelme, J.A.. (2010). Unsteady mixed convection in a vented enclosure partially filled with two non-Darcian porous layers. *Numerical Heat Transfer, Part A,* Vol.57, pp.1-23, ISSN 1521-0634.

Moraga, N.O. & Lemus-Mondaca, R. (2011). Numerical conjugate air mixed convection/non-Newtonian liquid solidification for various cavity configurations and rheological models". *International Journal of Heat and Mass Transfer,* Vol. 54, (23), pp. 5116-5125.

Nield, D. &. Bejan, A. (1992). Convection in Porous Media, Springer-Verlag, New York, ISBN 0387984437.

Patankar, S. (1980). Numerical Heat Transfer and Fluid Flow, Hemisphere, Washington, ISBN 0891165223.

Rada, E.C.; Taiss, M.; Ragazzi, M.; Panaitescu, V. & Apostol, T. (2007). Lower heating value dynamics during municipal solid waste bio-drying, *Environmental Technology,* Vol.28, No.4, pp.463-469, ISSN 0959-3330.

Richards, L.A. (1931). Capillary Conduction of Liquids Through Porous Mediums. *Journal of Applied Physics,* Vol.1, pp. 318-333, ISSN 0021-8979.

Serrano, S. (2004). Modeling Infiltration with Approximated Solution to Richards Equation. Journal of Hydrologic Engineering, Vol.9, No.5, pp. 421-432, ISSN 1084-0699.

Sidhu, H.S.; Nelson, M.I. & Chen, X.D. (2007). A simple spatial model for self-heating compost piles. *ANZIAM J. (CTAC2006),* Vol.41, pp.C135-C150, ISSN 1446-8735.

Versteeg, H.K. & Malalasekera, W.. (1995). An Introduction to Computational Fluid Dynamics. The Finite Volume Method. John Wiley & Sons, (Ed. 1), New York, ISBN 0-47023515-2.

Zambra, C.E.; Moraga, N.O. & Escudey, M. (2011). Heat and Mass Transfer in Unsaturated Porous Media: Moisture Effects in Compost Piles Self-Heating. *International Journal of Heat and Mass Transfer,* Vol.54, pp.2801-2810, ISSN 0017-9310.

Zambra, C.E.; Rosales, C.; Moraga & N.; Ragazzi, M. (2011). Self-Heating in a bioreactor: Coupling of heat and mass transfer with turbulent convection. International Journal of Heat and Mass Transfer, Vol.54, pp.5077-5086, ISSN 0017-9310.

Zhdanok, S.A.; Kennedy, L.A. & Koester, G. (1995). Superadiabatic combustion of methane-air mixtures under filtration ina a packed bed. *Combustion and Flame,* Vol. 100, pp.221-131.

Finite Volume Method for Streamer and Gas Dynamics Modelling in Air Discharges at Atmospheric Pressure

Olivier Ducasse, Olivier Eichwald and Mohammed Yousfi
University of Toulouse
France

1. Introduction

Electrical discharges in air at atmospheric pressure like corona or dielectric barrier discharges are generally crossed by thin ionized filament called streamers (about 100μm diameter). The streamer develops and propagates inside the background gas with a high velocity (around 10^6 m/s) higher than the electron drift velocity (around 10^5 m/s). During the transport of charged particles within the filaments under the action of the electric field, the energetic charged particles undergo many collisions with the background gas (neutral particles). The interactions between charged and neutral particles generate in turn a gas dynamics characterized by gas temperature and density gradients. The variation of density, momentum transfer and energy of the different particles, present within the ionized filaments, are governed by the fluid conservation laws (or continuity equations) coupled, in the charged particles case, to the electric field or Poisson equation.

It is very important to well known the electro-dynamics characteristics of these atmospheric pressure non thermal plasma generated by streamer or micro-discharge dynamics for an efficient use in the associated applications such as the pollution control of flue gases (Kim, 2004; Marotta et al., 2007), the combustion and ignition improvement (Starikovskaia, 2006), the airflow control (Eichwald et al., 1998; Moreau, 2007) and the biomedical fields (Laroussi, 2002; Fridman, 2008).

In fact, up to now, the optimal use of the atmospheric non thermal plasma sources needs further experimental research works and also modelling investigations in order to better understand the electro-dynamics processes and phenomena induced by the micro-discharges (Ebert&Sentman, 2008; Eichwald et al., 2008). In the frame work of the micro-discharge modelling, the obtained results (the streamer morphology and velocity, the production of charged and radical particles, the dissipated power) depend on the hydrodynamics physical model (Eichwald et al., 2006; Li et al., 2007) the discretisation method (Finite Difference Method: FDM, Finite Element Method: FEM, or Finite Volume Method: FVM) and the numerical solver used (Ducasse et al., 2007, 2010; Soria-Hoyo et al., 2008). Indeed, the solution of the micro-discharge fluid models requires high resolution numerical schemes in order to be able to consider the strong coupling between both the transport and field equations and the steep gradients of the charged particles evolution in a sharp and very fast ionizing wave (Soria-Hoyo et al., 2008). Therefore, the streamer dynamics modelling does not only depend on the selected physical model but also on the accuracy and the stability of the numerical algorithm. Furthermore, the parametric analysis of non thermal plasma discharge requires less time consuming and optimized numerical algorithms.

The present work is dedicated to the use for Finite Volume Method through the streamer discharge simulation and the gas dynamics simulation. We start with an overview on streamer and gas dynamics modelling followed by the model and numerical algorithms for streamers and gas dynamics; in this main part, we explicitly discuss how the model equations are discretized with the help of FVM. Finally, some results about both the streamer discharge and the gas dynamics simulations are shown; in the case of the streamer discharge we also discuss the validation of the present models from comparison between the experiment and the simulation.

2. Bibliographic overview on streamer discharge and gas dynamics modelling

Initials attempts at the numerical treatment of the electro-hydrodynamic model, in the case of gas discharges at atmospheric pressure, began in the 1960's with Davies et al. (Davies et al., 1964) and Ward (Ward, 1971). They used a first order method of characteristics in the context of the Finite Difference (FD) method. However, poor spatial resolution restricted their study to qualitative results. Towards the late 1970's, Davies improved the method of characteristics by introducing an iterative counterpart that increased the overall accuracy of the algorithm to second order (Davies et al., 1971, 1975, 1977). This method was adopted by several research teams (Kline, 1974), (Yoshida & Tagashira, 1976) and (Abbas & Bayle, 1980) and, as a result, it became the dominant method until the early 1980's. In 1981, Morrow and Lowke (Morrow & Lowke, 1981) presented a work that numerically integrated the system of continuity equations with the two-step Lax-Wendroff method of Roach (Roach, 1972). However, due to numerical dispersion and numerical instability, calculations were restricted to low density plasmas. Such restrictions were overcame by the introduction of the Finite Difference (FD) - Flux Corrected Transport (FCT) technique, originally developed by Boris and Book (Boris & Book, 1973) and extended to two dimensions by Zalezak (Zalesak, 1979). The FCT technique adds an optimal amount of diffusion and is remarkably stable in presence of sharp density gradients. In this context, FCT has become the most frequently used numerical method in streamer discharge modelling since Morrow introduced it for the first time to the gas discharge community in the 1980's (Morrow, 1981, 1982). Thus, Morrow was the first to offer an analysis (Morrow, 1982) for high density plasmas (up to electron density of 10^{13} cm^{-3}) and particular attention was paid to the selection of the Courant–Friedrichs–Lewy condition (CFL) (Courant et al., 1928). According to Morrow (Morrow, 1982) the CFL has to take values lower than 0.1 in order to quickly damp out any numerical oscillations resulting from steep density gradients. However, until the early 1980's, the FD based models did not exceed 1.5D description (1D for the continuity equations, and 2D for the electric field calculation in order to take into account the filamentary structure of the streamer: radial extension). By the middle of the 1980's, Dhali and Williams (Dhali & Williams, 1985) had launched the first fully two-dimensional simulation, using the FCT technique to solve the continuity equations. Thus, they elucidated several aspects of both positive and negative streamer phenomena. Subsequently, Kunhardt and Wu (Kunhardt & Wu, 1987) improved the FCT method and described a self-consistent numerical simulation of the formation and propagation of streamers in electropositive (N_2) and electronegative (N_2-SF_6) gases. Finally, an implicit version of FCT for gas discharge problems was presented by Steinle and Morrow (Steinle & Morrow, 1989). This new algorithm gave a threefold increase in the overall simulation speed because it was able to use a CFL ~ 1 while maintaining the scheme accuracy. However, this new method never became popular among the scientific community. In the mid-1990's, using the same model as Dhali and Williams (Dhali & Williams, 1985), Vittelo et al (Vittelo et al., 1993, 1994) reported a more accurate analysis of the negative streamer in N_2 with a quite small spatial resolution (2.5μm - 10μm). They also made the first simulations for streamer propagation in non-uniform gaps (point-to-plane electrode configuration) using a fully two-dimensional model. More systematic work on non-uniform gaps was also performed by Babaeva and Naidis (Babaeva & Naidis, 1996) (using the FCT technique), Kulikovsky (Kulikovsky, 1995a, 1995b) (using an optimized second-order Shurfetter-Gummel scheme) or Pancheshnyi and Starikovskii (Pancheshnyi & Starikovskii, 2003) (using a first-order upwind scheme). Another efficient second-order numerical scheme was introduced by Van Leer (Van Leer, 1979) and named the second order Monotonic Upwind-Centered Scheme For Conservation Laws (MUSCL) scheme. This algorithm was used through the Finite Volume Method (FVM) in the 3D modelling of high pressure micro-discharges in micro-cavities (Eichwald et al., 1998) in the 1.5D (Eichwald et al., 2006) and the 2D (Ducasse et al., 2007) modelling of the positive streamer propagation. At the beginning of the century (2000) a new approach to gas discharge modelling was presented in the works of Georghiou et al. (Georghiou et al., 1999, 2000) and Min et al. (Min et al., 2000, 2001) in which they used the Finite Element Method (FEM) to solve the electro-hydrodynamic model for parallel plate and wire-plate gaps. Based on a Finite Element Flux Corrected Transport algorithm (FEM-FCT) the simulations maintain the ability to handle steep gradients through the use of FCT, but also allow for the use of unstructured triangular cells. This method significantly reduces the number of unknowns, consequently reducing the computing time. Fine resolution is used only where it is necessary, enabling the model to be extended to a fully two-dimensional form and thereby making it possible to model complex geometries. Moreover, the first works that adapted the FEM to the charged carrier conservation equations were the papers of Yousfi et al. (Yousfi et al., 1994) and Novak and Bartnikas (Novak & Bartnikas, 1987) but the former restricted its application to 1.5D problems. Finally, a few works on 3D streamer discharge modeling using either FVM or FEM numerical schemes have been performed by Kulikovsky (Kulikovsky, 1998), Park et al. (Park et al., 2002), Akyuz et al. (Akyuz et al., 2003) Georghiou et al. (Georghiou et al., 2005) Pancheshnyi (Pancheshnyi, 2005) and Papageorghiou et al. (Papageorghiou et al., 2011).

3. Model and numerical algorithms for streamer discharge and gas dynamics

3.1 Model and equation discretisations with finite volume method (FVM) for streamer discharge and gas dynamics

In this work, streamer formation and propagation (streamer dynamics) is modelled using the first order electro-hydrodynamic model in the framework of the drift-diffusion approximation (Eichwald, 2006). Moreover, neutral dynamics is not taken into account; only the charged particle dynamics is considered. Thus, the equations involved in this model are the following:

$$\mathrm{div}\left(\varepsilon\vec{E}\right) = e\,(n_p - n_e - n_n)\,,\ \varepsilon = \varepsilon_0\varepsilon_r \tag{1}$$

$$\vec{E} = -\overrightarrow{\mathrm{grad}}V \tag{2}$$

$$\frac{\partial n_s}{\partial t} + \mathrm{div}\left(\vec{j}_s\right) = \sigma_s\left(E/n_g\right),\ \vec{j}_s = n_s\vec{v}_s\left(E/n_g\right),\ s{=}e,\ n,\ p \tag{3}$$

$$\vec{v}_s\left(E/n_g\right) = \vec{v}_{s-drift}\left(E/n_g\right) + \vec{v}_{s-diff}\left(E/n_g\right) = \mu_s\left(E/n_g\right)\vec{E} - \frac{D_s\left(E/n_g\right)}{n_s}\overrightarrow{\mathrm{grad}}n_s\,,\ s{=}e,\ n,\ p \tag{4}$$

In (1) (Maxwell-Gauss) $\vec{E}$ is the total electric field (due to geometry and space charge) e the absolute value of the electronic charge, n_e, n_p and n_n the electron, positive and negative ion densities, ρ_c the space charge density and ε_0 and ε_r the free space and relative permittivities; here, the relative permittivity is equal to 1. With regard to the continuity equations (3) of the density, subscript "s" stands for the electrons (e) or the positive (p) or negative (n) ions. Moreover, $\vec{v}_s$ is the velocity and σ_s the source term; both are functions of the reduced electric field E/n_g (n_g is the gas density) according to the local electric field approximation. The electron velocity is calculated using the classical drift-diffusion approximation (4), where μ_s and D_s are the mobility and the diffusion coefficients respectively.

The Maxwell-Gauss equation (1) is discretised with finite volume method (FVM). Thus, the equation is integrated on an elementary volume $\mathcal{V}_{ij}$ (cell) of the three-dimensional (3D) space. In addition, the Gauss-Ostrogradsky theorem (or divergence theorem) is used to transform the volume integration in surface integration. The Gauss-Ostrogradsky theorem is a mathematical statement of the physical fact that, in the absence of the creation or destruction of matter, the density within a region of space can change only by having it flow into or away from the region through its boundary. After resolution, the following equation is obtained:

$$S_{i+1/2}E_{i+1/2,j} - S_{i-1/2}E_{i-1/2,j} + S_{j+1/2}E_{i,j+1/2} - S_{j-1/2}E_{i,j-1/2} = \frac{\rho_{cij}\mathcal{V}_{ij}}{\varepsilon_0} \tag{5}$$

In the case of a z-axis of symmetry (Fig. 1), the elementary volume and surfaces in cylindrical coordinates are:

$$\mathcal{V}_{ij} = \pi\left(r_{i+1/2}^2 - r_{i-1/2}^2\right)\cdot dz_{cell}\,,\ \text{with}\ r_{i+1/2} = r_i + dr_i/2\,,\ r_{i-1/2} = r_i - dr_{i-1}/2\,,\ dz_{cell} = \frac{\left(dz_{j-1} + dz_j\right)}{2};$$

$$S_{i+1/2} = 2\pi r_{i+1/2}dz_{cell}\,,\ S_{i-1/2} = 2\pi r_{i-1/2}dz_{cell}\ \text{et}\ S_{j+1/2} = S_{j-1/2} = \pi\left(r_{i+1/2}^2 - r_{i-1/2}^2\right) \tag{6}$$

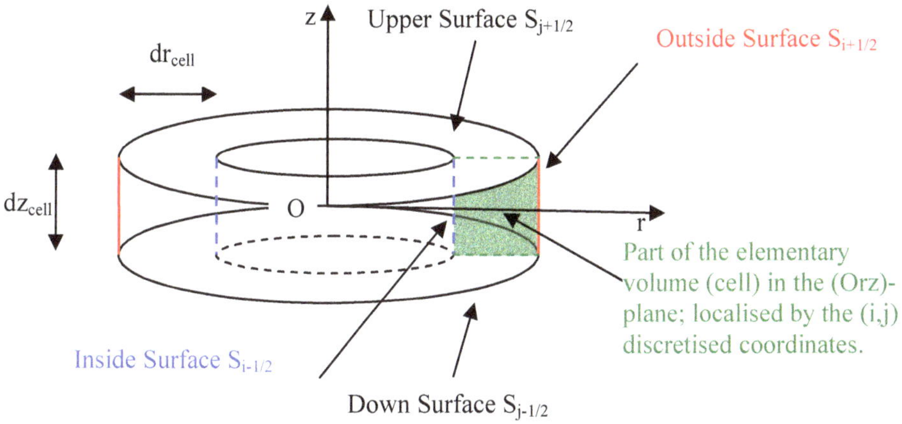

Fig. 1. Schematic representation of an elementary cell in the three-dimensional space. Due to the z-axial symmetry, the calculation on the half space is sufficient; thus computing time and memory size are saved.

In equation (5), the electric field components are expressed in function of the potential (7) to obtain equation (8) (FVM discretised Poisson equation); (7) is determined by first order finite difference of equation (2). Finally, (8) is ordered to generate the linear equation (9) for the (i,j) cell.

$$E_{i+1/2} = -\frac{V_{i+1} - V_i}{dr_i},\ E_{j+1/2} = -\frac{V_{j+1} - V_j}{dz_j} \tag{7}$$

$$\frac{S_{i+1/2}}{\mathcal{V}_{ij} dr_i}\left(V_{i+1} - V_{ij}\right) - \frac{S_{i-1/2}}{\mathcal{V}_{ij} dr_{i-1}}\left(V_{ij} - V_{i-1}\right) + \frac{S_{j+1/2}}{\mathcal{V}_{ij} dz_j}\left(V_{j+1} - V_{ij}\right) - \frac{S_{j-1/2}}{\mathcal{V}_{ij} dz_{j-1}}\left(V_{ij} - V_{j-1}\right) = -\frac{\rho_{cij}}{\varepsilon_0} \tag{8}$$

$$a_{ij} V_{i-1} + b_{ij} V_{i+1} + c_{ij} V_{ij} + d_{ij} V_{j-1} + e_{ij} V_{j+1} = f_{ij} \tag{9}$$

where $$a_{ij} = \frac{S_{i-1/2}}{\mathcal{V}_{ij} dr_{i-1}},\ b_{ij} = \frac{S_{i+1/2}}{\mathcal{V}_{ij} dr_i},\ c_{ij} = -\left(a_{ij} + b_{ij} + e_{ij} + d_{ij}\right),$$

$$d_{ij} = \frac{S_{j-1/2}}{\mathcal{V}_{ij} dz_{j-1}},\ e_{ij} = \frac{S_{j+1/2}}{\mathcal{V}_{ij} dz_j} \text{ and } f_{ij} = -\frac{\rho_{cij}}{\varepsilon_0}$$

For the whole elementary volumes of the 3D space, the equations rearrange in a matrix way as Ax=b, with A, x and b, respectively of dimension $n_r^2 \times n_z^2$, $n_r \times n_z$ and $n_r \times n_z$; x is the solution of the linear equation system. As regards the boundary conditions, we applied a Dirichlet or Neumann condition, following an electrode or an open space is considered.

The continuity equation for density of each charge species (3) is also discretised with FVM (like Poisson equation) and an explicit scheme. The explicit scheme calculates the state of a system at a later time from the state of the system at the current time. Thus, the discretised equation is the following:

$$\left(\overline{n_{i,j}}^{t+\Delta t} - \overline{n_{i,j}}^{t}\right) \mathcal{V}_{ij} = \left(\overline{n_{i-1/2,j} v_{r_{i-1/2,j}}}^{t} S_{i-1/2} - \overline{n_{i+1/2,j} v_{r_{i+1/2,j}}}^{t} S_{i+1/2}\right) \Delta t$$
$$+ \left(\overline{n_{i,j-1/2} v_{z_{i,j-1/2}}}^{t} S_{j-1/2} - \overline{n_{i,j+1/2} v_{z_{i,j-1/2}}}^{t} S_{j+1/2}\right) \Delta t + \sigma_{ij}^{t} \Delta t \tag{10}$$

The above continuity equations (10) and Poisson equation (9) are coupled through the source terms and the reduced electric field, which depends on the space charge density. Thus, the algorithms used have to be robust in order to prevent the development of non-physical oscillations or diffusion phenomena within the electro-hydrodynamic model.

The streamer development is described in a two-dimensional cylindrical (Orz) geometry, where (Oz) is associated with the streamer propagation axis, and (Or) with its radial extension. The next section is devoted to validating and comparing the efficiency of the algorithms to solve the continuity equations (10) and Poisson equation (9).

As regards the gas dynamic model, the system of equation bellow is used. The energy (11), momentum (12) and mass (13) continuity equations compose the model. The energy (thermal and kinetic), momentum and mass densities are respectively E, $\rho\vec{v}$, and ρ; moreover, P, $\vec{v}$, $\vec{j}_{th}$ and $\bar{\sigma}_{\mathcal{E}-\text{discharge}}$ are the pressure, velocity, thermal flux density, and the mean energy source term.

$$\frac{\partial \mathcal{E}}{\partial t} + \text{div}(\mathcal{E}\vec{v}) = -\text{div}(P\vec{v}) - \text{div}\left(\vec{j}_{th}\right) + \bar{\sigma}_{\mathcal{E}-\text{discharge}} \tag{11}$$

$$\frac{\partial(\rho v_r)}{\partial t} + \text{div}(\rho\vec{v}v_r) = -\frac{\partial P}{\partial r} \tag{12}$$

$$\frac{\partial(\rho v_z)}{\partial t} + \text{div}(\rho\vec{v}v_z) = -\frac{\partial P}{\partial z}$$

$$\frac{\partial \rho}{\partial t} + \text{div}\rho\,\vec{v} = 0 \tag{13}$$

The mean energy source term $\bar{\sigma}_{\mathcal{E}-\text{discharge}}(\vec{r}) = \frac{1}{t_1}\int_0^{t_1} f(E/n)\,\vec{j}_T(\vec{r},t)\cdot\vec{E}(\vec{r},t)dt$ is obtained via the streamer discharge simulation (previous model); f(E/n) is the distribution function of translation processes (depending on the reduced electric field) and $\vec{j}_T(\vec{r},t)$ the current density vector within the streamer discharge simulation. The mean energy is evaluated (for each cell of the calculation domain) on 150ns; the duration is negligible compared with the neutral gas dynamic duration process >1μs (shock wave). The thermal flux density $\vec{j}_{th} = -\lambda\,\overrightarrow{\text{grad}}(T)$ is expressed in function of the thermal conductivity coefficient and temperature gradient of the gas (air). The gas viscosity is not taking into account, there is no reactivity of the gas, the electrodes are at ambient temperature, and gliding conditions are applied to the electrodes. Finally, if the thermal transfer is not taking into account, the simulation diverges and there is no propagation wave.

One can notice the three equations (11) to (13) are based on the same structure as (3), the transport term in one side (left term) and source term on the other side (right term). Thus, the algorithm used to solve the equations is the same as (10).

3.2 Algorithm tests to solve the energy, momentum, density continuity and poisson equations

Several kinds of algorithms are presented to solve the two equation types of the model: continuity and Poisson equations. The algorithms have been tested in accuracy and computing time on special tests. For the continuity equations (or transport equations) we used the Davies' test (Davies & Niessen, 1990; Davies, 1992; Yousfi et al., 1994; Ducasse et al., 2010) in two directions of a cylindrical coordinate system (r, z). Six algorithms are tested: Upwind, Superbee Monotonic Upstream-centred Scheme for Conservation Law (MUSCL Superbee), Piecewise Parabolic Method PPM, ETBFCT, and Zalesak Peak Preserver (Ducasse et al., 2010). For the Poisson equation we compare the analytic solution to the numeric ones given by MUMPS (Direct method; not iterative) and SOR (Iterative method) (Amestoy, 2001, 2006; Fournié, 2010; Press, 2nd edition).

The Davies' test was first introduced by Davies and Niessen in order to compare the algorithm efficiency to solve one-dimensional continuity equations (14) ((15) is the FVM discretised form) without source term; what we write here is valid for energy, momentum and density continuity equations. The test is interesting for streamer modelling since it reproduces mathematically the behaviour of the streamer head propagation which is a fast ionizing wave that propagates steep density gradients in a sharp velocity field. Thus, the test is performed along a normalized z-axis [0, 1] divided into N=100 regular cells, and consists in propagating a square density profile (wave) $n(z,t)$ in a stationary oscillating velocity field, as shown in Fig. 2. Equations (16) and (17) give respectively the mathematical expressions for the density and the velocity profiles: This gives a velocity peak value at $z=0.5$ au (arbitrary unit) which is ten times greater than the values at the beginning and the end of the domain. The initial density of the square wave distribution $n(z,t=0)$ is enclosed between $z=0.05$ and $z=0.25$ with a constant value of 10au. In addition, the time step is chosen equal to 10^{-5}au which corresponds to a Courant-Friedrich-Levy number (CFL) equal to 10^{-2} ($\mathrm{CFL}=\frac{v_{max}\Delta t}{\Delta z}$).

$$\frac{\partial n}{\partial t}+\frac{\partial (n v_z)}{\partial z}=0 \tag{14}$$

$$\left(\overline{n_j}^{t+\Delta t}-\overline{n_j}^{t}\right)\mho_j=\left(\overline{n_{j-1/2}v_{z_{j-1/2}}}^{t}S_{j-1/2}-\overline{n_{j+1/2}v_{z_{j+1/2}}}^{t}S_{j+1/2}\right)\Delta t \tag{15}$$

$$\begin{cases} n(z,t=0)=10 \text{ if } 0.05\le z\le 0.25 \\ =0 \text{ elsewhere} \end{cases} \tag{16}$$

$$v_z(z) = 1 + 9\sin^8(\pi z) \quad (17)$$

Finally, the boundaries at z=0 and z=1 are periodic in the sense that any particle leaving the right side boundary enters at the left side; so that, after the period $T = \int_0^1 \frac{dz}{v_z(z)} \approx 0.59$, the transport density profile solution should be identical to the initial distribution $n(z,t=0)$. The comparison of the transported density profile with the exact solution at time t=T will determine the accuracy of the algorithms in handling discontinuities and steep gradients (for t grater than T see (Ducasse, 2010)). In order to quantify the algorithm accuracy, the mean absolute error (18) is calculated.

$$AE = \frac{1}{N}\sum_{j=1}^{N}\left|n_j^{analytic} - n_j\right| \quad (18)$$

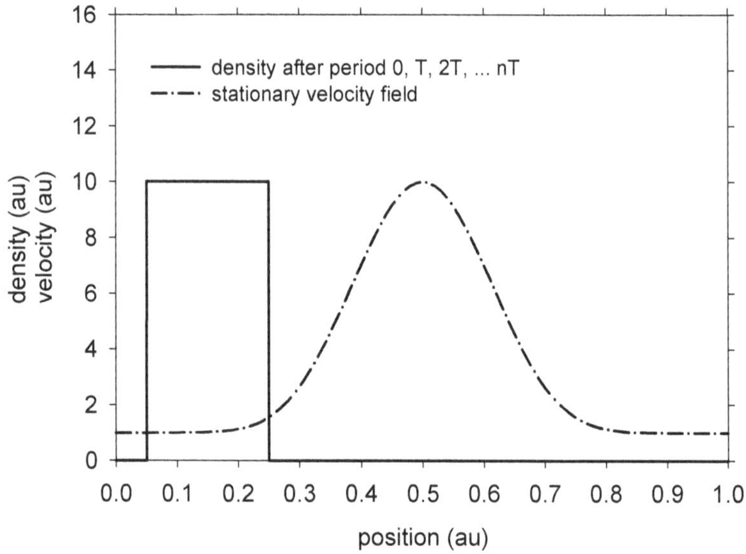

Fig. 2. Initial conditions in the Davies-test case for the density and the velocity field with the densities at times t=T.

Fig. 3 shows the numerical results obtained after one period, whereas Table 1 quantifies the performances of the algorithms in term of accuracy and time consumption. Moreover, with a CFL number equal to 10^{-2}, one period T of the square wave evolution corresponds to 59 070 iterations or time steps, which means that the flux correction is applied 59 070 times at the edges of each cells.

MUSCL Superbee, PPM, ETBFCT, Zalesak without Peak Preserver (ZNOPP) and Zalesak Peak Preserver (ZPP) algorithms generate similar results and nearly preserve the solutions from numerical diffusion and dispersion (Fig. 3). Nevertheless, after one period, the results clearly indicate that the PPM algorithm generates the most accurate solution since both the

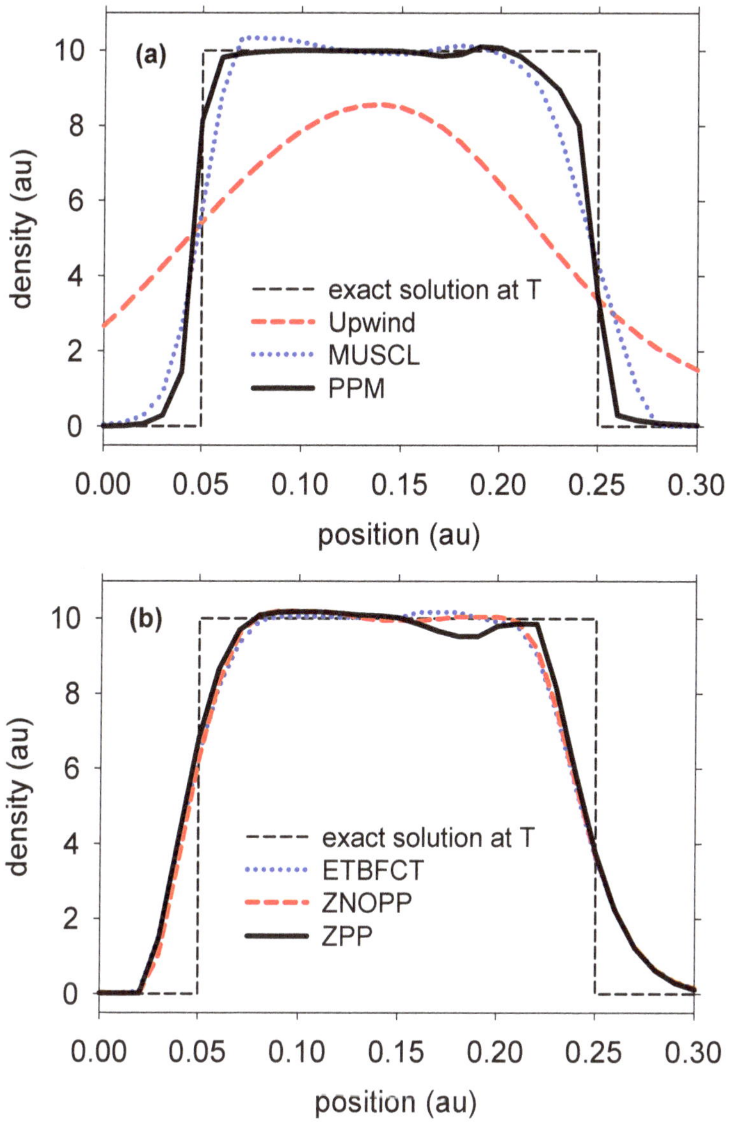

Fig. 3. Solutions obtained after one period (59 070 iterations) in the case of (a) Godunov-type schemes and (b) FCT technique. PPM is the most efficient.

steep gradients and the floor of the square wave are better reproduced (Fig. 3a). Furthermore, the PPM-AE is roughly two times lower in comparison with the other tested algorithms (Table 1); this is in accordance with the previous observations. For Upwind, we observe on Fig. 3a it introduces a large amount of numerical diffusion (comparable to physical diffusion) and its AE is ten times higher than PPM-AE.

The conservation criterion of the algorithms has been tested too. We observed that the particle conservation is verified for all the algorithms except for ZPP. Indeed, the particle number associated with ZPP increases of 1.1% as it was already emphasized by Morrow (Morrow, 1981).

The last two columns of Table 1, specify the absolute and relative computation time of the six algorithms. The processors used for this comparison are a 3GHz Intel® Pentium® IV with 512Ko of cache memory (768Mo of RAM) and a 2.8GHz Intel® quad-core Nehalem® EX with 8 Mo of cache memory for each processor (18Go of RAM). The results indicate ETBFCT is the fastest but also the less accurate (if Upwind is omitted). Therefore, by taken the ETBFCT values as the reference, it becomes possible to compare the gain of precision relatively to the computing time rise. For example, PPM is 2.32 times more accurate than ETBFCT but the computing time is multiplied by a factor 2.6 (2.5 with Nehalem®). In addition, ZPP is the less efficient since the precision increases by a factor 1.05 only, while the computation time increases by a factor 3.6 (4.3 with Nehalem®). MUSCL and ETBFCT show similar behaviours in term of computing time and accuracy. Moreover, we notice an important computing time fall from Pentium® IV to Nehalem® with a factor 5 for ETBFCT and about 3.5 to 4 for the others. In the particular case of Upwind we see a time consumption divided by 4 compared to ETBFCT, but more than 4 times less accurate than ETBFCT; some author still use the Upwind algorithm with a high space resolution to compensate the numerical diffusion (Pancheshnyi & Starikovskii, 2003; Urquijo et al., 2007).

Algorithm	AE after one period T (59070 iterations)	Computing time Intel-PentiumIV® 3GHz, 512Ko cache memory		Computing time Intel-Nehalem® 2.8GHz, 8Mo cache memory	
		Per iteration (μs)	Relative CPU time	Per iteration (μs)	Relative CPU time
Upwind	1.23	1.9	0.25	0.54	0.36
MUSCL	0.265	9.0	1.2	2.1	1.4
PPM	0.124	13	2.6	3.7	2.5
ETBFCT	0.288	7.5	1	1.5	1
ZNOPP	0.274	19	2.5	4.2	2.8
ZPP	0.275	22	3.6	6.5	4.3

Table 1. Absolute Error after one period (59 070 iterations) and mean computing time per iteration (no compilation option) for six numerical schemes.

Afterwards, the MUSCL Superbee algorithm is selected to be tested on the Kreyszig radial test (Kreyszig, 1999). Indeed, MUSCL with ETBFCT is the most interesting in terms of computing time and accuracy, with boundaries conditions simpler to implement than ETBFCT.

The Kreyszig radial test (Kreyszig, 1999) was used to observe the behaviour of the MUSCL Superbee algorithm for a physical quantity movement along the r-axis, both towards and away from the z-axis (symmetry axis). The test consists of the advection of a normalized square profile along the normalized radial direction, with a constant speed of $\pm 10^8$cm.s^{-1}, a mesh of 100 uniform cells, and a time step fixed at 10^{-11}s, which corresponds to a CFL equal to 0.1. In these conditions, Fig. 4 compares the numerical and analytical

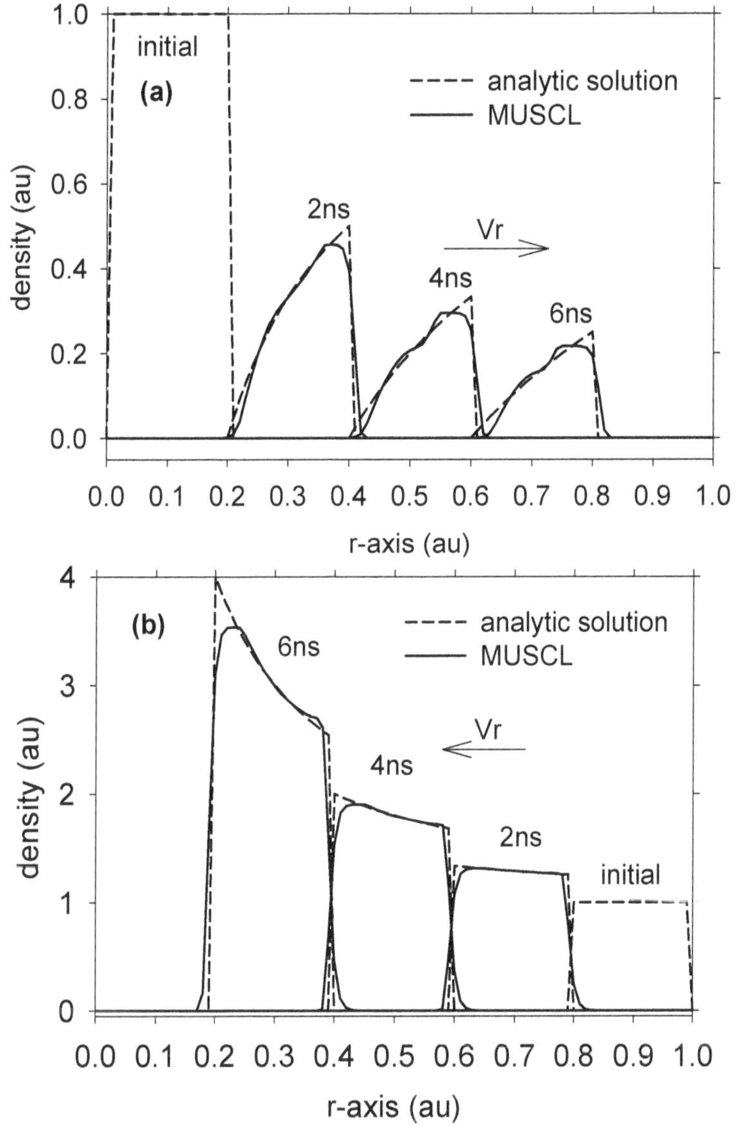

Fig. 4. Radial density solution given by the FVM-MUSCL algorithm, (a) for a positive radial velocity, and (b) for a negative radial velocity.

solutions, in a positive constant velocity field and a negative constant velocity field respectively. We note the correct behaviour of the algorithm, which introduce relatively small amounts of diffusion. The sharp corners are determined quite well despite the low spatial resolution. Thus, the solutions can be considered as satisfactory, all the more satisfactory the numerical solution tend to the analytic one with 1000 points and more accurate if a CFL=10^{-2} is added.

We conclude the MUSCL algorithm gives interesting results for both the absolute error of the solution and computing time. Moreover, we noted that the numerical solutions of the continuity equation tend to the analytic solutions in both axial and radial directions when the mesh step and (or) the time step are decreased (CFL).

At this stage we examine the numerical behaviour of SOR and MUMPS algorithms (Amestoy et al., 2001, 2006; Fournié et al., 2010; Press, 2nd edition) used to solve the Poisson equation (elliptic partial differential equation) without charge (i.e. Laplace's equation). We adopted a hyperbolic point-to-plane configuration for which the analytical potential field is known; the analytical solution was initially proposed by Eyring and first used by Morrow for streamers simulation (Eyring et al., 1928; Morrow & Lowke, 1997). Thus, the analytical solution is compared to the numerical one and the algorithm efficiency is quantified thanks to the relative error.

The curvature radius of the tip is 20μm and the inter-electrode space is 10mm; the applied voltage on the tip is equal to 9kV. The computational domain consists of a structured grid with none constant space cells in each direction. The limits of the domain are 19×19mm in z and r directions (cylinder of 19mm height and 19mm radius) and the number of nodal points in this domain is $n_r \times n_z = 307 \times 1186 = 364102$ (Δr_{min}=1μm and Δz_{min}=1μm). The spatial resolution along the z-axis is Δz=10μm from the plan until 200μm of the point, decrease down to Δz=1μm on the point, and increase again until the upper boundary; the radial step is progressively increased from Δr=1μm at the centre until the lateral boundary. Because of the symmetrical axis (Oz), the radial derivatives along the z-axis are set equal to zero. To perform the potential field comparisons, we set at each nodal point of the open boundaries (r=19mm and z=19mm) the analytical solution (Dirichlet conditions); we also performed the comparisons with a zero Neumann condition, since the simulation use this boundary condition.

The isopotential maps in Fig. 5 compare the analytic field with the MUMPS and SOR solutions. Very good agreements are observed between all results, as well as around the tip than in the whole domain; it can be quantitatively discussed using Fig. 6 for the solution along the z-axis. Thus, the relative error shows the direct method MUMPS gives the closest solution to the analytical one with a value less than 0.1%. For SOR, the error depends on the tolerance chosen for the convergence and the spectral radius (sr influence the speed convergence and solution accuracy); the convergence tolerance is defined as the maximum of the relative difference between the solution at iteration k and k-1. Indeed, with a 10^{-5} tolerance and an optimised convergence for the same tolerance (specific sr), we observe an error distribution lower than 0.5%, totally different from the MUMPS one, whereas with a 10^{-5} tolerance and an optimised convergence at 10^{-10} tolerance (other sr) the error tends to error distribution is constant on a large part of the inter-electrode space and decreases the MUMPS one; the result with a 10^{-10} tolerance and an optimised convergence for the same

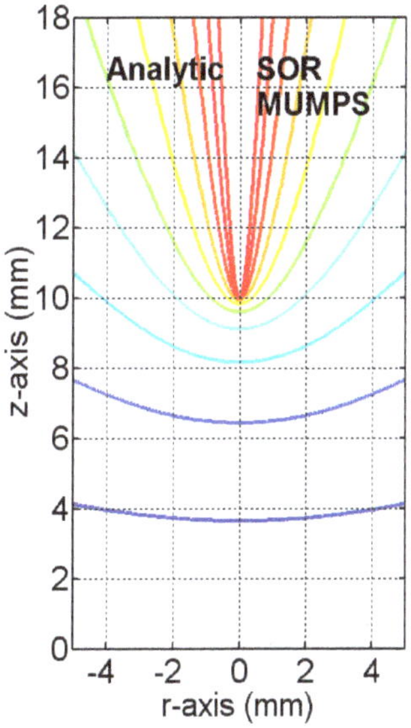

Fig. 5. Isopotentials given by the analytic solution (left side) and the numerical solutions (right side) which are identical for the two methods (SOR and MUMPS). NB: Zoom with a size of 5mm×18mm centred over the symmetry axis. The tip isopotential is 9000V and the interval between each isopotential is 900V

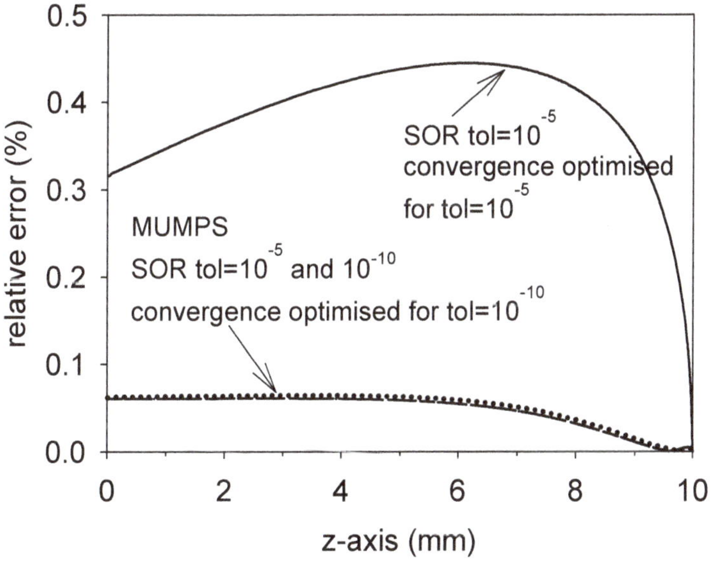

Fig. 6. Relative error between the analytic, and the SOR and MUMPS solutions along the z- axis.

tolerance is superimposed to the MUMPS one. In addition, for the three accurate results the reaching the point; it is due to the mesh, constant at the beginning and that starts to decrease close to the point. The contrary is observed if a constant step mesh is used: the error increases from the plan to the point (Kacem et al., 2011).

Table 2 gives the mean relative error calculated in the whole domain, the computing time and the number of iterations performed to satisfy the convergence tolerance fixed at 10^{-5}; these quantities are given for each method, several domain dimensions, several tolerance conditions, Dirichlet and Neumann conditions. Thus, SOR shows the highest mean relative error (0.30%) even if the value is acceptable. Moreover, one can notice MUMPS has the smallest mean relative error (4.2 10^{-2} %) since this direct method gives the nearest solution compared to the analytic one; SOR generate the same solution accuracy with both a 10^{-10} tolerance, or a 10^{-5} tolerance with an optimised spectral radius (convergence speed optimised for a 10^{-10} tolerance). Concerning the computing time, the SOR method needs between 17 and 34s to reach the specific tolerance criterion. MUMPS needs 830s to perform the direct calculation; thus, 830s are needed to construct the main matrix, analyse and performed the LU decomposition (Kacem et al., 2011), whereas only 0.23s are necessary to calculate the final product of matrices (if the LU decomposition is known, than only 0.23s is necessary to know the potential field). As regards the iteration number, SOR needs between 2400 and 5900 iterations in order to converge; the number depending on the tolerance.

Methods	Tolerance	Domain size (mm²)	Iterations	Mean Relative Error (%)	Computing Time (s) Intel-Nehalem® 2.8GHz, 8Mo cache memory
Dirichlet condition					
MUMPS	-	20×20	-	4.2×10^{-2}	8.3×10^{2} first time 0.23 (if a next time step)
SOR	10^{-5}		2438	0.30 (sr optimised at tolerance 10^{-5})	17
	10^{-5}		3510	4.4×10^{-2} (sr optimised at tolerance 10^{-10})	20
	10^{-10}		5902	4.2×10^{-2} (sr optimised at tolerance 10^{-10})	34
Neumann condition					
SOR	10^{-10}	10×20	10830	40	62
		20×20	13894	11	85
		30×30	10937	3.9	72
		40×40	14353	2.6	1.0×10^{2}

Table 2. Quantitative criteria to compare the method efficiency for Dirichlet and Neumann conditions on the open boundaries (r=rmax and z=zmax).

Afterwards, the point used is not hyperbolic anymore, so the potential at the boundaries of the domain is unknown (a Dirichlet condition is not possible) that is why we impose Neumann conditions. Thus, it is important to check the impact of a Neumann condition on the previous hyperbolic point and compare the numerical solution to the analytic one; the test is performed with the SOR method. Table 2 shows the boundary position in the previous tests (Dirichlet) is $2x2mm^2$. If we choose a $1x2mm^2$ domain, than we observe the mean relative error is too high (40%). But with a $2x2mm^2$ domain, the error decreases down to 11%; the value is acceptable. With a $3x3mm^2$ domain the error is still improved (3.9%) but at $4x4mm^2$ the improvement starts to slow down (2.6%).

For the boundary positions of the next simulation we use the $2x2mm^2$ domain with a tolerance of 10^{-5} and a spectral radius obtain for an optimised convergence at a tolerance of 10^{-10}; it is a compromise between the solution accuracy and the computing time.

To finish, the code was compared to another one developed with finite element method. We found a very good agreement with less than 10% of difference (Ducasse et al., 2007).

4. Simulation results

The first part presents the results obtain with the streamer discharge simulation and the second the results obtained with the gas dynamic simulation. The streamer ionizing wave and the shock wave involved by the streamer discharge are simulated via a PRHE MPI parallelised streamer code; the simulator is able to reproduce both phenomena thanks to efficient algorithms we previously studied (no commercial software is able to do it). Both the streamer and gas dynamics simulations are 3D simulations with axial symmetry; cylindrical coordinates are used. Thus, only half space of a 2D domain (plane) is solved.

The electric discharge is obtained with a point-to-plane electrode system (see the algorithm test part above). The tip curvature radius is 20µm, the gap is 10mm, and the discharge occurs in air at atmospheric pressure; a time varying positive potential (reaches a 9kV maximum on 60ns about) is applied to the point (Fig. 9). The transport of charged particles, their reactivity, their influence on the electric field and the photoionisation phenomenon are taken into account; the air neutral particles are fixed. Moreover, the reaction scheme is composed of 28 reactions the reader can find in (Bekstein et al., 2010) plus 10 ionic recombination reactions and 15 reactions with metastables (Table 3); so the reaction scheme is composed of 53 reactions. Finally, the reaction scheme is composed of electrons, two negative ions (O^- and O_2^-) seven positive ions (N_2^+, O_2^+, N^+, O^+, N_4^+, O_4^+ and $N_2O_2^+$) two radical atoms (O, N) and three metastables ($N_2(A)$, $N_2(a')$ and $O_2(a)$).

Electronic density and electric field are shown; the simulation results are compared with experimental ones for several inter-electrode space values. The simulation code has been parallelised and the computing time is given varying several parameters.

Fig. 7 and Fig. 8 show the electric field and electronic density evolutions from 20ns up to 200ns. We observe the primary streamer starts its propagation at 20ns about, i.e. when the applied voltage reaches 3.8kV (Fig. 9). At this time, the space charge formation near the positive point is responsible for the little current peak which appears in the current curve in Fig. 9. The streamer arrives at the plane at 87.5ns about, and corresponds to the maximum calculated current peak. Moreover, a secondary front propagates from the point (the mechanism is different from an ionising wave) in the same time than the streamer propagation and after, during the relaxation phase (beyond 87.5ns); but the speed of this second front is definitely slower. More details about the streamer mechanism, the radical production are available in (Eichwald et al., 2008, 2011).

Ion-Ion recombination (Kossyi et al., 1992)
$N_2^+ + O^- \rightarrow N_2 + O$ $N_2^+ + O_2^- \rightarrow N_2 + O_2$ $O_2^+ + O^- \rightarrow O_2 + O$ $O_2^+ + O_2^- \rightarrow 2\,O_2$ $N_4^+ + O^- \rightarrow 2\,N_2 + O$ $N_4^+ + O_2^- \rightarrow 2\,N_2 + O_2$ $O_4^+ + O^- \rightarrow 2\,O_2 + O$ $O_4^+ + O_2^- \rightarrow 3\,O_2$ $N_2O_2^+ + O^- \rightarrow N_2 + O_2 + O$ $N_2O_2^+ + O_2^- \rightarrow N_2 + 2\,O_2$
Metastable reactions (Yousfi & Benabdessadok, 1996; Kossyi et al., 1992)
$e^- + N_2 \rightarrow e^- + N_2(A)$ $e^- + N_2 \rightarrow e^- + N_2(a')$ $e^- + O_2 \rightarrow e^- + O_2(a)$ $N_2(A) + N_2 \rightarrow 2N_2$ $N_2(A) + O_2 \rightarrow N_2 + O_2$ $N_2(A) + O_2 \rightarrow O_2(a) + N_2$ $N_2(A) + O_2 \rightarrow 2\,O + N_2$ $N_2(A) + N \rightarrow N + N_2$ $2\,N_2(a') \rightarrow N_4^+ + e^-$ $N_2(a') + N_2(A) \rightarrow N_4^+ + e^-$ $N_2(a') + O_2 \rightarrow 2\,O + N_2$ $O_2(a) + N_2 \rightarrow N_2 + O_2$ $O_2(a) + O_2 \rightarrow 2O_2$

Table 3. Ion-Ion recombinaison and metastable reactions taken into account in the streamer reaction scheme.

A first comparison between simulation and experiment (through the current) was done at one specific potential and inter-electrode space (Eichwald at al., 2008). Here, the simulated currents are compared with the experimental ones at 9, 10, 11mm inter-electrode space and a 9kV DC applied voltage (the power is increased by hand to reach 9kV, afterward we observe a streamer discharge quite periodically). Thus the experimental applied potential shape is different, but the streamer channel is filiform in both cases (simulation and experiment; no branching phenomenon is observed).

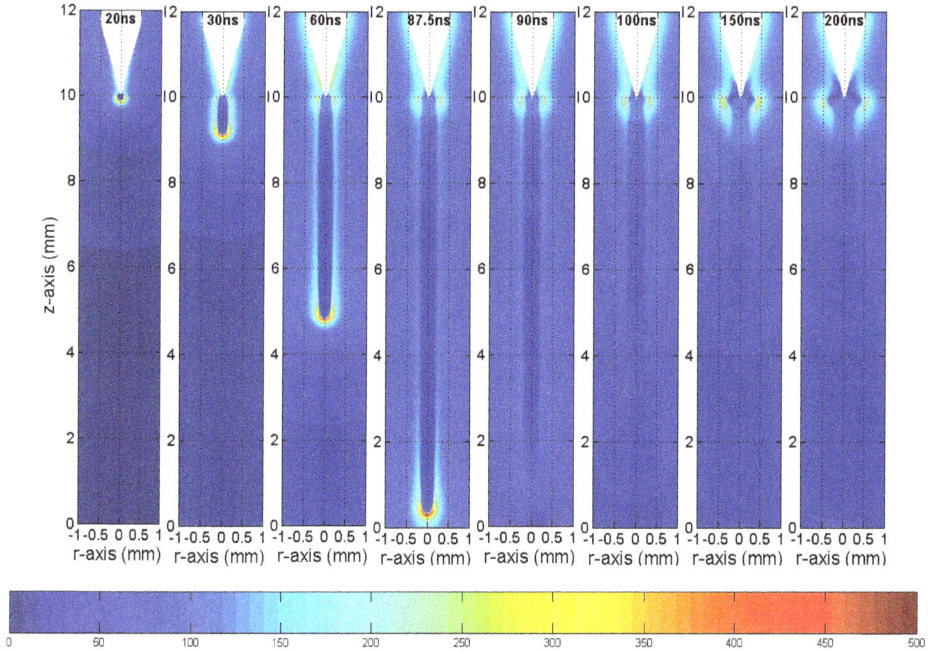

Fig. 7. Reduced electric field (Td) distribution as a function of time.

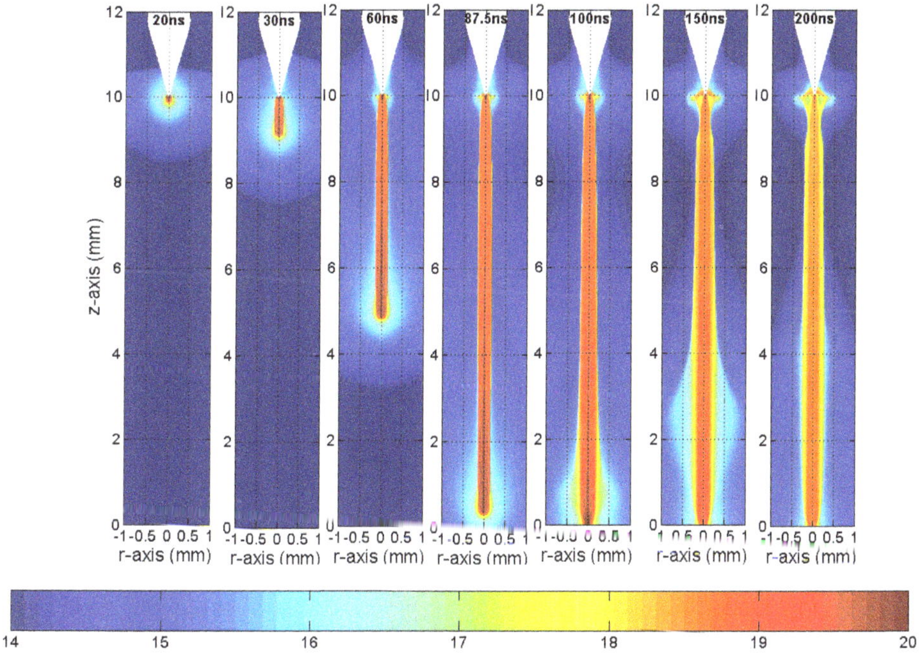

Fig. 8. Electron density distribution ($\log_{10}$ scale; m^{-3}) as a function of time.

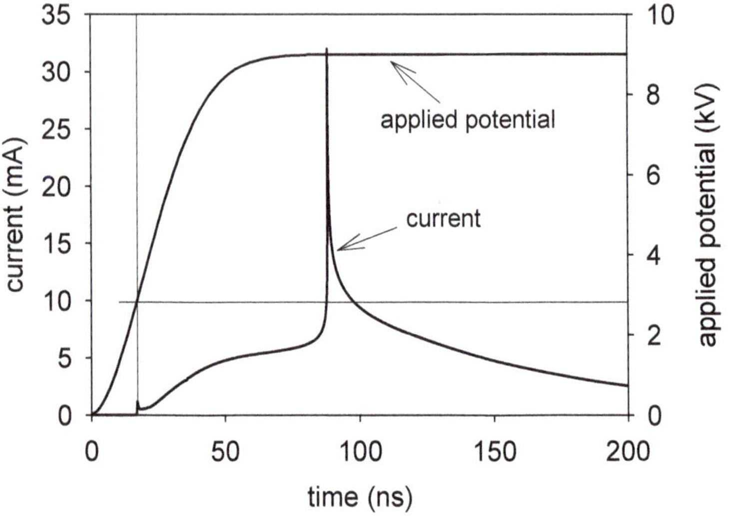

Fig. 9. Current and applied potential as a function of time; the streamer propagation starts at 20ns with a 2.8kV applied potential.

Fig. 10 shows there is a difference on the main peak of 10mA about at 9mm, and increases with the inter-electrode space; nevertheless the orders of magnitude are the same. Moreover, the bump observed on the experimental curve at 9mm (due to the secondary streamer) is not visible at 10 and 11mm; it seems the phenomenon is of the same amplitude as the primary streamer, but not at all visible on the simulated curves (even at 9mm); it could explain the

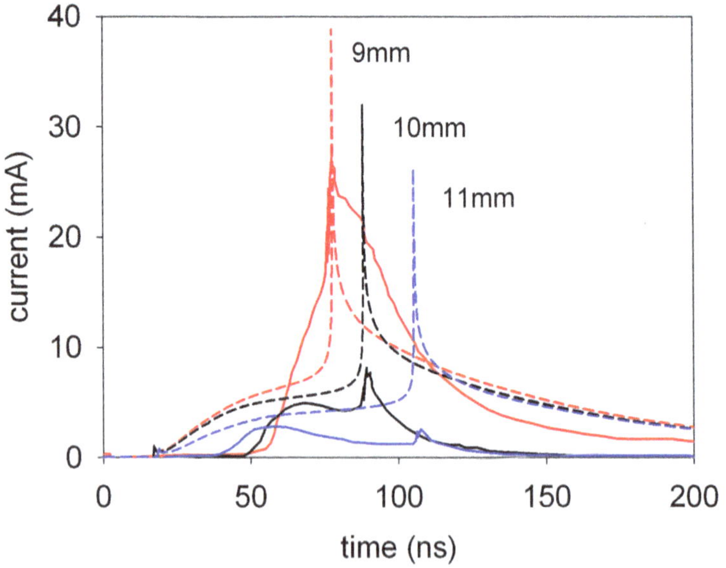

Fig. 10. Comparison (--) simulated - (-) experimental results for three inter-electrode spaces.

important current gap observed between 10 and 9mm we do not see on the simulated current. In addition, at 10 and 11mm we observe an important bump during the streamer propagation phase not present at 9mm; whereas the simulation always shows the same light bump (not depends on the gap size). Concerning the phenomenon duration, the configurations at 11mm are closer. Nevertheless, the fact the applied potential conditions are different for both the simulation and the experiment are responsible (may be in part) for the differences we observe.

The simulation results are generated via a MPI parallelised streamer code developed by the PRHE group (Laplace laboratory). Table 3 shows the calculation performances obtained on a Altix cluster ICE 8200 of 352 nodes named Hyperion (Toulouse University); each node is composed of two Nehalem EX quad-core processors at 2.8 GHz with 8 Mo of cache memory per processor, and 36 Go of RAM. Thus, the computing time increases from 9 to 11mm, but not linearly. Indeed, from 9 to 10, 10 hours more are necessary whereas from 10 to 11mm less than 2 hours are necessary; this has to be correlated with the time iteration numbers. Indeed, when the inter-electrode gap increases it is like the applied potential decreases; consequently the physical quantity gradients are lower and the iterative parts of the code converge faster. We do the same observation with SOR: SOR is faster than MUMPS for one iteration; if the gradients are lower enough to make SOR converge at one iteration than the calculation time will be certainly lower.

Inter-electrode Size (mm)	Mesh Definition $n_r \times n_z$ & Point Applied Potential (kV)	Potential Solver	Processor Number	Time Iteration Number	Computing Time to Generate the Result Intel-Nehalem® 2.8GHz, 8Mo cache memory	
					Total (h:min:s)	Per time iteration (s)
9	307×1190; 9	MUMPS	16	742174	69:20:29	0.33
10	307×1186; 9	MUMPS		726190	79:31:58	0.39
10	307×1186; 9	SOR		726771	41:53:55	0.21
11	307×1298; 9	MUMPS		728497	81:17:39	0.40

Table 4. Computing time necessary to generate the result with a PRHE MPI parallelised Streamer code at 16 Processors, for three inter-electrode dimensions, and MUMPS and SOR potential solvers. The calculations were made on the Hyperion supercomputer. The simulation generates a faster result when SOR solver is implemented.

The streamer discharge effect on the air gas dynamics is shown through the pressure Fig. 11; complementary information can be found in (Eichwald et al., 2011). The local air heated locally on the point by the electric discharge reaches a temperature of thousands of Kelvin. Thus, the thermal shock generates pressure gradients (Fig. 11) that induces wave propagations by a successive local compression - expansion mechanism. The gas expansion is characterised by a spherical and cylindrical shock wave. Indeed, the streamer discharge start to heat locally the air on the tip, forming afterwards a spherical wave, superimpose to the heat in the channel, forming a cylinder wave. Such spherical waves were already experimentally observed using the laser Schlieren technique (Ono & Oda, 2004). In addition, the simulation shows the spherical shock wave propagates at the air sound speed: between 1μs and 4μs the wave front propagates on 1200μm so a velocity of 400m/s.

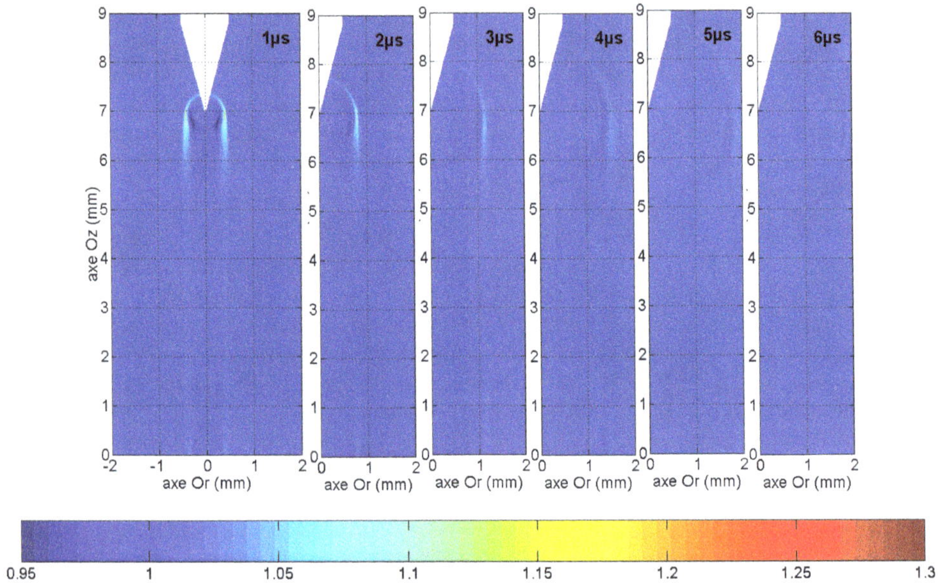

Fig. 11. Shock wave pressure distribution ($\times 10^5$ Pa) as a function of time.

5. Conclusion

In this chapter we have shown how the Finite Volume Method can be used for the discretisation of the transport and Poisson equations, allowing the simulations of streamer discharge development and the associated gas dynamics. It is clear that FVM is attractive since we work directly on elementary volumes that make sense from a physical point of view. Moreover, through a very carefully study, we showed important results as regards the algorithm accuracy, the algorithm convergence (SOR iterative method) the boundary conditions, and the computing time. In the case of the algorithms tested in our research group we have shown first that MUSCL Superbee is the most efficient to treat the conservation laws (Energy, momentum and density); we have also shown that SOR and MUMPS are both interesting in term of computing time. In fact SOR is efficient from a time step to another when the space charge varies slowly (it is the case at relatively low applied potential); MUMPS is efficient if the space charge varies rapidly (the computing time remaining practically the same). In addition, the PRHE-MPI-parallelised code is efficient with 16 processors on the Hyperion HPC system (2.8GHz Intel-Nehalem®; 8Mo cache memory). At this stage, it is possible to do parametric studies since the calculation is fast enough (around three days at 10mm inter-electrode distance and 200ns for the duration). Nevertheless, some improvements on the discharge model still have to be performed from a physical point of view. Indeed, we have shown the experimental current behaves differently when varying the inter-electrode space, whereas the simulation always showed the same shape. May be one of the improvement would be to take into account the local modifications of the air gas properties due to the streamer discharge by the direct coupling of gas dynamics and streamer discharge dynamics.

6. Acknowledgment

This work was granted access to the HPC resources of CALMIP under the allocation 2011-[p0604]. We thanks the calculation engineer Nicolas Renon for the helpful collaboration.

7. References

Abbas, I., & Bayle, P. (1980). A critical analysis of ionising wave propagation mechanisms in breakdown. *J. Phys D: Appl. Phys.*, vol. 13, pp. 1055–1068

Akyuz, M., Larsson, A., Cooray, V., & Strandberg, G. (2003). 3D simulations of streamer branching in air. *J. Electrostat.*, vol. 59, p. 115, 2003.

Amestoy, P. R., Duff, I. S., Koster, J., & L'Excellent, J. Y. (2001). A fully asynchronous multifrontal solver using distributed dynamic scheduling, *SIAM Journal of Matrix Analysis and Applications*, Vol 23, No 1, pp 15-41

Amestoy, P. R., Guermouche, A., L'Excellent, J. Y., & Pralet, S. (2006). Hybrid scheduling for the parallel solution of linear systems. *Parallel Computing* Vol 32 (2), pp. 136-156

Babaeva, Yu. N., & Naidis, G. V. (1996). Two-dimensional modelling of positive streamer dynamics in non-uniform electric fields in air. *J. Phys. D: Appl. Phys.*, vol. 29, pp. 2423–2431

Bekstein, A., Yousfi, M., Benhenni, M., Ducasse, & O., Eichwald, O. (2010). Drift and reactions of positive tetratomic ions in dry, atmospheric air: Their effects on the dynamics of primary and secondary streamers. *Journal of Applied Physics*, Vol. 107, N 10., pp. 103308 - 103308, ISSN 0021-8979

Boris, J. P., & Book, D. L. (1973). Flux-Corrected Transport I. SHASTA, A Fluid Transport Algorithm That Works. *J. Comp. Phys.*, vol. 11, pp. 38–69

Courant, R., Friedrichs, K., & Lewy, H. (1928). Über die partiellen Differenzengleichungen der mathematischen Physik. *Mathematische Annalen* vol. 100, no. 1, pp. 32–74

Davies, A. J., Evans, C. J., & Llewellyn Jones, F. (1964). Electrical Breakdown of gases: the spatio-temporal growth of ionization in fields distorted by space charge. *Proc. Roy. Soc.*, vol. 281, pp. 164–183

Davies, A. J., Davies, C. S., & Evans, C. J. (1971). Computer simulation of rapidly developing gaseous discharges. *Proc. IEE Sci. Meas. Technol.*, vol. 118, pp. 816–823

Davies, A. J., Evans, C. J., & Woodinson, P. M. (1975). Computation of ionization growth at high current densities. *Proc. IEE Sci. Meas. Technol.*, vol. 122, pp.765–768

Davies, A. J., Evans, C. J., Townsend, P., & Woodinson, P. M. (1977). Computation of axial and radial development of discharges between plane parallel electrodes. *Proc. IEE Sci. Meas. Technol.*, vol. 124, pp.179–182

Davies, A. J., & Niessen, W. (1989). The solution of the continuity equations in Ionization and Plasma growth. *Physics and Applications of Pseudosparks*, M. A. Gundersen and G. Schaefer, Eds. New York: Plenum, p. 197

Davies, A. J. (1992). Numerical solutions of continuity equations and plasma growth. Workshop, plasma chaud et modélisation des décharges, CIRM, Luminy, pp. 45–53

Dhali, S. K., & Williams, P. F. (1985). Numerical simulation of streamer propagation in nitrogen at atmospheric pressure. *Phys. Rev. A.*, vol. 31, p. 1219

Ducasse, O., Papageorghiou, L., Eichwald, O., Spyrou, N., & M. Yousfi (2007). Critical analysis on two-dimensional point to plane streamer simulations using the finite element and finite volume methods. *IEEE transactions on plasma science*, 35, issue 5, part 1, pp1287-1300

Ducasse, O.; Eichwald, O. & Yousfi, M. (2010). High order eulerian schemes for fluid modelling of streamer dynamics. *Advances and Applications in Fluid Mechanics*, vol.8, No.1,pp. 1-31, ISSN 0973-4686

Ebert, U., & Sentman, D. (2008). Streamers, sprites, leaders, lightning: from micro- to macroscales. *J. Phys. D: Appl. Phys.* 41: 230301

Eichwald, O., Bayle, P., Yousfi, M., Jugroot, M. (1998). Modeling and three-dimensional simulation of the neutral dynamics in an air discharge confined in a microcavity. Part I and II. *J. Appl. Phys.*, 84 (9), pp. 4704-4726

Eichwald, O., Ducasse, O., Merbahi, N., Yousfi, M., & Dubois, D. (2006). Effect of fluid order model on flue gas streamer dynamics, *J. Phys. D: Appl. Phys.*, vol. 39, pp. 99–107

Eichwald, O., Ducasse, O., Dubois, D., Abahazem, A., Merbahi, N., Benhenni, M., & Yousfi, M. (2008). Experimental analysis and modelling of positive streamer in air: towards an estimation of O and N radical production. *J. Phys. D: Appl. Phys.* 41, doi:10.1088/0022-3727/41/23/234002

Eichwald, O., Yousfi, M., Ducasse, O., Merbahi, N., Sarrette, J. P., Meziane, M., & Benhenni, M. (2011). Electro-Hydrodynamics of Micro-Discharges in Gases at Atmospheric Pressure. *InTech*

Eyring, C. F., Mackeown, S. S, & Millikan, R. A. (1928). Fields currents from points. *Physical Review*, vol 31, pp. 900-909

Fournié, M., Renon, N., Renard, Y., & Ruiz, D. (2010). CFD parallel simulation using Getfem++ and Mumps. Euro-Par 2010 - *Parallel Processing*, 16th International Euro-Par Conference

Fridman, G., Friedman, G., Gutsol, A., Shekhter, A. B., Vasilets, V. N., & Fridman, A. (2008). Applied plasma medicine. *Plasma Process. Polym.* 5, pp. 503-533

Georghiou, G. E., Morrow, R., & Metaxas, A. C. (1999). An improved finite-element flux corrected transport algorithm. *J. Comp. Phys.*, vol. 148, pp. 605–620

Georghiou, G. E., Morrow, R., & Metaxas, A. C. (2000). A two-dimensional finite element flux corrected transport algorithm for the solution of gas discharge problems. *J. Phys. D: Appl. Phys.*, vol. 33, pp. 2453–2466

Georghiou, G. E., Papadakis, A. P., Morrow, R., & Metaxas, A. C. (2005). Numerical modelling of atmospheric pressure gas discharges leading to plasma production. *J. Phys. D : Appl. Phys.*, vol. 38, pp. 303–328

Kacem, S., Eichwald, O., Ducasse, O., Renon, N., Yousfi, M., & Charrada, K. (2011). Full multi grid method for electric field computation in point-to-plane streamer discharge in air at atmospheric pressure. *Journal of Computational Physics*. doi:10.1016/j.jcp.2011.08.003

Kim, H. H. (2004). Non thermal plasma processing for air pollution control: A historical review, current issues and future prospects. *Plasma Process. Polym.*, pp. 191-110

Kline, L. (1974). Calculations of discharge initiation in overvolted parallel-plane gaps. *J. Appl. Phys.*, vol. 45, pp. 2046–2054

Kossyi, I. A.; A. Yu. Kostinsky, A. A. Matveyev, and V. P. Silakov (1992). Kinetic scheme of the non-equilibrium discharge in nitrogen-oxygen mixtures. *Plasma Sources Sci. Technol.* 1, pp. 207-220.

Kreyszig, E. (1999). *Advanced engineering mathematics*. John Wiley

Kulikovsky, A. A. (1995). Two-dimensional simulation of the positive streamer in N_2 between parallel-plate electrodes. *J. Phys. D: Appl. Phys.*, vol. 28, pp. 2483–2493

Kulikovsky, A. A. (1995). A More Accurate Scharfetter-Gummel Algorithm of Electron Transport for Semiconductor and Gas Discharge Simulation. *J. Comp. Phys.*, vol. 119, p149

Kulikovsky, A. A. (1987). Three-dimensional simulation of a positive streamer in air near curved anode. *Phys. Lett. A*, vol. 245, pp. 445–452

Kunhardt, E. E., & Wu, C. (1987). Towards a more accurate flux corrected transport algorithm. *J. Comp. Phys.*, vol. 68, pp. 127–150

Laroussi, M. (2002). Nonthermal decontamination of biological media by atmospheric-pressure plasmas: review, analysis and prospects. *IEEE Trans. Plasma. Sci.* 30 (4) pp. 1409-1415

Li, C., Brok, W. J. M., Ebert, U., & Van Der Mullen, J. J. A. M. (2007). Deviation from local field approximation in negative streamer head. *J. Appl. Phys.* 101: 123305

Marotta, E., Callea, A., Xianwen Ren, Rea, M., Paradisi, C. (2007). Mechanistic study of pulsed corona processing of hydrocarbons in air at ambient temperature and pressure. *International Journal of Plasma Environmental Science & Technology* 1 (1), pp. 39-45

Min, W. G., Kim, H. S., Lee, S. H., & Hahn, S. Y. (2000). An investigation of FEM-FCT method for Streamer Corona Simulation. *IEEE Trans. Magn.*, vol. 36, pp. 1280-1284

Min, W. G., Kim, H. S., Lee, S. H., & Hahn, S. Y. (2001). A study on the streamer simulation using adaptive mesh generation and FEM-FCT. *IEEE Trans. Magn.*, vol. 37, pp. 3141-3144

Morrow, R., & Lowke, J. J. (1981). Space-charge effects on drift dominated electron and plasma motion. *J. Phys D: Appl. Phys.*, vol. 14, pp. 2027–2034

Morrow, R. (1981). Numerical solution of hyperbolic equations for electron drift in strongly non-uniform electric fields. *J. Comput. Phys.* 43, pp. 1-15

Morrow, R. (1982). Space-charge effects in high-density plasmas. *J. Comp. Phys.*, vol. 46, pp. 454–461

Morrow, R., & Lowke, J. J. (1997). Streamer propagation in air. *J. Phys. D, Appl. Phys.*, vol. 30, no. 4, pp. 614–627

Moreau, E. (2007). Airflow control by non-thermal plasma actuators. *J. Phys. D: Appl. Phys.* 40, pp. 605-636

Novak, J. P., & Bartnikas, R. (1987). Breakdown model of a short plane-parallel gap. *J. Appl. Phys.*, vol. 62, p. 3605

Ono, R. & Oda, T. (2004). Visualization of Streamer Channels and Shock Waves Generated by Positive Pulsed Corona Discharge Using Laser Schlieren Method. *Japanese Journal of Applied Physics*, vol. 43, No. 1, pp. 321–327

Pancheshnyi, S. V., & Starikovskii, A. Yu. (2003). Two-dimensional numerical modelling of the cathode-directed streamer development in a long gap at high voltage. *J. Phys. D: Appl. Phys.*, vol. 36, p. 2683.

Pancheshnyi, S. V. (2005). Role of electronegative gas admixtures in streamer start, propagation and branching phenomena. *Plasma Sources Sci Technol.*, vol. 14, pp. 645–653

Papageorgiou, L., Metaxas, A. C. & Georghiou, G. E. (2011). Three-dimensional numerical modelling of gas discharges at atmospheric pressure incorporating photoionization phenomena. *Journal of Physics D: Applied Physics*, vol. 44, 045203

Park, J. M., Kim, Y. H., & Hong, S. H. (2002). Three-dimensional numerical simulations on the streamer propagation characteristics of pulsed corona discharge in a wire-cylinder reactor. *8th Int. Symp. On High Pressure Low Temperature Plasma Chemistr* , Estonia, p. 104

Press, H. W., Teukolsky, S. A., Vetterling, W. T., & Flannery, B. P. (1972). *Numerical recipes in fortran, the art of scientific computing*. Cambridge university press, second edition

Roache, P. J. (2008). *Computational Fluid Dynamics*. Albuquerque, NM: Hermosa

Soria-Hoyo, C., Pontiga, F., Castellanos, A. (2008). Two dimensional numerical simulation of gas discharges: comparison between particle-in-cell and FCT techniques. *J. Phys. D: Appl. Phys.* 41: 205206

Starikovskaia, S. M. (2006). Plasma assisted ignition and combustion. *J. Phys. D: Appl. Phys.* 39, pp. 265-299

Steinle, P. & Morrow, R. (1989). An implicit flux-corrected transport algorithm. *J. Comp. Phys.*, vol. 80, pp. 61–71

deUrquijo, J., Juarez, A. M., Rodriguez-Luna, J. C., & Ramos-Salas, J. S. (2007). A Numerical Simulation Code for Electronic and Ionic Transients From a Time-Resolved Pulsed Townsend Experiment. *IEEE transactions on plasma science* 35, issue 5, part 1, pp1204-1209, ISSN 0093-3813

Van Leer, B. (1979). Toward the ultimate conservative difference scheme. V. A second order sequel to Godunov's method. *J. Comp. Phys*, vol. 32, pp. 101–136

Vitello, P. A., Penetrante, B. M., & Bardsley, J. N. (1993). Multi-dimensional modelling of the dynamic morphology of streamer coronas. *Non Thermal Techniques for Pollution Control*, NATO ASI Series, vol. G34, part A, ed. B. M. Penetrante and S. E. Schultheis, Springe-Verlag Berlin Heidelberg, pp. 249–272

Vitello, P. A., Penetrante, B. M., & Bardsley, J. N. (1994). Simulation of negative streamer dynamics in nitrogen. *Phys. Rev. E*, vol. 49, pp. 5574–5598

Ward, A. L. (1965). Calculation of electrical breakdown in air at near atmospheric pressure. *Phys. Rev.*, vol. 138, pp. 1357–1362

Yoshida, K., & Tagashira, H. (1976). Computer simulation of a nitrogen discharge at high overvoltages. *J. Phys D: Appl. Phys.*, vol. 9, pp. 491–505

Yousfi, M., Poinsignon, A., Hamani, A. (1994). Finite element method for conservation equation in electrical discharges areas. *J. Comp. Phys.*, vol. 113, 2, pp. 268–278

Yousfi, M., & Benabdessadok, M. D. (1996). Boltzmann equation analysis of electron-molecule collision cross sections in water vapor and ammonia. *J. Appl. Phys.*, 80 (12), pp. 6619-6631

Zalezak, S. T. (1979). Fully multidimensional flux-corrected transport algorithms for fluids. *J. Comp. Phys.*, 31, pp. 335–362

Mass Conservative Domain Decomposition for Porous Media Flow

Jan M. Nordbotten, Eirik Keilegavlen and Andreas Sandvin
University of Bergen
Norway

1. Introduction

Understanding flow in subsurface porous media is of great importance for society due to applications such as energy extraction and waste disposal. The governing equations for subsurface flow are a set of non-linear partial differential equations of mixed elliptic-hyperbolic type, and the parameter fields are highly heterogeneous with characteristic features on a continuum of length scales. This calls for robust discretization methods that balance the challenges in designing efficient and accurate methods. In this chapter we focus on a class of linear solvers for elliptic systems that aims at providing fast approximate solutions, preferably in one iteration, but fall back to being iterative methods with good convergence properties if higher accuracy is needed.

We consider flow of flow of a single fluid in a porous media, transporting a passive particle. This can for instance represent flow of a pollutant as a result of groundwater contamination. Governing equations for the flow will be presented in the next section. Analytical solutions to the flow problem can only be found in very special cases, and in general, numerical approximations must be sought. The primary numerical schemes for commercial simulations are control volume methods. These methods are formulated such that conservation of mass is ensured, which is considered crucial in applications. After discretizing, an elliptic equation needs be solved for the pressure. This process is computationally expensive and may constitute the majority of the simulation time.

The permeability (fluid conductivity of the rock) in subsurface porous media has a truly multiscale nature, with highly permeable pathways with significant correlation lengths. Hence the elliptic pressure equation will experience strong non-local effects, posing a challenge for linear solvers. Moreover, the permeability field constructed by geologists is highly detailed; the number of cells in the geo-model can easily be several orders of magnitude higher than what is feasible to handle in a flow simulation. The traditional approach to this problem has been to upscale the permeability, e.g. to compute a representative permeability on a coarser grid. For the pressure equation this gives a linear system that is much smaller and computationally cheaper to solve. The drawback is of course that details in the geological characterization may be lost during upscaling, and these details are known to have significant impact on transport. An alternative approach is offered by the so-called multiscale methods, which have been a highly popular research field in the last decade (Tchelepi & Juanes, 2008). Like upscaling, multiscale methods perform a coarsening to end up with a relatively small linear system to solve. However, a multiscale method also provides a mapping from the coarse solution onto the fine grid. This projected solution will not be equal to a direct solution on the fine grid, but the two solutions will share many properties; in particular, many multiscale methods provide a velocity field that is mass conservative on the fine scale. Hence it can be used to solve fine scale transport equations. Numerical experiments have shown that this strategy can be extremely effective and highly accurate when measured in metrics that are important for applications (Kippe, et al., 2008; Efendiev & Hou, 2009).

Despite the success of multiscale methods in porous media flow, the strategy has certain weaknesses. In this chapter we highlight the quality of the coarse operator: If this does not represent essential features of the flow field, the quality of the fine scale velocity field may be poor. In particular, long and high permeable pathways are difficult to capture in the coarse scale operator. A natural approach would therefore be to introduce a scheme that allows for iterations on the multiscale solution. The idea of a multi-level iterative method resembles domain decomposition, and in Nordbotten & Bjørstad, 2008, the equivalence between the multiscale finite volume method (Jenny, et al., 2003) and a special domain decomposition strategy was shown. The resulting iterative scheme was termed mass conservative domain decomposition (MCDD), and it can be classified as an additive Schwarz preconditioner with minimal overlap. Contrary to classical domain decomposition methods, MCDD will produce solutions that are mass conservative at any iteration step, thus it is not necessary to reduce the pressure residual to a very low value before solving transport equations. Various aspects of MCDD have been tested for two-dimensional problems (Kippe, et al., 2008; Sandvin, et al., 2011; Lunati, et al., 2011). However, to formulate multiscale methods for three-dimensional problems has turned out to be considerably more difficult in general, and to our knowledge, no applications of MCDD-type methods within an iterative setting have been reported in three dimensions.

In this chapter, we consider multiscale methods and preconditioners defined for arbitrary number of spatial dimensions. We show how the multiscale method can be formulated both as a top-down and as a bottom-up method, and that these formulations give rise to different interpretations of the resulting approximations and preconditioners. Numerical examples illustrate the main strengths and weaknesses of the approach. Moreover, the numerical examples highlight the capabilities of the framework in terms of producing quick calculations when possible, but also producing more accurate results when needed.

2. Governing equations and discretization

The primary focus of the current chapter is linear solvers. The particular linear solvers we discuss are designed to preserve certain properties from the physical problem. Therefore, the linear solvers cannot be discussed without first specifying both the governing equations and the particular discretizations we are concerned with.

2.1 Governing equations

We consider flow of an incompressible fluid in a porous medium. For an introduction to flows in porous media, see e.g. (Bear 1972); for a reference focusing on appropriate numerical methods for this problem, confer (Chen, et al., 2006). Here we will only provide a brief review of the main ideas of importance to this chapter. Conservation of mass (volume) for each phase can be modeled by the equation:

$$\nabla \cdot \boldsymbol{u} = \upsilon.$$

Here the flux of each phase is represented by $\boldsymbol{u}$, and υ denotes the volumetric source / sink terms. The flux is usually assumed to be given by Darcy's law, which reads

$$\boldsymbol{u} = -\boldsymbol{k}\nabla p,$$

where the permeability is denoted $\boldsymbol{k}$, and p is the fluid potential. Additionally, we consider a dissolved concentration c which is passively advected with the velocity field,

$$\phi \frac{\partial c}{\partial t} + \nabla \cdot (c\boldsymbol{u}) = \psi, \tag{1}$$

where ϕ is the fraction of the void space available for fluid flow, referred to as porosity, and the material source term is given by ψ. We note that by introducing the particle velocity $\boldsymbol{v} = \phi^{-1}\boldsymbol{u}$ the advection of the dissolved concentration can be written in terms of the material derivative on Lagrangian form

$$\frac{Dc}{Dt} = \frac{\psi}{\phi} - c\frac{\nabla \cdot \boldsymbol{u}}{\phi}.$$

By eliminating the fluid flux from the statement of volume conservation we obtain an elliptic equation for pressure

$$-\nabla \cdot (\boldsymbol{k}\nabla p) = \upsilon. \tag{2}$$

In the sequel, we will study numerical solution techniques for Eq. (2), while keeping in mind that the methods must also be applicable for complex situations. Specifically, by integrating the Lagrangian version of the transport equation, we see that volume balance errors lead to exponential growth of errors in the dissolved concentration. Thus, it is of importance for the problems we consider that the velocity field must always be mass conservative in order to be suitable for use with most transport schemes.

2.2 Discretization

To discretize Eq. (1), we consider scalar discretizations, and in particular control volume schemes as they are particularly well suited for an exact representation of the conservation equation. Introducing the usual L_2 inner product, we can write the elliptic equation on a weak form as: Find $p \in H_0^1$ such that

$$(\boldsymbol{k}\nabla p, \nabla w) = (\upsilon, w) \quad \text{for all } w \in W. \tag{3}$$

Here, W represents a suitably chosen space, and we have for simplicity assumed zero Dirichlet boundary conditions to simplify the exposition. From this equation, we obtain control volume methods by choosing the finite subset of $W_h \in W$ to be the piece-wise constants forming a partition of unity on a cell-based grid, from which we obtain

$$\int_{\partial\omega} \boldsymbol{u} \cdot \boldsymbol{n} \, d\sigma = \int_{\omega} \upsilon \, d\tau,$$

for each (primal) cell ω. Note that $\boldsymbol{n}$ is the unit normal vector pointing out from the cell, and the product $\boldsymbol{u} \cdot \boldsymbol{n}$ is the normal flux over the boundary. Various control volume methods can now be defined by their approximation to the (flux) boundary integrals, most of which can be interpreted as particular choices of the finite space for p_h. We will in the following assume that such a choice has been made (for concreteness, one may consider the control-volume finite element method which is defined by p_h lying in the space spanned by piece-wise linears with nodes forming a dual grid to the partition induced by W_h). Furthermore, we assume for simplicity that the choice of flux approximation leads to a local approximation of the flux, in the sense that fluxes can be explicitly represented as a combination of fluid potentials in near-by cells.

We have now described a general setting for discrete representations of volume balance and Darcy's law which lead to a sparse linear system for the scalar variable p_h, which can be given on vector-matrix form as

$$\boldsymbol{A}\boldsymbol{p} = \boldsymbol{v}. \tag{4}$$

Remark 1: The control-volume finite element method, while attractive for educational purposes, is not very accurate in practice. Therefore, in reservoir simulation, the flux over a face has traditionally been approximated as driven by the pressure difference in the two adjacent cells only, giving rice to two-point schemes for the flux (Aziz & Settari, 1979). For logically Cartesian grids, this gives a classical 5- or 7-point cell stencil in 2 and 3 dimensions, respectively. However, in situations when the principal axis of the permeability tensor deviates considerably from the orientation of the grid, two-point schemes are known to produce inaccurate results. As a remedy, so-called multi-point schemes have been introduced (Edwards & Rodgers, 1998; Aavatsmark, 2002). These produce more accurate results to the price of a larger computational stencil, for Cartesian grids, the resulting linear system will have 9 and 27 bands in 2 and 3 dimensions, respectively. As we will see, the reduced accuracy of two-point schemes for rough grids is not only important for discretizing Eq. (2); similar considerations are also important when constructing fast linear solvers.

Remark 2: If a method is defined by choosing both p_h and w_h to lie in the same finite-dimensional space, the classical finite element method is recovered. In particular, the simplest choice, the piece-wise (multi)-linear functions give a system of equations that have a similar algebraic structure to the control-volume methods discussed above, but do not explicitly represent conservation.

3. Mass conservative domain decomposition

Here we describe the ingredients for the mass conservative domain decomposition (MCDD). We will start our presentation with describing the development of Schur-complement systems for N-dimensional problems. Readers may of course chose to disregard this generality, and consider only the special cases $N = 2, 3$. From this, we will see how we can form classical domain decomposition methods as well as multiscale control-volume methods. Note that due to the large number of matrices and vectors involved in the presentation, these will no longer be marked as bold as long as there is no room for confusion.

3.1 Schur complement systems

Being consistent with the discretization outlined above, we assume that computational domain is partitioned into a fine scale grid, and that control volume discretization enforces conservation of mass on the fine grid cells. In order to proceed in the construction of a two-level method, we need to introduce the notion of coarse grids.

Consider a continuous collection of cells, referred to as internal boundary cells, which partition the domain into isolated subdomains. By isolated, we mean in the sense of the discretization of the elliptic operator, such that no cell in one subdomain is dependent on any cells of any other subdomains. We interpret the subdomains as cells of the coarse dual grid, and the internal boundary cells thus form the nodes, edges, faces etc. of the dual coarse grid. We will identify cells and variables with a numerical subscript dependent on what part of the dual coarse grid they form part of: 0 indicates dual coarse nodes, 1 indicates dual coarse edges, 2 indicates faces, and so on, until N denotes cells that lie in the subdomains (where N is the dimension of the problem). Confer Fig. 1 for an illustration. When considering only the internal boundary cells, we will refer with subscript B to all subscripts less than N. Note that we have not yet introduced a coarse primal grid; this will not be needed before a later section.

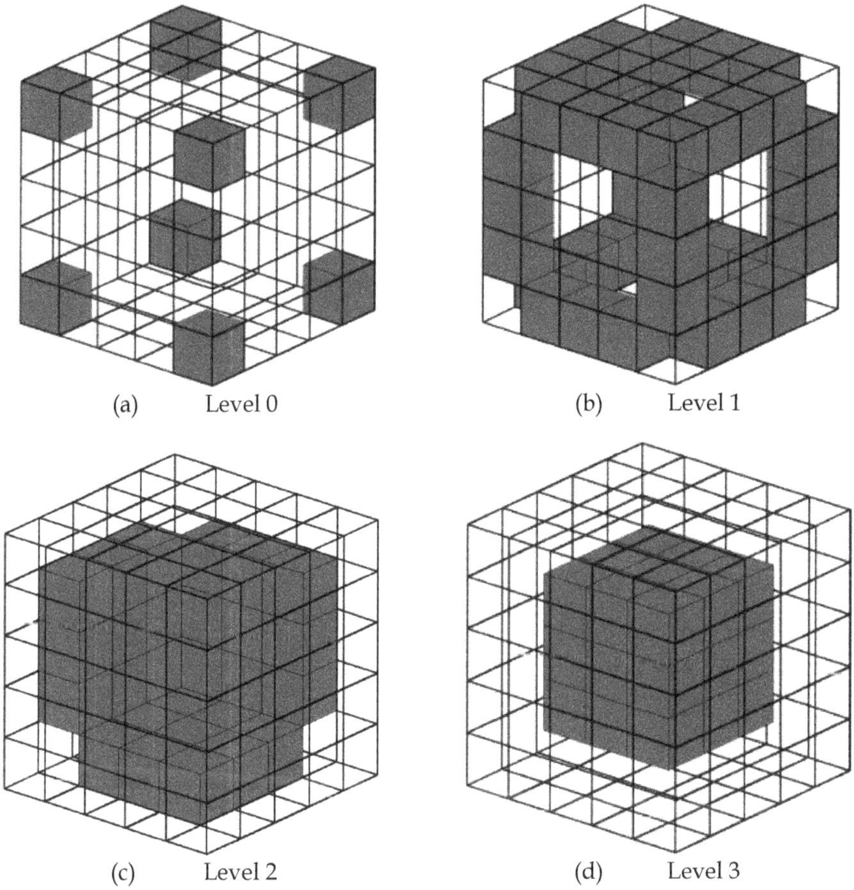

Fig. 1. Illustration of cells on different levels in a three-dimensional Cartesian grid. For clarity of visualization, only some of the cells on level 2 are indicated.

We now start to manipulate the linear system of equations (4), with the ultimate goal of obtaining a coarse linear system that captures non-local structures. By a reordering of the unknowns based on the dual coarse grid, Eq. (4) can then be written as

$$\begin{pmatrix} A_{00} & \cdots & A_{0N} \\ \vdots & \ddots & \vdots \\ A_{N0} & \cdots & A_{NN} \end{pmatrix} \begin{pmatrix} p_0 \\ \vdots \\ p_N \end{pmatrix} = \begin{pmatrix} v_0 \\ \vdots \\ v_N \end{pmatrix}. \tag{5}$$

In the last row, by construction, A_{NN} is a sparse block diagonal matrix, with each block representing the interactions within each isolated subdomain. This implies that we can find the values p_N by a local calculation given the variables on the internal boundary cells. We write these local calculations as

$$p_N = A_{NN}^{-1}(v_N - A_{NB}p_B).$$

We use this expression to formally eliminate internal cells from our system of equations. Thus, by substitution into Eq. (5) we have the *Schur Complement* system

$$\begin{pmatrix} S_{00} & \cdots & S_{0(N-1)} \\ \vdots & \ddots & \vdots \\ S_{(N-1)0} & \cdots & S_{(N-1)(N-1)} \end{pmatrix} \begin{pmatrix} p_0 \\ \vdots \\ p_{(N-1)} \end{pmatrix} = \begin{pmatrix} \tilde{v}_0 \\ \vdots \\ \tilde{v}_{(N-1)} \end{pmatrix},$$

where the Schur complement matrices S_{ij} are defined as

$$S_{ij} \equiv A_{ij} - A_{iN}A_{NN}^{-1}A_{Nj},$$

and the right hand side has been updated to reflect the elimination of the internal nodes by

$$\tilde{v}_i \equiv v_i - A_{iN}A_{NN}^{-1}v_N.$$

We make a few comments about the Schur complement system.

Remark 3: By the Schur complement formulaiton the number of unknowns has been reduced, from what was essentially an N dimensional problem to an $(N-1)$ dimensional problem. This significant reduction in model complexity comes at the cost of the Schur complement system being in general much denser than the original system. Furthermore, the computational cost of calculating the full Schur complement matrices is frequently prohibitive. As such, the Schur complement formulation by itself is seldom used.

Remark 4: For local discretizations the direct coupling between variables p_i and p_j, where i and j are more than one integer apart, is usually small (and indeed there is no coupling for the two-point flux approximation methods). This implies that the matrices A_{ij} and thus also S_{ij} are for many practical problems essentially zero for $|i-j| \geq 2$, and the full Schur complement system is therefore essentially block tri-diagonal. Furthermore, we see that S_{ij} only differs significantly from A_{ij} in the case where $i = j = N-1$.

In the particular case of two spatial dimensions, the matrix S describes interaction between edge and vertex nodes only. In the case of three dimensions, it describes the interaction between faces, edges and vertexes, where we expect that the interactions between vertexes and faces are weak.

While the Schur complement system itself may be prohibitive to form and solve, it provides the framework for developing approximate solvers. Classically, these fall in the category of domain decomposition preconditioners (Smith, et al., 1996; Quateroni & Valli, 1999; Toselli & Widlund, 2005). In this chapter, we see how this framework also gives us both multiscale methods and preconditioners based on them.

Recall that the Schur complement system is (essentially) tridiagonal. The main approximation strategies to this system fall in two categories: The top down strategy gives a low-rank approximation to S based on only the degrees of freedom associated with vertexes of the coarse dual grid, which are then identified as the coarse degrees of freedom. This essentially forms a multiscale subspace based on the lower-diagonal component of S, and is the approach we will emphasize in the following. The bottom-up strategy goes the other way, successively applying Schur complement strategies to eliminate all variables until only a system for p_0 remains. Since the Schur complement matrices themselves are too expensive to calculate, the bottom-up approach requires introducing low-rank approximations to the Schur complement (e.g. probing based techniques (Chan & Mathew, 1992)) at every stage in the succession. The class of domain decomposition methods known as substructuring methods is often formulated in terms of the bottom-up framework.

3.2 Multiscale basis approximations

The multiscale basis approximations to the Schur complement system use the (block) lower diagonal component of S. Retaining the dependence on p_0, which we hereafter identify as the coarse variable, we then see that we obtain an explicit expression for the remaining degrees of freedom. In the block tri-diagonal case, this can be written compactly as:

$$p_i = \left(\prod_{j=1}^{i} -\hat{S}_{jj}^{-1} \hat{S}_{j(j-1)} \right) p_0 + \sum_{k=1}^{i} \left(\prod_{j=k+1}^{i} -\hat{S}_{jj}^{-1} \hat{S}_{j(j-1)} \right) \tilde{v}_k.$$

In this expression, the matrix products are ordered right to left, and we have marked the Schur complement matrices with a hat, indicating that approximate choices of these matrices can be used in order to define different multiscale bases. In the general case, where a block tri-diagonal system is not assumed, the above expression is defined recursively. Either way, for conciseness, we denote the linear operator associated with the reconstruction of the full approximation p by its homogeneous and heterogeneous parts,

$$p = \Psi p_0 + \Upsilon \tilde{v}.$$

At this point we make the following remarks.

Remark 5: The space spanned by the projection of p_0 to the full set of variables defined by Ψ is termed the *multiscale space* W_H^{MS}. It can be characterized by the basis functions obtained by setting $p_0 = e_i$, where e_i is the elementary vectors. The resulting product allows us to define the *multiscale basis function* ψ_i^{MS} as columns of Ψ,

$$\psi_i^{MS} \equiv \Psi e_i.$$

Given suitable choices of $\hat{S}_{ij}$, this gives various multiscale basis functions from literature, as seen in the following remarks. In the terminology of domain decomposition, these basis functions are often referred to as prolongation operators.

Remark 6: The natural interpretation of $\hat{S}_{jj}^{-1}\hat{S}_{j(j-1)}$ is to solve local problems at the level j using level $j-1$ as boundary conditions. This motivates the usual approximations to these Schur complements. Three important alternatives exist.

1. $\hat{S}_{jj}$ can be chosen as a discretization of the original differential operator restricted to the part of the internal boundary associated with j. This is the original multiscale basis functions of Hou & Wu 1997, and this is also the strategy we will apply in our numerical experiments.
2. For arbitrary operators, the differential operator restricted to a lower dimension may not be a good approximation to the problem, and this approximation is unstable. For such cases, a simple linear interpolation on internal boundaries can be suggested (Lunati & Jenny, 2007), and $\hat{S}_{jj}$ is then chosen as any matrix which admits the relevant (multi-)linear solutions.
3. Both the preceding operators require knowledge about the original geometry of the problem, and can thus be seen as geometric methods. If it is desired to implement multiscale methods strictly algebraically, then it is possible to construct algebraic approximations $\hat{S}_{jj}$ based on the information in S_{jj}, as was explored in Sandvin, et al., 2011.

Remark 7: It is common to not approximate the last Schur complement S_{NN}. Note that this does not imply that this Schur complement matrix needs to be computed, as we only need to know its action on the elements of the multiscale basis and on the right hand side. If this component is retained exactly, then the method becomes residual-free on the subdomains, which is an important aspect that can be exploited at later stages.

Keeping in mind that we now have an explicit representation of the solution covering the domain given the knowledge of the coarse nodes, we can use this representation to obtain a coarse system of equations.

3.3 Coarse scale equations retaining conservation form

From the last section, we see that we can use the Schur complement system to obtain a multi-scale basis. This is essentially a low-dimensional approximation of the solution space for the homogeneous part of the discrete differential operator. What remains in order to get an approximate solution is to consider coarse equations that constrain the remaining degrees of freedom in the multiscale space.

The original system of equations provides us with the first option for a set of coarse equations, since the equations associated with the coarse variables are simply our fine-scale discretization. Recalling the notation Ψ that indicates the linear operator that reconstructs the homogeneous part of the solution from the coarse basis we see that our original system of equations is simply

$$\begin{pmatrix} A_{00} & \cdots & A_{0N} \end{pmatrix}\Psi\, p_0 = v_0 - \begin{pmatrix} A_{00} & \cdots & A_{0N} \end{pmatrix}\Upsilon\tilde{v}.$$

This is however a poor choice of constraint for our coarse variables, as it physically represents only the differential operators locally around the coarse node. From the perspective of the variational derivation of the control volume method, this solution thus satisfies Eq. (3) with $p \in W_H^{MS}$ and w in the space of piecewise constants with support around the *local cells associated with* p_0. From this understanding, we are motivated to think of reformulating the system such that the test functions for the coarse equations have a larger support, and in particular also form a partition of unity.

To be precise, consider a coarse partition of the domain, referred to as the primal coarse grid, which has the following properties: Each cell in the primal coarse grid consists of a set of cells from the fine grid, and contains exactly one vertex (cell on level 0) of the dual coarse grid, see Fig. 2. Since the primal coarse grid is a subset of the fine-scale grid, we know that the space of piecewise constant functions W_H on the primal coarse grid is a sub-space of the space of piecewise constant functions W_h on the fine grid. Therefore, by a change of representation, we could write the original discretization such that the discrete equations for the coarse variables satisfied Eq. (3) for all piecewise constant functions of the primal coarse grid. This leads to our desired coarse equations.

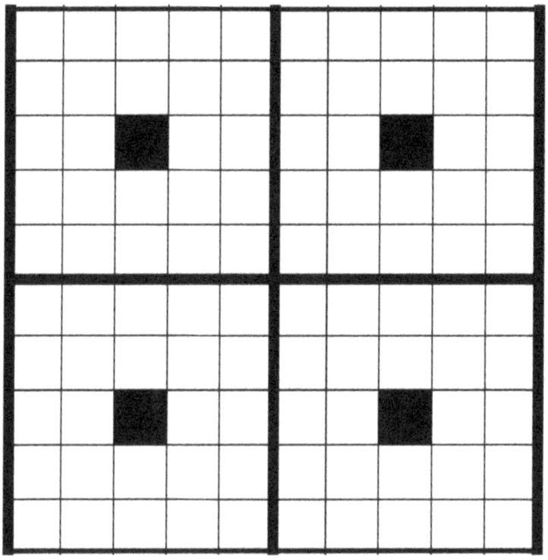

Fig. 2. A two-dimensional fine scale grid with a primal coarse grid imposed on it (bold lines). Black cells denotes center cells in the primal cell, these are on level 0 in the dual topology.

More practically, let the A, as before, represent a standard control volume discretization. Then let R_i be the restriction matrix to primal coarse cell i, and let $A_i = R_i A$. If furthermore M is an integration matrix that sums all rows in A_i into the row of the center cell, that is

$$M_i = I + e_{i0}(\mathbf{1} - e_{i0})^T, \tag{6}$$

where I is an identity matrix, e_{i0} is a unit vector identifying the center of the coarse cell, and $\mathbf{1}$ is a vector of ones. Multiplication with M_i for all primal coarse cells, and mapping the result back to the whole domain gives a linear system

$$Cp = v, \tag{7}$$

where $C = \sum_i R_i^T M_i R_i A$, and $v \longrightarrow \sum_i R_i^T M_i R_i v$. The linear system (7) is the original conservation of mass on the fine scale for all variables p_i where $i \geq 1$, however it represents conservation on the coarse scale for variables p_0. Note in particular that this means that $C_{ij} = A_{ij}$ for $i \geq 1$, and that this linear transformation does not change the solution p.

We now see that the coarse equations, as given by

$$(C_{00} \quad \cdots \quad C_{0N})\Psi p_0 = v_0 - (C_{00} \quad \cdots \quad C_{0N})\Upsilon\tilde{v},$$

solve the problem given by Eq. (3) for with $p \in \text{span}\,\psi^{MS}$ and $w \in W_H$, the space of piecewise constant functions on the dual coarse grid. We have thus derived a coarse control volume discretization, utilizing exactly a multiscale basis function to represent the solution. As a direct method, this is the so-called Multiscale Control Volume (Finite Element) Method as was first discussed (assuming $v_i = 0$ for $i \geq 1$) in Jenny, et al., 2003. The multiscale control volume methods described in the context of linear preconditioners are the Mass Conservative Domain Decomposition preconditioners derived in Nordbotten & Bjørstad, 2008.

Remark 8: In *Remark 1* at the end of the discretization section we saw that the standard finite element method is obtained by choosing test functions that are in the same multiscale space W_H^{MS} as the solution space. One may ask if the same is the case for multiscale methods. The answer is that yes, in the sense that if the integration on the primal grid defined in Eq. (7) is replaced by a weighted sum, using the multiscale basis itself as weights, the classical Multiscale Finite Element method of Hou & Wu, 1997 is recovered.

3.4 Recovering a conservative fine-scale flux field

The method as outlined so far constructs a two-level set of control volume methods. This can be seen from several perspectives: Either as the basis for a multi-level method, as the basis for a preconditioner in an iterative method, but also from the perspective of deriving a new, (coarse) single-scale control-volume method. We will consider the third perspective in this section.

When discussing the coupled set of equations outlined in Section 2.1, we pointed out the importance of retaining local mass conservation. This property is often necessary to consider (almost) point-wise, while by construction, the control-volume methods consider this only on the primal cells of the grid. It is therefore natural to consider whether a post-processing can be performed to extend this cell-wise property to a more local property, and whether this operation can be conducted locally.

For a local post-processing, it is natural to use the (cell-wise conservative) fluxes over boundaries of the primal grid as the basis of solving Neumann boundary problems inside each cell. The Neumann problem for the elliptic problems we consider is well-known to only admit solutions if the compatibility condition is satisfied, which is to say that the boundary conditions exactly integrate to the sum of all internal sources or sinks. The control-volume methods satisfy the compatibility condition by construction. Note that after post-processing, we will obtain a flux field that is everywhere conservative, but as a consequence will not everywhere satisfy Darcy's law.

In the case of single-scale control-volume methods, the permeability coefficient k is usually considered constant inside each primal cell, and locally post-processed fluxes can be calculated analytically for some cell shapes. While this is not used much from the perspective of practical simulation, it is an invaluable tool in the derivation of error estimates.

For the multiscale control-volume method, the permeability is of course possibly heterogeneous inside each coarse primal cell, and a numerical calculation must be performed as a post-processing step. This can be achieved using the same grid and discretization as used when obtaining the multi-scale basis functions, and leads to an approximation with the following important properties: A post-processed flux which is conservative on the fine-scale primal grid. This post-processed flux allows for transport simulations to be performed on a significantly finer grid than the coarse control-volume scheme that was derived.

It is important to note that the possibility of post-processing the fluxes is the most important property of the multiscale control-volume method. Moreover, the construction of the MCDD preconditioner explicitly preserves this property, such that at any iteration of an iterative approach, the approximate solution to the fine-scale problem can also be post-processed in an identical manner.

3.5 Multiscale methods as iterative solvers

The domain decomposition method formulated in Section 3.3 can be applied as a stand-alone solver for the pressure system (4). This was the approach advocated in the early multiscale papers (Hou & Wu, 1997; Jenny, et al., 2003; Aarnes, 2004). Since the action of the method on a vector can be evaluated solving local systems related to the Schur complements, as well as a (relatively small) coarse linear system, we understand that the method offers an efficient way to obtain a pressure approximation and a mass conservative fine scale velocity field. Indeed, simulations of petroleum recovery indicate that in some cases, this strategy provides a fairly accurate and very cheap alternative to traditional approaches.

However, the above strategy is insufficient for more challenging problems. A particular weakness of the multiscale methods is the reliance on somewhat arbitrary approximations to the Schur complements $\hat{S}_{ii}$. Indeed, since the approximate Schur complements determine the subspace W_H^{MS}, we understand that for any approximate Schur complement, cases exist where the solution to the fine-scale problem lies in a space orthogonal to the multiscale space. Thus multiscale methods as direct solvers will always have problems with robustness. The practical performance of multiscale methods unfortunately deteriorates with the number of spatial dimensions; to be specific, multiscale methods have turned out to perform significantly worse in three spatial dimensions than in 2D.

When faced with these issues, there are a few techniques that can be applied to improve the solution. One is to consider sophisticated ways to construct $\hat{S}_{ii}$, using in particular non-local information and information about the right hand side. Another approach, to be described next, is to apply the MCDD preconditioners in an iterative setting to improve the approximation. The simplest such strategy is a Richardson scheme, where we, equipped with an initial guess p^0, define the iterative scheme

$$p^{l+1} = p^l + \tau B_{MS}(b - Ap^l),$$

where B_{MS} represent one application of the multiscale method and τ is a damping factor. We observe that when the multiscale method is applied as a stand-alone solver, this corresponds to applying a single Richardson iteration with the MCDD preconditioner and $\tau = 1$. The Richardson scheme will in general exhibit poor convergence for our problem. A better utilization of the multiscale method is as a preconditioner inside an iterative solver such as GMRES (Saad & Schultz, 1986). Since the problem is likely to be more difficult in some parts of the domain, the application of the preconditioner can be restrained to those parts, if they can be identified by error estimates.

An important and often time consuming ingredient of GMRES is to ensure orthogonality of the basis vectors for the Krylov subspace in which the approximated solution lies. When GMRES is preconditioned with the multiscale method, this computational cost can be reduced considerably by exploiting a special feature of the solution: If the internal nodes p_N are eliminated using an exact solver, e.g. introducing no approximation to S_{NN}, the residual in the interior will be zero after one application of the preconditioner as discussed in *Remark* 7. This does not mean the pressure is exact for those cells, but rather that the influence of nodes on level N on the residual is lumped into the higher levels. This also means that GMRES does not need to minimize the residual for cells on level N, the orthogonalization needs only consider levels $0, \ldots, N-1$, leading to an often significant reduction of the computational cost. Note that level N cannot be totally ignored, since some nodes there are a part of the flux expression for level $N-1$.

We now realize that the MCDD applied as a preconditioner in an iterative setting possess several advantageous features in comparison to standard preconditioners:

1. For relatively simple problems, where standard multiscale methods are applicable, the iterative procedure can be terminated after a single iteration.
2. For moderately complex problems, the iterative method can be terminated at any point where the solution is deemed accurate enough, and a locally conservative flux field can be recovered.
3. For truly challenging problems, the MCDD preconditioner is comparable to standard non-overlapping domain decomposition based preconditioners for these problems.

Thus we see, for applications where the exact solution to the linear system is not necessary, the current methodology allows for a substantial savings in number of iterations. This is of great practical importance, since the error introduced by a discrete approximation to (2) can frequently be orders of magnitude larger than the tolerance used in traditional linear solvers.

3.6 Computational cost

While a full assessment of the computational cost is beyond the scope of the chapter, we will make some brief comments that allow the reader to get a general impression of the cost of both the multiscale methods as well as their application as preconditioners.

The computational cost of the MCDD preconditioner is composed of three components. First, the approximate Schur complement system involves approximating the action of S_{NN} on a (small) set of vectors. Physically, this corresponds to solving the local problems inside the internal subdomains for given boundary conditions. Denoting the number of internal subdomains as N_{SD}, a naïve estimate of cost would be $N_{SD} \cdot \dim p_0$. However, by construction, most approximate Schur complements will be local, such that each subdomain typically only has non-zero boundary conditions associated with the variables in p_0 that are associated with cells on the boundary of the subdomain. For Cartesian coarse subdomain, this is identified as the corners, such that the computational cost is proportional to $N_{SD} \cdot 2^d \cdot C_F$. Here C_F is defined as the coarsening factor, which is the ratio of degrees of freedom in the fine and coarse spaces, $C_F \equiv \frac{\dim p}{\dim p_0}$. The multiscale basis is only calculated once.

Secondly, there is a cost associated with the right-hand side, which needs to evaluated at every iteration. As seen in Section 3.2, the right hand side is also associated with local calculations, forced by source terms in contrast to the multiscale basis functions. The cost is thus proportional to $N_{SD} \cdot C_F \cdot N_I$, where N_I is the number of iterations.

Finally, there is the cost associated with solving the coarse set of equations. Here, there are two contrasting strategies. The domain-decomposition strategies argue for aggressive coarsening, where the coarse problem has (almost) negligible size and cost. This has the advantage that the cost of the coarse solve can be neglected, at the expense of more costly construction of the multiscale basis. However, as the multiscale basis calculation is trivially parallel, this may be a good strategy on some computational architectures, and in particular if the selection of coarse grids is hard to automate. A contrasting strategy is in the multi-grid flavor, where a much less aggressive coarsening is applied, which leads to a non-negligible cost in the coarse problem. However, since the coarse problem has the same control-volume structure as the fine-scale discretization, the multiscale method can be called recursively. The resulting algorithm has a better performance from the perspective of computational cost, but may be more difficult to implement as the problem is no longer trivially parallel. Note that for a conservative approximation to be obtained, the reconstruction of the flux field must also be conducted recursively.

In general, the multiscale methods are designed for problems where there is a coupling between the permeability k and the concentration field c. As c evolves locally, the multiscale basis functions may only need to be updated locally in space, allowing for further computational savings compared to a generic linear solver that is not adapted to these features. These aspects have been carefully highlighted in a suite of 2D test cases (Kippe, et al., 2008).

4. Numerical examples

In this section, we show numerical examples illustrating the properties of the domain decomposition method. For these examples, we have chosen the permeability field defined according to the SPE 10th comparative benchmark study much used to study upscaling and multiscale methods (Christie & Blunt, 2001). This test case involves a Cartesian 60 x 220 x 85 grid. The permeability in the upper 35 layers have a somewhat smooth distribution (consistent with a shallow marine depositional system), whereas the 50 lower layers are characterized by sharp permeability contrasts and highly permeable channels with long correlation length (consistent with a fluvial depositional system). The lower layers are expected to pose challenges for linear solvers. Representative layers from the upper and lower parts of the formation are shown in Fig. 3. The permeability field spans more than 10 orders of magnitude, rendering a challenging test problem for our methods.

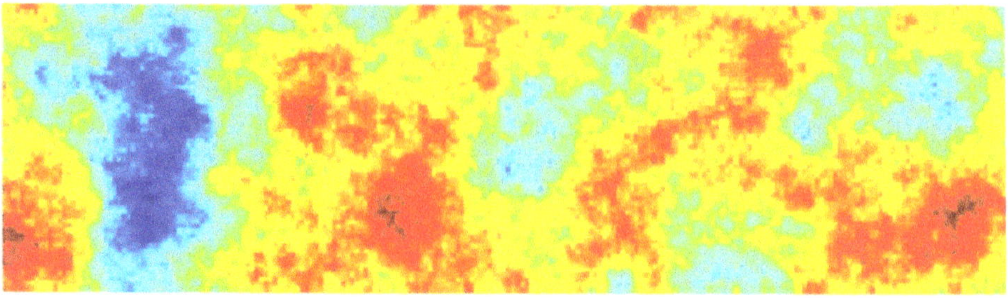

(a) Uppermost layer

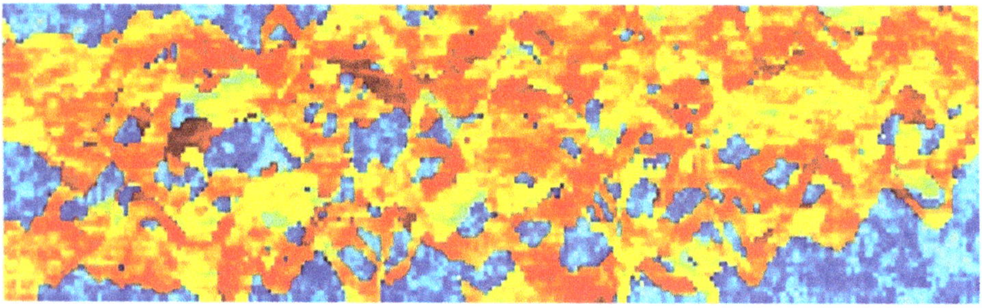

(b) Lowermost layer

Fig. 3. The base-10 logarithm of the permeability from the uppermost (a) and lowermost (b) layers in the SPE 10 test case. These are used in the 2D tests, and they are representative for the upper and lower part of the 3D formation, respectively. Blue and red corresponds to high and low-permeable regions, respectively. For convenience, the figures are rotated **90°**.

On this grid, we will consider simple setups, with one injection well and one producer. Both for 2D and 3D tests the injector is located along the boundary (the position differs somewhat between the tests, as we avoid injecting into low-permeable cells), and the producer is located in the middle of the domain. The pressure equation (2) is discretized using a two-point scheme, and for simplicity, periodic boundary conditions are assumed. For all test cases, post-processing of the flux field as discussed in Section 3.4 will be applied to ensure mass conservation on the fine scale.

4.1 2D examples

We start with two instructive examples in 2D, using permeability from the uppermost and lowermost layer of the SPE10 dataset, as pictured in Fig. 3. Thus the fine scale grid has 60 x 220 cells, and we use a coarse grid with 4x20 cells, rendering a coarsening factor of 165. Fig. 4 shows the pressure profiles obtained by a fine scale solution and the multiscale solver. For the upper layer, the multiscale solution is similar to that of the true solution; and has a quality that is as good as can be expected keeping in mind that the multiscale method is essentially a coarse discretization. In both solutions the pressure contours clearly indicate flow from injector to the producer, although again, the resolution of the local flow around the producer is better refined on the fine-scale grid. For the lower layer, the multiscale solution is highly oscillatory with false local minima in the solution. This can be interpreted as a case where the approximation to the Schur complement $\hat{S}_{ii}$ is not good enough, where a better approximation, or iterations, are needed to produce a pressure profile that resembles that of the fine scale solution.

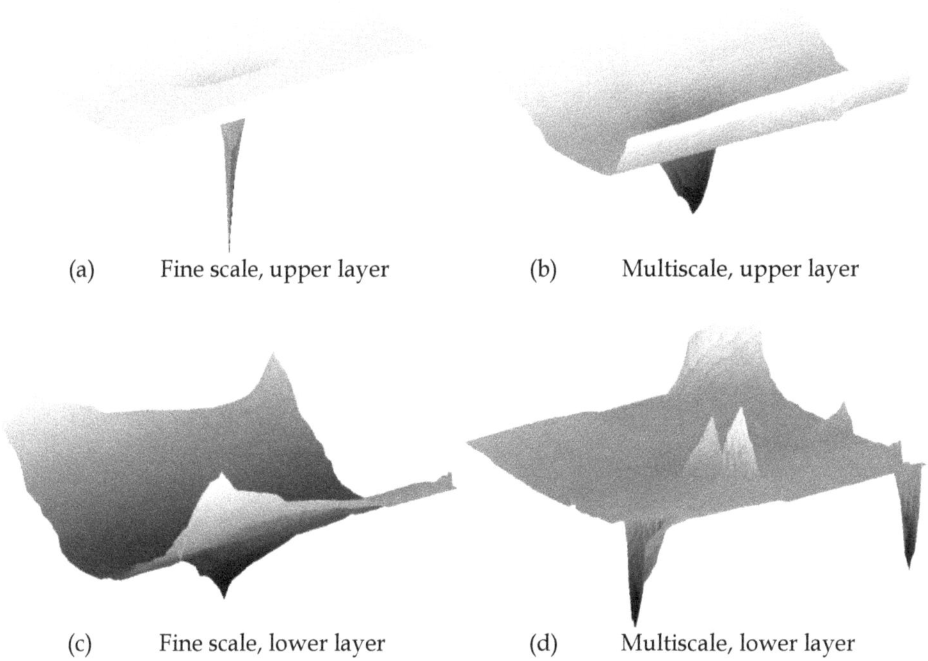
(a) Fine scale, upper layer
(b) Multiscale, upper layer
(c) Fine scale, lower layer
(d) Multiscale, lower layer

Fig. 4. Pressure solutions obtained by a fine scale and a multiscale solution for the uppermost and lowermost permeability layers. The injection well is located along the left boundary for all plots, while the producer is associated with the downward spike visible in the middle of the domain visible in all figures except (d).

For the uppermost layer of SPE10, the relatively good MS approximation to pressure is reflected in the post-processed fluxes. We illustrate this by the solution to the transport equation (1), as displayed in Fig. 5 (a) and (b). Note that despite the relatively coarse grid used for the multiscale control-volume approximation, the reconstruction of the fine-scale fluxes leads to a flow field with no visible artifacts. From the perspective of practical simulation, the solutions are indistinguishable.

Surprisingly, despite the relatively poor approximation to the pressure field, quite satisfactory fluxes can be obtained also for the lowermost layer as shown in Fig. 5 (c-d). This illustrates that the coarse scale conservation of mass combined with post-processing of the velocity field, leads to a multiscale approximation that is applicable to transport problems also for highly challenging problems. Note however that in these lower layers, the multiscale approximation leads to some cases where flow-channels are either suppressed or exaggerated.

The results from the concentration maps in Fig 5 are further confirmed by considering time series of the concentration in the production well, as shown in Fig. 6. For the upper layer, the curves corresponding to the fine scale and multiscale solutions are almost identical, and the differences are relatively small also for the lower layer.

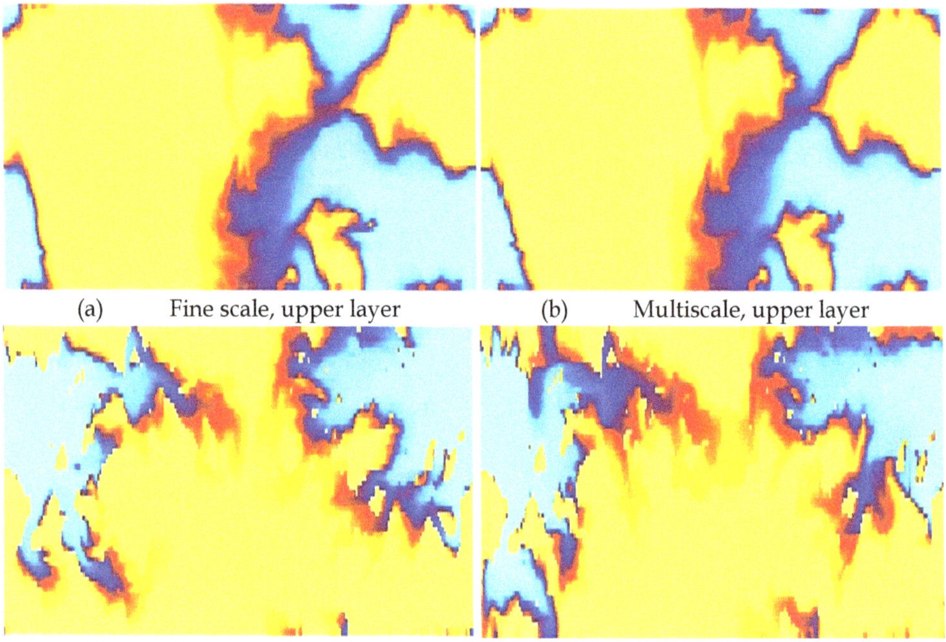

(c) Fine scale, lower layer (d) Multiscale, lower layer

Fig. 5. Concentration profiles obtained by solving the transport equation based on the post-processed pressure solutions. High concentration of the injected species is indicated with blue. We emphasis that for the multiscale solution, the velocity field is post processed to achieve local conservation of mass.

Remark 9: The appearance of oscillatory behavior in the multiscale solution is not unexptected. Again, we can analyze the multiscale control-volume method as simply being a single-scale control volume method on the coarse primal grid. It is known that for problems where the anisotropy in k is not aligned with the grid, local control-volume methods (and indeed this also holds for some other discretization families) in general cannot be constructed that are both consistent, as well as oscillation-free (Nordbotten, et al., 2007; Keilegavlen, et al., 2009). The channelized features that are shown in Fig. 3b are clearly not aligned with the general directions of the domain, and therefore they will lead to an effective permeability on the coarse grid that is also not aligned with the grid. The argument from the single-scale methods can thus be lifted to the multi-scale setting, which then informally may be states as: *No approximation $\hat{S}$ can be defined that leads to a local coarse-level control-volume method that is monotone for general channelized media.*

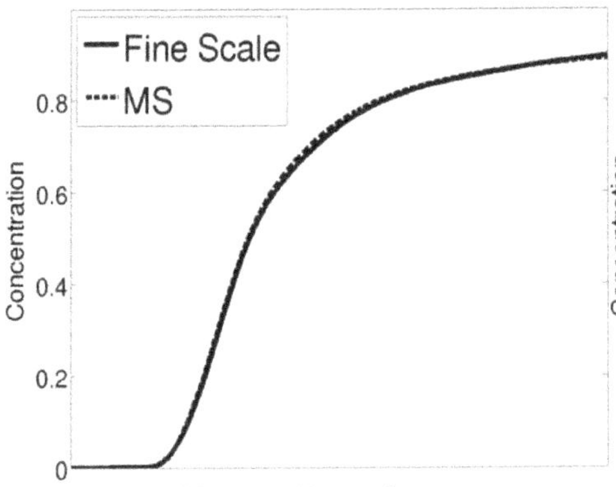

(a) Upper layer

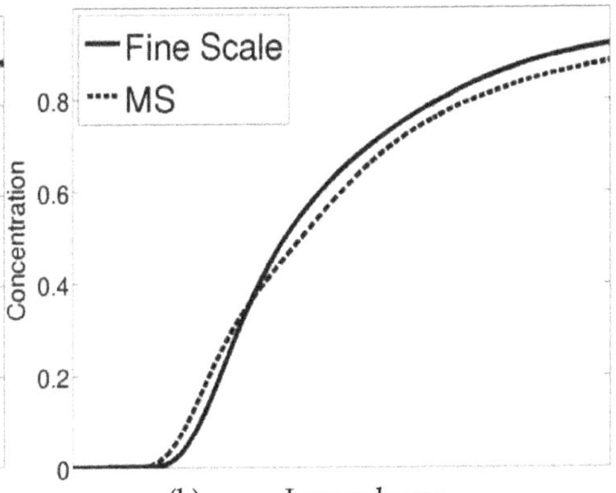

(b) Lower layer

Fig. 6. Time series of the concentration in the production well in the upper and lower layer.

4.2 3D examples

The 2D examples showed that the multiscale method can provide reliable solutions for challenging permeability fields and relatively high coarsening ratios. As previously mentioned, the performance of the multiscale method deteriorates significantly when going from 2D to 3D. As we will see, the multiscale solution may be insufficient for transport purposes, and the application as a MCDD preconditioner inside an iterative solver is essential in order to recover accuracy. For all 3D simulations, we consider coarse grid cells composed of 15 x 11 x 5 fine cells, rendering a coarsening ratio of 825.

4.2.1 Multiscale method as preconditioner

We first consider simulations in the 10 uppermost and lowermost layers of the SPE10 formation, extending the two cases considered in the 2D case. Again there is an injection well in a corner of the domain, and a producer in the middle of the domain. We consider transport solutions based on a fine scale solution, a pure multiscale solution, and from MCDD preconditioned GMRES iterations. Since visualization is more difficult in 3D than in 2D, we will in 3D only give the time-series type plots similar to Figs. 6. The time series of concentration in the production cells are shown in Fig. 7 both for the upper and lower layers. For the upper layers, we observe that in contrast with the 2D examples, the multiscale solution now deviates significantly from the fine scale solution. Applying some GMRES iterations improves the quality of the solution somewhat, until after a sufficient number of iterations renders a curve that is indistinguishable from the fine scale. For the lower layers, the stand-alone multiscale solver produce a time series that is vastly different from the fine scale solution, and it is therefore not shown in the figure. For this difficult problem, it takes more iterations to produce a time series that resembles that of the fine scale solve.

The above test shows that in 3D the multiscale method does not reproduce concentration curves that are comparable to the fine-scale curves even when for relatively easy case of the upper layers of SPE10. Thus the present test shows the utility of having a framework that, when the fast multiscale solution is insufficient, can fall back to an iterative scheme, and fairly quickly recover a velocity field that is good enough for transport purposes. The increased difficulty in approximating the solution is also shown in the development of the residual during the corresponding GMRES iterations, see Fig. 8. In the lower layers the residual decreases slower and more iterations are needed to obtained what might be deemed a satisfactory solution. Based on these two simple tests, we observe that the relative residual error in the linear solver of 10^{-3} to 10^{-4} is needed in order to reproduce a good transport solution. This value is many orders of magnitude higher than the typical residual errors used in iterative solvers for the linear system. Indeed, the residual error as a function of iterations is shown in Fig. 8, where we see that more than 600 iterations are needed to obtain a converged iteration for the upper layers. The ability to truncate the iterations early when using MCDD as a preconditioner will therefore in this case represent a savings of about 90% in terms of number of iterations in the iterative solver. Also for the lower layers, an early truncation saves a majority of the computational effort.

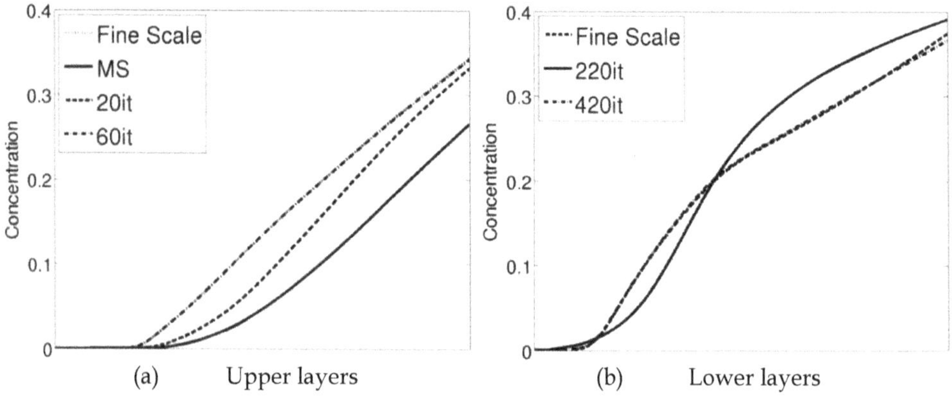

Fig. 7. Time series of the concentration in the production well for simulations in the upper and lower part of the SPE10 formation. The solutions obtained from the fine scale are located on top of those from 60 and 420 GMRES iterations for the upper and lower parts, respectively. A multiscale solution is not shown for the lower part, due to the low quality of the results produced.

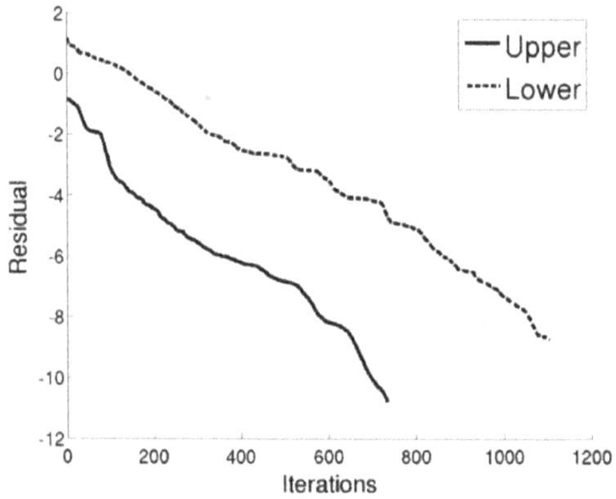

Fig. 8. The residual as a function of GMRES iterations for parts of the upper and lower part of the formation. The iterations terminates when the relative residual is reduced to a factor $\mathbf{10^{-10}}$.

At this point, it is appropriate to mention a third option to improve the multiscale solution, in addition to advanced approximations of the Schur complements $\hat{S}_{ii}$ and increasing the number of iterations: By increasing the number of coarse variables (in essence moving variables to level 0), the range of the multiscale basis functions ψ^{MS} can be increased to capture more of the solution. These ideas are exploited in (Sandvin, et al., Submitted), and show promising results in that the number of iterations needed can be reduced significantly with only a minor increase in the computational cost.

4.2.2 Quality control of MCDD solution

Returning to the original formulation of the system, we realize that an MCDD-based solution can be interpreted as the *exact* solution of Equations (2), for a modified permeability k^*. Using the pressure solution from the iterative solver together with the post-processed fluxes, we can calculate k^*. As the permeability is typically a value associated with great uncertainty for geological applications, we can compare the difference between k and k^* as a metric on the quality of the MCDD-based approximation. The most basic version of this comparison is to recall that from the physical motivation of the problem, the original permeability k is symmetric positive definite, and we can thus assess the quality of the approximate solution based on whether the modified permeability k^* also satisfies this physical constraint.

In Table 1, iteration counts and the number of sign changes are shown for a series of residual tolerances for GMRES. The grid consists of 60x220x10 cells, and the permeability is found from the channelized part of the SPE10 formation. Note that for the SPE10 dataset, the permeability tensor is diagonal, so positive definiteness is equivalent to positive diagonal elements. The table shows that for a high residual tolerance, more than a third of the fluxes change sign during post processing. Moreover, even for high accuracy of the GMRES solution there are some sign changes during flux post-processing.

$\log_{10}$(Tolerance)	-2	-4	-6	-8	-10
Iterations	224	429	717	915	1079
Negative elements in k^*	37%	9.4%	1.8%	.17%	.0018%

Table 1. The relative residual in GMRES, together with numbers for iterations and percentage of negative elements in the modified permeability.

The deviation of the flux and potential from a physical flow field, as measured by k^*, represents one attractive metric for assessing the approximation quality. However, more classical *a posteriori* error bounds and estimates are also applicable in this setting, and may be of equal importance for practical applications.

5. Concluding remarks

The present chapter has reviewed the construction of multiscale control volume methods in arbitrary dimensions from an algebraic perspective, allowing for a completely decoupled implementation of the fine-level (control volume) discretization and the multi-scale framework. We have emphasized several important aspects, including the points where key approximations are made, together with both their algebraic and physical interpretations. By bringing attention to the formulation of multiscale methods in this general setting, we have been able to highlight aspects of how multiscale control volume methods relate to classical single-scale discretizations, iterative preconditioners, and multi-level approximations. Through carefully chosen numerical examples, we have sought to illustrate both the quality of the multiscale approximation to the primary variable (pressure), but more importantly the role of the multiscale approximation in the setting of a coupled system of equations. These examples clearly illustrate the increasing complexity faced with problems in 3D over 2D, and the care with which one needs to deal with notions of approximate solvers and multiscale numerics.

In closing this chapter we wish to take the opportunity to discuss some of the main obstacles and benefits of multiscale methods as one considers more challenging problems.

As a stand-alone solver for a single elliptic problem, it is difficult for multiscale methods and preconditioners to compete with multigrid methods. The advantage of the methodology lies therefore in different aspects.

- A coarse discretization is obtained directly, with explicit coarse flux expressions, leading to an understanding of the nature of the effective coarse-scale operator for the system.
- For time-dependent, where multiple (similar) problems need to be solved in succession, a large amount of calculations can be re-used from previous time-steps.
- For (locally) spatially periodic problems, sub-domain problems may be identical and computational savings can be obtained through re-use again.
- For problems with scale-separation (where homogenization is applicable), the multiscale method gives a good approximation to both the homogenized and true solutions after a single iteration.

Despite the initial promise, and the evidence that the advantages can be realized for model problems, several challenges remain before multiscale methods attain the robustness required for practical applications. Some of the major limitations, together with their potential remedies are:

- For irregular grids (both on the fine and coarse scale) and for anisotropic media, the multiscale approximation is again less robust, especially when local Schur approximations are applied. To some extent, this can be overcome by oversampling, through enriching the coarse space, or by bottom-up approaches such as matrix probing, although as noted in *Remark 9*, a local and consistent coarse operator can in general never be designed.
- For non-linear elliptic equations (e.g. if the permeability coefficient is a function of the pressure or its gradients), the method is no longer residual-free in the interior if the multiscale basis functions are re-used. Recalculating multiscale basis functions in an iterative setting is prohibitively expensive, and it remains unclear if good multiscale approximations can be constructed.
- For higher-dimensional problems (more than 3), the quality of multiscale approximations has yet to be addressed at all.

With these perspectives in mind, it is clear that multiscale methods and preconditioners are still a topic of very active research. As such, there will most certainly be aspects of the current chapter that later research will both clarify and improve upon. Nevertheless, we hope that the present text succeeds in giving a current perspective on multiscale methods that will have value for both the general and specialized reader.

6. Acknowledgement

This research was financed in part through the Norwegian Reseach Grant #180679, "Modelling Transport in Porous Media over Multiple Scales". The book chapter is written in part based on lectures at the Radon Institute for Computational and Applied Mathematics (RICAM) Special Semester on Multiscale Simulation & Analysis in Energy and the Environment, Linz, October 3-December 16, 2011

7. References

Aarnes, J., 2004. On the use of a mixed multiscale finite element method for greater flexibility and increased speed or improved accuracy in reservoir simulation. *SIAM Multiscale Model. Simul.*, 2(3), pp. 421-439.

Aavatsmark, I., 2002. An introduction to the multipoint flux approximations for quadrilateral grids. *Comput. Geosci.*, 6(3-4), pp. 405-432.

Aziz, K. & Settari, A., 1979. *Petroleum Reservoir Simulation.* s.l.:Chapman & Hill.

Bear, J., 1972. *Dynamics of Fluids in Porous Media.* s.l.:Elsevier.

Chan, T. & Mathew, T., 1992. The interface probing technique in domain decomposition. *SIAM J. Matrix Anal. Appl.*, 13(1), pp. 212-238.

Chen, Z., Huan, G. & Ma, Y., 2006. *Computational methods for multiphase flows in porous media.* s.l.:SIAM.

Christie, M. & Blunt, M., 2001. Tenth SPE Comparative Solution Project: A Comparison of Upscaling Techniques. *Proc. of SPE Reservoir Simulation Symposium.*

Edwards, M. & Rodgers, C., 1998. Finite volume discretization with imposed flux continuity for the general tensor pressure equation. *Comput. Geosci.*, 2(4), pp. 459-490.

Efendiev, Y. & Hou, T., 2009. *Multiscale finite element methods: theory and applications.* s.l.:Springer.

Hou, T. & Wu, X.-H., 1997. A multiscale finite element method for elliptic problems in composite materials and porous media. *J. Comput. Phys.*, 134(1), pp. 169-189.

Jenny, P., Lee, S. & Tchelepi, H., 2003. Multi-scale finite-volume method for elliptic problems in subsurface flow simulations. *J. Comput. Phys.*, 187(1), pp. 47-67.

Keilegavlen, E., Nordbotten, J. & Aavatsmark, I., 2009. Sufficient criteria are necessary for monotone control volume methods. *Appl. Math. Lett.*, 22(8), pp. 1178-1180.

Kippe, V., Aarnes, J. & Lie, K.-A., 2008. A comparison of multiscale methods for elliptic problems in porous media flow. *Comput. Geosci.*, 12(3), pp. 277-298.

Lunati, I. & Jenny, P., 2007. Treating highly anisotropic subsurface flow with the multiscale finite-volume method. *SIAM Multiscale Model. Simul.*, 6(1), pp. 308-318.

Lunati, I., Tyagi, M. & Lee, S., 2011. An iterative multiscale finite volume algorithm converging to the exact solution. *J. Comput. Phys.*, Volume 230, pp. 1849-1864.

Nordbotten, J., Aavatsmark, I. & Eigestad, G., 2007. Monotonicity of control volume methods. *Numer. Math.*, 106(2), pp. 255-288.

Nordbotten, J. & Bjørstad, P., 2008. On the relationship between the multiscale finite-volume method and domain decomposition preconditioners. *Comput. Geosci.*, 12(3), pp. 367-376.

Quateroni, A. & Valli, A., 1999. *Domain decomposition methods for partial differential equations.* s.l.:Oxford University Press.

Saad, Y. & Schultz, H., 1986. A generalized minimal residual algorithm for solving nonsymmetric linear systems. *SIAM J. Sci. Stat. Comput.*, 7(3), pp. 856-869.

Sandvin, A., Keilegavlen, E. & Nordbotten, J., Submitted. Auxiliary variables in multiscale simulations. *J. Comput. Phys.*.

Sandvin, A., Nordbotten, J. & Aavatsmark, I., 2011. Multiscale mass conservative domain decomposition preconditioners for elliptic problems on irregular grids. *Comput. Geosci.*, 15(3), pp. 587-602.

Smith, B., Bjørstad, P. & Gropp, W., 1996. *Domain Decomposition.* s.l.:Cambridge University Press.

Tchelepi, H. & Juanes, R., 2008. *Special issue on multiscale methods for flow and transport in heterogeneous porous media.* 3 ed. Comput. Geosci.: s.n.

Toselli, A. & Widlund, O., 2005. *Domain Decomposition Methods - Algorithms and Theory.* s.l.:Springer.

Part 3

Application of FVM in Medicine and Engineering

Wood Subjected to Hygro-Thermal and/or Mechanical Loads

Izet Horman, Dunja Martinović, Izet Bijelonja and Seid Hajdarević
Mechanical Engineering Faculty, University of Sarajevo
Bosnia and Herzegovina

1. Introduction

The FV method was originally developed for fluid flow, heat and mass transfer calculations (Patankar, 1980), and later generalized for stress analysis in isotropic linear and non-linear bodies (Demirdžić & Muzaferija, 1994; Demirdžić et al., 1997; Demirdžić & Martinović, 1993). For the purpose of the stress analysis in the wood, the method is modified to take into account the anisotropic nature of the wood and influence of the moisture content and the temperature on the deformation and stresses (Horman, 1999). Also, performance of the wood is found to be very sensitive to the moisture content and the temperature. Thus, it is of a great importance to be able to predict behavior of such materials under different hygro-thermo-mechanical loads. In order to demonstrate the methods capabilities, a transient analysis of fields of temperature, moisture, and stresses and displacement in the wood subjected to hygro-thermal or mechanical loads is performed.

2. Theory

2.1 Governing equations

The behaviour of an arbitrary part of a solid, porous body at any instant of time can be described by the following energy, mass and momentum balance equations which, when written in a Cartesian tensor notation, read:

$$\frac{\partial}{\partial t}\left(\rho c_q T\right) = -\frac{\partial q_j}{\partial x_j} + \rho s_q \tag{1}$$

$$\frac{\partial}{\partial t}\left(\rho c_m M\right) = -\frac{\partial \dot{m}_j}{\partial x_j} + \rho s_m \tag{2}$$

$$\frac{\partial}{\partial t}\left(\rho \frac{\partial u_i}{\partial t}\right) = \frac{\partial \sigma_{ij}}{\partial x_j} + \rho b_i , \qquad i = 1,2,3 \tag{3}$$

In these equation, t is time, x_i is the Cartesian coordinate, ρ is the mass density, c_q and c_m are the specific heat and the specific moisture, T is the temperature, M is the moisture potential and u_i is the displacement, s_q and s_m are the heat and mass source, b_i is the body force, and q_j, $\dot{m}_j$ and σ_{ij} are the heat and mass flux vector, and stress tensor components, respectively.

2.2 Constitutive relations

In order to close the system of Eqs. (1)-(3) the constitutive relations for heat and mass flux based on the theory of Luikov (1966) which takes into account both the Soret and Duffort effect, together with the constitutive relation for a solid body are used:

- for Eqs. (1) and (2) heat and mass flux vector are

$$q_j = -k_{jl}^q \frac{\partial T}{\partial x_l} + \varepsilon r \dot{m}_j = -\left(k_{jl}^q + \varepsilon r \delta k_{jl}^m\right)\frac{\partial T}{\partial x_l} - \varepsilon r k_{jl}^m \frac{\partial M}{\partial x_l} \tag{4}$$

$$\dot{m}_j = -k_{jl}^m \frac{\partial M}{\partial x_l} - \delta k_{jl}^m \frac{\partial T}{\partial x_l} \tag{5}$$

- for an elastic, porous, orthotropic material for Eqs. (3) is

$$\sigma_{ij} = C_{ijkl}\,\varepsilon_{kl} - \alpha_{ij}\Delta T - <\beta_{ij}\Delta M> = \frac{1}{2}C_{ijkl}\left(\frac{\partial u_k}{\partial x_l} + \frac{\partial u_l}{\partial x_k}\right) - \alpha_{ij}\Delta T - <\beta_{ij}\Delta M> \tag{6}$$

Here k_{ij}^q and k_{ij}^m are the heat and mass conduction coefficient tensor components, respectively, ε is the ratio of the vapour diffusion coefficient to the coefficient of total diffusion of moisture, r is the heat of the phase change, δ is the temperature-gradient coefficient, ε_{ij} are the strain tensor components, C_{ijkl} are the elstic constant tensor components, α_{ij} are the coefficients of thermal expansion, β_{ij} are the shrinkage (contraction) coefficients, $\Delta T = T - T_u$, $\Delta M = M - M_h$ and T_u is the temperature at an udeformed state and M_h is the moisture potential at the fiber saturation point. For an orthotropic material and the coordinate axes aligned with the symmetry axes, Eqs. (4)-(6) can be written in the following matrix form:

$$\begin{bmatrix} q_1 \\ q_2 \\ q_3 \end{bmatrix} = -\begin{bmatrix} k_{11}^q + \delta\varepsilon r k_{11}^m & 0 & 0 \\ 0 & k_{22}^q + \delta\varepsilon r k_{22}^m & 0 \\ 0 & 0 & k_{33}^q + \delta\varepsilon r k_{33}^m \end{bmatrix} \begin{bmatrix} \frac{\partial T}{\partial x_1} \\ \frac{\partial T}{\partial x_2} \\ \frac{\partial T}{\partial x_3} \end{bmatrix} - \varepsilon r \begin{bmatrix} k_{11}^m & 0 & 0 \\ 0 & k_{22}^m & 0 \\ 0 & 0 & k_{33}^m \end{bmatrix} \begin{bmatrix} \frac{\partial M}{\partial x_1} \\ \frac{\partial M}{\partial x_2} \\ \frac{\partial M}{\partial x_3} \end{bmatrix} \tag{7}$$

$$\begin{bmatrix} \dot{m}_1 \\ \dot{m}_2 \\ \dot{m}_3 \end{bmatrix} = -\begin{bmatrix} k_{11}^m & 0 & 0 \\ 0 & k_{22}^m & 0 \\ 0 & 0 & k_{33}^m \end{bmatrix} \begin{bmatrix} \frac{\partial M}{\partial x_1} \\ \frac{\partial M}{\partial x_2} \\ \frac{\partial M}{\partial x_3} \end{bmatrix} - \delta \begin{bmatrix} k_{11}^m & 0 & 0 \\ 0 & k_{22}^m & 0 \\ 0 & 0 & k_{33}^m \end{bmatrix} \begin{bmatrix} \frac{\partial T}{\partial x_1} \\ \frac{\partial T}{\partial x_2} \\ \frac{\partial T}{\partial x_3} \end{bmatrix} \tag{8}$$

$$
\begin{bmatrix} \sigma_{11} \\ \sigma_{22} \\ \sigma_{33} \\ \sigma_{12} \\ \sigma_{23} \\ \sigma_{31} \end{bmatrix} = \begin{bmatrix} A_{11} & A_{12} & A_{31} & 0 & 0 & 0 \\ A_{12} & A_{22} & A_{23} & 0 & 0 & 0 \\ A_{31} & A_{23} & A_{33} & 0 & 0 & 0 \\ 0 & 0 & 0 & A_{44} & 0 & 0 \\ 0 & 0 & 0 & 0 & A_{55} & 0 \\ 0 & 0 & 0 & 0 & 0 & A_{66} \end{bmatrix} \begin{bmatrix} \varepsilon_{11} - \alpha_{11}\Delta T - < \beta_{11}\Delta M > \\ \varepsilon_{22} - \alpha_{22}\Delta T - < \beta_{22}\Delta M > \\ \varepsilon_{33} - \alpha_{33}\Delta T - < \beta_{33}\Delta M > \\ \varepsilon_{12} \\ \varepsilon_{23} \\ \varepsilon_{31} \end{bmatrix} \tag{9}
$$

where the terms in $\langle\ \rangle$ brackets are „active" only for $M < M_h$, while the nine non-zero orthotropic elastic constants A_{ij} are related to the Young's moduli E_i, the Poisson's coefficients ν_{ij} and the shear moduli G_{ij} by the following relations:

$$
A_{11} = \frac{E_1^2 E_2 E_3 - \nu_{23}^2 E_3^2 E_1^2}{E_1 E_2 E_3 (1 - 2\nu_{12}\nu_{23}\nu_{31}) - E_1^2 E_2 \nu_{31}^2 - E_2^2 E_3 \nu_{12}^2 - E_3^2 E_1 \nu_{23}^2},
$$

$$
A_{22} = \frac{E_1 E_2^2 E_3 - \nu_{31}^2 E_1^2 E_2^2}{E_1 E_2 E_3 (1 - 2\nu_{12}\nu_{23}\nu_{31}) - E_1^2 E_2 \nu_{31}^2 - E_2^2 E_3 \nu_{12}^2 - E_3^2 E_1 \nu_{23}^2},
$$

$$
A_{33} = \frac{E_1 E_2 E_3^2 - \nu_{12}^2 E_2^2 E_3^2}{E_1 E_2 E_3 (1 - 2\nu_{12}\nu_{23}\nu_{31}) - E_1^2 E_2 \nu_{31}^2 - E_2^2 E_3 \nu_{12}^2 - E_3^2 E_1 \nu_{23}^2},
$$

$$
A_{44} = 2G_{12}, \quad A_{55} = 2G_{23}, \quad A_{66} = 2G_{31}, \tag{10}
$$

$$
A_{12} = \frac{\nu_{23}\nu_{31} E_1^2 E_2 E_3 + \nu_{12} E_1 E_2^2 E_3}{E_1 E_2 E_3 (1 - 2\nu_{12}\nu_{23}\nu_{31}) - E_1^2 E_2 \nu_{31}^2 - E_2^2 E_3 \nu_{12}^2 - E_3^2 E_1 \nu_{23}^2},
$$

$$
A_{23} = \frac{\nu_{31}\nu_{12} E_1 E_2^2 E_3 + \nu_{23} E_1 E_2 E_3^2}{E_1 E_2 E_3 (1 - 2\nu_{12}\nu_{23}\nu_{31}) - E_1^2 E_2 \nu_{31}^2 - E_2^2 E_3 \nu_{12}^2 - E_3^2 E_1 \nu_{23}^2},
$$

$$
A_{31} = \frac{\nu_{12}\nu_{23} E_1 E_2 E_3^2 + \nu_{31} E_1^2 E_2 E_3}{E_1 E_2 E_3 (1 - 2\nu_{12}\nu_{23}\nu_{31}) - E_1^2 E_2 \nu_{31}^2 - E_2^2 E_3 \nu_{12}^2 - E_3^2 E_1 \nu_{23}^2}.
$$

Note that the pair of constitutive Equations (4) and (5) can be extended to take into account the effect of the pressure gradient on the heat and mass transfer.

- for a thermo-elasto-plastic isotropic material for Eqs (3) and (1) are

$$
\delta\sigma_{ij} = 2G\delta\varepsilon_{ij} + \lambda\delta_{ij}\delta\varepsilon_{kk} - (3\lambda + 2G)\alpha\delta_{ij}\delta T - \left\langle \frac{3G\sigma_{ij}^{d}\sigma_{kl}^{d}\delta\varepsilon_{kl}}{\bar{\sigma}^2\left(\dfrac{H}{3G} + 1\right)} \right\rangle \tag{11}
$$

and constitutive relation (4) $\dot{m}_j = 0$.

Here

$$\sigma_{ij}^{d} = \sigma_{ij} - \frac{1}{3}\delta_{ij}\sigma_{kk} \tag{12}$$

is the stress deviator and

$$\bar{\sigma} = \left(\frac{3}{2}\sigma_{ij}^{d}\sigma_{ij}^{d}\right)^{1/2} \tag{13}$$

is the effective stress (in the case of Von Mises yield criterion), G and λ are Lame's constants, α is the thermal expansion coeffifient, G is the shear modulus, H' is the plastic modulus, and δ_{ij} is the Kronecker delta.

Lame's constants are related to the more commonly used elastic modulus E and Poisson's coefficient ν by the following relationships:

$$\lambda = \frac{\nu E}{(1+\nu)(1-2\nu)}, \quad G = \frac{E}{2(1+\nu)} \tag{14}$$

In the case of elastic conditions, the expression within the brackets $\langle\ \rangle$ vanishes, and the constitutive realtion (11) reduces to the Duhamel-Neumann form of Hooke's law.

2.3 Initial and boundary conditions

In order to complete the mathematical model, initial and boundary conditions have to be specified. As initial conditions, the temperature, the moisture potential, and the displacement and velocity components have to be specified at all points of the solution domain.

For a wood heat treatment process, boundary conditions can be either of Dirichlet or Von Neuman type, i.e. temperature and/or heat flux and displacements and/or forces (surface tractions) have to be specified at all boundaries.

For a convective wood drying process, the following boundary conditions are normally appropriate:

$$k_{jl}^{q}\frac{\partial T}{\partial x_l}n_j + h_q(T-T_a) + (1-\varepsilon)rh_m(M-M_a) = 0$$

$$k_{jl}^{m}\frac{\partial M}{\partial x_l}n_j + h_m(M-M_a) + \delta k_{jl}^{m}\frac{\partial T}{\partial x_l}n_j = 0 \tag{15}$$

$$\sigma_{ji}n_j = f_{si}$$

where h_q and h_m are the (convective) heat and mass transfer coefficients, respectively, f_{si} is the surface traction, and all quantities are calculated at the solution domain boundary, except for those with subscript a which correspond to the ambient air.

3. Numerical method

3.1 Generic transport equation

Before the construction of a numerical algorithm is started, it is important to notice that the governing Eqs. (1)-(3) or Eqs. (1) and (3) when combined with constitutive Eqs. (7)-(9), or Eqs. (4) ($\dot{m}_j = 0$) and (11) can be written in the form of the following generic transport equation:

$$\frac{\partial}{\partial t}\left(\rho B_\psi\right) - \frac{\partial}{\partial x_j}\left(\Gamma_{jl}^{\psi}\frac{\partial \psi}{\partial x_l}\right) - S_\psi = 0, \quad \left(\Gamma_{jl}^{\psi} = 0 \quad \text{for} \quad j \neq l\right) \tag{16}$$

which can be integrated over an arbitrary solution domain V bounded by the surface A, with unit outer normal vector n_j to yield:

$$\int_V \frac{\partial}{\partial t}\left(\rho B_\psi\right) dV - \int_A \Gamma_{jl}^{\psi}\frac{\partial \psi}{\partial x_l} n_j \, dA - \int_V S_\psi dV = 0, \quad \left(\Gamma_{jl}^{\psi} = 0 \text{ for } j \neq l\right) \tag{17}$$

The generic variable ψ stands for T, M or u_i.

The maeaning of the coefficients B_ψ, Γ_{jj}^{ψ} and S_ψ for the wood drying process is given in Table 1.

ψ	B_ψ	Γ_{11}^{ψ}	Γ_{22}^{ψ}	Γ_{33}^{ψ}	S_ψ
T	$c_q T$	$k_{11}^{q} + \varepsilon r \delta k_{11}^{m}$	$k_{22}^{q} + \varepsilon r \delta k_{22}^{m}$	$k_{33}^{q} + \varepsilon r \delta k_{33}^{m}$	$\frac{\partial}{\partial x_j}\left(\varepsilon r k_{jl}^{m}\frac{\partial M}{\partial x_l}\right) + \rho s_q$
M	$c_m M$	k_{11}^{m}	k_{22}^{m}	k_{33}^{m}	$\frac{\partial}{\partial x_j}\left(\delta k_{jl}^{m}\frac{\partial T}{\partial x_l}\right) + \rho s_m$
u_1	$\frac{\partial u_1}{\partial t}$	A_{11}	$\frac{A_{44}}{2}$	$\frac{A_{66}}{2}$	$\frac{\partial}{\partial x_1}\left(A_{12}\frac{\partial u_2}{\partial x_2} + A_{31}\frac{\partial u_3}{\partial x_3}\right) + \frac{\partial}{\partial x_2}\left(\frac{A_{44}}{2}\frac{\partial u_2}{\partial x_1}\right) + \frac{\partial}{\partial x_3}\left(\frac{A_{66}}{2}\frac{\partial u_3}{\partial x_1}\right) - \frac{\partial}{\partial x_1}\left[A_{11}\left(\alpha_{11}\Delta T + <\beta_{11}\Delta M>\right) + A_{12}\left(\alpha_{22}\Delta T + <\beta_{22}\Delta M>\right) + A_{31}\left(\alpha_{33}\Delta T + <\beta_{33}\Delta M>\right)\right] + \rho b_1$
u_2	$\frac{\partial u_2}{\partial t}$	$\frac{A_{44}}{2}$	A_{22}	$\frac{A_{55}}{2}$	$\frac{\partial}{\partial x_1}\left(\frac{A_{44}}{2}\frac{\partial u_1}{\partial x_2}\right) + \frac{\partial}{\partial x_2}\left(A_{12}\frac{\partial u_1}{\partial x_1} + A_{23}\frac{\partial u_3}{\partial x_3}\right) + \frac{\partial}{\partial x_3}\left(\frac{A_{55}}{2}\frac{\partial u_3}{\partial x_2}\right) - \frac{\partial}{\partial x_2}\left[A_{12}\left(\alpha_{11}\Delta T + <\beta_{11}\Delta M>\right) + A_{22}\left(\alpha_{22}\Delta T + <\beta_{22}\Delta M>\right) + A_{23}\left(\alpha_{33}\Delta T + <\beta_{33}\Delta M>\right)\right] + \rho b_2$
u_3	$\frac{\partial u_3}{\partial t}$	$\frac{A_{66}}{2}$	$\frac{A_{55}}{2}$	A_{33}	$\frac{\partial}{\partial x_1}\left(\frac{A_{66}}{2}\frac{\partial u_1}{\partial x_3}\right) + \frac{\partial}{\partial x_2}\left(\frac{A_{55}}{2}\frac{\partial u_2}{\partial x_3}\right) + \frac{\partial}{\partial x_3}\left(A_{31}\frac{\partial u_1}{\partial x_1} + A_{23}\frac{\partial u_2}{\partial x_2}\right) - \frac{\partial}{\partial x_3}\left[A_{31}\left(\alpha_{11}\Delta T + <\beta_{11}\Delta M>\right) + A_{23}\left(\alpha_{22}\Delta T + <\beta_{22}\Delta M>\right) + A_{33}\left(\alpha_{33}\Delta T + <\beta_{33}\Delta M>\right)\right] + \rho b_3$

Table 1. The meaning of $B_\psi, \Gamma_{jj}^{\psi}$ and S_ψ in Eqs. (16) and (17)

3.2 Finite volume discretisation

As all numerical methods, the present one consists of time, space and equations discretisation. The time interval of interest is subdivided into a number of subintervals δt, not necessarily of the the same length. The space is discretised by a number of contiguous, non-overlapping hexahedral control volumes (CV), with the computational points at their centres (Fig.1). Then the integrals in generic Equation (17) are calculated by employing the midpint rule, the gradients are evaluated by assuming a linear variation of the dependent variable ψ between the computational points, and a fully implicit temporal scheme is employed. As a result a non-linear algebraic equation of the following form for each CV is obtained:

$$a_P \psi_P - \sum_K a_K \psi_K = b \qquad (K = W, E, S, N, B, T) \tag{18}$$

where the coefficients a_K and b are defined as:

$$a_E = \left(\Gamma_{11}^{\psi}\right)_e \frac{A_e}{\delta x_{1e}} \text{ and similar expressions for other cell faces,}$$

$$a_P^o = \frac{V}{\delta t}\left(\rho \frac{B_\psi}{\psi}\right)_P^o, \qquad a_P = \sum_K a_K + a_P^o \ , \qquad b = S_{\psi P} V + a_P^o \psi_P^o \tag{19}$$

where the subscripts P and e denote values at the centre of the CV and at the centre of the east cell-face, respectively, A_e is the area of the east cell face, V is the volume of the CV, δx_e is the distance between points P and E, and all quantities refer to the current time level, except for those with the superscript o which refer to the previous, „old" time level.

3.3 Solution algorithm

After assembling Eqs. (18) for all CVs and for all transport equations, five (four in 2D case) sets of N mutually coupled non-linear algebraic equations are obtained, where N is the number of CVs. Those equations are solved by employing the following segregated iterative procedure.

First, all dependent variables are given their initial values. Then the boundary conditions which correspond to the first time step are applied, and the sets of equations for each individual dependent variable (T, M, u_i) are linearised and temporarily decoupled by assuming that coefficient a_K and source terms b are known (calculated by using depedent variable values from the previous iteration or the previous time step), resulting in a system of linear algbraic equations of the form:

$$A_\psi \boldsymbol{\psi} = \mathbf{b}_\psi \tag{20}$$

for each dependent variable, where A_ψ is an N x N matrix, vector $\boldsymbol{\psi}$ contains values of depedent variable ψ at N nodal points and $\mathbf{b}_\psi$ is the source vector.

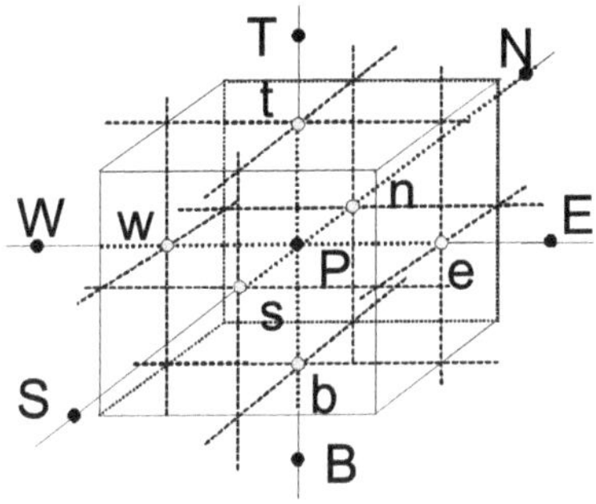

- Computational points
- o Cell-face centres

Fig. 1. A typical control volume and the compass labelling scheme

Systems Eqs. (20) are then solved sequentially in turn until a converged solution is obtained. The procedure is assumed converged when the following conditions are satisfied for all five (four in 2D case) sets of equations:

$$\sum_{i=1}^{N}\left| a_P\psi_P - \sum_K a_K\psi_K - b \right| < pR_\psi \tag{21}$$

$$\left|\psi_i^m - \psi_i^{m-1}\right| < q\left|\psi_i^m\right|, \qquad i = 1,2,....,N$$

where p and q are typically of the order 10^{-3}, R_ψ is a suitable normalisation factor and superscripts m and $m-1$ denote values at two successive iterations.

In the next time step the whole procedure is repeated, except that the initial values are replaced by the values from the previous time step.

The present discretisation procedure ensures that the matrix A_ψ has the folowing desirable properties: it is seven (five in 2D case) - diagonal, symmetric, positive definite and diagonally dominant, which makes Eq. (20) easily solvable by a number of iterative methods which retain the sparsity of the matrix A_ψ. Note that it does not make sense to solve Eq. (20) to a tight tolerance since its coefficients and sources are only approximate (based on the values from the previous iteration/time step). Normally, reduction of the absolute residuals for one order of magnitude suffices.

The segregated solution strategy employed enables re-use of the same storage for the matrix A and vestor **b** for all depedent variables ψ, thus requiring only $8N$ storage locations ($6N$ in a 2D case). It is also important to mention that the fully implicit time differencing used, avoids stability-related time step restrictions. In principle, it allows any magnitude of the time step to be used, and in practice it is limited only by the required temporal accuracy.

When constitutive Eqs. (11) for a thermo-elasto-plastic isotropic material are applied, an elastic deformation is assumed at the beginning of iterations of each (load increment) time step (the expression within the brackets $\langle\ \rangle$ in Eqs. (11) is omitted). In the next iteration step in CVs in which the effective stress has reached the yield stress an elasto-plastic deformation is assumed and the expression within the brackets $\langle\ \rangle$ in Eqs. (11) is activated.

After each time step (load increment) displacements and stresses are updated adding displacement and stress increments in the current time step to the total displacements and total stresses from the previous time step. This procedure is repeated until the prescribed number of time steps (or load increments) is completed.

4. Application of the method

The method described in the previous sections has been applied to a number of both linear and non linear solid body deformation problems, few of which will be presented.

4.1 Numerical predictions of the wood drying process

The wood drying process is an important step in the manufacturing of wood products. During that process a non-uniform distribution of moisture content and temperature causes deformation and stresses in the wood and may result in a deformed and/or cracked end-product.

A wood drying process can be described as an unsteady process of heat, mass and momentum transfer in an orthotropic continuum with variable physical properties. The method solves a coupled set consisting of energy, moisture potential and momentum equations (1-3) with the constitutive relations (4-6).

Beech-wood beams (600x50x50 mm³) are exposed to the uniform, unsteady flow of hot air in a laboratory dryer with an automatic control of the ambient air parameters (Horman, 1999).

The temperature and/or moisture dependent physical properties of the wood, obtained by fitting available experimental data, are given in Table 2. The others are considered constant and are given in Table 3. The timber is known to be cylindrically orthotropic. However, the wood samples used in this study are taken from the outer region of a cylindrical timber log and the rectilinear isotropy of samples is a reasonable assumption.

$E\,(Pa)$	$C < 30$ %	$C \geq 30$ %
E_{11} (Pa)	$\left(6{,}69 - 4{,}66 e^{-1{,}1 \cdot 10^{7} C^{-6{,}3}}\right)(1{,}8 - 0{,}02T)10^8$	$2{,}05(1{,}8 - 0{,}02T)10^8$
E_{22} (Pa)	$\left(13{,}22 - 9{,}3 e^{-2{,}5 \cdot 10^{6} C^{-5{,}75}}\right)(1{,}8 - 0{,}02T)10^8$	$4{,}04(1{,}8 - 0{,}02T)10^8$
E_{33} (Pa)	$\left(81{,}11 - 57{,}03 e^{-2{,}5 \cdot 10^{6} C^{-5{,}75}}\right)(1{,}8 - 0{,}02T)10^8$	$24{,}79(1{,}8 - 0{,}02T)10^8$
ρ (kg/m³)	$\dfrac{559(100 + C)}{100 - 0{,}47(30 - C)}$	$559\left(1 + \dfrac{C}{100}\right)$
c_q (J/kg K)	$467[C(100 + T)]^{0,2}$	
k_{11}^{q} (W/m K)	$1{,}36(0{,}088 + 0{,}000709T + 0{,}00181C)$	
k_{22}^{q} (W/m K)	$1{,}15k_{11}^{q}$	

Table 2. Temperature and/or moisture dependent physical properties of wood ($C = c_m M$ (%) is the moisture content)

At the beginning of the drying process the wood samples had a uniform distribution of temperature, moisture, displacement and velocity:

$$T = 21\ °\text{C},\ M = 75\ °\text{M},\ u_i = \dot{u}_i = 0 \text{ for } t = 0 .$$

Property	Value	Property	Value	Property	Value
r (J/kg)	$2{,}3 \cdot 10^6$	ν_{12}	0,36	α_{11} (1/K)	$37{,}6 \cdot 10^{-6}$
c_m (kg$_m$/kg°M)	0,01	ν_{21}	0,71	α_{22} (1/K)	$28{,}4 \cdot 10^{-6}$
k_{11}^m (kg$_m$/ms°M)	$4{,}5 \cdot 10^{-9}$	ν_{13}	0,043	α_{33} (1/K)	$4{,}16 \cdot 10^{-6}$
k_{22}^m (kg$_m$/ms°M)	$1{,}15\ k_{11}^m$	ν_{31}	0,52	β_{11} (1/°M)	$36{,}8 \cdot 10^{-4}$
G_{12} (Pa)	$3 \cdot 10^8$	ν_{23}	0.073	β_{22} (1/°M)	$18{,}0 \cdot 10^{-4}$
δ (°M/K)	2	ν_{32}	0,45	β_{33} (1/°M)	$1{,}8 \cdot 10^{-4}$

Table 3. Constant physical properties of wood

The coefficients of convective heat and mass transfer, based on the ambient air velocity of $v_a = 2$ m/s and moisture of $M_a = 10{,}5$ °M, were taken as:

$$h_q = 40 \text{ W/m}^2\text{K}, \qquad h_m = 1{,}8 \cdot 10^{-6} \text{ kg/ m}^2 \text{ s°M},$$

while the ambient air temperature and the ratio of the vapour diffusion coefficient to the coefficient of total diffusion of moisture were assumed to vary during the drying process according to the following schedules:

$$T_a = \begin{cases} 28\ °\text{C} \\ 0{,}42t + 23.8 \\ 49\ °\text{C} \end{cases} \quad \text{for} \quad \begin{cases} 0 \le t \le 10 \\ 10 < t \le 60 \\ t > 60 \end{cases} \text{ min} \qquad C_a = 10{,}5\ \%$$

$$\varepsilon = \begin{cases} 0{,}1 \\ 0{,}5 \\ 1{,}0 \end{cases} \quad \text{for} \quad \begin{cases} 0 \le t < 60 \\ 60 \le t < 3660 \text{ min} \\ t \ge 3660 \end{cases}$$

Zero surface tractions are assumed and boundary conditions Eqs. (15) are applied.

For the purpose of the numerical calculations the problem is considered to be a 2D plane strain problem. Due to the double symmetry, only one quarter of the cross-section is taken as the solution domain. For all calculations presented in this study, a uniform numerical mesh consisting of 20 x 20 CV was employed, while the time step was varied from 10 to 100 min (first seven time steps of 10 min, 31 time steps of 30 min, and finally 140 time steps of 100 min). These results are found to be grid and time independent by performing a systematic grid and time-step refinement (difference between the results on the 20 x 20 CV mesh differ from ones obtained on a 40 x 40 CV mesh for less than 1%, while the results obtained with $\delta t = 3$ h practically coincide with results obtained with $\delta t = 1{,}5$ h).

During the initial phase of drying ($0 \le t < 2$ h) the moisture content is above the fiber saturation point and the deformation is a consequence of the thermal stress only. Figure 2 shows the calculated fields at $t = 70$ min. One can see that an increase in temperature (Fig. 2a) causes the expansion of wood sample (Fig. 2b) and that the outer region is subjected to compressive and the inner region to extensive stresses (Fig. 2c, d).

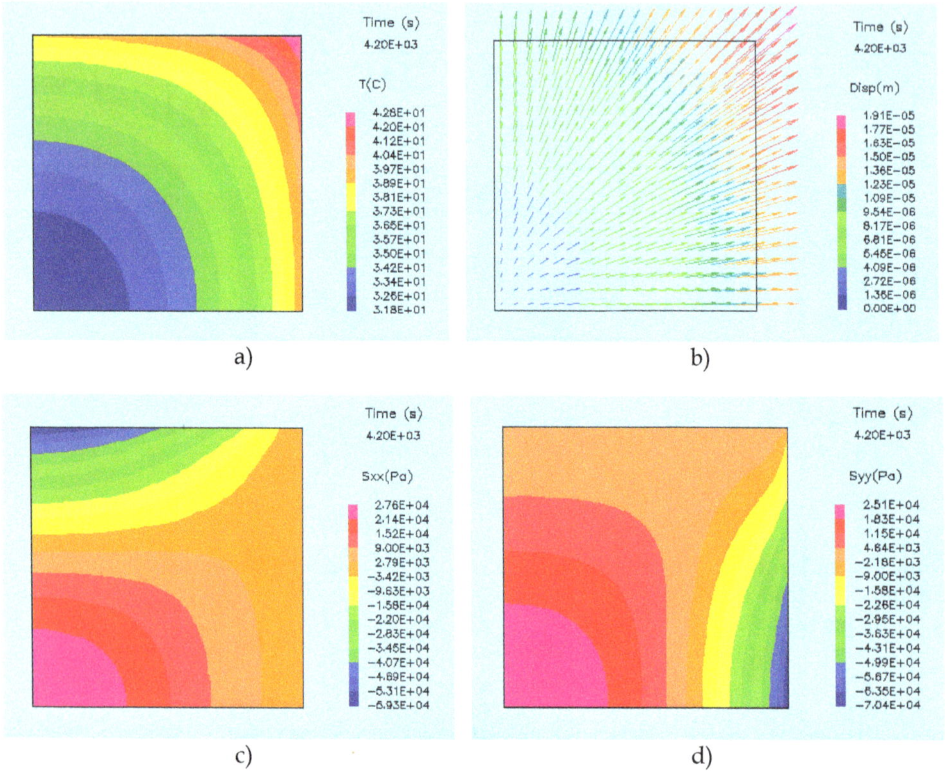

Fig. 2. Temperature (a), displacement (b), and normal stresses (c) and (d) at $t = 70$ min

During the period of intensive drying ($60 \leq t < 190$ h) the deformation and stresses due to hygroscopic loads dominate. Figure 3 shows that at $t = 108$ h the moisture content has fallen below the fiber saturation point (Fig. 3a) and that this causes the shrinking of the wood sample (Fig. 3b). Around $t = 100$ h the stresses reach their maximum values and are extensive in the outer region and compressive in the interior of the sample (Fig. 3c, d). By comparing the values of stresses at $t = 70$ min and $t = 108$ h, it can be seen that the thermal stresses are around 200 times smaller than the stresses caused by the drop in the moisture content below the fiber saturation point.

If one plots the contours of the effective stress at $t = 108$ h, when it is at its maximum (Fig. 4), one can see that the effective stress is greater than the yield stress ($\sigma_y = 10$ MPa at 10% moisture; $\sigma_y = 6$ MPa at 30%) only in a very narrow surface region (1mm deep), which indicates that the plastic defomation did not take place in the interior of the sample, and that the drying schedule is well designed.

At the end of the drying process ($t = 246$ h), the moisture content in the sample varies from 11,1 to 14,4 % (Fig. 5a), while Fig. 5b and 5c illustrate the anisotropy of the wood sample, the contraction is 1,3 mm in the x and 0,6 mm in the y direction, or 6,5% and 3,3% (axis x and y).

In order to confirm the validity of the FV predictions, the calculated temperature, moisture and displacements are compared with experimental data (Horman, 1995., Institut für Holzphysik und mechanische Technologie des Holzes, Hamburg) at reference points (Fig. 6). Figures 7 and 8 show temperature and moisture content histories at two reference points. It can be seen a good agreement between calculations and experiment: maximum difference for both temperature and moisture was 8%, and the average difference was less than 2% (Martinović et al., 2001).

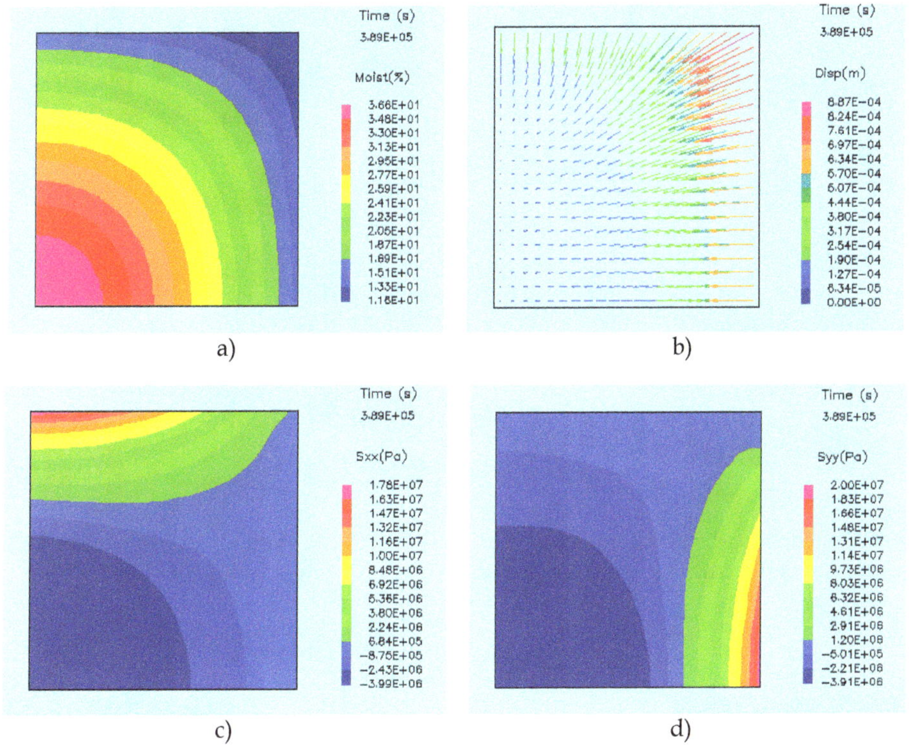

Fig. 3. Moisture (a), displacement (b), and normal stresses (c) and (d) at $t = 108$ h

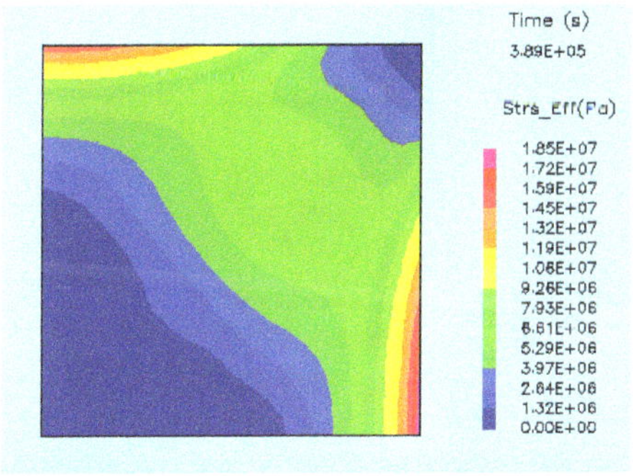

Fig. 4. Effective stress at $t = 108$ h

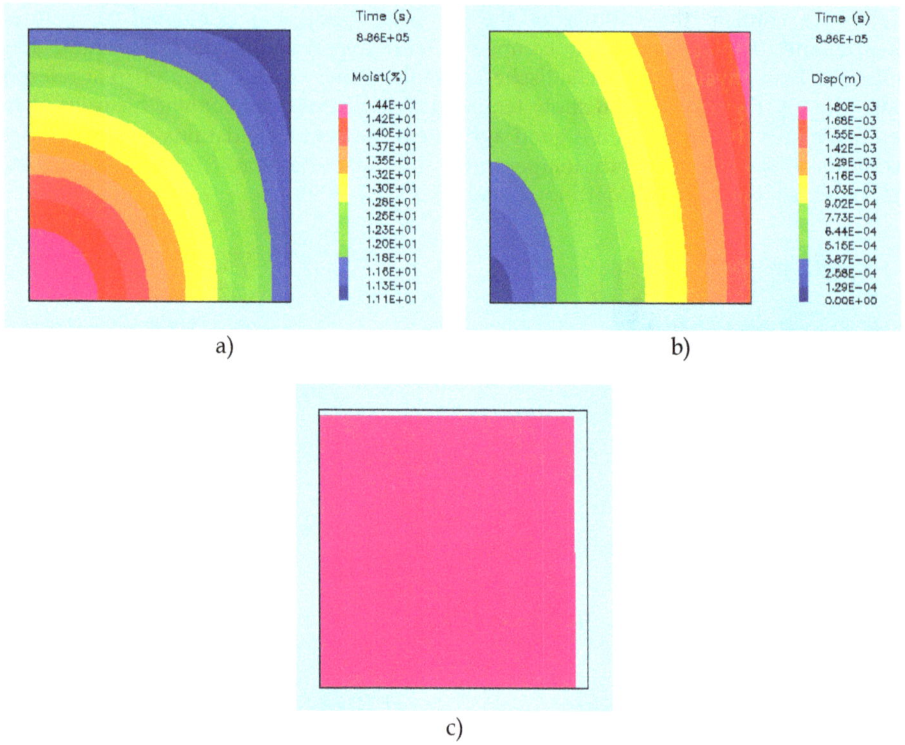

Fig. 5. Moisture (a), displacement (b), and cross section shape of deformed wood sample (one quarter of the cross section) contours at the end of drying schedule ($t = 246$ h)

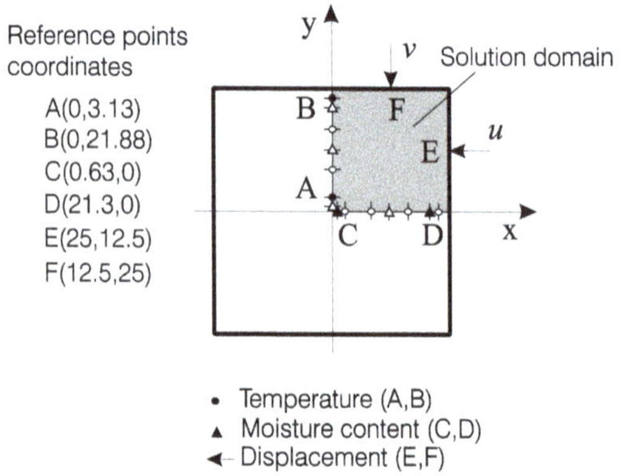

Fig. 6. Solution domain and reference points

Figure 9 shows how the displacements at two points on the surface of the sample vary during the drying process. One can see very little deformation during the initial phase ($t \leq 1000$ min) and a considerable shrinking of the sample afterwards, and that predictions closely follow experimental data (maximum difference 15%, average difference 5%).

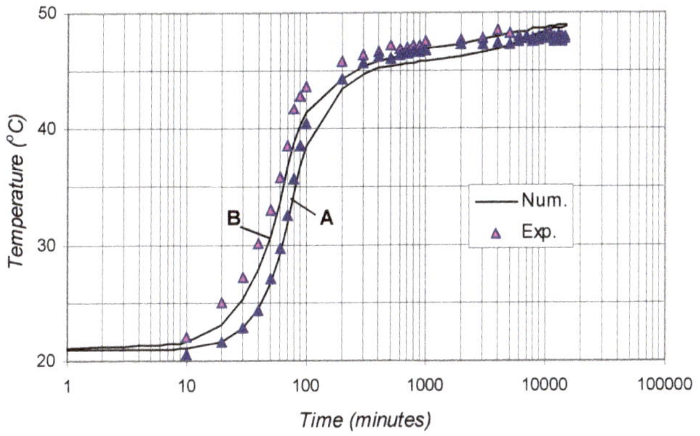

Fig. 7. Temperature history at reference points A and B

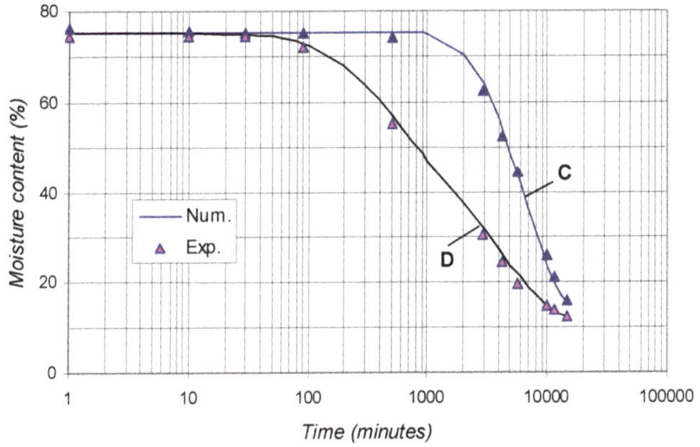

Fig. 8. Moisture content history at reference points C and D

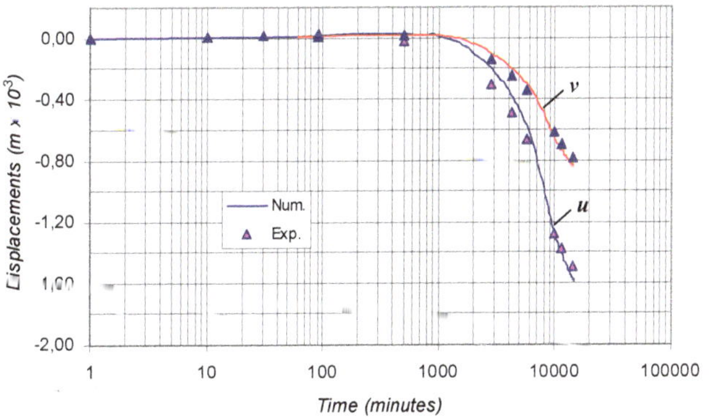

Fig. 9. u displacement at reference point E and v displacement at reference point F

4.2 Numerical predictions of the wood heat treatment process

The prediction of temperature, stresses and displacements in logs during their thermal preparation in the veneers production (wood steaming) is an important step for designing satisfying heating regime of logs preparation, without damaging in wood. The equations governing heat and momentum balance (Eqs. (1) and (3)) with corresponding constitutive relations (Eq. 11) in thermo-elasto-plastic material are solved.

For a mathematical description of a thermo-elasto-plastic deformation of the body the incremental plasticity theory is applied. The problem is considered to be a 2D plane strain problem (Horman et al., 2003).

A beech log with a diameter of 0,42 m and length of 5,1 m was exposed to steam, which temperature history during the phases of heating up, through-heating and cooling down is in Fig. 10 depicted. For numerical calculations the heat transfer coefficient $h_q = 7840\ W/m^2K$, and thermal and mechanical properties of the wood given in Table 4 are used.

ρ	c	k	E	G	ν	α	σ_y
kg/m^3	J/kgK	W/mK	Pa	Pa	-	1/K	Pa
950	2950	0,54	$4{,}3 \cdot 10^8$	$1{,}6 \cdot 10^8$	0,35	$3{,}2 \cdot 10^{-5}$	$1{,}2 \cdot 10^6$

Table 4. Thermal and mechanical properties of wood ($c = 70\%$, $T = 80^o C$)

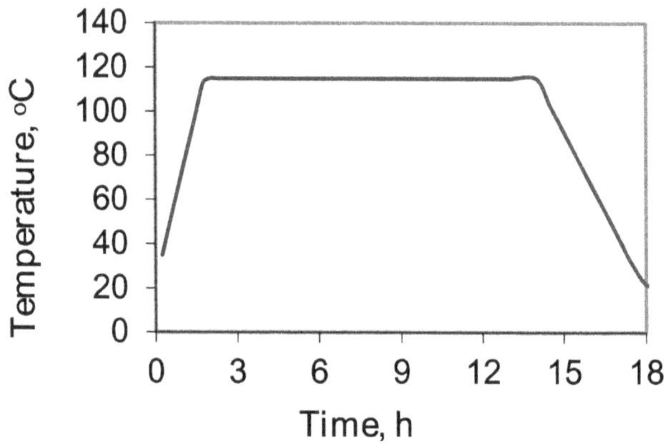

Fig. 10. Temperature history of the steam during the phases of heating up, through-heating and cooling down

Temporal temperature, radial displacement, and stress distributions at three points of the log cross section which is used for veneer production ($0{,}08m < r < 0{,}21m$) are shown in Figs. 11a-11d. In Figs. 12a and 12b effective stress distributions at three cross sections, and at four time values are depicted.

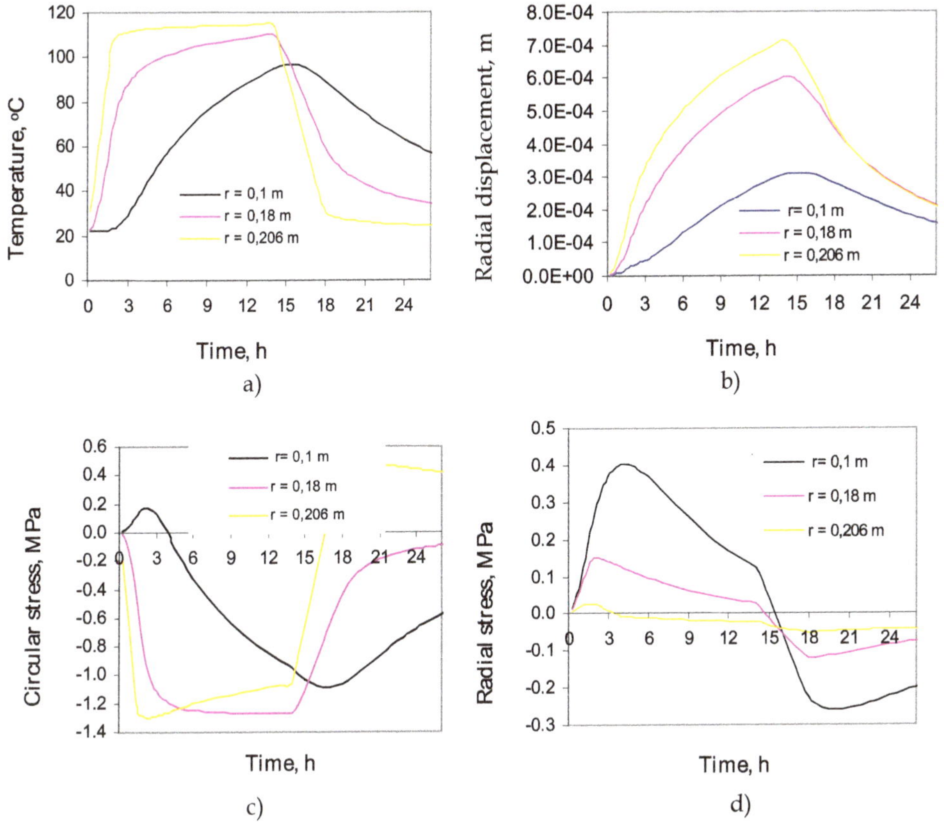

Fig. 11. Temporal a) temperature, b) radial displacement, c) circular stress, d) radial stress at three points of the log cross section $r_1 = 0,1$ m, $r = 0,18$ m i $r = 0,206$ m ($\varphi = const$) which is used for veneer production

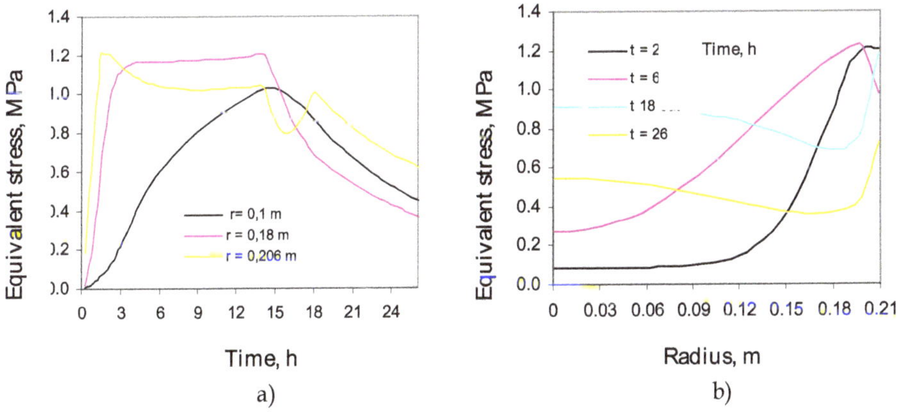

Fig. 12. Eeffective stress distributions a) at three cross sections, and b) at four time values

4.3 Numerical analysis of stress and strain conditions of a three-dimensional furniture skeleton construction and its joints

At the design stage of some pieces of furniture, their complex skeleton construction is subjected to stress and strain analysis. That allows them to satisfy all the functional demands (comfort), aesthetic demands, but also the strength and stiffness both by their shape and their dimensions. To achieve that, it is necessary to carry out a numerical simulation of the stress of a complex construction.

The finite volume method is used in the calculation. Orthotropy of the wood material is accounted for by approximating it with an isotropic material whose elastic modulus E and Poisson's ratio v are calculated by employing the least-square method. The functional Q is minimized by E and v (Martinović et al., 2008)

$$Q = \int_{sphere} \int_0^{2\pi} \Big[\left(\sigma_{xx}^{ort} - \sigma_{xx}^{izo}\right)^2 + \left(\sigma_{yy}^{ort} - \sigma_{yy}^{izo}\right)^2 + \left(\sigma_{zz}^{ort} - \sigma_{zz}^{izo}\right)^2 + \left(\sigma_{xy}^{ort} - \sigma_{xy}^{izo}\right)^2 + \left(\sigma_{xz}^{ort} - \sigma_{xz}^{izo}\right)^2 + \left(\sigma_{yz}^{ort} - \sigma_{yz}^{izo}\right)^2 \Big] d\alpha_x dA_{sphere} \tag{22}$$

and the obtained expressions are

$$E = \frac{(1+\nu)(1-2\nu)}{15(1-\nu)} \left[3(A_{xx} + A_{yy} + A_{zz}) + 2(A_{xy} + A_{xz} + A_{yz}) + 4(A_{kk} + A_{ll} + A_{mm}) \right] \tag{23}$$

$$\nu = \frac{A_{xx} + A_{yy} + A_{zz} + 4(A_{xy} + A_{xz} + A_{yz}) - 2(A_{kk} + A_{ll} + A_{mm})}{2\left[2(A_{xx} + A_{yy} + A_{zz}) + 3(A_{xy} + A_{xz} + A_{yz}) + (A_{kk} + A_{ll} + A_{mm}) \right]} . \tag{24}$$

The coefficients of the stiffness matrix A_{ij} are given in Eqs. (10).

The physical model is angle 3D joint and skeleton construction chair (Fig. 13.)

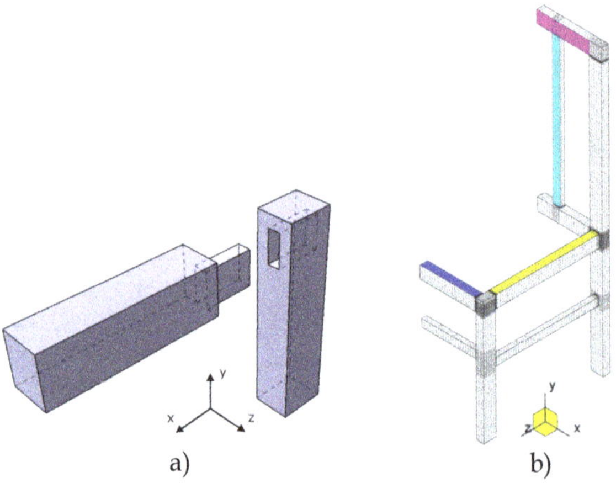

Fig. 13. a) The angle joint, b) the model of an examined chair

Mechanical properties of wood, spruce, for temperature 20°C and moisture content 9,8 %, are given in Table 5. Mass density is 0,44 g/cm³.

E_t	E_r	E_l	G_{rt}	G_{lr}	G_{lt}	ν_{tr}	ν_{rt}	ν_{rl}	ν_{lr}	ν_{tl}	ν_{lt}
GPa	GPa	GPa	GPa	GPa	GPa	-	-	-	-	-	-
0,392	0,686	15,916	0,0392	0,618	0,765	0,24	0,42	0,019	0,43	0,013	0,53

Table 5. Mechanical properties of wood, spruce

Elastic modulus and Poisson's ratio for the simulated isotropic material for 3D model (Eqs. (23) and (24)) are $E = 3,98$ GPa, and $\nu = 0,192$. The following assumptions and boundary conditions are used:

- angle joint is simplified; the glue line is neglected in a space,
- the force on the angle joint is exchanged with a uniform load.

Stress σ_{xx} and effective stress σ_{eff} contours are presented in Fig. 14.

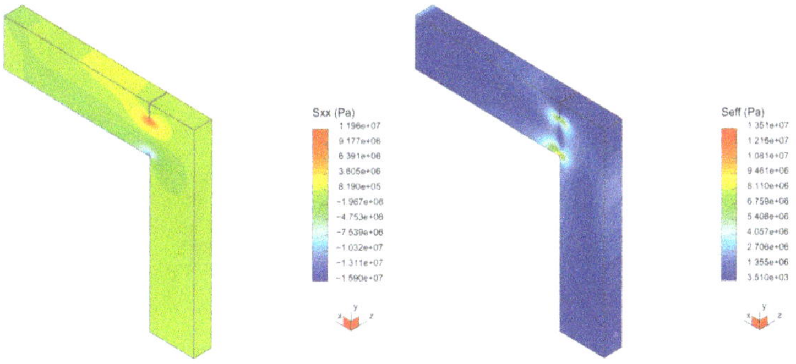

a) b)

Fig. 14. a) Normal stress σ_{xx} contours in the angle joint, b) effective stress σ_{eff} contours in the angle joint

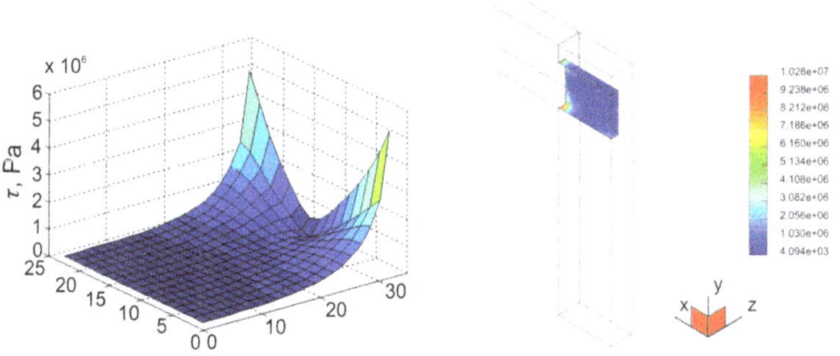

a) b)

Fig. 15. a) Tangential stress at the plane xy, at the distance of 4,6 mm from the symmetry plane b) tangential stress at the plane xy and resulting stress at the planes xz (the place of osculation of the planes of the tenon)

The highest value of the compressive stress σ_{xx} is in the symmetry plane, at undermost point of the tenon (~15,9 MPa), and the highest value of the tensile stress is at upper point of the tenon (~11,9 MPa). The place of the highest value of the effective stress is at the place of the highest compressive stress ($\sigma_{eff\,\max} = 13{,}5$ MPa). Tangential stress is presented on the plane xy at the distance of 4,6 mm from the symmetry plane. Figure 15a shows that the places of maximal stress ($\tau_{\max}$ ~5 MPa) are at $x = 35$ mm. At the same plane xy and the plane xz, in the place of osculation of the planes of the tenon, tangential stress and resulting stress (normal σ_{yy} and tangential stress τ_{yx}) are calculated, respectively and it are presented in Figure 15b. The maximal stress is ~ 10 MPa and it can be seen at the under part of the tenon.

In the end, the stress - strain analysis is done for the symmetrical half of the loaded chair (Horman et al., 2010). Mass load of the horizontal underframe of the whole chair is 100 kg and of the vertical frame is 22 kg. Effective stress contours at the elements of the chair frame and at the joints of the highest stresses ($\sigma_{eff\,\max}$ ~14 MPa) are presented in Figure 16.

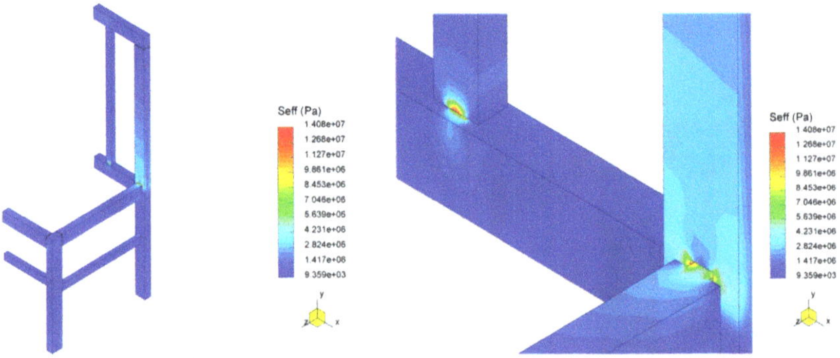

a) b)

Fig. 16. a) Distribution of effective stress at the skeleton chair, b) the joints of the highest stresses

Deformation of the chair is presented in Figure 17. The highest displacement is 13,3 mm.

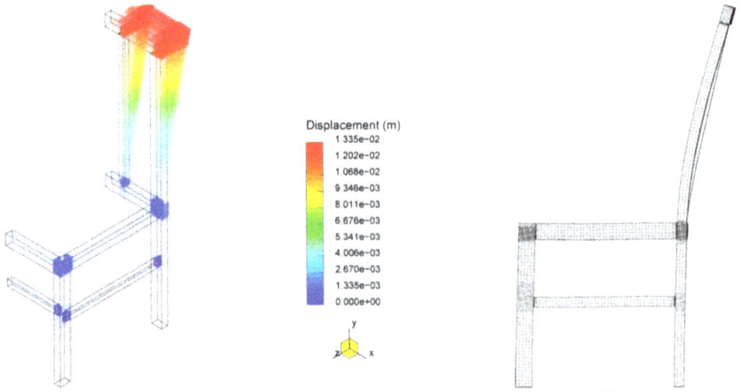

a) b)

Fig. 17. a) Distribution of displacements at the skeleton chair, b) deformed skeleton chair

5. Conclusion

The presented finite volume method for solution of the problems of energy, mass and momentum balance in conjugation with heat and mass transfer in an anisotropic, elasto-plastic, porous body is successful applied. Predictions of temperature, moisture content, strain and stress field in the wood drying as well as wood heat treatment process show high accurate results for course numerical grids due the second order accurate fully conservative spatial differencing scheme. The fully implicit unconditionally stable temporal differencing scheme enable large time steps during heat treatment processes. The applied finite volume discretisation procedure results in the diagonal dominant system of algebraic equations which are suitable for an iterative solution algorithm. The segregated iterative solution algorithm comprising the linearization and temporary decoupling of the system of equations for each dependent variable shows efficiency as well robustness solving highly nonlinear system of equations.

6. References

Demirdžić, I. & Martinović, D. (1993). Finite volume method for thermo-elasto-plastic stress analysis,*Comput. Methods Appl. Mech. Engrg.*, Vol 109, Issues. 3-4, (November 1993), pp. 331-349, ISSN 0045-7825

Demirdžić, I. & Muzaferija, S. (1994). Finite volume mtehod for stress analysis in complex domains, *Int. J. Numer. Methods Engrg.*, Vol 37, Issues. 21, (November 1994), pp. 3751-3766, ISSN 0029-5981

Demirdžić, I.; Muzaferija, S. & Perić, M. (1997). Benchmark solutions of some structural analysis problems using finite-volume mtehod and multigrid acceleration, *Int. J. Numer. Methods Engrg.*, Vol 40, Issues. 10, (May 1997), pp. 1893-1908, ISSN 0029-5981

Horman, I. (1999). *Finite volume method for analysis of timber drying,* PhD Thesis, University of Sarajevo (In Bosnian)

Horman, I.; Hajdarević, S.; Martinović, S. & Vukas, N. (2010). Numerical Analysis of Stress and Strain in a Wooden Chair, *Drvna industrija.*, Vol 61, No. 3, (September 2010), pp. 151-158, ISBN 0012-6772

Horman, I.; Martinović, D. & Bijelonja, I. (2003). Numerical Analysis Process of Wood Heat Treatment, *Proceedings of 4th International Scientific Conference of Production Engineering "RIM 2003"*, pp. 443-450, ISBN 9958-624-16-8, Bihać, Bosnia and Herzegovina, September 25-27, 2003

Luikov, A. V. (1966). *Heat and Mass Transfer in Capillary Porous Bodies,* Pergamon Press, Oxford, UK

Martinović, D.; Horman, I. & Demirdžić, I. (2001). Numerical and Experimental Analysis of a Wood Drying Process, *Wood Science and Technology,* Vol 35, No. 1-2, (April 2001), pp. 143-156, ISSN 0043-7719

Martinović, D.; Horman, I. & Hajdarević, S. (2008). Stress Distribution in Wooden Corner Joints, *Strojarstvo,* Vol 50, No. 4, (July-August 2008), pp. 193-204, ISBN 0562-1887

Patankar, S. V. (1980). *Numerical Heat Transfer and Fluid Flow,* McGraw-Hill, ISBN 0070487405, New York, USA

Integrated Technology for CAD Modeling and CAE Analysis of a Basic Hydraulic Cylinder

Radostina Petrova[1] and Sotir Chernev[2]
[1]*Technical University of Sofia, Faculty of Engineering and Education*
[2]*"HES" PLC, Yambol*
Bulgaria

1. Introduction

Usually solution of different engineering problems requires design of various objects or systems. Basically, there are three general approaches to solving engineering problems: an experimental approach, a computational approach and a computational-experimental approach, which combines both of the formentioned. Each of the first two approaches has advantages and disadvantages, while the last one joins the advantages and avoids the disadvantages of the other two. Complex engineering problems, including the presented one, are solved mainly in this way.

When selecting the most suitable computational code for solving a problem, it is obligatory to mind that each computational code is based on a mathematical model of the governing physical processes, expressed in the form of a set of equations derived from physical laws, including semi-empirical and empirical constants or relationships. Consequently, an appropriate method for solving these equations is also required.

For problems which solution is based on Finite Volume Method (FVM) (Versteeg & Malalasekera, 2007; Wendt & Anderson, 2009) the equations of the mathematical model are solved in a discrete form on a computational mesh. The solution of the mathematical problem is obtained with a certain degree of accuracy, depending on the method of discretising the differential and/or integral equations and on the method of solving the obtained discrete equations. Of course, the solution also depends on the introduced initial data. It is known that higher accurate solution requires finer computational mesh, provided through rather substantial computer memory and CPU time.

2. Theoretical background of the problem

The authors of this chapter provide and discuss an example of modeling and simulation of an engineering problem realized through SolidWorks+SW Flow Simulation+SW Motion software correctly and adequately from a physical viewpoint. The solution of this design challenge is a result of the cooperation between industry and science. Coperating company "HES" PLC, Yambol is among the best producers of hydraulic cylinders in Bulgaria. Most of its products are produced in limited series and are result of an individual design work. The idea of the study is to develop user-frendly and adaptable applicable environment, suitable for investingation on different firm designer products. It has been decided that the study should contain the following modules:

- *Technological design of a hydraulic cylinder, specified by its structural scheme.* This scheme is a part of the production plan of "HES" PLC, Yambol and describes manufacturing of each of the cylinder's components. The module is implemented through 3D CAD system - in our case SolidWorks (SW) (Lombard, 2011).
- *Numerical simulation investigating the fluid flow inside the hydraulic cylinder.* The investigation aims to calculate the active force that pushes the Piston Rod Kit (PRK) precisely (see fig. 1 and 2), to visualize the flow trajectories and consequently, to outline the vulnerable domains in cylinder structure. This module uses SolidWorks Flow Simulation (SW Flow …, 2011; Solving …, 2011).
- *Dynamic numerical simulation on Piston Rod Kit motion.* The aim is to calculate the friction force between the Double Acting Seal (DAS) and Cylinder Tube (CT) (see fig. 1), which influences strongly on maintenance, reliability and life of parts. This module is realized through SolidWorks Motion (SW Motion) software package (SW Motion ..., 2011).

2.1 Governing equations in investigating the flow work, (SW Flow..2011; Solving..,2011)

The Navier-Stokes equations, which formulate mass, momentum and energy conservation laws for fluid flows are used. They are supplemented by fluid state equations defining the nature of the fluid, including its density, viscosity and thermal conductivity of temperature. The particular problem is specified by the necessity of defining flow's geometry, boundary and initial conditions. The authors rely entirely on software capacity of predicting laminar and turbulent flows.

Generally, the state equation of a fluid has the following form:

$$P=f(p,T,y) \tag{1}$$

where $y=(y_1, \dots y_M)$ is the concentration vector of the fluid mixture components. In our case the gas is treated as an ideal one. The mass transfer is calculated under the following specific equation:

$$\frac{\partial \rho y_m}{\partial t}+\frac{\partial}{\partial x_i}\left(\rho u_i y_m\right)=\frac{\partial}{\partial x_i}\left(\left(D_{mn}+D_{mn}^t\right)\frac{\partial y_n}{\partial x_i}\right)+S_m, \tag{2}$$
$$m=1,2,\dots M$$

wherein D_{mn} and D_{mn}^t are molecular and turbulent matrices of diffusion and S_m is the rate of production or consumption of the *m-th* component.

In simulating the flow's motion we use calculating of local mean age (LMA). This is the average time τ for fluid to travel from the selected inlet opening to the pointed outlet, considering both the velocity and diffusion. It is determined by solving the following equation:

$$\sum_{i=1}^{3}\frac{\partial}{\partial x_i}\left(\rho\tau u_i-\left(\frac{\mu}{\sigma}+\frac{\mu_t}{\sigma_t}\right)\frac{\partial \tau}{\partial x_i}\right)=\rho, \tag{3}$$

where x_i is the *i-th* coordinate, ρ is the density, u_i is the *i-th* velocity component, μ is the dynamic viscosity coefficient, μ_t is the turbulent eddy viscosity coefficient, σ and σ_t are the laminar and turbulent Schmidt numbers. The equation is solved under the $\tau = 0$ boundary condition on the inlet opening. Dimensionless LMA is chosen in the provided example. This is LMA divided by the V/Q ratio, where V is the volume of the computational fluid domain and Q is the volume flow rate of the fluid entering the fluid volume.

The employed in this example numerical solution technique is standart, robust and reliable. Hence, it does not require any user knowledge about the computational mesh and the numerical methods employed. The used software package solves the governing equations through the finite volume (FV) method on a spatially rectangular computational mesh, designed in Cartesian coordinate system with the planes orthogonal to its axes and refined locally at the solid/fluid interface and, if necessary, additionally in specified fluid regions, at the solid/solid surfaces, and in the fluid region during calculation. Values of all the physical variables are stored at the mesh cell centers. Due to the FV method, the governing equations are discretized in a conservative form. The spatial derivatives are approximated with implicit difference operators of second-order accuracy. The time derivatives are approximated with an implicit first-order Euler scheme. The viscosity of the numerical scheme with respect to the fluid viscosity is negligible.

2.2 Theoretical background of dynamic simulation of RPK motion, (SW Motion.., 2011)

Static studies assume that loads are constant or applied very slowly until they reach their full values. Thus, the velocity and acceleration of each particle of the model are assumed to be zero. As a result, static studies neglect inertial and damping forces. For many practical cases, loads are not applied slowly or they change in time. Generally, if the frequency of a load is larger than 1/3 of the lowest (fundamental) modal frequency, a dynamic study should be used. Objectives of the dynamic analysis include: design of structural and mechanical systems that ought to operate without failure in dynamic environments and modifying system's characteristics (i.e. geometry, damping mechanisms, material properties, etc.) in order to reduce vibration effects. A dynamic simulation is also known as a kinetic simulation.

Many of the engineering products contain moving assemblies of components. For their analysis and correct design, it is necessary to perform a dynamic simulation of the mechanism. This is a time-history solution of all displacements, velocities, accelerations and internal reaction forces in the model driven by a set of external forces and excitations. Unlike kinematic and static simulations which involve the solution of only algebraic equations, dynamic simulations are more complex because they involve the solution of differential and algebraic equations (DAEs). It is the most complex and computationally demanding type of simulation and is meant to be used with models that have one or more degrees of freedom.

The basic algorithm available in SW Motion solver performs the numerical integration required for dynamic analyses based on stiff solution methods that use implicit, backward difference formulations (BDF) to solve the DAEs. It sets coupled differential and algebraic equations to define the functions of motion of the model. A numerical solution to these equations is obtained by integrating the differential equations while satisfying algebraic constraint equations at each time step. The set of differential equations is numerically stiff when there is a wide interval between high and low frequency eigenvalues. The possible solution of the equations of motion depends on the numerical stiffness of the equations - the stiffer the equations are, the slower the solution is. Numerically stiff differential equations require stiff integration methods to compute the solution's efficiency because other types of methods for solving differential equations perform poorly and are too slow.

There are three stiff integration methods for computing motion by SW Motion solver:

- Gear (GSTIFF)
- Modified Gear (WSTIFF)
- Stabilized Index-2 method (SI2_GSTIFF), which is a modification of the GSTIFF method.

GSTIFF integration method, developed by C. W. Gear (Gear, 1971a; 1971b), is the most widely-used and tested integrator. It is a variable-order, variable-step and multi-step integrator with a maximum integration order of six. Among benefits of GSTIFF are high speed, high accuracy of the system displacements and robust in handling a variety of analysis problems. A limitation of the procedure is that velocities and especially accelerations can have errors. An easy way to minimize these errors is to control the maximum time step that the integrator is allowed to take, so that the integrator runs at a constant step size and runs consistently at a high order (three or more). Additionally, corrector failures can be encountered at small step sizes. These failures occur because the Jacobian matrix is a function of the inverse of the step size and becomes ill-conditioned at small steps. However, GSTIFF method is a fast and accurate method for computing displacements for a wide range of motion analysis problems. It ensures that the solution satisfies all constraints, although it does not ensure that the velocities and accelerations calculated satisfy all first- and second-time derivatives. The solver monitors an integration error only in system displacements, not in velocities.

WSTIFF (Brenan, 1996) is another variable order, variable step size stiff integrator. GSTIFF and WSTIFF are similar in formulation and behavior. Both use BDF. They differ in that GSTIFF coefficients are mostly calculated assuming a constant step size, whereas WSTIFF coefficients are a function of the step size. If the step size changes suddenly during integration, GSTIFF introduces a small error, while WSTIFF can handle step size changes without loss of accuracy. Sudden step size changes occur whenever there are discontinuous forces, discontinuous motions etc.

SI2_GSTIFF, a Stabilized Index-2 method, is a modification of the GSTIFF method. This integration method provides better error control over the velocity and acceleration terms in the equations of motion. Provided the motion is sufficiently smooth, SI2_GSTIFF velocity and acceleration results are more accurate than those computed with GSTIFF or WSTIFF and that is true even for motions with high frequency oscillations. SI2_GSTIFF is also more accurate with smaller step sizes, but is still significantly slower.

All of the three integrators (GSTIFF, WSTIFF, and SI2_GSTIFF) use Newton-Raphson iterations to solve the DAEs of motion. This iteration process is referred to as correcting the solution. The adaptivity value modifies the corrector error tolerance to include a term that is inversely proportional to the integration step size. This is intended to loosen the corrector tolerance when the step size gets small. If the integration step size is equal to *h*, *Adaptivity/h* is added to the corrector tolerance. The *Adaptivity* value affects the GSTIFF, WSTIFF and SI2_GSTIFF integrators.

The control over the convergence of the calculations can be done by adjusting the values of several dynamic parametres, such as: accuracy, number of iterations, maximum step size, recalculation the Jacobian matrix.

3. Basic description of the used software technics

3.1 Detailed description of the used by Flow Simulation technics, (SW Flow..2011; Solving..,2011)

Since Flow Simulation is based on solving time-dependent Navier-Stokes equations, steady-state problems are solved through a steady-state approach. To obtain the steady-state solution quicklier a method of local (over the computational domain) time steps is employed. A multigrid method is used for accelerating the solution convergence and suppressing parasitic oscillations. The computational mesh is built by dividing the computational domain into parallelepiped cells whose sides are orthogonal to the Global coordinate system axes. Procedures of the computational mesh refinement are used to better resolve the model features, such as high-curvature surfaces in contact with fluid, narrow flow passages (gaps) and the specified insulators' boundaries. During the subsequent calculations while solving the problem the computational mesh can be refined additionally to better resolve the high-gradient flow and solid regions revealed in the calculations.

Most of the fluid flows encountered in engineering practice are turbulent, so Flow Simulation was mainly developed to simulate and study turbulent flows. To predict turbulent flows, the Favre-averaged Navier-Stokes equations are used where time-averaged effects of the flow turbulence on the flow parameters are considered. Through this procedure the Reynolds stresses appear in the equations for which additional information must be provided. To close the system of equations, Flow Simulation employs transport equations for the turbulent kinetic energy and its dissipation rate. Flow Simulation employs one system of equations to describe both laminar and turbulent flows. Moreover, transition from a laminar state to turbulent one and vice versa is possible. Flows in models with moving walls are computed by specifying the corresponding boundary conditions. Thus, for choosing model's characteristics it is necessary to remember how important the right choice of boundary conditions is. For internal flows, i.e., flows inside models, Flow Simulation offers the following two options of specifying the flow boundary conditions: manually at the model inlets and outlets (i.e. model openings), or to specify them by transferring the results obtained in another Flow Simulation calculation in the same coordinate system. With the first option, all the model openings are classified into "pressure" openings, "flow" openings and "fans", depending on the flow boundary conditions, which are intended to be specified on them. A "flow" opening boundary condition is imposed when dynamic flow properties (i.e., the flow direction and mass, volume flow rate or velocity) are known at the opening. The pressure at the opening is determined as a part of the solution. In Flow Simulation the default velocity boundary condition at solid walls corresponds to the no-slip condition and the solid walls are also considered to be impermeable. In addition to this, the wall surface's translation and/or rotation can be specified.

3.2 Detailed description of the used by SW Motion integration technics, (SW Motion.., 2011)

SW Motion uses complete kinematic modeling to compute component motion and dynamic calculations to analyze forces, excited in mechanisms by springs, dampers, motors, gravity, friction, etc. The large and diverse number of kinematic connections in the program's libraries enables the creation of mechanisms with varying degrees of complexity. They can be designed through setting one or more drives or through defining a trajectory or a set of trajectories of a particular point from the mechanism's parts. Tracking the motion parameters (linear and angular speeds and accelerations, etc.) as a function of time or of another chosen by the user parameter is also enabled.

For obtaining the final results SW Motion solver provides three types of integration methods: GSTIFF (default), SI2_GSTIFF and WSTIFF.

The user directly controls the following integration options:

- *Maximum Iterations* - Default value is 25. It specifies the maximum number of times the numeric integrator iterates in the search of a solution for a given time step. If the program exceeds this limit, a convergence failure is recorded.
- *Initial Integrator Step Size* – The command enters the first integration step size used by the variable step integrator. The initial integrator step size controls the speed at which the integration method starts and its initial accuracy. The user can run the simulation quicklier in subsequent runs by increasing this value.
- *Minimum Integrator Step Size* - This is the value of the lower bound of the integration time step. The simulation time can be decreased by increasing this value.
- *Maximum Integrator Step Size* – This enters the upper bound of the integration time step. This is important if the integration method does not detect short-lived events such as impacts. Otherwise, it is recommended this value be of the same order as the short-lived events. If the user sets this value too large, some events can be ignored by the simulation.
- *Jacobian Re-evaluation* – It enables the frequency of matrix re-evaluation to be controlled. More frequent re-evaluation gives better simulation accuracy at the cost of simulation time. If the model does not change significantly over time, a smaller Jacobian re-evaluation option can be used.

3.3 Friction phenomenon in SW Motion, (SW Motion..., 2011)

Friction is a resistive force that occurs in joints and between parts in contact. When parts are in contact, friction is calculated based on the static and dynamic coefficients of friction and the normal force acting on the part. The static and dynamic friction properties applied to the contact calculation are basically derived from the material properties. SW Motion incorporates dynamic friction into the contact calculation. Contact friction is the friction that occurs between bodies in contact. The following friction parametres can be monitored:

- *Dynamic Friction Velocity* - Specifies the velocity at which dynamic friction becomes constant.
- *Dynamic Friction Coefficient* - Specifies the constant used to calculate forces due to dynamic friction.
- *Static Friction* - Includes static friction in the contact calculation.

- *Static Friction Velocity* - Specifies the velocity at which the static frictional force is overcomed so that a stationary component begins to move.
- *Static Friction Coefficient* - Specifies the constant used to calculate the force necessary to overcome forces between two touching bodies at rest.

In the presented example the contact friction force is calculated. The velocities and coefficients of friction used are assigned automatically based on the predefined for each contact set materials. However, they may not be the most appropriate parameters based on the dynamics of the model. Then, if necessary, these coefficients can be set manually. SW Motion uses the Coulomb friction method and fits a smooth curve to the friction parameters to solve the friction force.

4. Basic description of the used SolidWorks software

4.1 CAD/CAE systems

The integration of CAD/CAE systems in developed manufacture gives experts the opportunity to put fewer resources and less energy into technical activities supporting the process of design. Thus they focus their effort and intellectual capability on creating innovative and optimal solutions or on generating new ideas for solving known or new problems and challenges. CAD (Computer Aided Design) by definition means a combination of hardware and software used for optimal solution of the geometric problems in product design. CAD systems solve mainly tasks related to describing the geometry of components, the assemblies and products as a whole. In recent years the capabilities of CAD systems have been expanded. Today they have been successfully integrated to other subsystems (CAE, CAM, etc.), aiming to avoid manual reformulation of data and to connect individual work environments. While the term CAD includes all the geometry-oriented tasks, CAE (Computer Aided Engineering) covers all the computing tasks that take place within the designing process. This involves both all calculations during the designing and all optimising procedures in achievement of constructive solutions. There is a steady interaction between CAD and CAE, since the geometry generated by the CAD system is often the basis for CAE and vice versa (Янакиев & Николов, 2010; Топалова & Бакърджиев, 2006).

The creation and processing of the product's digital model is the basis of the engineering design methodology using CAD/CAE systems. It includes geometric characteristics, data about the material and the properties of the product, some additional specific information for the designed product whereas the main component is the geometrical model. It is created and displayed as a graphic image on the screen display. The principles of operation in geometric modeling are independant from the software environment. The impact of CAE technologies on engineering has been increasing since the second half of the previous century so that it could encounter the enforced requirements for higher precision and suitability of the model and environment. Modern development of computing devices and software allows the users to model the process of deformation of complex-shaped components and structures made of different materials.

4.2 SolidWorks (SW) CAD environment

SolidWorks is also used as a software platform for many other programs. SolidWorks is an integrated CAD/CAE system, providing a unified and interconnected environment for design, for structural and dynamic analyses, for data management of engineered products. It was developed in 1993 and since 1997 has been a trade brand of French company Dassault Systèmes. SolidWorks is known to be among the most widely used standard softwares for 3D designing. The system incorporates modules for preparing technical documentation, kinematic, dynamic, thermal and strength analysis of structures, flow simulation technics, design of specific product (sheet metal products, mold products, etc.) and so on.

Based on a multifunctional interface, SW has a parametric, 100% associative and hybrid modeler. The large software libraries provide a lot of parametric 3D parts, ready to be implemented in the model. SolidWorks enables joining of various parts together and developing different structures and mechanisms. It enables establishement of complex mechanical relations, checking the assemblies for interference, collisions and alignment of the parts. SolidWorks imports/exports files to AutoCAD, DraftSight (a free product developed by Dassault Systèmes) and some other well-known software products.

4.3 CAD modeling of a hydraulic cylinder in SW environment

For the establisment of a CAD model, design drawings of all parts, constituting the hydraulic cylinder as a part of the production program of "HES" PLC, Yambol are used. The developed hydraulic cylinder is a double acting cylinder of a piston type. Both piston courses (forward and backward) are carried out under the influence of a working fluid (hydraulic oil). The components of the hydraulic cylinder are shown in fig.1.

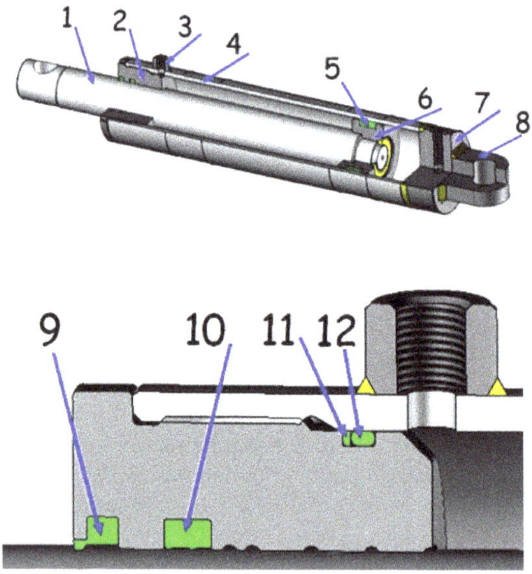

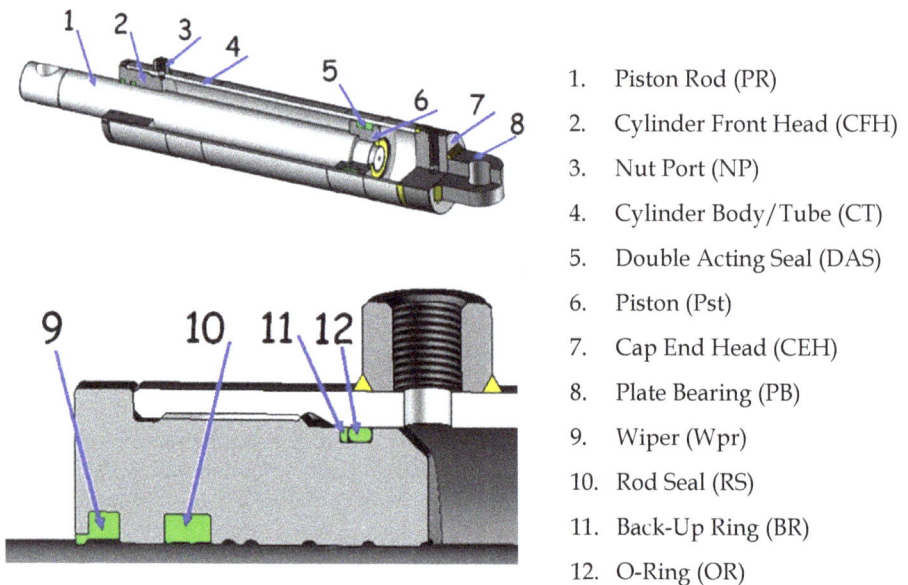

1. Piston Rod (PR)
2. Cylinder Front Head (CFH)
3. Nut Port (NP)
4. Cylinder Body/Tube (CT)
5. Double Acting Seal (DAS)
6. Piston (Pst)
7. Cap End Head (CEH)
8. Plate Bearing (PB)
9. Wiper (Wpr)
10. Rod Seal (RS)
11. Back-Up Ring (BR)
12. O-Ring (OR)

Fig. 1. Components of a hydraulic cylinder

The parts of the hydraulic cylinder are constructed in a precise manner. Fig. 2 shows the structural block diagram of hydraulic cylinder's manufacturing. Each block of the structure incorporates the abbreviations of the parts and the digital identification numbers, by which they are classified in the product management system. These numbers are unique for each object. The given block diagram shows that the PTHC (Hydraulic Cylinder of Piston Type) is located at the highest organizational level. It represents the final product which incorporates all components of the hydraulic cylinder structure. In the production management system PTHC exists under the digital identification number 311113597, which specifies its particular classification. Through it, the cylinder is controlled by the design and operation systems.

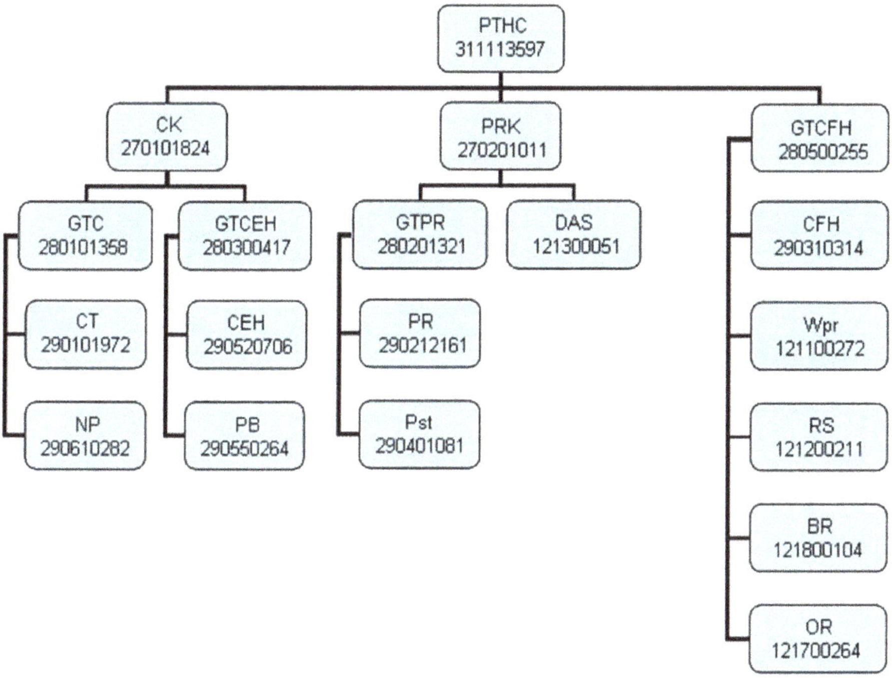

DAS - Double Acting Seal
CT - Cylinder Tube
PR - Piston Rod
Pst - Piston
CFH - Cylinder Front Head
CEH - Cap End Head
PB - Plate Boaring
NP - Nut Port
OR - O-Ring
BR - Back-Up Ring

RS - Rod Seal
Wpr - Wiper
GTC - General Type Cylinder
GTPR - General Type Piston Rod
GTCFH - General Type Cylinder Front Head
GTCEH - General Type Cap End Head
CK - Cylinder Kit
PRK - Piston Rod Kit
PTHC -Hydraulic Cylinder of a Piston Type

Fig. 2. Structural block diagram constituting a hydraulic cylinder

- The studied PTHC is assembled through the following few steps:

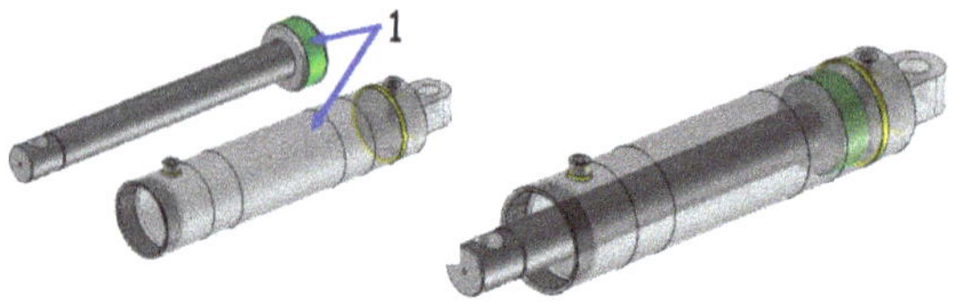

Step 1 -setting the mates between the axes of the Piston Rod Kit and of the cylinder. Thus, the relative degrees of freedom of the PRK to cylinder body are set to 2: longitudinal translation and rotation around its axis.

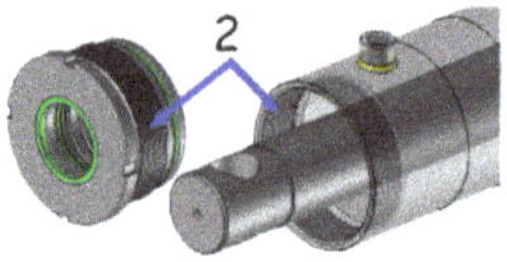

Step 2 - setting mates between the axis of the cylinder front head and the axis of the cylinder. The axis of the cylinder front head is automatically alined towards the cylinder axis.

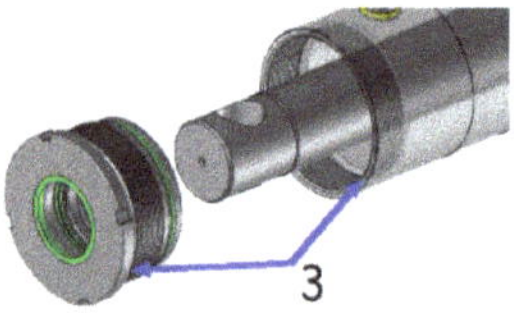

Step 3 - setting „Face to Face"the front surface of the cylinder front head and the front surface of the cylinder. The cylinder front head automatically touches the front of the cylinder.

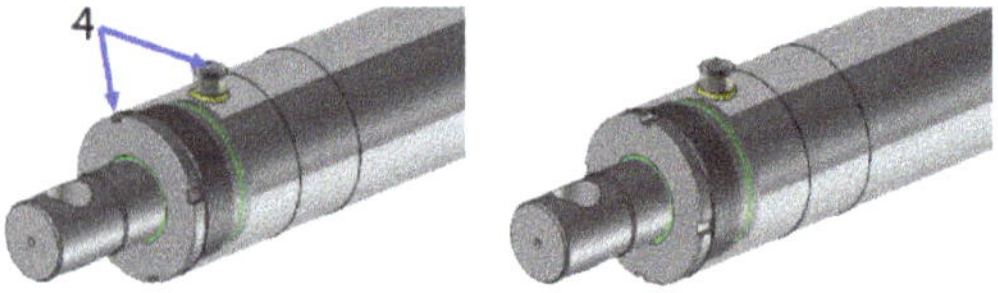

Step 4 - orientating the cylinder front head towards the nut port opening and establishing the final PTHC appearance.

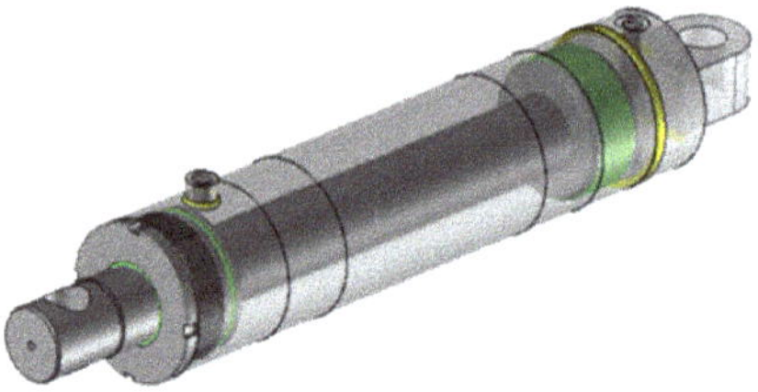

Fig. 3. Assembling of a hydraulic cylinder of a piston type (PTHC)

4.4 SolidWorks (SW) Flow Simulation environment

SW Flow Simulation package is designed for simulation of fluid and gas phenomena. It enables development of various scenarios for modeling different fluids and simulating their floating, including calculating of speed, acceleration, turbulence, etc. of fluid particles, of the forces on the surrounding walls, of the heat exchange effects, etc. The library set of examined fluids varies starting from air and water through gas and liquid chemicals to ice cream paste, honey, plastic melts, blood, etc. The user can also create user-defined fluids. Some of the main features of SW Flow Simulation enable modeling of: heat exchange effects with conduction or convection; heat exchange through radiation, including calculations on solar radiation; fluid simulation of gas and fluid flows in valves, regulators and pipes;

simulation of complex rotating flows in mechanisms. Additional SW Flow Simulation options enable: investition on externally wrapped moving objects; analysis of turbulent flows and illustration of trajectories of turbulating particles; simulation of real gasses for more precise analysis of high pressure or low temperature problems for example identification of domains where cavitation is possible and so on.

4.5 SW Flow Simulation analysis of the modeled hydraulic cylinder

Through this simulation the design of the real working environment of the hydraulic cylinder under a specified variety of boundary conditions, real time monitoring the motion of the fluid inside the cylinder, examination of the magnitude of the force between the plunger and the cylinder walls for different working volumes of the two cylinder chambers and systematizing the numerical and graphical results of all performed numerical experiments are done.

Depending on a predefined scenario of hydraulic cylinder work SW Flow Simulation software tracks the fluid flow trajectories inside the volumes of the hydraulic cylinder, regarding their development with time. All mathematical procedures are influenced by the boundary conditions and the scenario goals and are described below.

Through the first stage a mathematical definition of the investigated volume is done. The local coordinate systems inside each examined chamber volume are introduced and the computing domain is outlined. The disposition of the fluid, particularly its location, form and type (viscosity, density, operating temperature, etc.) is also input.

The boundary conditions and the type of mathematical calculation, including choosing the solver, are adjusted at the second stage. Thus, working environment is described.

The numerical and visual presentations (graphical or as a movie) are the focus of the third stage. All obtained data is systematized by the software and can be analized by the user.

To enable SW Flow Simulation to run the simulation and to set goals correctly, it is necessary to specify the material of each part of the model. Since the provided hydraulic cylinder is a product system of "HES" PLC, Jambol, all material characteristics correspond to the technical data given in its technical documentation. The structured parts are made of carbon steel. The materials of all non-structural parts are also defined.

To perform the research a new project is started. The SI measuring system is chosen.

After that the following characteristics of the "internal" fluid are specified: location in a bounded space; no cavities; positive gravity direction - along +Y axis; axis of motion - Y. Other applicable options like heat conduction, radiation, etc are not considered in this case.

Next, the type of the fluid is chosen. It is oil of type ISO-L-HL, according to standart system ISO 6743/4, also known as MH-L32. It posesses the following characteristics: viscosity class - 32; density at 20°C - 0,874 g/ml; kinetic viscosity at 40°C - 32 mm^2/s; viscosity index - 96; liquefaction temperature - -21°C. As there is no such type of fluid in SW database, a new one is assigned.

The next few steps implement computational domain function (fig 4-left), fluid subdomains (fig. 4-right), boundary conditions (fig. 5) and computational goals. The precise introduction of computational domain is of significant importance for its insufficient volume may cause inaccurate results, while its unreasonable increase "aggravates" the model and extends the necessary CPU time. The surrounding surfaces of inside chambers are highlighten to outline the fluid subdomains. The two volumes should be totally separate, should have no common overlap and should be completely bounded. Among the defined goals are the factors, which influence the system operation, such as the pressure magnitude, including its dynamic and static components. „Flow Trajectories"command is used to track the motion of the fluid particles. Obtained data can be exported to files of the following types avi, excell or pictures (*.jpg or others).

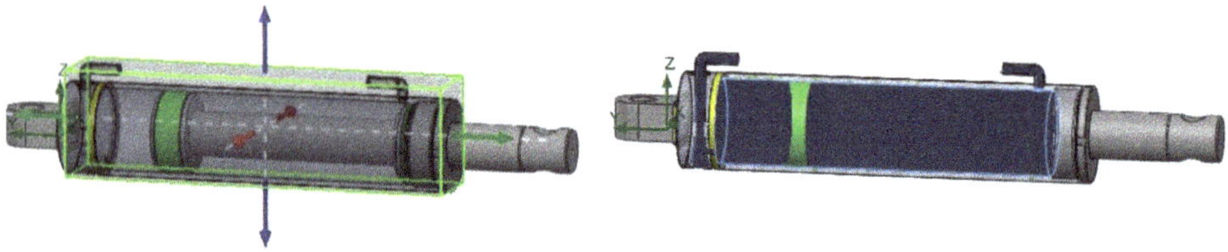

Computational domain Fluid sub-domain

Fig. 4. Introducing the of the examined fluid domains

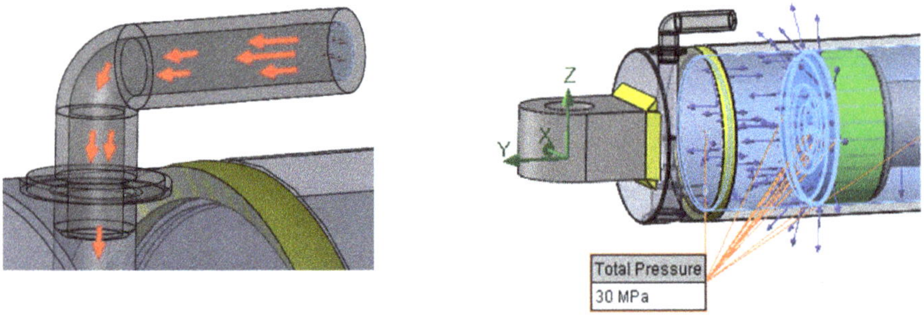

„Pipe" domain „Chamber" domain

Fig. 5. Defining the boundary conditions of the incoming fluid

During the first stage of the simulation the piston rod kit (PRK) moves forward.

To define the boundary conditions of the incoming fluid the following steps are passed through:

- For the the „pipe" domain (fig. 5 - left):
  1. Specifying the opening (area, surface) through which the fluid fills the pipe.
  2. The direction of fluid motion is selected.
  3. "Flow Openings" function is introduced by specifying properties (weight, volume or speed) of the fluid filling the pipe. In our case this is the specific flow rate. The required flow rate for the designed hydraulic cylinder is 35 l/min or 0,00058 m^3/s.
- For the the „chamber" domain (fig. 5 - right):
  4. Highlightening all the inside walls which outline the first volume and which are exposed to pressure.

5. Highlightening the direction of the rod motion - Y.
6. Selecting "Pressure Openings" function and specifying surfaces, exposed to pressure. For this project "Total Pressure" option is chosen and the pressure value is 30MPa. This is the test pressure, according to hydrauluc cylinder's technical documentation.
7. The temperature of the fluid is set at 50°C.

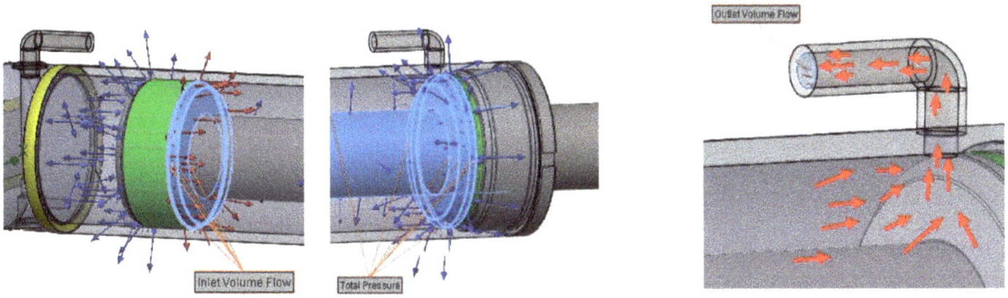

„Chamber" domain „Pipe" domain

Fig. 6. Defining of boundary conditions for the outgoing fluid

Boundary conditions of the outgoing fluid (fig. 6) are set, depending on following requirements:

1. To visualize the simulation of the fluid going out of the right chamber, it is obligatory to define the boundary conditions. While moving, the piston pushes the fluid out of the right chamber. The pressure resulting from the fluid expulsion is introduced throught highlightening the affected surfaces. All these actions are similar to the ones described above.
2. The boundary conditions at the fluid exit are input about the inner surface of the pipe. The procedure "Outlet Volume Flow" is used.

Thus all boundary conditions are precisely defined.

The next step is to define the "Goals" of the output, i.e. the results that are going to be analysied by the user later. The aim of the study is to find the value of the force, raised by fluid and moving the piston.

5. Numerical data and results obtained through SW Flow Simulation for the modeling hydraulic cylinder

5.1 Moving forward of the piston rod kit

5.1.1 Basic assumptions

The course of the piston of the hydraulic cylinder is 310 mm (in both directions). This means that the distance L varies from 0 ÷ 310mm (fig. 7). As the used software is unable to solve the system for the three processes (filling the left chamber, moving the piston and empting the right chamber) going on simultatiously the following assumption is made: The system is solved for some close chain steady state situations while the initial boundary conditions for each situation coincide with the final boundary conditions of the previous situation.

To enable calculation of the magnitude of force $\vec{F}$ (fig.7) depending on the piston's position, the operating volume of the left chamber of the hydraulic cylinder increases as distance L increases and the simulation is run for each particular position of the piston.

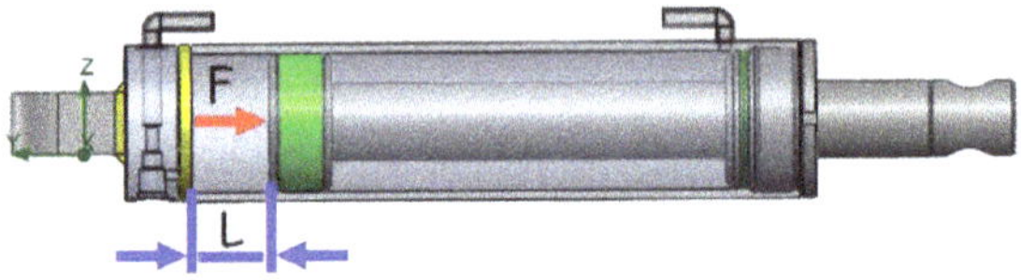

Fig. 7. Operating volume of the hydraulic cylinder

As the formulation and the goals of the problem are defined the calculation should be run.

5.1.2 Visualizing the results and comments on the simulation process

A preview of operating hydraulic cylinder during forward piston course is shown in fig. 8. The PRK tries to move forward in *Vrod* direction, while squeezing forward the outgoing fluid. Consequently the volume of the right chamber shrinks.

Visualized as separate vectors flow trajectories enable easier tracking of the motion of fluid particles (fig. 8 and 9).

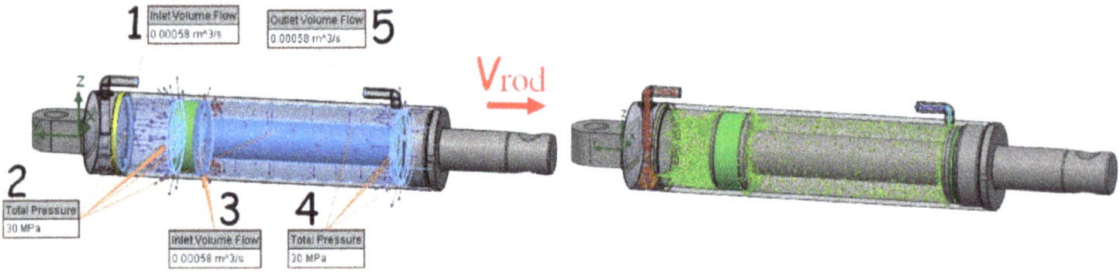

Boundary conditions Flow trajectories

Fig. 8. A preview of an operating hydraulic cylinder - PRK moving forward

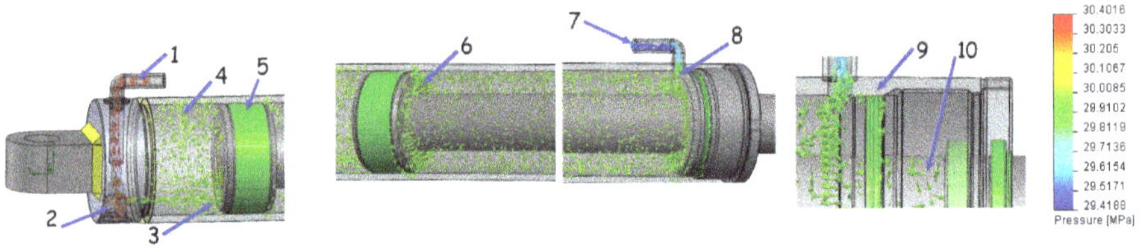

Fluid filling the chamber Fluid empting the right chamber Legend

Fig. 9. Flow trajectories of fluid particles inside the hydraulic cylinder for L= 65mm.

- Position 1 - The working fluid fills the volume through a pipe of a diameter of 8 mm with a flow rate Q = 35 l/min.
- Position 2 - The fluid particles form a small vortex at the bottom and track forward through an opening of 6mm diameter. For the presented situation the volume of the left cylinder chamber is $V = 0{,}000413m^3$ and is filled for t=0,71s.
- Position 3 - The velocity of the fluid filling the left cylinder chamber as a strongly directed stream, is about 20,6 m/s.

- Position 4 - Some of the fluid particles re-bounce off the piston and vortex around. According to software calculation, their speed is approximaely 10m/s.
- Position 5 - PRK separating the two volumes.
- Position 6 - Moving forward the piston initiates the fluid expulsion from the right chamber. The pressure of the fluid is about 29,95 MPa. The fluid particles try to squeeze the piston head and press the DAS.
- Position 7 -The fluid particles slide throught the chamfer of the cylinder front head and float up the pipe with a velocity of about 11,6 m/s and pressure of about 29,7MPa.
- Position 8 - The fluid leaves the computational domain through the pipe, which diameter is 8 mm. The pressure inside this domain decreases slightly to about 29,4 MPa.
- Position 9 - Some of the fluid particles have passed through the joint between the cylinder front head and the cylinder and have disposed around the ring.
- Position 10 - Some other particles of the fluid have disposed in the volume surrounded by the rod surface and the screw channel of the cylinder front head, re-bouncing off the rod seal.

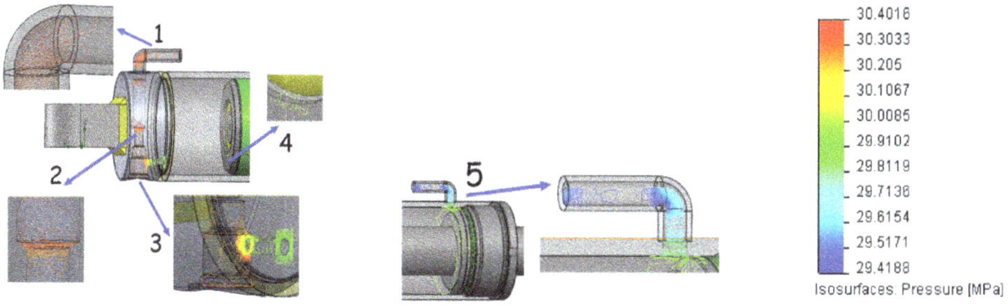

Fig. 10. Dangerous domains in the hydraulic cylinder chambers

To define the dangerous domains during hydraulic cylinder operation, i.e. the domains where the flows pressure is high, "Isolines" function is used - fig. 10. This figure enables user to find the potential crack originators in the cylinder structure.

- Position 1 - The fluid fills the pipe with high velocity. It is possible for cracks to be originated at the corner of the pipe.
- Position 2 - The fluid floats through a zone chamfering pipes with different diametres.
- Position 3 - The high pressure squeezing the fluid inside the hydraulic left cylinder chamber can cause vulnerability around this zone.
- Position 4 - The fluid particles re-bounce off the piston of the highest velocity. The chamfered piston profile is exposed to random time-varying pressure.
- Position 5 - The fluid is released at a high speed, cracks are possible to occur in the joint between the cylinder tube and the pipe.

5.1.3 Numerical data and verification of the results

All numerical results of any importance to the study are exported to MS Excel file. Some of them are systematized in table 1 and graphically presented in fig. 11. During the second half of the piston's forward course the magnitude at first decreases slightly and then remains constant.

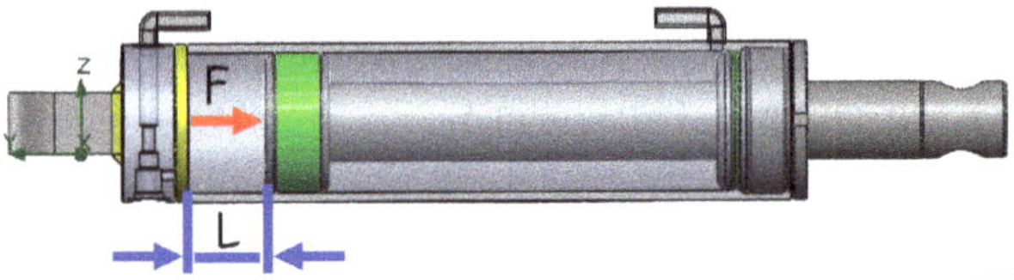

№	L, [mm]	V, [m³]	t, [s]	Vr, [mm]	F, [kN]	Legend
1	5	3,179E-05	0,055	5,13	190,7552	**L** - the distance defining PRK position inside the cylinder; **V** - the volume of the left cylinder chamber; **t** - the time for filling the chamber of the hydraulic cylinder with fluid; **Vr** - the displacement of PRK in forward direction during its motion; **F** - the calculated by the SW Flow Simulation magnitude of the tracked force during the forward motion of PRK;
2	20	1,272E-04	0,218	20,32	190,7474	
3	35	2,225E-04	0,382	35,62	190,7397	
4	50	3,179E-04	0,545	50,81	190,7321	
5	65	4,133E-04	0,709	66,10	190,7245	
6	80	5,087E-04	0,872	81,30	190,7181	
7	95	6,041E-04	1,036	96,59	190,7111	
8	110	6,994E-04	1,199	111,79	190,7067	
9	125	7,948E-04	1,363	127,08	190,7048	
10	140	8,902E-04	1,526	142,27	190,7041	
11	155	9,856E-04	1,690	157,56	190,7039	
12	170	1,081E-03	1,853	172,76	190,7037	
13	185	1,176E-03	2,017	188,05	190,7038	
14	200	1,272E-03	2,180	203,25	190,7036	
15	215	1,367E-03	2,344	218,54	190,7041	
16	230	1,462E-03	2,507	233,74	190,7038	
17	245	1,558E-03	2,671	249,03	190,7036	
18	260	1,653E-03	2,834	264,22	190,7041	
19	275	1,749E-03	2,998	279,51	190,7039	
20	290	1,844E-03	3,161	294,71	190,7036	
21	305	1,939E-03	3,325	310,00	190,7038	

Table 1.

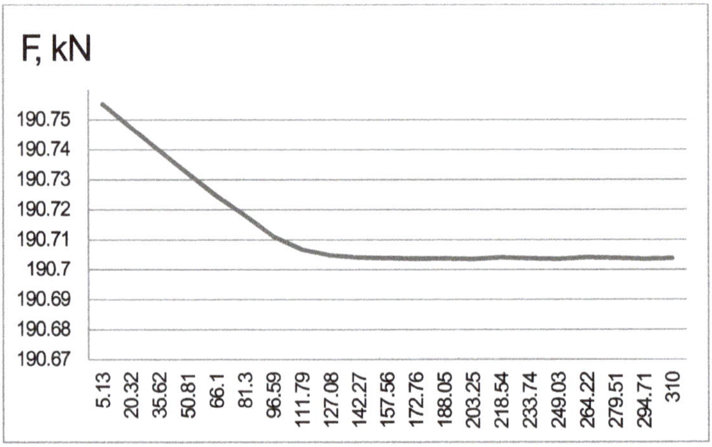

Fig. 11. Graph of the magnitude of force $\vec{F}$ versus forward displacement of the piston **Vr**

The calculated magnitude of the force $\vec{F}$ is verified through the basic methods of Fluid Mechanics, as follows:

- test pressure of the hydraulic cylinder - $P = 30MPa$;
- the area at "face" surface of the piston - $A = \frac{\pi.D^2}{4} = \frac{3,14.90^2}{4} = 6358,5mm^2$
- calculated force - $F = 30.6358,5 = 190755N \Rightarrow 190,755kN$
- error - $\Delta\% = \frac{190,755 - 190,713}{190,755} 100\% = 0,022\%$

The error is under 0,025%, where the value 190,755 is the theoretically calculated force magnitude and the value 190,713 is the average value of the results run by the software.

5.2 Moving backwards of the piston

5.2.1 Basic assumptions and boundary conditions

The method and the stages of simulating the backward motion of PRK are the same as they are during its forward movement simulation but the set boundary conditions are different.

Fig. 12 shows the sequence of setting the input boundary conditions.

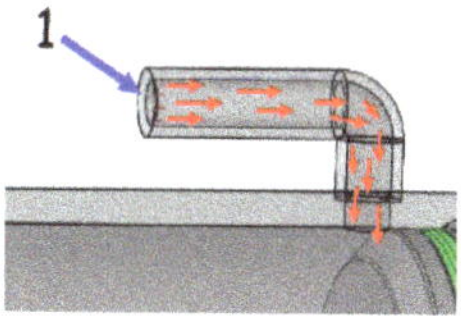

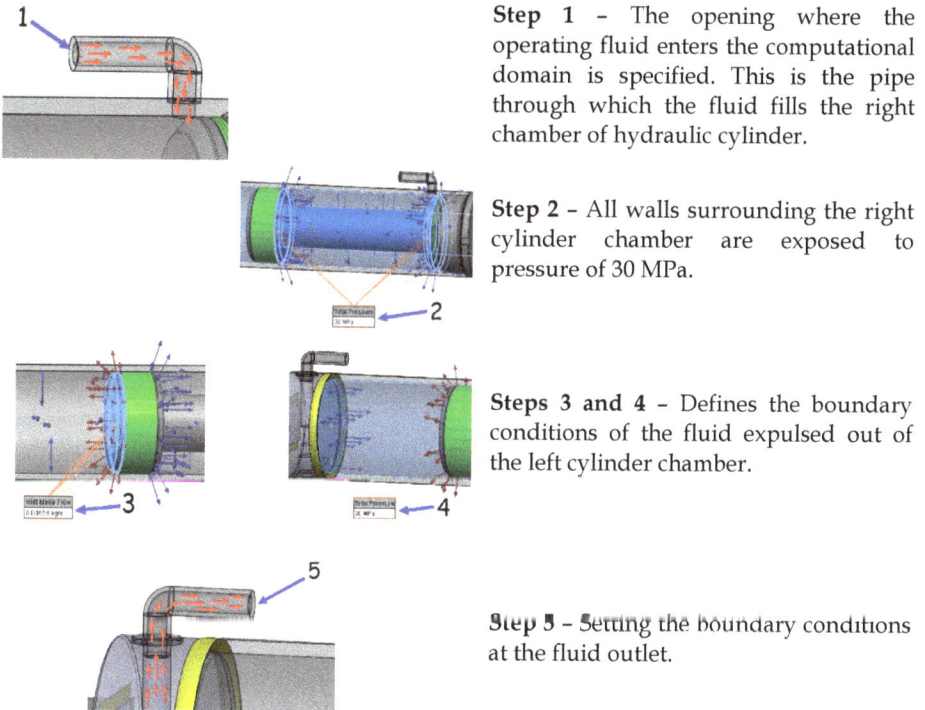

Step 1 - The opening where the operating fluid enters the computational domain is specified. This is the pipe through which the fluid fills the right chamber of hydraulic cylinder.

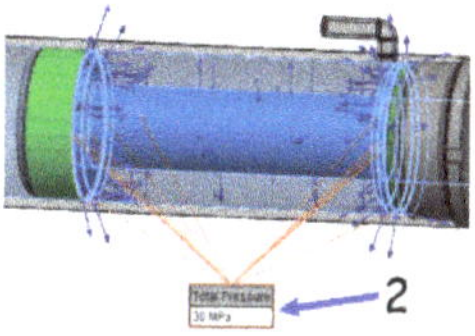

Step 2 - All walls surrounding the right cylinder chamber are exposed to pressure of 30 MPa.

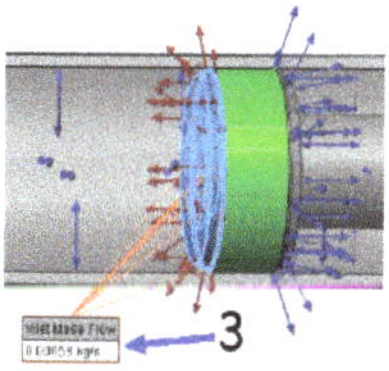

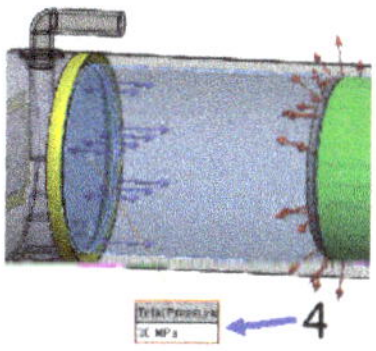

Steps 3 and 4 - Defines the boundary conditions of the fluid expulsed out of the left cylinder chamber.

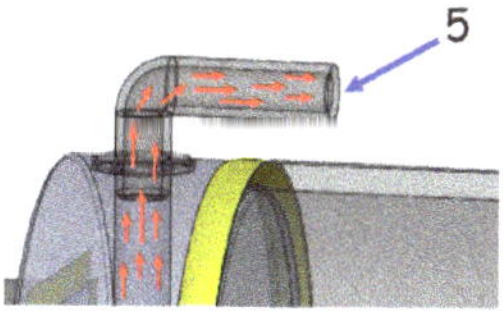

Step 5 - Setting the boundary conditions at the fluid outlet.

Fig. 12. Setting the boundary conditions during PRK backward motion

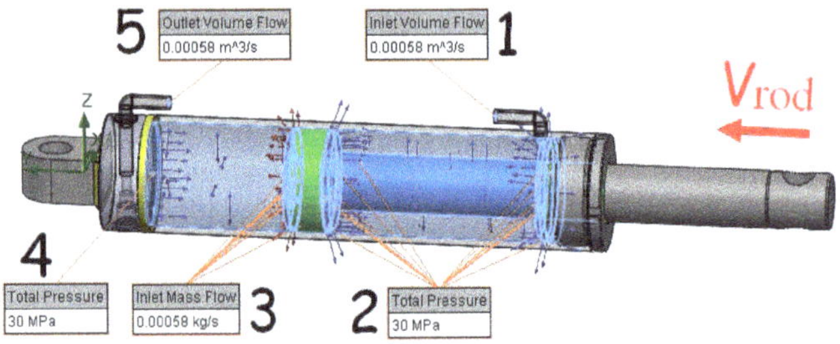

Fig. 13. Scheme of the operating cylinder during piston's backward course

Scheme of the operating cylinder during piston's backward course is given in fig. 13.

- Position 1 - The fluid fills the right chamber of the hydraulic cylinder throught the pipe opening at a flow rate of $0.00058m^3/s$
- Position 2 - The inside pressure of 30 MPa is established.
- Position 3 - PRK moves backwards in Vrod direction, the volume of the left chamber shrinks while the piston expulses the fluid.
- Position 4 - The inside pressure of the left cylinder chamber is 30 MPa.
- Position 5 - The fluid inside the left cylinder chamber is expulsed.

5.2.2 Flow trajectory and isoline results of the run simulation

Fig. 14-right shows the flow trajectories in the right chamber of the hydraulic cylinder at L = 185 mm (see table 2).

- Position 1 - The operating fluid fills the domain through an 8mm pipe at a flow rate of Q= 35l/min ($0,0005833m^3/s$).
- Position 2 - The fluid fills the cylinder chamber as a highly concentrated stream. Some of its particles rebounce off the chamber of the cylinder front head and the piston rod. The velocity of the fluid at that domain is about 11,6 m/s.
- Position 3 - The fluid fills the chamber, while some of its particles reboूce off the surface of the piston. The right cylinder chamber is filled for about t = 1,39s.
- Position 4 - The fluid flows around the piston rod and fills the entire right chamber.

The motion of the fluid particles inside the left chamber of the hydraulic cylinder is shown in fig. 14-left.

- Position 1 - The back side of the piston pushes the fluid forward with a pressure of about 29,94MPa. The fluid is attempting to drain through the DAS to the right cylinder chamber.
- Position 2 - Some of the fluid rebounce off the caped head.
- Position 3 -The fluid empties the chamber of the hydraulic cylinder through a 6mm opening. The pressure in this domain is about 29,5MPa and the velocity is around 20,6m/s.
- Position 4 -The fluid goes out through the hole at the bottom.
- Position 5 - The fluid leaves the computational domain.

The general view of the flow trajectories inside the hydraulic cylinder is given in fig. 15.

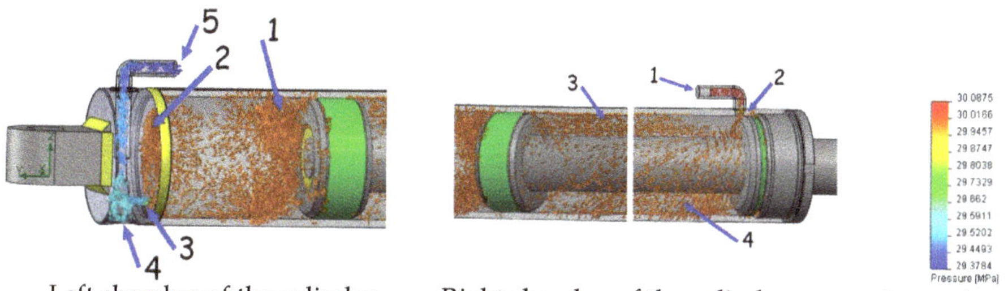

Left chamber of the cylinder | Right chamber of the cylinder | Legend

Fig. 14. Flow trajectories inside the hydraulic cylinder at L = 185 mm (see table 2).

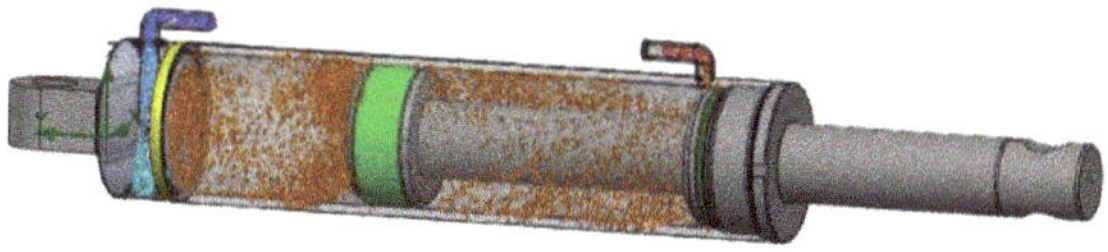

Fig. 15. General view of the flow trajectories inside the hydraulic cylinder

"Isolines" function helps the designers to study in details the vulnerable zones of the hydraulic cylinder structure (fig. 16).

- Position 1 - The fluid fills the right chamber of the hydraulic cylinder at the highest velocity and occurrence of cracks in the pipe corner is possible.
- Position 2 - The fluid empties the cylinder through the left chamber pipe.
- Position 3 - The joint between the cylinder chamber and the pipe is among the vulnerable cylinder zones too. Because of the high pressure inside this domain and the chamfered joints crack occurence is possible.

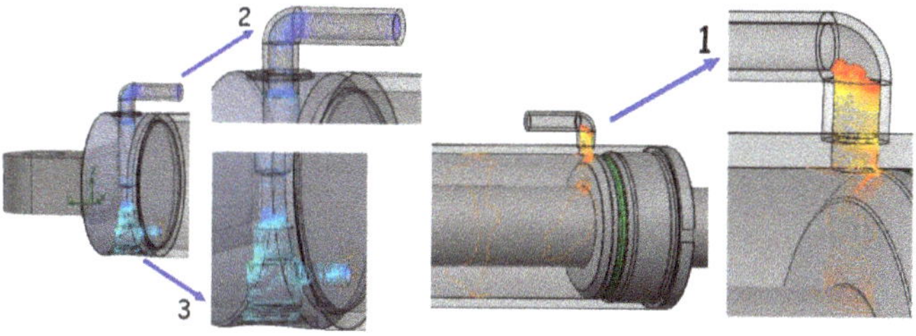

Fig. 16. Vulnerable domains of the hydraulic cylinder structure during PRK backward motion

5.2.3 Numerical data and verification of the results

The numerical and graphical presentations of the magnitude of the studied force $\vec{F}$ during piston's backward motion are given below (table 2 and fig. 17)

The calculated magnitude of the force $\vec{F}$ is verified through the methods of Fluid Mechanicsm, once more:

- test pressure of the hydraulic cylinder - $P = 30MPa$;
- the area of the piston's "face" surface - $A_{piston} = \frac{\pi.D^2}{4} = \frac{3,14.90^2}{4} = 6358,5mm^2$

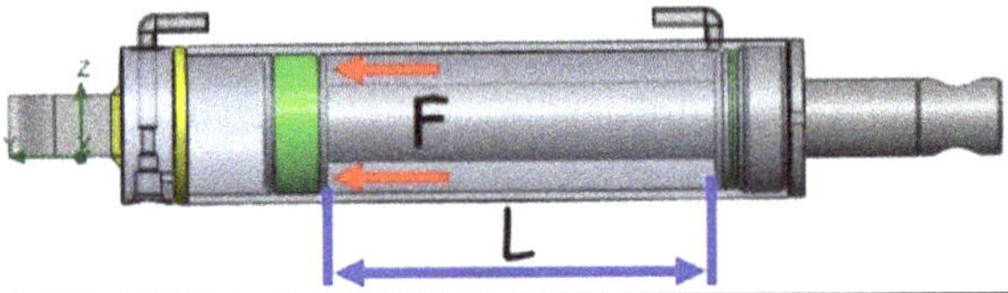

№	L, [mm]	V, [m³]	t, [s]	Vr, [mm]	F, [kN]	Legend
1	5	2,198E-05	0,038	5,13	131,8805	
2	20	8,792E-05	0,151	20,37	131,8668	L - the distance defining PRK position inside the cylinder;
3	35	1,539E-04	0,264	35,61	131,8576	
4	50	2,198E-04	0,377	50,86	131,8505	
5	65	2,857E-04	0,490	66,10	131,8466	V - the volume of the right cylinder chamber;
6	80	3,517E-04	0,603	81,34	131,8453	
7	95	4,176E-04	0,716	96,59	131,8448	
8	110	4,836E-04	0,829	111,83	131,8445	t - the time for filling the chamber of the hydraulic cylinder with fluid;
9	125	5,495E-04	0,942	127,08	131,8447	
10	140	6,154E-04	1,055	142,32	131,8448	
11	155	6,814E-04	1,168	157,56	131,8446	
12	170	7,473E-04	1,281	172,81	131,8447	Vr - the displacement of PRK in backward direction during its motion;
13	185	8,133E-04	1,394	188,05	131,8446	
14	200	8,792E-04	1,507	203,29	131,8448	
15	215	9,451E-04	1,620	218,54	131,8445	
16	230	1,011E-03	1,733	233,78	131,8447	F - the calculated by the SW Flow Simulation magnitude of the tracked force during the backward motion of PRK;
17	245	1,077E-03	1,846	249,03	131,8445	
18	260	1,143E-03	1,959	264,27	131,8446	
19	275	1,209E-03	2,072	279,51	131,8448	
20	290	1,275E-03	2,185	294,76	131,8447	
21	305	1,341E-03	2,298	310,00	131,8445	

Table 2.

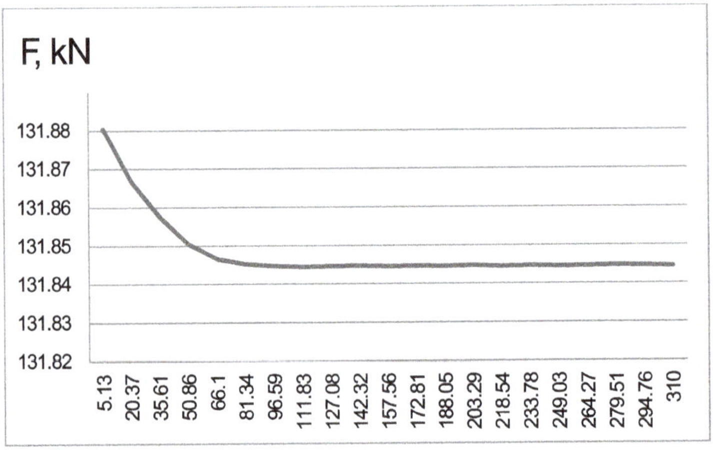

Fig. 17. Graph of the magnitude of force $\vec{F}$ versus backward displacement of the piston **Vr**

- the area of the rod - $A_{rod} = \frac{\pi.D^2}{4} = \frac{3,14.50^2}{4} = 1962,5mm^2$.
- the "working" area of the piston - $A = A_{piston} - A_{rod} = 6358,5 - 1962,5 = 4396mm^2$
- calculated force - $F = 30.4396 = 131880N \Rightarrow 131,880kN$
- error - $\Delta\% = \frac{131,88 - 131,85}{131,88}100\% = 0,023\%$

The error is under 0,025%. It is calculated as the theoretical force magnitude is equal to 131,88 kN and 131,85 is the average value of the results run by the software.

Based on the given in fig. 17 graph it can be concluded that the magnitude of the force $\vec{F}$ decreases slightly in the second half of the course motion. The tendency in madnitude decrease corresponds to the one observed during the pistons course forward. The difference in magnitude values is due to gravity impact and to the difference in the "working" area of the piston.

6. CAE simulation of the operational movement of a hydraulic cylinder using the SW Motion environment

6.1 Developing a SW motion scenario

The SW Motion software enables performing a real time dynamic investigation on the phonomena in the hydraulic cylinder. The input data is predefined in our CAD model and SW Flow Simulation research.

The objectives of this part of the research are:

- to simulate the dynamic effects of the piston motion under predefined input data and various boundary conditions;
- to calculate through a numerical experiment and to analyse the friction force arising between the inner surface of hydraulic cylinder and the double acting seal (DAS).

At first a design scenario, including all facts, which influence the cylinder operation and the calculating steps is defined. The direction and the motion velocity are introduced. All necessary data is taken from performed SW Flow Simulation research. The intermediate values are calculated automatically through cubic interpolation.

The second step is to define the contact conditions between the inner surface of the cylinder and the outer surface of the DAS. Here the characteristics of the materials are input. A greasy environment is chosen. Based on user's choice, the program itself calculates coefficients of static and kinetic friction between the materials.

During the third step a time dependent calculating intervals are input. After starting the "Solve" procedure, all kinematic and dynamic parameters of the motion are automatically calculated. During this last step all results selected by the user are systematized to enable easier analysis.

It is necessary to add some additional explanations about the second and the third steps of the model development.

The materials which form the seal (fig. 18) are different, but they all belong to polyester elastomers'group. Elastomers are polymeric materials that can be subjected to reversible high-elasticity deformations in extreme operating conditions. They possess an extreme abrasion and extrusion resistance. Most often the piston seals are made of Turcon materials or Zurcon polyurethane. In our case, the material of the whole seal is defined as "Greasy", i.e. working in viscous environment. Thus, the coefficients of static and kinetic friction (μ_s and μ_k) are automatically found by the software.

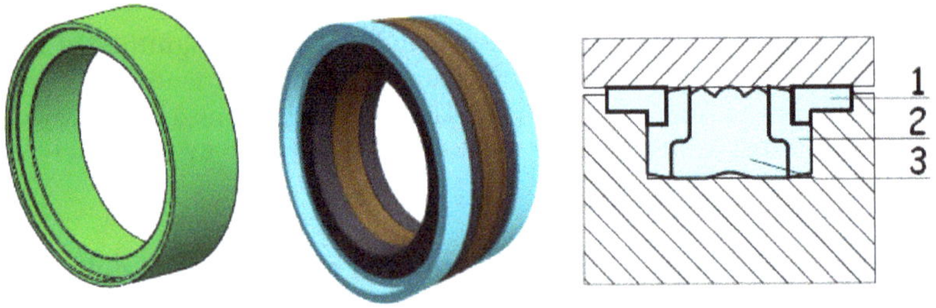

position 1 position 2 legend

Fig. 18. Double acting seal

Real time interval of the simulation for forward motion of PRK is equal to 3,325s (see the previous item). The backward motion scenario developes analogically, but the process lasts 2,298s.

6.2 Analysis of the results obtained through the developed SW Motion Scenario

The motion of the piston is stored in visual file of the type *. avi, which is easily transferred from one working environment to another. Some snapshots of the forward and backward motion of PRK are given in fig. 19.

In the final simulation stage the graphs of some studied parameter versus time are displayed. The software offers a lot of types of results, such as "Force", "Displacement", "Velocity", "Acceleration", "Momentum", "Energy", "Power", etc. The most important for the investigation function is classified in "Force" group. It includes "Applied Force", "Reaction Force", "Reaction Momentum", etc. The authors are interested in "Contact Force", which tracks the magnitude of the friction force between the cylinder and the DAS (fig. 20) during the simulation. The friction force graph versus time is given in fig. 21.

As the authors could not find any research on the topic of how the contact force between the cylinder and the DAS can be calculated, they have decided to verify the proposed methodology of calculating the friction force as follows: there are some measurements made by the producer "HES" PLC, Jambol of the pulling/pushing force on the piston in hydraulic cylinder of that size and construction. The contact force, arisen during PRK motion due to its dead weight, varied in-between 20-30N. The simulated force varied in the range of 22-28N (see fig. 21). The time duration of rod piston course has been chosen to be the same as if the hydraulic cylinder is filling with a fluid of a pressure of 30MPa.

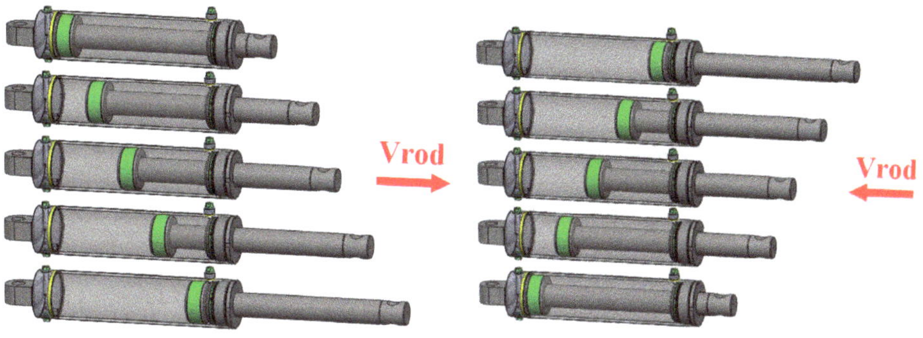

PRK forward motion PRK backward motion

Fig. 19. Some snapshots of the motion of the piston rod kit

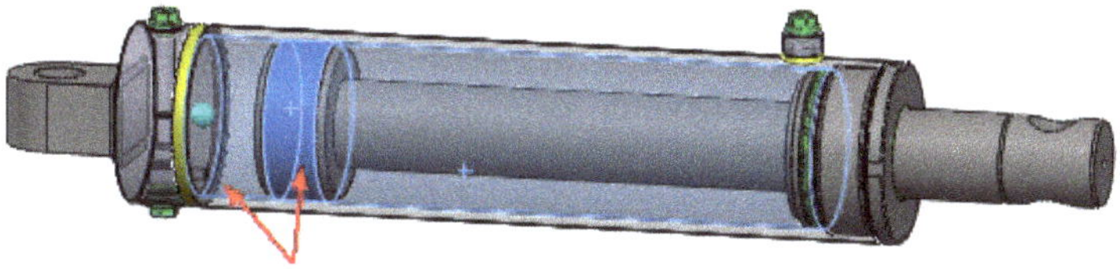

Fig. 20. The surfaces between which the friction force is tracked

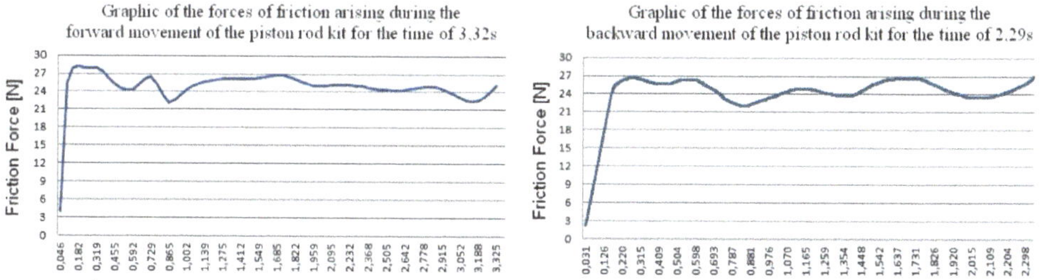

Friction force during forward PRK motion Friction force during backward PRK motion

Fig. 21. Graph of the friction force during the motion of the piston rod kit

7. Conclusions

The developed integrated technology of CAD modeling and CAE analysis of a basic hydraulic cylinder is a part of the long-lasting strategy of company "HES" PLC, Jambol for increase of quality and production control, for bettering of the working environment, etc. Among the basic objectives of this strategy are:

- Improvement of the productivity of three-dimensional modeling of hydraulic cylinders as a result of the established structural scheme and easier and quicker design of hydraulic cylinders' particular elements.
- Development of reliability and overall structural and technological optimizing procedures of the designed hydraulic cylinders, supported by integrated technology for CAE analysis of the fluid flows through FVM.

- Increase of quality and control of the produced products, throughout the use of competitive technology targeting stronger integration of CAD/CAE/CAM systems for product design.

All these stages result in a new working environment, which provokes designers' creativity, optimises technology, increases the competitiveness of the company and finally, strengthens its positions on the market of hydraulic cylinders.

8. Acknowledgment

The presented research has been supported by "HES" PLC company and Erasmus bilateral agreement between Technical University of Sofia, Bulgaria and University of Technology of Vienna, Austria, to whom the authors express their great acknowledgment.

9. References

Brenan, K.E., Campbell, S.I., and Perzold, L.R. (1996). Numerical Solution of Initial Value Problems in Differential-Algebraic Equations, Classics in Applied Mathematics, ISBN 0-89871-353-6

Gear, C.W. (1971a). The Simultaneous Solution of Differential Algebraic Systems. IEEE Transactions on Circuit Theory, CT-18, No.1, 89-95.

Gear, C.W. (1971b). Numerical Initial Value Problems in Ordinary Differential Equations. New Jersey: Prentice-Hall.

Lombard Matt (2011). SolidWorks 2011 Assemblies Bible, John Wiley and Sons, ISBN10: 1118002768

Lombard Matt (2011). SolidWorks 2011 Parts, John Wiley and Sons, ISBN10: 111800275x

SolidWorks Flow Simulation 2011 Technical Reference, (2011), Dassault Systèmes SolidWorks Corp.

Solving Engineering Problems with Flow Simulation 2011, (2011), Dassault Systèmes SolidWorks Corp.

SolidWorks Motion Studies, on-line reference guide, (2011), Dassault Systèmes SolidWorks Corp.

Versteeg Henk Kaarle, Weeratunge Malalasekera (2007). An introduction to computational fluid dynamics: the Finite Volume Method, Pearson Education Limited, ISBN 978-0-13-127498-3

Wendt John F., John David Anderson (2009). Computational fluid dynamics: an introduction, Springer-Verlag, Berlin, ISBN 978-3-540-85055-7

Топалова М., Ст. Бакърджиев (2006). 3D моделиране на технологични единици, Известия на СУБ - Сливен, т. 10, 2006, с. 49-51, ISSN 1311-2864.

Янакиев И., Ст. Николов (2010). CAD/CAM/CAE системи в машиностроенето, ISBN 978-954-737-802-5

Conjugate Gradient Method Applied to Cortical Imaging in EEG/ERP

X. Franceries[1,2,4], N. Chauveau[1,2,*], A. Sors[3], M. Masquere[4] and P. Celsis[1,2]
[1]*Inserm, Imagerie Cérébrale et Handicaps Neurologiques UMR 825, Toulouse*
[2]*Université de Toulouse, UPS, Imagerie Cérébrale et Handicaps Neurologiques UMR 825, CHU Purpan, Place du Dr Baylac, Toulouse Cedex 9*
[3]*LU 48 LERISM Laboratoire d'Etudes et de Recherche en Imagerie Spatiale et Médicale, UPS, Toulouse Cedex 4*
[4]*Université de Toulouse, UPS, INPT, LAPLACE (Laboratoire Plasma et Conversion d'Energie), Toulouse Cedex 9*
France

1. Introduction

Electroencephalography (EEG) and/or Event Related Potentials (ERP) are powerful non-invasive techniques which have broad clinical applications for epilepsy (Gloor et al., 1977; Hughes, 1989; Jaseja, 2009; Myatchin et al., 2009). It is also the case for psychiatric and developmental disorders (Pae et al., 2003; Ruchsow et al., 2003; Youn et al., 2003). There are developments in brain cognition research as for dyslexia (Horowitz-Kraus&Breznitz, 2008; Nuwer, 1998; Russeler et al., 2007), for visual treatment in face recognition (Chaby et al., 2003; George et al., 1996). In all these situations, specific brain areas are activated, and inverse techniques based on ERP treatment can help to estimate them. Techniques based on EEG/ERP are known to be incontestably inoffensive and cheap. This explains why they are often used and are still of great interest in medicine. The optimization of such medical tools, in research on brain cognition and/or as clinical tools, often requires knowledge of the intra-cerebral current sources. In EEG/ERP, this information can be obtained by solving of the so-called "inverse" problem consisting in finding the localization of the spatio-temporal intra-cerebral activity from scalp potential recordings. Various methods have been proposed in the EEG/ERP literature for computing this inverse problem.

Although scalp potentials were first recorded by Hans Berger in 1929 (Berger, 1929), the first inverse problem approach was introduced by Cuffin et al. (Cuffin&Cohen, 1979) in both MEG and EEG, followed by Hämäläinen et al. in 1984 (Hämäläinen&Ilmoniemi, 1984) in MEG. They later extended and increased the performance of the inverse approach applied to MEG (Hämäläinen&Ilmoniemi, 1994). The method was based on the Euclidean norm, which estimates the shortest vector solution in the source-current space (Hämäläinen&Ilmoniemi, 1994). This so-called Minimum Norm Estimate (MNE) is close to Tikhonov regularization (Tikhonov&Arsenin, 1977). However, the MNE solution is

* Corresponding Author

known to misreport actual deep sources as being in the outermost cortex (Pascual-Marqui, 1999; Pascual-Marqui et al., 2002). In order to compensate for the tendency of MNE to favour weak and surface sources, some authors have introduced a "weighting" matrix, calling this inverse method the Weighted Minimum Norm Estimate (WMNE) (Ding, 2009). Then, derived from this reasoning, many inverse methods have been used and/or improved specifically for EEG/ERP, modifying and/or reducing the solution space. Baillet introduced a priori to the solution which can be seen as a weighting matrix using a Bayesian probability, based on anatomical or functional knowledge (Baillet&Garnero, 1997). A Weighted Resolution Optimization (WROP), extending the Backus-Gilbert inverse method (Backus&Gilbert, 1968), has been developed (Grave de Peralta Menendez et al., 1997). The same technique has been modified, using biophysical and psychological a priori to the method called "Local Auto Regressive Average" (LAURA) (De Peralta-Menendez&Gonzales-Andino, 1998). Other authors have considered that restricting the potential solution to the cortical surface is sufficient to make the brain localization, and that the potential maps on the cortex surface must be significantly smooth, which has given rise to the inverse methods called "LOw Resolution brain Electromagnetic Tomography" (LORETA) (Pascual-Marqui et al., 1994), sLORETA (Pascual-Marqui, 2002) which is close to the "Variable Resolution Electrical Tomography" (VARETA) method (Bosch-Bayard et al., 2001). The above list of inverse methods is not exhaustive; a wide range of techniques exist for deriving inverse methods for use in EEG/ERP and new developments continue to be relevant today.

It should be noted that another type of inverse method has been developed at the same time. The main assumption is that the number of intra-cerebral current sources is limited (<10) and each source is punctual. Examples of such inverse methods are implemented in Brain Electric Source Analysis (BESA) (Scherg&Berg, 1991), using the so-called "simplex method" developed by Nelder and Mead (Nelder&Mead, 1965) and the Multiple Signal Classification (MUSIC) algorithm (Mosher et al., 1999). This type of method will not be discussed in our study, which only takes all cortex surfaces into account as possible locations of brain activity.

Inverse methods are numerous and cover many domains, especially in physics and medicine. Recent research can be found that uses the CGM for problems such as the determination of local boiling heat fluxes (Egger et al., 2009), the spatial distribution of Young's modulus (Fehrenbach et al., 2006), 3D elastic full-waveform seismic inversion (Epanomeritakis et al., 2008). Other applications can be found in other journals e.g., thermal diffusivity in plasma (Perez et al., 2008; Yang et al., 2008) and conductivity changes in impedance tomography (Zhao et al., 2007), proving, if it were necessary, the wide use of CGM in many different fields of application. Nevertheless, despite some attempts to use inverse methods such as the CGM in EEG/ERP, there is a lack of studies on the application of CGM to inverse problems in electroencephalography and/or event related potentials. Our contribution is to study the interest of applying CGM in EEG or ERP inverse problems.

In this article, the dependence of the reconstruction quality on the number of electrodes and the noise level are studied using CGM in numerical simulation. The main goal of this work is to evaluate the quality of intra-cerebral source reconstruction using CGM and to compare these results to the Cortical Imaging Technique (CIT). The model parameters and the CGM are described in Sec. 2. Then, in Sec. 3, the theoretical reconstruction of cortical potentials, as if they had been solved from experimentally recorded scalp potentials, are presented and discussed, considering various numbers of electrodes and noise levels. In Sec. 4, previous results are compared to those obtained by CIT, using the comparison tools MAG and RDM factors. The conclusions of this work are given in Sec. 5.

2. Material and method

2.1 Head model

To localize brain activity from recorded scalp potentials in EEG/ERP, mathematical/physical models that describe the geometrical and electrical properties of the head and the intra-cerebral current sources are needed. Generally, the head is described as a conductive volume with piecewise constant conductivity to represent the conductivity of each of its different parts. (Chauveau et al., 2004; He et al., 2002; Zhang et al., 2003). In our study (Figure 1), five compartments were used to construct the head model for the simulation, using the ICBM-152 (http://packages.bic.mni.mcgill.ca/tgz/) T1 template from Montreal Neurological Institute.

Resolution was 2 mm. Conductivities were those used in our previous study on CIT (Chauveau et al., 2008).

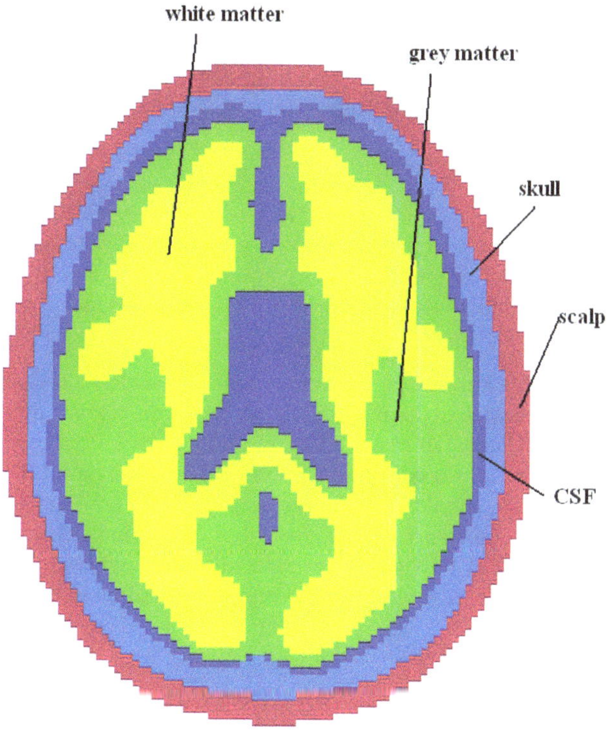

(Chauveau et al., 2005)

Fig. 1. **The five tissues** (white and grey matter, cerebrospinal fluid, skull and scalp), after segmentation of a realistic head geometry (e.g. The T1 Montreal head template).

2.2 Method

Determining scalp potentials from the simulation of intra-cerebral sources, called the forward problem, was an initial step towards the solution of the inverse problem, which aimed to find the sources at the origin of scalp potentials. Various numerical methods (Chauveau et al., 2005; 2005; Darvas et al., 2006; Franceries et al., 2003) have been proposed in the literature for computing the forward problem, including finite difference (FDM) ((Mattout, 2002; Vanrumste, 2001), boundary element (BEM) (Crouzeix, 2001; Kybic et al., 2005; Yvert et al., 1995) and finite element (FEM) (Darvas et al., 2006; Thevenet et al., 1991) methods, the last two being the most widely used. FEM with inclusion anisotropic conductivities have also been developed (Wolters et al., 2007). Although the simulations are usually time consuming, all give rise to numerical solutions and most of them are adequate to simulate brain activation. The Resistor Mesh Model (RMM) (Chauveau et al., 2005; 2005; Franceries et al., 2003), close to Finite Volume Method (FVM) first proposed by Patankar (Patankar, 1980), gives very stable results and is easy to set up. The RMM is made of 2 mm size voxel elements. A sparse square symmetric admittance matrix Y describes the model. Each element represents a resistor, completely determined by its geometry and its conductivity (Franceries et al., 2003). The elements are assembled at the nodes of the model. The forward solution for a vector of currents I is a vector of potentials V so that I = Y x V. The resolution is obtained by using a numerical technique as Newton Raphson algorithm. It should be noted that, in some very special and simple cases (e.g. spherical models), an analytical solution is available but this is not the case for realistic head geometry (de Munck&Peters, 1993; Yvert et al., 1997; Zhou&van Oosterom, 1992)

2.3 Source configuration

In EEG/ERP, brain activation was first simulated by one or several dipolar current sources. The brain activation of each source was modelled by a current dipole, as introduced in 1953 by Plonsey (Plonsey&Barr, 1988). Other types of extended brain activity model have been proposed, e.g. ring extended sources to mimic the gamma frequency range EEG (Tallon-Baudry et al., 1999) .

In our study, we chose the current dipole model, which is widely used and well suited to the RMM. The intra-cerebral activation was simulated by four current dipoles, as used in a previous study on CIT (Chauveau et al., 2008), in order to make the comparison between CGM and CIT. A complex source configuration was used with 2 radial (RR = Radial Right, RL = Radial Left) and 2 tangential (TR = Tangential Right ant TL = Tangential Left) dipoles, placed on or close to the cortex surface. We chose these four dipoles because we wanted to test the inverse technique on two major points: radial or tangential dipoles (EEG being known to be most sensitive to radial), and symmetric dipoles (is the technique able to separate right and left activity?).

2.4 Forward solution

The RMM was applied to solve the forward problem with the previously described source configuration and a sample head model of five tissue compartments, using 2 mm voxels. The method computes potentials at all nodes (the RMN model contains 486,850 nodes and 1,413,720 elements) inside the head model and at the head surface where the electrodes are placed (i.e. on the patient's scalp surface). The use of a large matrix made solving this complete forward problem time consuming. In EEG/ERP, in order to reduce time of resolution and to minimize hard disk space, a lead field (*LF*) matrix, linking the electrode potentials and the currents at the cortex surface is constructed, using the Helmholtz reciprocity principle (Helmholtz, 1853). This new matrix is smaller, reducing calculation time, but the potentials are computed only at the electrode nodes of the scalp surface. The forward problem of computing scalp potential at electrode position (V_e) from a source configuration (I) thus becomes a reduced linear system as follows:

$$Ve = LF.I \tag{1}$$

2.5 Inverse problem

Generally, the inverse problem is solved by using the same matrix as the one for the numerical forward method (i.e. LF), but using inversion. The LF matrix is not square and so cannot be inverted directly. Many methods exist to solve this ill-posed inverse problem, detailed mathematically by Tikhonov (Tikhonov&Arsenin, 1977). Depending on the physical problem, the matrix conditioning and the optimal inverse method have to be adapted. Up to now, CGM have not been applied to EEG/ERP and the most widely used inverse method is the pseudo-inverse matrix, e.g. the Moore-Penrose technique:

$$I = LF^{+}.Ve = (LF^{T}.LF)^{-1}.LF^{T}.Ve \tag{2}$$

where LF^{+} and LF^{T} are respectively the pseudo-inverse matrix of LF by Moore-Penrose and the transpose matrix of LF.

In real measurements, data are corrupted by noise and a regularization technique has to be used in the inversion procedure. Zero-order Tikhonov regularization permits this problem to be solved:

$$I = (LF^{T}.LF + \lambda.I)^{-1}.LF^{T}.Ve \tag{3}$$

λ is a regularization factor depending on noise level, the optimal value of which is obtained at the angle of the associated L-curve (Carthy, 2003; Hansen, 2000; Tikhonov, 1963).

In EEG/ERP, the Cortical Imaging Technique (CIT) is one of the possible inverse methods, which limits the space of solutions for current dipoles to the cortex surface. This method has been described and evaluated (Chauveau et al., 2008; He et al., 2002) and it provided the comparison technique used in our study.

2.6 Conjugate gradient method (CGM)

CGM is an iterative technique. Other iterative techniques have been proposed (Gorodnitsky et al., 1995; Hansen, 1994; Ioannides et al., 1990). Ioannides proposes continuous probabilistic solutions to the biomagnetic inverse problem, very efficient for deep sources. Gorodnitsky describes a recursive weighted minimum norm algorithm (FOCUSS). Hansen has developed regularization tools for Matlab: he describes the iterative regularization methods, and presents CGM as a process which has some inherent regularization effect where the number of iterations plays the role of regularization parameter.

CGM was first designed for solving linear equations thanks to a square symmetric matrix. The application of CGM can be extended to rectangular non symmetric matrix as lead fields are, for the inverse solution. That is this way we use CGM in this study. The following equations do not need specific hypothesis on the properties of the linear matrix.

When using a gradient method (GM) in EEG/ERP, the inverse problem is replaced by an estimation problem in which the unknown source configuration I_k is varied iteratively until the difference between the measured and calculated scalp potentials is as small as possible:

$$R_k = Ve - Ve_k = Ve - LF.I_k \tag{4}$$

R_k is the residual of the measured scalp potentials, Ve, minus the computed ones, Ve_k, at the kth iteration and LF the lead field matrix.

$$R_{k+1} = R_k + \beta.LF.P_k \tag{5}$$

The simple gradient method (Amari, 1977) is based on a local derivative function, in order to minimize the error. At each step of a gradient method, a trial set of values for the variable is used to generate a new set corresponding to a lower value of the error function.This was improved in the steepest gradient method (Curry, 1944), where the descent method takes the direction of the maximum gradient of the error function, which reduces the number of iterations. A further improvement is CGM, in which the previous (k) and the next (k+1) search directions are defined to be orthogonal in the residual associated error space, so that CGM explores a maximum of R_k space. The CGM (Press et al., 1992) is an iterative method which computes:

$$I_{k+1} = I_k - \beta.P_k \tag{6}$$

$$F_k = R_k^T.R_k = (Ve - Ve_k)^T.(Ve - Ve_k) \Leftrightarrow F_k = (Ve - LF.I_k)^T.(Ve - LF.I_k) \tag{7}$$

where P_k is a vector of search direction at the kth iteration and β is a scalar of optimal step of descent obtained by finding the minimal argument of the objective function F_k, defined by the norm of residual R_k:

$$\beta = ArgMin\ (F_{k+1}(I_{k+1}))$$

This is equivalent to looking for the β value which cancels the derivative.

$$\beta \rightarrow \frac{\partial F_{k+1}}{\partial \beta} = 0$$

$$\frac{\partial F_{k+1}}{\partial \beta} = \frac{\partial (R_{k+1}^T.R_{k+1})}{\partial \beta} = 0 \tag{8}$$

$$\frac{\partial\left[(Ve - LF.I_{k+1})^T.(Ve - LF.I_{k+1})\right]}{\partial \beta} = 0 \tag{9}$$

Replacing I_{k+1} by its value in equation 6 and developing equation 9 gives:

$$\beta = -\frac{P_k^T.LF^T.R_k}{P_k^T.LF^T.LF.P_k} \tag{10}$$

The new iterative direction P_{k+1} is computed from the previous one P_k using:

$$P_{k+1} = LF^T.R_{k+1} - \gamma.P_k \tag{11}$$

and by imposing that the previous P_k and the next P_{k+1} search direction are orthogonal

$$P_{k+1}^T LF^T.LF.P_k = 0 \tag{12}$$

Replacing the value of P_{k+1} of equation 11 in equation 12 gives the new conjugation factor γ given by:

$$\gamma = \frac{R_{k+1}^T.LF.LF^T.LF.P_k}{P_k^T.LF^T.LF.P_k} \tag{13}$$

The solution I_k is obtained at the kth iteration when the value of the chosen stopping criterion C of CGM is reached:

$$\sqrt{\frac{R_k^T.R_k}{V_e^T.V_e}} < C \tag{14}$$

C corresponds to a value, chosen by the user: it must be higher or equal to 0.01, from our experience and with our model. The root mean square of the relative error is compared to that value to stop the iterations.

In real conditions, data are corrupted by noise. In ERP/EEG protocols, noise vector No can be simply estimated on the pre-triggering time interval before events. Then the smaller criterion to reach becomes:

$$C_{noise} = \sqrt{\frac{No^T.No}{V_e^T.V_e}} \tag{15}$$

Then we stop the iterations when

$$\sqrt{\frac{R_k^T.R_k}{V_e^T.V_e}} < C_{noise} \tag{16}$$

CGM does not need a priori conditions for solving the inverse problem in EEG/ERP, especially on the number of possible current sources in the cerebral volume. Moreover, CGM is faster than the classical Gradient Method because it needs less iteration to converge.

The main particularity of CGM is that the variation of the vector current obtained at each program loop is made orthogonal to the previous one. This permits to explore more quickly the space of solutions. In our application, CGM uses a lead field matrix, which is never inverted. The minimization is achieved on the square difference between measured and estimated electrode potentials. To stop the process, two methods are reported:

- A precision criterion chosen by the user (equation 14) which can generate oscillations.
- A precision criterion estimated from noise and signal (equation 15) avoiding oscillations, stopping the iterations when equation 16 is validated.

3. Results

3.1 CGM results without noise

3.1.1 Single dipoles

Figure 2 shows cortical potentials obtained by direct simulation in comparison with cortical potentials computed at all nodes in the RMM by CGM (with 107 electrodes). All the cortical potentials reconstructed show a good localization for each single dipole, even though individual dipoles are smoother at the cortex cerebral surface. In order to quantify the results, we used the accuracy measures described in the appendix: magnification factors (MAG) and relative difference measure (RDM).

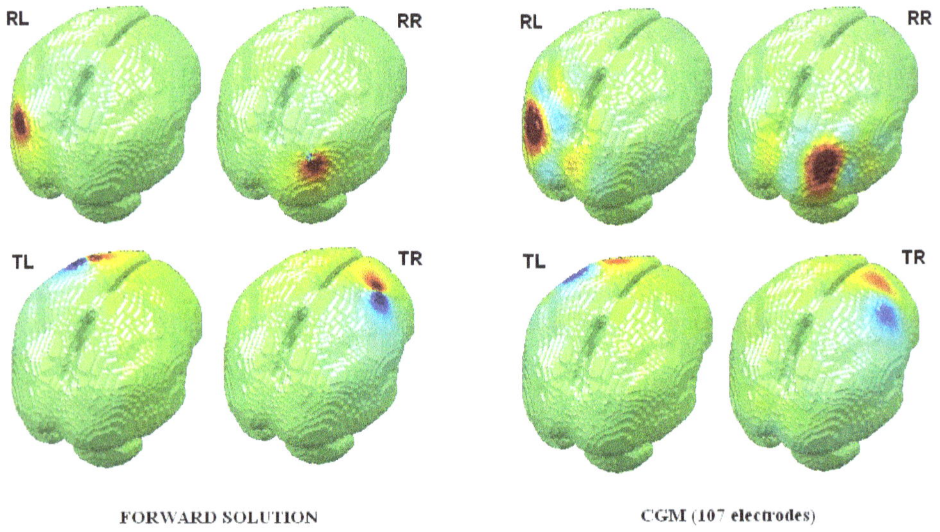

Fig. 2. **Forward solution and CGM** left part: forward solution of cortical potentials in 2 mm voxels for each dipole (RL, radial left ,RR, radial right, TL tangential left, TR tangential right), right part: CGM without noise for each dipole (right part) [-5e-5 +5e-5 volts] from 107 electrode potentials.

CGM performance is given for single and multiple dipoles in Table 1. It appears that the cortical potentials obtained by CGM underestimate the dipole amplitudes in comparison with potentials obtained by the forward solution. The worst result is observed for the dipole RR, which is correctly located on the cortex, but with a spread cortex area (Fig. 2) and with a local maximum much lower than for the forward solution, which explains the low value of MAG.

	Cortex		Scalp		Electrodes	
107 electrodes	MAG	RDM	MAG	RDM	MAG	RDM
1 dipole						
RR	0.31	1.21	1.23	0.28	1.08	0.06
RL	1.57	0.98	1.06	0.16	1.05	0.03
TR	0.92	0.47	1.09	0.14	1.04	0.03
TL	0.72	0.77	1.03	0.06	1.01	0.01
4 dipoles	0.48	1.15	1.05	0.20	1.02	0.07

Table 1. MAG and RDM of cortical potentials, scalp potentials and electrode potentials obtained by CGM for simulation of single dipole and the 4 dipoles with 107 electrodes in 2 mm voxel, without noise

3.1.2 Effect of number of electrodes on CGM with the 4 dipoles

As EEG is only recorded at a limited number of electrodes, it is important to estimate the role of this number on the quality of the inverse solution.

Figure 3 shows the cortical potentials obtained by CGM for 60 and 107 electrodes without noise. CGM was used to compute cortical potentials from the electrode potentials of the forward solution. As we can see on the figure, the CGM solution with 107 electrodes is more accurate than the solution obtained with 60 electrodes (tangential dipoles are better defined: red and blue areas are closer). We also observe, taking the potentials of the forward solution as a reference, that CGM with 60 and 107 electrodes underestimates the potentials at the cortical surface, especially for tangential dipoles. MAG reported in Table 1, and Table 2 and 3 (for noise 0%) confirms lower potential estimation at the cortex, whereas high RDM indicates mismatch on the shape or position.

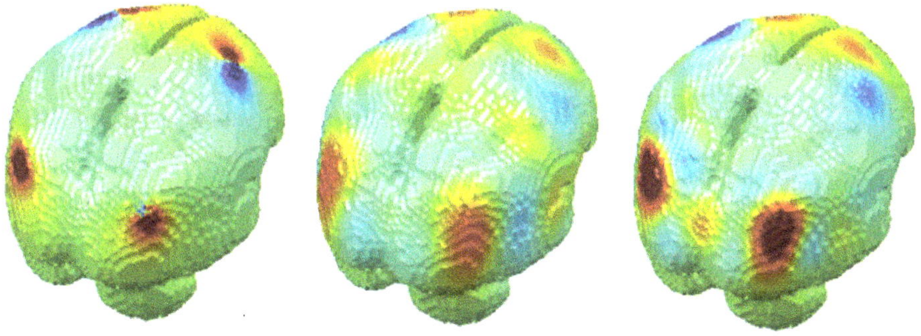

Forward solution — CGM (60 electrodes) — CGM (107 electrodes)

Fig. 3. Forward solution of cortical potentials in 2 mm voxels (left part) for 4 dipoles, CGM with 60 electrodes (central part) and CGM with 107 electrodes without noise for 4 dipoles [-5e-5 +5e-5 volts]

MAG	criterion			MAG	criterion			MAG	criterion		
60 elec.	0.02	0.05	0.10	scalp	0.02	0.05	0.10	cortex	0.02	0.05	0.10
0%	0.99	0.98	0.98	0%	1.11	1.14	1.14	0%	0.42	0.40	0.36
2%	1.00	1.00	0.99	2%	1.08	1.06	1.14	2%	0.45	0.43	0.39
5%	1.08	1.08	1.07	5%	1.18	1.17	1.20	5%	0.67	0.60	0.50
10%	1.13	1.10	1.07	10%	1.23	1.26	1.28	10%	0.98	0.92	0.79

RDM	criterion			RDM	criterion			RDM	criterion		
60 elec	0.02	0.05	0.10	scalp	0.02	0.05	0.10	cortex	0.02	0.05	0.10
0%	0.07	0.08	0.13	0%	0.31	0.37	0.39	0%	1.18	1.18	1.19
2%	0.13	0.13	0.14	2%	0.29	0.28	0.39	2%	1.19	1.19	1.19
5%	0.25	0.24	0.23	5%	0.40	0.37	0.39	5%	1.25	1.23	1.21
10%	0.50	0.50	0.50	10%	0.68	0.62	0.61	10%	1.32	1.32	1.30

Table 2. Results of MAG and RDM of electrode, scalp and cortical potentials, obtained by CGM for different values of stopping criterion for simulation of 4 dipoles with 60 electrodes in 2 mm voxel and noise level varying from 0% to 10%. Values in grey indicate cases where criterion is lower than noise, which is not valuable.

MAG	criterion			MAG	criterion			MAG	criterion		
107 elec	0.02	0.05	0.10	scalp	0.02	0.05	0.10	cortex	0.02	0.05	0.10
0%	1.02	1.04	1.04	0%	1.05	1.12	1.20	0%	0.48	0.43	0.38
2%	1.03	1.02	1.02	2%	1.03	1.11	1.17	2%	0.62	0.49	0.40
5%	1.06	1.05	1.01	5%	1.12	1.14	1.02	5%	1.14	0.97	0.72
10%	1.06	1.05	1.05	10%	1.71	1.48	1.27	10%	3.03	2.86	2.49

RDM	criterion			RDM	criterion			RDM	criterion		
107 elec	0.02	0.05	0.10	scalp	0.02	0.05	0.10	cortex	0.02	0.05	0.10
0%	0.07	0.09	0.13	0%	0.20	0.26	0.38	0%	1.15	1.17	1.19
2%	0.10	0.09	0.11	2%	0.25	0.27	0.36	2%	1.21	1.17	1.19
5%	0.24	0.23	0.22	5%	0.54	0.45	0.36	5%	1.29	1.27	1.23
10%	0.56	0.57	0.58	10%	0.92	0.90	0.92	10%	1.35	1.34	1.34

Table 3. Results of MAG and RDM of electrode, scalp and cortical potentials, obtained by CGM for different values of stopping criterion C for simulation of 4 dipoles with 107 electrodes in 2 mm voxel and noise level varying from 0% to 10%.

3.2 CGM results with noise

A recent review on solving the inverse problem in EEG (Grech et al., 2008) presents the techniques in non-parametric and parametric methods, depending on the fixed number of dipoles (assumed a priori or not). No specific technique appears to give much better results than the others, and research in this field is continuing. For simulation studies, EEG noise must be taken into account, and Gaussian White Noise (GWN) is often used (Chauveau et al., 2008; He et al., 2002). We tested the CGM with 3 different GWN levels: 2%, 5% and 10% of the maximum electrode potential.

Figure 4 shows the cortical potential distribution obtained by CGM with 60 and 107 electrodes for noise levels varying from 0% to 10% (criterion defined in equation 14). Oscillations increase with the level of noise with 60 and 107 electrodes. So, the higher the noise level, the less correctly the cortical potential cartography is reconstructed. These results also show that oscillations in cortical potential distributions increase relatively faster when the noise level is higher than 5% with 107 electrodes, and 10% with 60 electrodes.

3.3 CGM versus CIT

Figure 5 shows the cortical potential distributions obtained with CGM, for different values of relative noise level and with 107 electrodes (criterion defined in equation 14), in comparison with the results of CIT. A Tikhonov regularization was used in CIT, but there are anyway some oscillations for high noise level. CGM presents oscillations when the criterion is too small compared to noise, but for each noise level it a correct estimation can be obtained.

In real conditions, the noise level can be easily estimated on the pre-stimuli interval before the triggers used for ERP. Taking into account the noise level, the criterion is then limited to C_{noise} (equation 15). Iterations are stopped when C_{noise} is reached. Results are reported in figure 6.

Qualitative factors have been calculated by means of MAG and RDM (see appendix) at the electrodes, at the scalp surface and at the cortex surface for 60 electrodes (Table 2) and for 107 electrodes (Table 3).

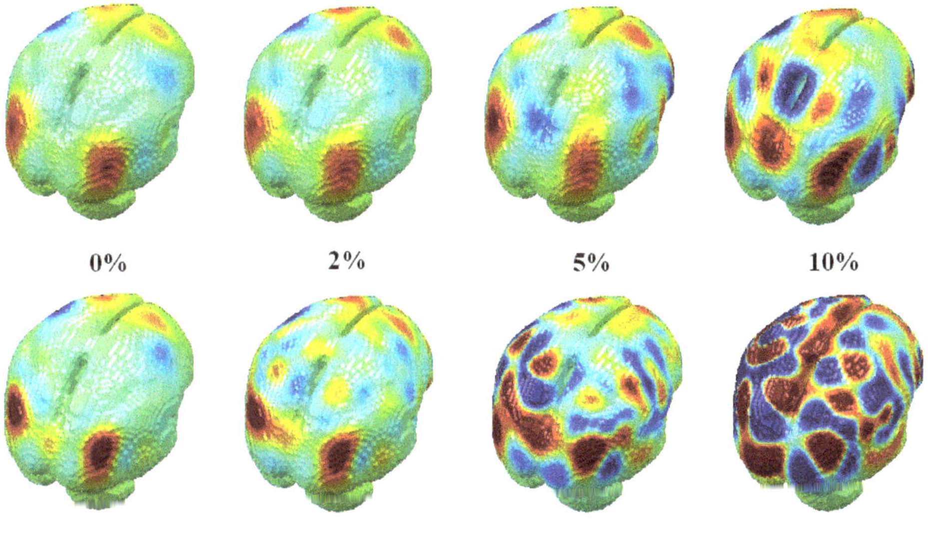

Fig. 4. CGM for 60 and 107 electrodes (criterion set to 0.01)
Cortical potentials in 2 mm voxels for 4 dipoles and different noise levels for CGM with 60 electrodes (top) and CGM with 107 electrodes (bottom) [-5e-5 +5e-5 volts]

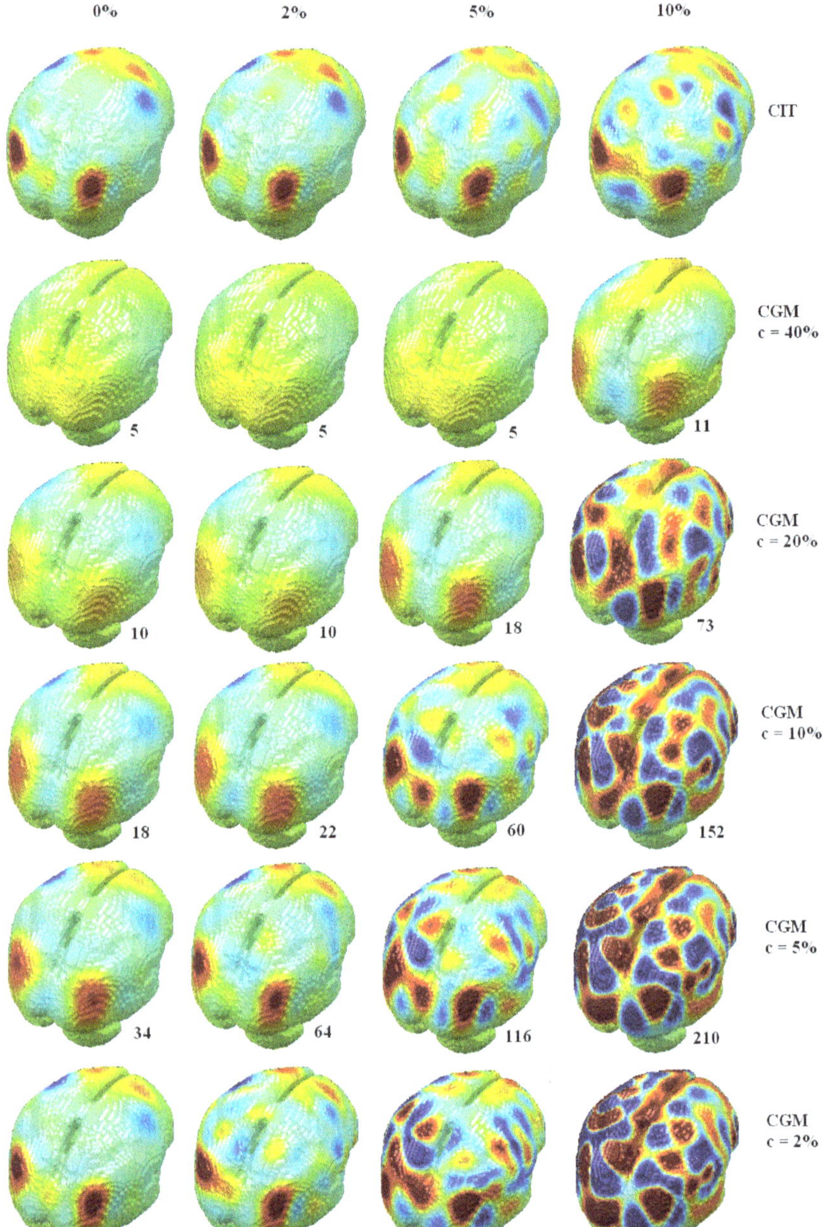

Fig. 5. CIT and CGM 107 electrodes (equation 14)
Cortical potentials in 2mm voxels for 4 dipoles with noise varying from 0% to 10% and 107 electrodes, CIT solutions (first line) and CGM (all other lines) for different criterion value c (equation 14), from 2% to 40%. For each CGM solution, the number of iterations is reported.

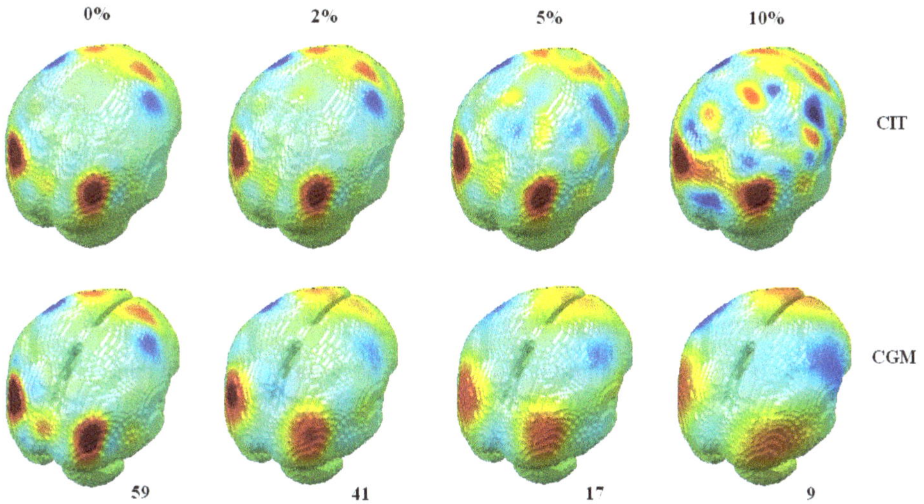

Fig. 6. CIT and CGM 107 electrodes (criterion depending on noise level)
Cortical potentials estimated from 107 electrodes in 2mm voxels for 4 dipoles with noise varying from 0% to 10% (criterion of equation 15): CIT solutions (first line) and CGM (last line). For each CGM solution, the number of iterations is reported.

4. Conclusion

This study by simulation has shown that CGM gives coherent results in the detection of simultaneous multiple dipoles (4 in our case). CGM solutions give satisfactory localization and estimation of cortical potentials even though the area of each dipole is increased. Symmetrical dipoles are well detected while tangential dipoles are more difficult to observe, as for any inverse technique in EEG/ERP.

We have shown that, without noise, CGM correctly localizes individual and simultaneous dipoles, with an underestimation of the cortical potentials. Moreover, the number of electrodes strongly conditions the quality of the solution obtained by CGM. So, without noise in the data, the higher the number of electrodes, the more accurate the dipole localization and the more correctly reconstructed the corresponding cortex potentials. So increasing the number of electrodes reduces the number of unknowns in the inverse problem in EEG/ERP. In consequence, cortical potentials are better evaluated.

With the addition of white Gaussian noise (WGN), this observation becomes partially true, because solutions obtained with high numbers of electrodes are less stable than those obtained with a smaller number when noise level increases. There is then an optimal number of electrodes for each simulated noise level. Solutions obtained by CIT and by CGM present oscillations which increase with the noise level. Cortical potential solutions of CGM are quite similar to the ones of CIT for a low noise level. When this level increases, CIT presents oscillation, still gives quite correct position for the sources but added potentials corrupt the result, while CGM presents solutions with less oscillation, but may be with less precision. The combination of CIT and CGM results permits to validate the source positions:

CGM permit to clearly identify there are 4 sources in our case (2 radial dipoles and 2 tangential dipoles), and CIT permits to point out where they are.

It is then possible to use CGM as a complementary tool to solve inverse problems in EEG/ERP. The advantage of CGM is that there is no need for matrix inversion and there is not a prior in the number of current sources or in their propagation direction in the cerebral volume. This iterative method avoids having to invert huge rectangular matrices which are time and memory consuming when the spatial resolution of the model is ambitious. The estimation of noise permits to calculate a realistic stopping criterion to use, avoiding oscillations.

5. Appendix: Comparison tools MAG and RDM

MAG is an index for potential magnitude comparison between two series of equivalent data, and RDM estimates the variation of spatial distribution between the two series.

MAG and RDM are given by:

$$MAG=\sqrt{\frac{\sum_{i=1}^{n} V_{Ci}^2}{\sum_{i=1}^{n} V_{Fi}^2}} \quad RDM=\sqrt{\sum_{i=1}^{n}\left(\frac{V_{Ci}}{\sum_{i=1}^{n} V_{Ci}^2}-\frac{V_{Fi}}{\sum_{i=1}^{n} V_{Fi}^2}\right)^2}$$

where V_{Ci} is a series of computed potential data points obtained with a specific technique from electrode potentials (CIT or CGM), V_{Fi} is the forward solution for the same points and n is the number of chosen points.

6. References

Amari, S. I. (1977). "Neural theory of association and concept-formation." Biol Cybern 26(3): 175-85.

Backus, G. and Gilbert, F. (1968). "The revolving power of gross earth data." Geophys., J.R. Astr. Soc. 16: 169-205.

Baillet, S. and Garnero, L. (1997). "A Bayesian approach to introducing anatomo-functional priors in the EEG/MEG inverse problem." IEEE Trans Biomed Eng 44(5): 374-85.

Berger, H. (1929). "das Elektrenkephalogram des Menschen." Arch. F. Psychiatr. 87: 527-570.

Bosch-Bayard, J.;Valdes-Sosa, P., et al. (2001). "3D statistical parametric mapping of EEG source spectra by means of variable resolution electromagnetic tomography (VARETA)." Clin Electroencephalogr 32(2): 47-61.

Carthy, P. J. M. (2003). "Direct analytic model of the L-curve for Tikhonov regularization parameter selection." Inverse Problems 19(3): 643.

Chaby, L.;George, N., et al. (2003). "Age-related changes in brain responses to personally known faces: an event-related potential (ERP) study in humans." Neurosci Lett 349(2): 125-9.

Chauveau, N.;Franceries, X., et al. (2008). "Cortical imaging on a head template: a simulation study using a resistor mesh model (RMM)." Brain Topogr 21(1): 52-60.

Chauveau, N.;Franceries, X., et al. (2004). "Effects of skull thickness, anisotropy, and inhomogeneity on forward EEG/ERP computations using a spherical three-dimensional resistor mesh model." Hum Brain Mapp 21(2): 86-97.

Chauveau, N.;Morucci, J. P., et al. (2005). "Resistor mesh model of a spherical head: part 1: applications to scalp potential interpolation." Med Biol Eng Comput 43(6): 694-702.

Chauveau, N.;Morucci, J. P., et al. (2005). "Resistor mesh model of a spherical head: part 2: a review of applications to cortical mapping." Med Biol Eng Comput 43(6): 703-11.

Crouzeix, A. (2001). "Méthodes de localisation des générateurs de l'activité électrique cérébrale à partir de signaux électro- et magnéto- encéphalographiques." Ph D Dissertation, INSA Lyon, France

Cuffin, B. N. and Cohen, D. (1979). "Comparison of the magnetoencephalogram and electroencephalogram." Electroencephalogr Clin Neurophysiol 47(2): 132-46.

Curry, H. B. (1944). "The method of steepest descent for nonlinear minimization problems." Quart. Appl. Math. 2: 258-261.

Darvas, F.;Ermer, J. J., et al. (2006). "Generic head models for atlas-based EEG source analysis." Hum Brain Mapp 27(2): 129-43.

de Munck, J. C. and Peters, M. J. (1993). "A fast method to compute the potential in the multisphere model." IEEE Trans Biomed Eng 40(11): 1166-74.

De Peralta-Menendez, R. G. and Gonzales-Andino, S. L. (1998). "A critical Analysis of Linear Inverse Solutions to the Neuroelectromagnetic Inverse Problem." IEEE Trans Biomed Eng 45(4): 440-448.

Ding, L. (2009). "Reconstructing cortical current density by exploring sparseness in the transform domain." Phys Med Biol 54(9): 2683-97.

Egger, H.;Heng, Y., et al. (2009). "Efficient solution of a three-dimensional inverse heat conduction problem in pool boiling." Inverse Problems 25: 095006 (19pp).

Epanomeritakis, I.;Akcelik, V., et al. (2008). "A Newton-CG method for large-scale three-dimensional elastic full-waveform seismic inversion." Inverse Problems 24: 034015 (26pp).

Fehrenbach, J.;Masmoudi, M., et al. (2006). "Detection of small inclusions by elastography." Inverse Problems 22: 1055-1069.

Franceries, X.;Doyon, B., et al. (2003). "Solution of Poisson's equation in a volume conductor using resistor mesh models: application to event related potential imaging." J Appl Phys 93(6): 3578-3588.

George, N.;Evans, J., et al. (1996). "Brain events related to normal and moderately scrambled faces." Brain Res Cogn Brain Res 4(2): 65-76.

Gloor, P.;Ball, G., et al. (1977). "Brain lesions that produce delta waves in the EEG." Neurology 27(4): 326-33.

Gorodnitsky, I. F.;George, J. S., et al. (1995). "Neuromagnetic source imaging with FOCUSS: a recursive weighted minimum norm algorithm." Electroencephalogr Clin Neurophysiol 95(4): 231-51.

Grave de Peralta Menendez, R.;Hauk, O., et al. (1997). "Linear inverse solutions with optimal resolution kernels applied to electromagnetic tomography." Hum Brain Mapp 5(6): 454-467.

Grech, R.;Cassar, T., et al. (2008). "Review on solving the inverse problem in EEG source analysis." J Neuroeng Rehabil 5: 25.
Hämäläinen, M. S. and Ilmoniemi, R. J. (1984). "Interpreting measured magnetic fields of the brain: estimates of current distributions." Tech. Rep. TKK-F-A559, Helsinki University of technology, Espoo.
Hämäläinen, M. S. and Ilmoniemi, R. J. (1994). "Interpreting measured magnetic fields of the brain: minimum norm estimates." Med. biol. Eng. & Comput. 32: 35-42.
Hansen, P. C. (1994). "REGULARIZATION TOOLS: A Matlab package for analysis and solution of discrete ill-posed problems." Numerical Algorithms 6: 1-35.
Hansen, P. C. (2000). "The L-curve and its use in the numerical treatment of inverse problems." Computational Inverse Problems in Electrocardiologyn, WIT Press, Southampton.
He, B.;Zhang, X., et al. (2002). "Boundary element method-based cortical potential imaging of somatosensory evoked potentials using subjects' magnetic resonance images." Neuroimage. 16(3 Pt 1): 564-76.
Helmholtz, H. (1853). "Über einig Gesetze des Verteilung elektrischer Strome in Korperlischen Leitern mit Anwendung auf die tierisch-elektrischen Versuche." Ann. Phys. Chem. 29: 211-233, 353-377.
Horowitz-Kraus, T. and Breznitz, Z. (2008). "An error-detection mechanism in reading among dyslexic and regular readers--an ERP study." Clin Neurophysiol 119(10): 2238-46.
Hughes, J. R. (1989). "The significance of the interictal spike discharge: a review." J Clin Neurophysiol 6(3): 207-26.
Ioannides, A. A.;Bolton, J. P. R., et al. (1990). "Continuous probabilistic solutions to the biomagnetic inverse problem." Inverse Problems 6: 523-42.
Jaseja, H. (2009). "Significance of the EEG in the decision to initiate antiepileptic treatment in patients with epilepsy: A perspective on recent evidence." Epilepsy Behav.
Kybic, J.;Clerc, M., et al. (2005). "A common formalism for the integral formulations of the forward EEG problem." IEEE Trans Med Imaging 24(1): 12-28.
Mattout, J. (2002). "Ph. D. thesis." Approches statistiques multivariées pour la localisation de l'activation cérébrale en magnétoencéphalographie et en imagerie par résonance magnétique fonctionnelle. University of Paris VI.
Mosher, J. C.;Baillet, S., et al. (1999). "EEG source localization and imaging using multiple signal classification approaches." J Clin Neurophysiol 16(3): 225-38.
Myatchin, I.;Mennes, M., et al. (2009). "Working memory in children with epilepsy: an event-related potentials study." Epilepsy Res 86(2-3): 183-90.
Nelder, J. A. and Mead, R. (1965). "A simplex method for function minimization." Computer Journal 7: 308-313.
Nuwer, M. R. (1998). "Fundamentals of evoked potentials and common clinical applications today." Electroencephalogr Clin Neurophysiol 106(2): 142-8.
Pae, J. S.; Kwon, J. S., et al. (2003). "LORETA imaging of P300 in schizophrenia with individual MRI and 128-channel EEG." Neuroimage 20(3): 1552-60.
Pascual-Marqui, R. D. (1999). "Review of Methods For Solving the EEG Inverse Problem." International Journal of Bioelectromagnetism 1: 75-86.

Pascual-Marqui, R. D. (2002). "Standardized low-resolution brain electromagnetic tomography (sLORETA): technical details." Methods Find Exp Clin Pharmacol 24 Suppl D: 5-12.

Pascual-Marqui, R. D.; Esslen, M., et al. (2002). "Functional imaging with low resolution brain electromagnetic tomography (LORETA): review, new comparisons, and new validation." Japanese Journal of Clinical Neurophysiology 30: 81-94.

Pascual-Marqui, R. D.; Michel, C. M., et al. (1994). "Low resolution electromagnetic tomography: a new method for localizing electrical activity in the brain." Int J Psychophysiol 18(1): 49-65.

Patankar, S. V. (1980). "Numerical heat transfer and fluid flow." (New York: Mc Graw Hill).

Perez, L.; Autrique, L., et al. (2008). "Implementation of a conjugate gradient algorithm for thermal diffusivity identification in a moving boundaries system." Journal of Physics: Conference Series 135: 12082.

Plonsey, R. and Barr, R. (1988). "Biolectricity: a quantitative approach." (New York: Plenum Press): 21-30.

Press, W. H.; Teukolsky, A. A., et al. (1992). "Conjugate Gradient Methods in Multidimensions." Numerical recipes in C. The Art of Scientific Computing Second Edition, Cambridge University Press, New York: 420-424.

Ruchsow, M.; Trippel, N., et al. (2003). "Semantic and syntactic processes during sentence comprehension in patients with schizophrenia: evidence from event-related potentials." Schizophr Res 64(2-3): 147-56.

Russeler, J.; Becker, P., et al. (2007). "Semantic, syntactic, and phonological processing of written words in adult developmental dyslexic readers: an event-related brain potential study." BMC Neurosci 8: 52.

Scherg, M. and Berg, P. (1991). "Use of prior knowledge in brain electromagnetic source analysis." Brain Topogr 4(2): 143-50.

Tallon-Baudry, C.; Bertrand, O., et al. (1999). "A ring-shaped distribution of dipoles as a source model of induced gamma-band activity." Clin Neurophysiol 110(4): 660-5.

Thevenet, M.; Bertrand, O., et al. (1991). "The finite element method for a realistic head model of electrical brain activities: preliminary results." Clin Phys Physiol Meas 12 Suppl A: 89-94.

Tikhonov, A. N. (1963). "Sov. Math.-Dokl." 4: 1035.

Tikhonov, V. L. and Arsenin, V. Y. (1977). Solutions of Ill-Posed Problems. New York, Wiley.

Vanrumste, B. (2001). "EEG dipole source analysis in a realistic head model." Ph. D. thesis, University of Ghent, Belgium

Wolters, C. H.; Kostler, H., et al. (2007). "Numerical Mathematics of the Subtraction Method for the Modeling of a Current Dipole in EEG Source Reconstruction Using Finite Element Head Models." SIAM J. Sci. Comput. 30(1): 24-45.

Yang, Y. C.; Chang, W. J., et al. (2008). "Modelling of thermal conductance during microthermal machining with scanning thermal microscope using an inverse methodology." Physics Letters A 373: 519-523

Youn, T.; Park, H. J., et al. (2003). "Altered hemispheric asymmetry and positive symptoms in schizophrenia: equivalent current dipole of auditory mismatch negativity." Schizophr Res 59(2-3): 253-60.

Yvert, B.; Bertrand, O., et al. (1995). "Improved forward EEG calculations using local mesh refinement of realistic head geometries." Electroencephalogr Clin Neurophysiol 95(5): 381-92.

Yvert, B.; Bertrand, O., et al. (1997). "A systematic evaluation of the spherical model accuracy in EEG dipole localization." Electroencephalogr Clin Neurophysiol 102(5): 452-9.

Zhang, X.; van Drongelen, W., et al. (2003). "High-resolution EEG: cortical potential imaging of interictal spikes." Clin Neurophysiol 114(10): 1963-73.

Zhao, B.; Wang, H., et al. (2007). "Linearized solution to electrical impedance tomography based on the Schur conjugate gradient method." Measurement Science and Technology 18: 3373-3383.

Zhou, H. and van Oosterom, A. (1992). "Computation of the potential distribution in a four-layer anisotropic concentric spherical volume conductor." IEEE Trans Biomed Eng 39(2): 154-8.

Permissions

The contributors of this book come from diverse backgrounds, making this book a truly international effort. This book will bring forth new frontiers with its revolutionizing research information and detailed analysis of the nascent developments around the world.

We would like to thank Dr. Radostina Petrova, for lending her expertise to make the book truly unique. She has played a crucial role in the development of this book. Without her invaluable contribution this book wouldn't have been possible. She has made vital efforts to compile up to date information on the varied aspects of this subject to make this book a valuable addition to the collection of many professionals and students.

This book was conceptualized with the vision of imparting up-to-date information and advanced data in this field. To ensure the same, a matchless editorial board was set up. Every individual on the board went through rigorous rounds of assessment to prove their worth. After which they invested a large part of their time researching and compiling the most relevant data for our readers. Conferences and sessions were held from time to time between the editorial board and the contributing authors to present the data in the most comprehensible form. The editorial team has worked tirelessly to provide valuable and valid information to help people across the globe.

Every chapter published in this book has been scrutinized by our experts. Their significance has been extensively debated. The topics covered herein carry significant findings which will fuel the growth of the discipline. They may even be implemented as practical applications or may be referred to as a beginning point for another development. Chapters in this book were first published by InTech; hereby published with permission under the Creative Commons Attribution License or equivalent.

The editorial board has been involved in producing this book since its inception. They have spent rigorous hours researching and exploring the diverse topics which have resulted in the successful publishing of this book. They have passed on their knowledge of decades through this book. To expedite this challenging task, the publisher supported the team at every step. A small team of assistant editors was also appointed to further simplify the editing procedure and attain best results for the readers.

Our editorial team has been hand-picked from every corner of the world. Their multi-ethnicity adds dynamic inputs to the discussions which result in innovative outcomes. These outcomes are then further discussed with the researchers and contributors who give their valuable feedback and opinion regarding the same. The feedback is then collaborated with the researches and they are edited in a comprehensive manner to aid the understanding of the subject.

Apart from the editorial board, the designing team has also invested a significant amount of their time in understanding the subject and creating the most relevant covers. They scrutinized every image to scout for the most suitable representation of the subject and create an appropriate cover for the book.

The publishing team has been involved in this book since its early stages. They were actively engaged in every process, be it collecting the data, connecting with the contributors or procuring relevant information. The team has been an ardent support to the editorial, designing and production team. Their endless efforts to recruit the best for this project, has resulted in the accomplishment of this book. They are a veteran in the field of academics and their pool of knowledge is as vast as their experience in printing. Their expertise and guidance has proved useful at every step. Their uncompromising quality standards have made this book an exceptional effort. Their encouragement from time to time has been an inspiration for everyone.

The publisher and the editorial board hope that this book will prove to be a valuable piece of knowledge for researchers, students, practitioners and scholars across the globe.

List of Contributors

Marek Brandner, Jirí Egermaier and Hana Kopincová
The University of West Bohemia, Czech Republic

J.H.M. ten Thije Boonkkamp and M.J.H. Anthonissen
Department of Mathematics and Computer Science, Eindhoven University of Technology, Eindhoven, the Netherlands

A. Ashrafizadeh, M. Rezvani and B. Alinia
K. N. Toosi University of Technology, Iran

Árpád Veress and József Rohács
Budapest University of Technology and Economics, Hungary

M. Ebrahim and R. Jalalabadi
K. N. Toosi University of Technology, Iran

Jure Mencinger
LFDT, Faculty of Mechanical Engineering, University of Ljubljana, Slovenia

A.M. Afonso, M.S.N. Oliveira, M.A. Alves and F.T. Pinho
Transport Phenomena Research Centre, Faculty of Engineering, University of Porto, Porto

P.J. Oliveira
Department of Electromechanical Engineering, Textile and Paper Materials Unit, University of Beira Interior, Covilhã, Portugal

Yousuf Alhendal and Ali Turan
The University of Manchester, United Kingdom

Accary Gilbert
Holy-Spirit University of Kaslik, Faculty of Engineering, Lebanon

H. C. Zhang, Y. Y. Guo, H. P. Tan and Y. Li
Harbin Institute of Technology, P. R. China

Nelson O. Moraga
Universidad de La Serena/Departamento de Ingeniería Mecánica, La Serena, Chile

Carlos E. Zambra
Centro de Investigación Avanzada en Recursos Hídricos y Sistemas Acuosos (CIDERH), CONICYT-REGIONAL GORE-TARAPACÁ, Iquique, Chile

Olivier Ducasse, Olivier Eichwald and Mohammed Yousfi
University of Toulouse, France

Jan M. Nordbotten, Eirik Keilegavlen and Andreas Sandvin
University of Bergen, Norway

Izet Horman, Dunja Martinović, Izet Bijelonja and Seid Hajdarević
Mechanical Engineering Faculty, University of Sarajevo, Bosnia and Herzegovina

Radostina Petrova
Technical University of Sofia, Faculty of Engineering and Education, Bulgaria

Sotir Chernev
"HES" PLC, Yambol, Bulgaria

N. Chauveau and P. Celsis
Inserm, Imagerie Cérébrale et Handicaps Neurologiques UMR 825, Toulouse, France
Université de Toulouse, UPS, Imagerie Cérébrale et Handicaps Neurologiques UMR 825, CHU Purpan, Place du Dr Baylac, Toulouse Cedex 9, France

A. Sors
LU 48 LERISM Laboratoire d'Etudes et de Recherche en Imagerie Spatiale et Médicale, UPS, Toulouse Cedex 4, France

M. Masquere
Université de Toulouse, UPS, INPT, LAPLACE (Laboratoire Plasma et Conversion d'Energie), Toulouse Cedex 9, France

X. Franceries
Inserm, Imagerie Cérébrale et Handicaps Neurologiques UMR 825, Toulouse, France
Université de Toulouse, UPS, Imagerie Cérébrale et Handicaps Neurologiques UMR 825, CHU Purpan, Place du Dr Baylac, Toulouse Cedex 9, France
Université de Toulouse, UPS, INPT, LAPLACE (Laboratoire Plasma et Conversion d'Energie), Toulouse Cedex 9, France